Basic Lessons on Isometries, Similarities and Inversions in the Euclidean Plane

A Synthetic Approach

Basic Lessons on Isometries, Similarities and Inversions in the Euclidean Plane

A Synthetic Approach

Ioannis Markos Roussos

Hamline University, USA

World Scientific

NEW JERSEY · LONDON · SINGAPORE · BEIJING · SHANGHAI · HONG KONG · TAIPEI · CHENNAI · TOKYO

Published by

World Scientific Publishing Co. Pte. Ltd.

5 Toh Tuck Link, Singapore 596224

USA office: 27 Warren Street, Suite 401-402, Hackensack, NJ 07601

UK office: 57 Shelton Street, Covent Garden, London WC2H 9HE

Library of Congress Control Number: 2021039757

British Library Cataloguing-in-Publication Data
A catalogue record for this book is available from the British Library.

BASIC LESSONS ON ISOMETRIES, SIMILARITIES AND INVERSIONS IN
THE EUCLIDEAN PLANE
A Synthetic Approach

ISBN 978-981-123-985-4 (hardcover)
ISBN 978-981-124-037-9 (paperback)
ISBN 978-981-123-986-1 (ebook for institutions)
ISBN 978-981-123-987-8 (ebook for individuals)

For any available supplementary material, please visit
https://www.worldscientific.com/worldscibooks/10.1142/12358#t=suppl

Typeset by Stallion Press
Email: enquiries@stallionpress.com

Dedication

To: All my Teachers

Contents

Biography

Ioannis Markos Roussos

Professor Ioannis Markos Roussos was born on November 5, 1954 at the village Katapola of Amorgos, Greece. After the primary and secondary education he studied mathematics at the National and Kapodistrian University of Athens and received his BSc Degree (1972–1977). Then, he study graduate mathematics and computer sciences at the University of Minnesota and received his Masters and PhD degrees (1978–1986). His specialization in mathematics was in Differential Geometry and Analysis. He has taught mathematics at the University of Minnesota, University of South Alabama and Hamline University. Besides this book, he has published 17 research papers, 10 expository papers and the book *Improper Riemann Integrals*. He has participated in meetings and has refereed papers and promotions of other professors. Other interests are classical music, history, international relations and travelling.

Acknowledgements

I would like to express my deepest thanks to my colleague, Dr. Arthur Guetter, Professor of Mathematics, who used his great expertise to make some of the figures in this book using the program *Mathematica*, and then taught me how to do the rest of them. He has also been my LaTex teacher. Without him, this work would not have been completed. I would also like to thank Hamline University for offering me sabbatical leave in 2017–2018, during which I had the time needed to start and complete this work. I must also thank my students for their questions, hints, moral support and encouragement.

Prologue

The aim of this book is to provide a complete exposition of plane isometries, similarities and inversions to students and teachers who need to study and teach classical Euclidean plane Geometry at a higher level. Many textbooks provide some information on these topics, but they are far from a concise and complete exposition. Therefore, this book is also an important addition to Mathematics Education. Experience has shown that students have difficulty with these topics, especially when they are briefly presented. Therefore, our aim is to provide teachers and students all the necessary details in an accessible and complete exposition using synthetic geometry only. Thus, we stay within the classical synthetic geometry and avoid analytic geometry and / or the geometry of the complex numbers. In two proofs only, we use some analysis and topology in order to preserve generality and also show how some results from higher theories could be conveniently used. Still, these two proofs can be bypassed by anyone who has become familiar with the whole corpus. The interplay of straight segments and free vectors as well as of angles with oriented angles is necessary in order to establish general results in a succinct and comprehensive way. Some basic knowledge of algebraic groups with the operation of composition is convenient and rather necessary even in a synthetic approach in order to render the material more concise and elegant. The topics exposed in this book can be used to prove many results and solve many problems of classical geometry, which are encountered with different proofs in the literature; their applications are numerous and some are useful to engineers. To emphasize this fact, we have included several good examples, important applications and numerous exercises of various level and difficulty. The reader must be familiar with the basic definitions, concepts and results of the classical plane geometry, such as, equality and similarity of triangles, Pythagoras' theorem, parallel lines, Thales' theorem, the theorems of the angle bisectors of a triangle, the radical axis and radical center of circles, and so on. After the preliminary Chapter 1, the exercises are written at the end of each chapter, as separate chapters in order not to interrupt the flow of the theory. They are classified in three groups: general exercises, geometrical constructions and geometrical loci. Some exercises or groups of exercises can be assigned as projects to students.

The author
Professor Ioannis M. Roussos, Ph.D.
Hamline University
Twin Cities, Minnesota, USA, Summer 2021.

Chapter 1

Some Preliminaries

1.1 Definitions

We consider the Euclidean[1] plane $\mathcal{P}$ in which we have (arbitrarily) designated a straight segment XY as the **unit segment**, i.e., we assign to it the unit length represented by the number 1. See **Figure 1.1**. If we pick any two points A and

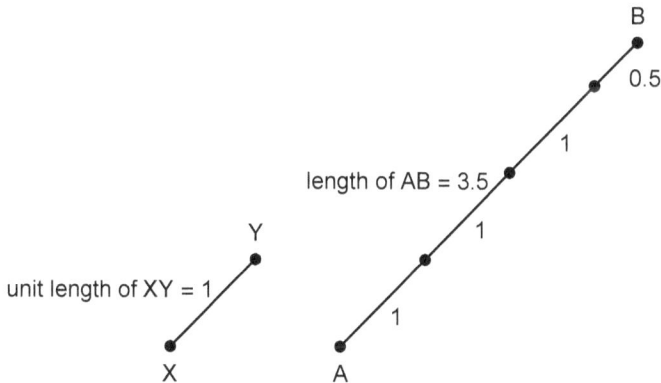

Figure 1.1: Length of a straight segment

B, we define the **length of the straight segment** AB or BA to be the number $\mu(AB) = r$ that represents how many times the segment XY fits in AB.[2] Then,

---

[1]Euclid of Alexandria, Egypt. Greek mathematician, 365?-3th century BCE. He wrote the book "*The Elements*" in which he founded Euclidean Geometry and exhibited all mathematics up to his time. This book is considered the most influential book in civilization.

[2]The strict and precise definition of length in the axiomatic classical geometry would take a whole chapter to develop that also involves the concept of number. A length can be represented by an integer, rational, algebraic and transcendental number. So, here, we pass over it by making this intuitive definition, since the foundation of geometry is not included in the scope of this book. The interested reader should refer to pertinent bibliography on this matter.

r is a real non-negative number, $r \geq 0$. We note that: $r = 0 \iff A = B$, i.e., the two points coincide.

Definition 1.1.1 *The **(Euclidean) distance function** or **metric** (function) in $\mathcal{P}$ is the function defined by*

$$d \;:\; \mathcal{P} \times \mathcal{P} \longrightarrow \mathbb{R}, \quad (A, B) \longmapsto d(A, B) = \mu(AB) = \mu(BA).$$

*The distance function (as any distance or metric function) is characterized by the following **properties**:*

1. $\forall$ A and B points in $\mathcal{P}$, $d(A, B) \geq 0$ **(non-negativity)**.

2. $d(A, B) = 0 \iff A = B$ **(non-degeneracy)**.

3. $\forall$ A and B points in $\mathcal{P}$, $d(A, B) = d(B, A)$ **(symmetry)**.

4. $\forall$ A, B and C points in $\mathcal{P}$, $d(A, B) \leq d(A, C) + d(C, B)$ **(triangle inequality)**.

The first three properties follow immediately from the definition of d. The fourth one, called the **triangle inequality**, follows from the Theorem of Plane Geometry that says: *"The sum of any two sides of a triangle is greater than the third side"*. See **Figure 1.2**. As written above, equality holds iff A, B, C are on the same straight line and C lies between A and B, or $A = C$, or $B = C$.

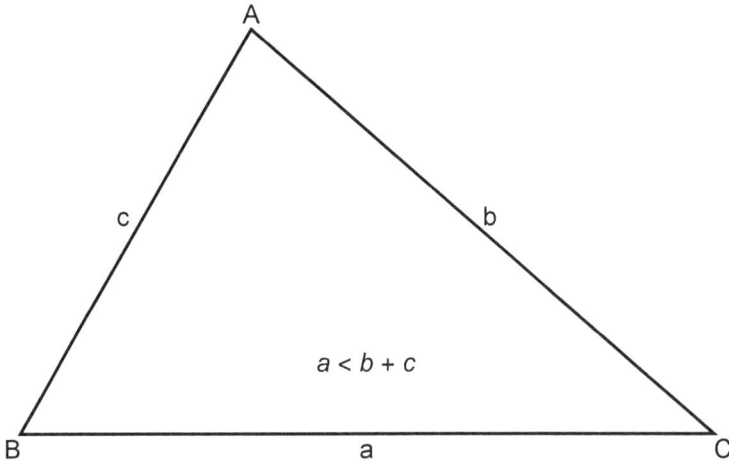

Figure 1.2: The triangle inequality

Note: In the sequel, the notation AB may mean the straight segment defined by the points A and B, or the length of AB, $\mu(AB)$, or the distance between A and B, $d(A, B)$. We also consider the equality $AB = CD$ as equivalent to $\mu(AB) = \mu(CD)$, or to $d(A, B) = d(C, D)$. We shall use these three notations interchangeably, but as it is customary, we mostly use the notation AB for all these three concepts, for the sake of a simpler notation.

If we introduce coordinate axes in $\mathcal{P}$, as we do in calculus, the points are represented by ordered pairs of numbers

$$A = (x_1, y_1) \qquad \text{and} \qquad B = (x_2, y_2).$$

See **Figure 1.3**. Then, with the help of the Pythagorean[3] Theorem, we prove that

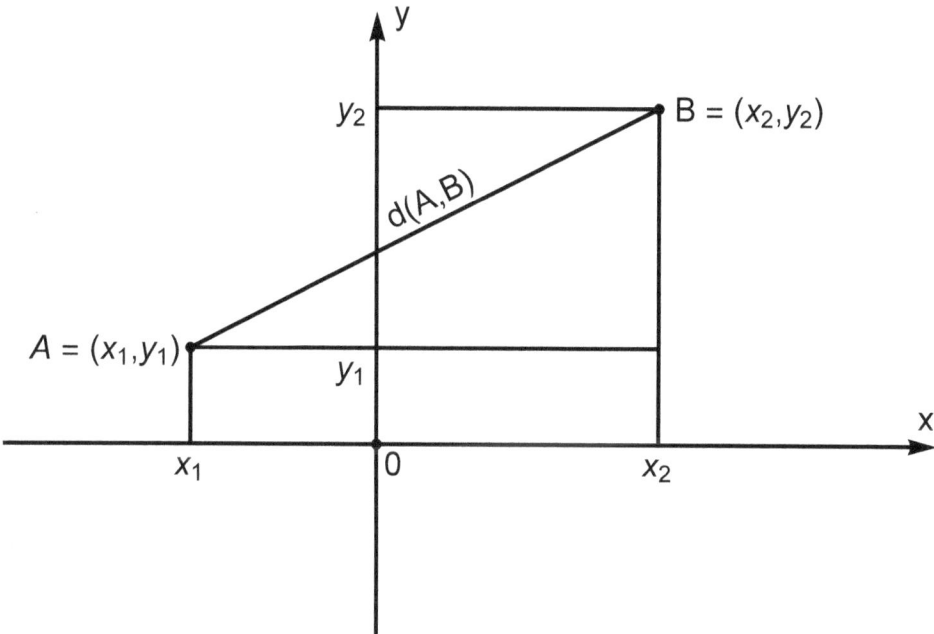

Figure 1.3: Distance of two points in the plane

$$d(A, B) = d[(x_1, y_1), (x_2, y_2)] = \sqrt{(x_2 - x_1)^2 + (y_2 - y_1)^2}$$

$$= d(B, A) = d[(x_2, y_2), (x_1, y_1)] = \sqrt{(x_1 - x_2)^2 + (y_1 - y_2)^2}.$$

A space which has a **metric function** is called **metric space**. Such a space is called **complete**, if every Cauchy[4] sequence (find the definition in analysis or topology books) converges in the space. The Euclidean plane with the above metric is a complete metric space.

---

[3]Pythagoras of Samos, Greek mathematician and philosopher, c.570-c.495 BCE.
[4]Augustin-Louis Cauchy, French mathematician, 1789-1857.

1.2   Free Vectors

Free vector in the plane is any oriented straight segment $\overrightarrow{AB}$ in which we consider motion from A to B on the straight segment AB. A is the initial point and B the terminal point of the free vector $\overrightarrow{AB}$. It is customary to denote a free vector by a single letter with an arrow on it, e.g., $\vec{u} = \overrightarrow{AB}$.

Two free vectors $\vec{u} = \overrightarrow{AB}$ and $\vec{v} = \overrightarrow{CD}$ are **equal** and we write $\vec{u} = \vec{v}$, if, by definition: (1) either both are zero-vectors, which happens when $A = B$ and $C = D$ and so $\vec{u} = \vec{v} = \vec{0}$, or otherwise, (2) they satisfy the following three conditions:

1. The straight segments AB and CD are parallel, $AB \parallel CD$, or they are on the same straight line.

2. The straight segments AB and CD have equal lengths, $\mu(AB) = \mu(CD)$.

3. The vectors $\vec{u} = \overrightarrow{AB}$ and $\vec{v} = \overrightarrow{CD}$ have the same direction or same orientation.

We must accurately define what we mean by **the same or opposite direction, or orientation, of two free parallel vectors** $\vec{u} = \overrightarrow{AB} \parallel \vec{v} = \overrightarrow{CD}$.

Case (1): The parallel segments $AB \parallel CD$ are on two different parallel lines. Then, $\vec{u} = \overrightarrow{AB}$ and $\vec{v} = \overrightarrow{CD}$ have the **same direction**, if the tip points B and D lie in the same closed half-plane that the whole plane $\mathcal{P}$ is separated by the straight line AC. See **Figure 1.4**.

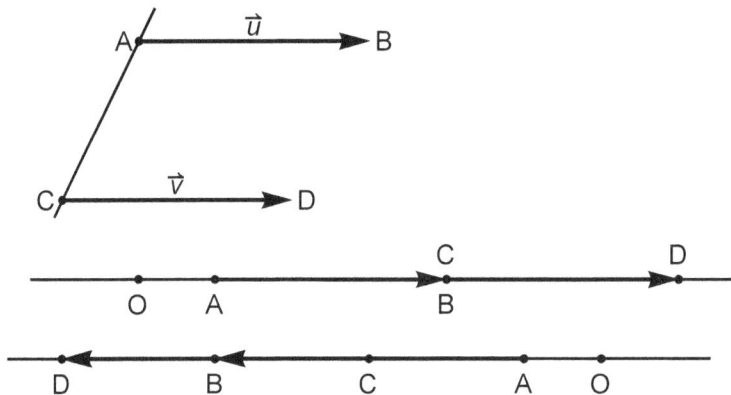

Figure 1.4: **Equal free vectors must have the same direction**

Case (2): The segments AB and CD are on the same straight line l. We pick a point O on the line l such that all four points A, B, C, and D are located on the same side of O. Then, $\vec{u} = \overrightarrow{AB}$ and $\vec{v} = \overrightarrow{CD}$ have the **same direction**, if either $OA \leq OB$ and $OC \leq OD$, or $OA \geq OB$ and $OC \geq OD$. See **Figure 1.4**.

Case (3): Otherwise [that is, cases **(1)** and **(2)** do not occur], we say that the **parallel** free vectors $\vec{u} = \overrightarrow{AB}$ and $\vec{v} = \overrightarrow{CD}$ have **opposite direction**. See **Figure 1.5**.

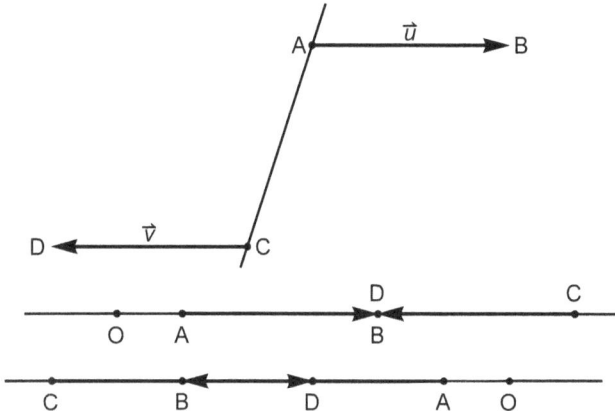

Figure 1.5: **Free vectors with opposite directions**

If $AB \nparallel CD$, then we do not compare the directions of $\overrightarrow{AB} \nparallel \overrightarrow{CD}$. See **Figure 1.6**. The zero vector may be considered having any direction.

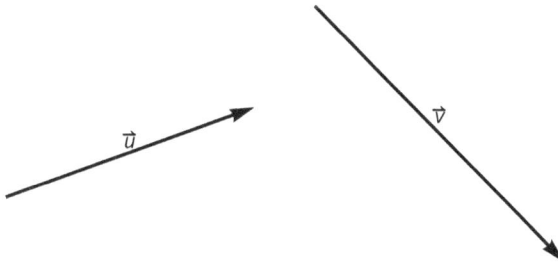

Figure 1.6: **Free vectors in free directions**

Since parallelism, same orientation and equality of lengths are equivalence relations, the **equality among free vectors** of the plane is an **equivalence relation**. That is, given $\vec{u}$, $\vec{v}$ and $\vec{w}$ free vectors in the plane $\mathcal{P}$, we have:

1. $\vec{u} = \vec{u}$ (**reflexive** property).

2. $\vec{u} = \vec{v} \implies \vec{v} = \vec{u}$ (**symmetric** property).

3. $\vec{u} = \vec{v}$ and $\vec{v} = \vec{w} \implies \vec{u} = \vec{w}$ (**transitive** property).

Any given free vector $\vec{u}$ in the plane, determines an **equivalence class** which is the set of all free vectors in the plane equal to $\vec{u}$.

We define the **oriented angle of two non-zero free vectors** $\angle(\vec{u}, \vec{v})$ in the following way: We pick any point $P \in \mathcal{P}$ and transport $\vec{u}$ and $\vec{v}$ equal to themselves at P, thus determining points A and B, such that $\overrightarrow{PA} = \vec{u}$ and $\overrightarrow{PB} = \vec{v}$. See **Figure 1.7**. Then, we define the oriented angle from $\vec{u}$ to $\vec{v}$ by

$$\angle(\vec{u}, \vec{v}) = \angle(\overrightarrow{PA}, \overrightarrow{PB}) = \angle(APB).$$

A value of this angle (positive, negative, or zero) is determined modulo 2π.

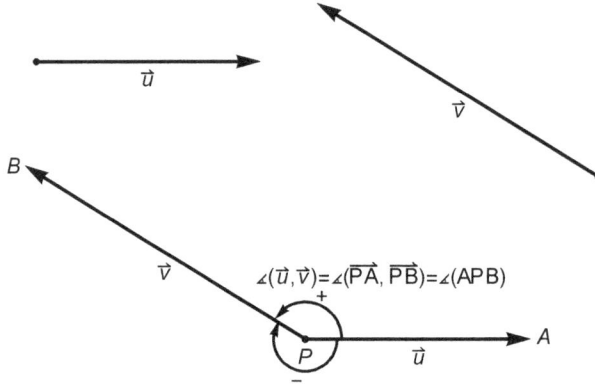

Figure 1.7: **Determination of the angle of two free vectors, (mod 2π)**

We work similarly if we want to determine the angle of two half-lines. The vectors $\vec{u}$ and $\vec{v}$ above could represent half lines starting at their initial points and having the same directions as the vectors.

We will use the following relation on vectors. Given any vector $\overrightarrow{AB}$ and any point O in the plane, then

$$\overrightarrow{AB} = \overrightarrow{OB} - \overrightarrow{OA} \qquad \text{or} \qquad \overrightarrow{OA} + \overrightarrow{AB} = \overrightarrow{OB}.$$

See **Figure 1.8**. This is an immediate consequence of the operation of addition on vectors. It simplifies proofs and computations in many situations.

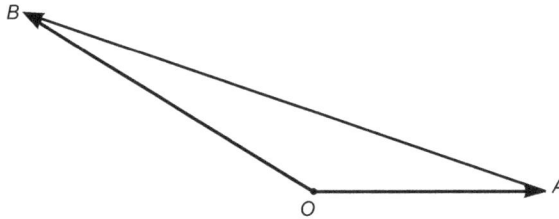

Figure 1.8: **Relation of free vectors**

We will use a similar relation on oriented angles. We are given any three (or more) rays (half lines) Ox, Oy, and Oz, all starting at a point O of the plane. See **Figure 1.9**. In any direction around O, positive or negative, we have the angle relations

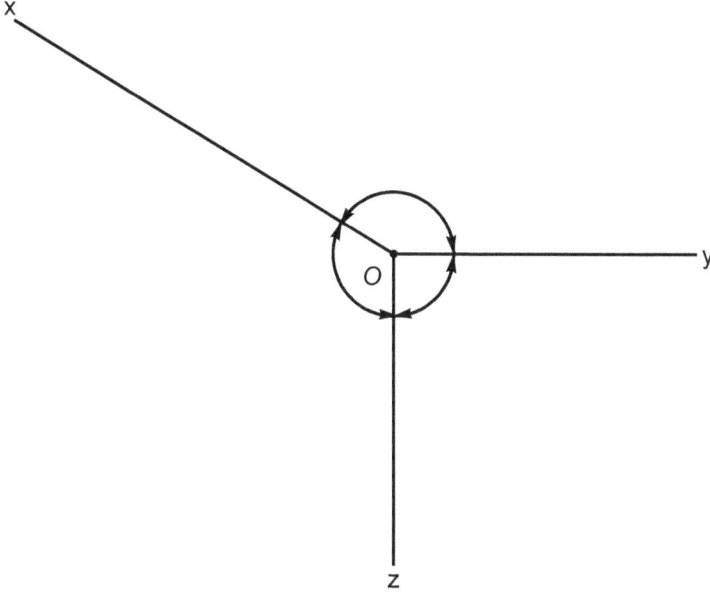

Figure 1.9: Relation of the angles

$$\sphericalangle(Ox, Oy) + \sphericalangle(Oy, Oz) = \sphericalangle(Ox, Oz), \quad (\text{mod } 2\pi),$$
$$\sphericalangle(Ox, Oz) + \sphericalangle(Oz, Oy) = \sphericalangle(Ox, Oy), \quad (\text{mod } 2\pi),$$
$$\sphericalangle(Oz, Oy) + \sphericalangle(Oy, Ox) = \sphericalangle(Oz, Ox), \quad (\text{mod } 2\pi),$$

and so on, (mod 2π). (Generalize these relations to the case of n concurrent rays at O.)

Relations with vectors and angles, as the above, simplify proofs and computations in many situations. We will also need the following definition:

Definition 1.2.1 *Axis*[5] *is a straight line l in which a **point of reference** or the **origin** $O \in l$ is designated (any point may be chosen), the positive and negative directions are assigned, and a unit-length positive vector $\vec{i} = \overrightarrow{OA}$ is chosen on it.*

We arbitrarily pick a point $A \neq O$ on l and assign the direction of the vector $\overrightarrow{OA} \neq \vec{0}$ to be the positive direction. Then, the direction of $\overrightarrow{AO} = -\overrightarrow{OA}$ is the

---

[5]The fundamental concept of **axis** was introduced by René Descartes, (French philosopher and mathematician, March 31, 1596 - February 11, 1650) and this was the beginning of Analytic Geometry in which we deal with the Cartesian coordinates.

negative. See **Figure 1.10**. We may also assign the unit length 1 to the vector $\overrightarrow{OA}$ and from this to determine the abscissas of all the other points on l relative to O. The abscissa of O is zero, the points in the positive direction of $\overrightarrow{OA}$ have abscissas the positive numbers and the points in the negative direction of $-\overrightarrow{OA}$ the negative numbers.

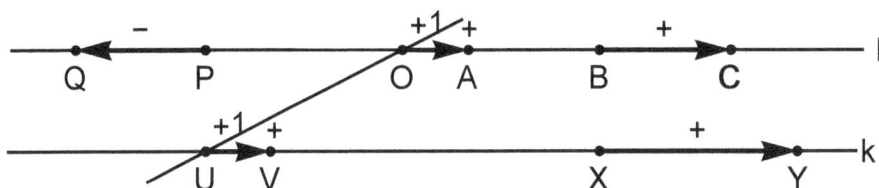

Figure 1.10: Two parallel axes and oriented segments

In such a case, when we have two points B and C on the axis l, the symbol $\overline{BC}$ means

$$\overline{BC} = \begin{cases} +\text{length of segment } BC, & \text{if } \overrightarrow{BC} \text{ has the positive direction,} \\ \\ -\text{length of segment } BC, & \text{if } \overrightarrow{BC} \text{ has the negative direction.} \end{cases}$$

In other words, $\overline{BC}$ means the **oriented or signed length** of BC. We may also call $\overline{BC}$ an **oriented segment** (instead of a vector) when the axis we refer to is given explicitly. If the abscissas of B and C are the real numbers b and c, respectively, then

$$\overline{BC} = c - b \quad \text{and} \quad \overline{CB} = b - c.$$

Then, notice

$$\overline{BB} = 0, \ \overline{BC} = -\overline{CB}, \ \overline{BC} = \overline{AC} - \overline{AB} = \overline{OC} - \overline{OB}, \ \overline{BC}^{\,2} = BC^{\,2},$$

etc., and

$$\overline{A_1 A_2} + \overline{A_2 A_3} + \overline{A_3 A_4} + \ldots + \overline{A_{n-1} A_n} = \overline{A_1 A_n}.$$

The last equation is analogous to

$$\overrightarrow{A_1 A_2} + \overrightarrow{A_2 A_3} + \overrightarrow{A_3 A_4} + \ldots + \overrightarrow{A_{n-1} A_n} = \overrightarrow{A_1 A_n},$$

with free vectors in succession in the plane. But, this equation is true for any successive free vectors in the plane (and so, on an axis too), whereas the previous equation is valid for successive oriented segments on the **same** axis.

Also, oriented segments can be compared and combined with the ordinary four operations if they are on the same axis, or on parallel axes, if all the parallel axes are given the same orientation (as it is done with parallel vector of the same direction) and the same unit vector (the same scale). See **Figure 1.10**. For example, the oriented segments $\overline{BC}$ and $\overline{PQ}$ and $\overline{XY}$ can be compared and combined using the numbers of their oriented lengths.

1.3 Preliminary Exercises

1. Consider three points O, A, B on an axis and M the midpoint of $\overline{AB}$. Prove that
$$\overline{OA} + \overline{OB} = 2\overline{OM}.$$

2. Consider the rays OX, OA, OB and OM the bisector of the angle $\sphericalangle(OA, OB)$. Prove that
$$\sphericalangle(OX, OA) + \sphericalangle(OX, OB) = 2\sphericalangle(OX, OM), \quad (\mathrm{mod}\ 2\pi).$$

3. Consider four points A, B, C, and D on an axis and M, N, P and Q the midpoints of $\overline{AC}$, $\overline{BD}$, $\overline{AD}$, and $\overline{BC}$, respectively. Prove that

   (a) $\qquad \overline{AB} + \overline{CD} = \overline{AD} + \overline{CB} = 2\overline{MN},$

   (b) $\qquad \overline{AB} - \overline{CD} = \overline{AC} - \overline{BD} = 2\overline{PQ}.$

4. Consider three axes $\overrightarrow{x}$, $\overrightarrow{y}$, and $\overrightarrow{z}$ (or three vectors). Prove the **Chasles**[6]-**Möbius**[7] **relation for three axes**
$$\sphericalangle(\overrightarrow{y}, \overrightarrow{z}) = \sphericalangle(\overrightarrow{x}, \overrightarrow{z}) - \sphericalangle(\overrightarrow{x}, \overrightarrow{y}), \quad (\mathrm{mod}\ 2\pi).$$

5. Consider three straight lines k, l, and m. First prove that
$$\sphericalangle(k, l) = \sphericalangle(\overrightarrow{k}, \overrightarrow{l}), \quad (\mathrm{mod}\ \pi)$$
and then prove the **Chasles-Möbius relation for three straight lines in the plane**
$$\sphericalangle(k, l) + \sphericalangle(l, m) = \sphericalangle(k, m), \quad (\mathrm{mod}\ \pi).$$

6. Consider two straight lines l_1 and l_2 and an axis $\overrightarrow{x}$. Prove
$$\sphericalangle(l_1, l_2) = \sphericalangle(\overrightarrow{x}, l_1) - \sphericalangle(\overrightarrow{x}, l_2), \quad (\mathrm{mod}\ \pi).$$

7. Consider two axes $\overrightarrow{x}$ and $\overrightarrow{y}$. We call bisector of the angle $\sphericalangle(\overrightarrow{x}, \overrightarrow{y})$, the axis $\overrightarrow{b}$ such that
$$\sphericalangle(\overrightarrow{x}, \overrightarrow{b}) = \sphericalangle(\overrightarrow{b}, \overrightarrow{y}).$$
Prove that if $\overrightarrow{z}$ is a third axis, then
$$\sphericalangle(\overrightarrow{b}, \overrightarrow{z}) = \frac{1}{2}\left[\sphericalangle(\overrightarrow{b}, \overrightarrow{x}) + \sphericalangle(\overrightarrow{b}, \overrightarrow{y})\right], \quad (\mathrm{mod}\ \pi).$$

---

[6] Michel Floréal Chasles, French mathematician, 1793-1880.
[7] August Ferdinand Möbius, German mathematician, 1790-1868.

8. Consider two straight lines l_1 and l_2. We call a bisector of the angle $\angle(l_1, l_2)$ a s.l. b through the common point of l_1 and l_2 that satisfies

$$\angle(l_1, b) = \angle(b, l_2), \pmod{\pi}.$$

Prove that the angle $\angle(l_1, l_2)$ has two bisectors b_1 and b_2, such that $b_1 \perp b_2$ and if $\overrightarrow{x}$ is any axis, then

$$\angle(\overrightarrow{x}, b_1) = \frac{1}{2}[\angle(\overrightarrow{x}, l_1) + \angle(\overrightarrow{x}, l_2)], \quad \pmod{\pi},$$

$$\angle(\overrightarrow{x}, b_2) = \frac{1}{2}[\angle(\overrightarrow{x}, l_1) + \angle(\overrightarrow{x}, l_2) + \pi], \quad \pmod{\pi}.$$

9. Consider four concurrent straight lines a, a', b and b' (i.e., they go through the same point). Prove that the pairs (a, a') and (b, b') are isogonal (determine equal smallest positive angles, or, equivalently, equal greatest negative angles) iff

$$\angle(a, b) = \angle(b', a'), \quad \pmod{\pi}.$$

10. Consider vectors $\overrightarrow{AB} = \overrightarrow{BC} = \overrightarrow{CD} \neq \overrightarrow{0}$. Compute the ratios:

$$\frac{\overrightarrow{AB}}{\overrightarrow{BD}}, \quad \frac{\overrightarrow{AD}}{\overrightarrow{DC}}, \quad \frac{\overrightarrow{BA}}{\overrightarrow{AD}}, \quad \frac{\overrightarrow{BC}}{\overrightarrow{DA}}, \quad \frac{\overrightarrow{BC}}{\overrightarrow{BA}}.$$

11. Consider four points X, Y, Z, and M on an axis. Prove:

(a) $\overline{MX} \cdot \overline{YZ} + \overline{MY} \cdot \overline{ZX} + \overline{MZ} \cdot \overline{XY} = 0$ (Euler).

(**Euler's**[8] **relation on an axis.**)

(b) $\overline{MX}^2 \cdot \overline{YZ} + \overline{MY}^2 \cdot \overline{ZX} + \overline{MZ}^2 \cdot \overline{XY} + \overline{ZX} \cdot \overline{XY} \cdot \overline{YZ} = 0$ (Stewart).

(See also **Stewart's**[9] **Theorem, 6.5.3**, for generalization.)

12. Consider three points A, B, and P on an axis. Let M be the midpoint of AB. Prove:

(a) $\overline{PA} \cdot \overline{PB} = \overline{PM}^2 - \overline{MA}^2 = \overline{PM}^2 - \overline{MB}^2,$

(b) $\overline{PA}^2 + \overline{PB}^2 = 2\overline{PM}^2 + 2\overline{MA}^2 = 2\overline{PM}^2 + 2\overline{MB}^2,$

(c) $\overline{PA}^2 - \overline{PB}^2 = 2\overline{AB} \cdot \overline{MP}.$

---

[8]Leonhard Euler, Swiss mathematician, 1707-1783.
[9]Matthew Stewart, Scottish mathematician, 1717-1785.

13. Consider two fixed points A and B on an axis, and two real numbers λ and μ. For any point P of the axis, we consider the point G of the axis determined by $\lambda \overrightarrow{GA} + \mu \overrightarrow{GB} = 0$. Prove:

$$\text{(a)} \qquad \lambda \overline{PA} + \mu \overline{PB} = (\lambda + \mu) \overline{PG}.$$
$$\text{(b)} \qquad \lambda \overline{PA}^2 + \mu \overline{PB}^2 = (\lambda + \mu) \overline{PG}^2 + \lambda \overline{GA}^2 + \mu \overline{GB}^2.$$

14. Consider three points A, B, and C on an axis of origin O. Prove:

$$\frac{\overline{OB} \cdot \overline{OC}}{\overline{AB} \cdot \overline{AC}} + \frac{\overline{OC} \cdot \overline{OA}}{\overline{BC} \cdot \overline{BA}} + \frac{\overline{OA} \cdot \overline{OB}}{\overline{CA} \cdot \overline{CB}} = 1.$$

15. Consider four points A, B, C, and D on an axis and let I and J be the midpoints of AB and CD, respectively. Prove:

(a) $\overline{AB} + \overline{CD} = \overline{AD} + \overline{CB}$.

(b) $I = J$ iff $\overline{AC} + \overline{BD} = \overline{AD} + \overline{BC} = 0$.

16. *We say that a point C (anywhere) on an axis that contains a vector $\overrightarrow{AB} \neq \overrightarrow{0}$ (and so $A \neq B$), divides the vector $\overrightarrow{AB} \neq \overrightarrow{0}$ in ratio λ, if*

$$\frac{\overline{AC}}{\overline{CB}} = \lambda.$$

*This ratio is also called the **simple ratio**[10] of the points A, B, C in this order, denoted by $\{A, B; C\}$.*

(Remember this definition in the pertinent exercises that follow.)

So, when $A \neq B$, prove:

(a) C is between A and B iff $\lambda > 0$.

(b) C is outside A and B iff $\lambda < 0$.

(c) $C = A$ iff $\lambda = 0$.

(d) If $C = B$, then λ is undefined.
Also, explain how we get $\lambda = +\infty$ or $\lambda = -\infty$ as a limit.

(e) $\overline{AC} = \dfrac{\lambda}{\lambda + 1} \overline{AB}$ or $\dfrac{\overline{AC}}{\overline{AB}} = \dfrac{\lambda}{\lambda + 1}$ or $\dfrac{\overline{CA}}{\overline{AB}} = -\dfrac{\lambda}{1 + \lambda}$.

(f) $\dfrac{\overline{BC}}{\overline{CA}} = \dfrac{1}{\lambda}$.

(g) $\dfrac{\overline{AB}}{\overline{BC}} = -(1 + \lambda)$.

---

[10] Other authors defined the **simple ratio** in different equivalent ways. For example,

$$\frac{\overline{AC}}{\overline{BC}} = \mu,$$

which is equal to $-\lambda$ defined above. Read carefully the exact definition of the simple ratio in the book that you study and check its equivalence with the definition we cite here.

(h) $\dfrac{\overline{BA}}{\overline{AC}} = -\left(1 + \dfrac{1}{\lambda}\right) = -\dfrac{1+\lambda}{\lambda}.$

(i) $\dfrac{\overline{CB}}{\overline{BA}} = -\dfrac{1}{1+\lambda}$ or $\overline{BC} = -\dfrac{1}{1+\lambda}\overline{AB}.$

(j) C is the midpoint of AB iff $\lambda = 1$.

(k) $\{A, B; C\} = \dfrac{\overline{AC}}{\overline{CB}}$ approaches -1 iff C approaches $+\infty$ or $-\infty$ on the axis.

(l) Draw the simple ratio in a system of axes with respect to c, the abscissa of C.

17. Given three points A, B, and C on an axis prove that:
    $A = B \neq C$ iff the simple ratio $\{A, B; C\} = -1$.
    (If $A = B = C$ then the simple ratio is not defined.)

18. Let A, B, C and C' be points on an axis such that

$$\{A, B; C\} = \frac{\overline{AC}}{\overline{CB}} = \lambda \qquad \text{and} \qquad \{B, A; C'\} = \frac{\overline{BC'}}{\overline{C'A}} = \lambda$$

    Prove:

    (a) The points C and C' are different, in general.

    (b) The points C and C' are the same iff $\lambda = 1$.
        In this case, the point $C = C'$ is the midpoint of AB.

19. On an axis we take $A = -2$ and $B = 1$. Find point C that divides $\overrightarrow{AB}$ in ratio $\lambda = 2$ or $\dfrac{-1}{2}$ or -2.

20. On an axis we take $A = -1$, $B = 2$ and $C = 5$.

    (a) Find the ratio λ that C divides $\overrightarrow{AB}$.

    (b) Find point D that divides $\overrightarrow{AB}$ in ratio $-\lambda$ [found in **(a)**].

    (c) If J is the midpoint of CD, find the ratio that J divides $\overrightarrow{AB}$.

21. Let $AB = 4$. Find the points that divide $\overrightarrow{AB}$ in ratios 3 and -3.

22. Let $AB = 4 + \sqrt{5}$. Find point C that divides $\overrightarrow{AB}$ in ratio $\dfrac{\sqrt{5}}{-1+\sqrt{5}}$ and compute AC and BC.

23. Let $AB = 5$ and points C and D divide the $\overrightarrow{AB}$ in ratios $\dfrac{3}{4}$ and $\dfrac{-3}{4}$.
    Compute CA, CB, DA, DB, and CD.

24. Construct the points that divide a straight segment of length $s \neq 0$ in ratios $\sqrt{2}$ and $-\sqrt{2}$.

25. Consider an axis ξ and any three successive non-zero vectors $\overrightarrow{AB}, \overrightarrow{BC}, \overrightarrow{CA}$ whose lengths are equal to each other. Let $\alpha = \measuredangle(\xi, \overrightarrow{BC})$, $\beta = \measuredangle(\xi, \overrightarrow{CA})$ and $\gamma = \measuredangle(\xi, \overrightarrow{AB})$. Using some geometry and trigonometry prove that

$$[4\sin(\alpha)\sin(\beta)\sin(\gamma)]^2 + [4\cos(\alpha)\cos(\beta)\cos(\gamma)]^2 = 1,$$

$$\text{or,} \quad [\sin(\alpha)\sin(\beta)\sin(\gamma)]^2 + [\cos(\alpha)\cos(\beta)\cos(\gamma)]^2 = \frac{1}{16}.$$

Chapter 2

Isometries

2.1 Definitions and Basic Isometries

Definition 2.1.1 *A function*

$$F \; : \; \mathcal{P} \longrightarrow \mathcal{P}$$

*is called a (**plane**) **isometry** if it satisfies the property:*

$$\forall \; A, \; B \text{ points of } \mathcal{P}, \text{ it holds that } d[F(A), F(B)] = d(A, B).$$

See **Figure 2.1**.

In this context, instead of function, we may use the terms: map, mapping, and transformation.

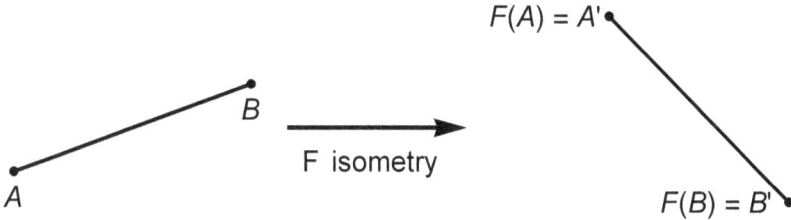

Figure 2.1: An isometry preserves distances,
$\mu(\mathbf{AB}) = \mu(\mathbf{A'B'})$ **(or simply AB = A'B')**

As we prove in the **next Section, [2.2, property (8)]**, the isometry maps a straight segment or a vector to a straight segment or a vector and the interior of a straight segment or a vector to the interior of a straight segment or a vector. So, here, the segment AB is mapped to the segment $A'B' = F(A)F(B)$ and the vector $\overrightarrow{AB}$ is mapped to the vector $\overrightarrow{A'B'} = \overrightarrow{F(A)F(B)}$. This mapping is also done in one to one and onto fashion, as we shall see **[2.2, properties (2) and (3)]**. But first we must give the examples of the basic isometries.

2.1.1   Basic Isometries

To show that isometries exist and thus their definition is not vacuous, we first study the basic isometries. These are the following:

(1) Parallel Translation by a Free Vector $\vec{u}$

The mapping of the parallel translation by a given free vector $\vec{u}$ (kept the same) is

$$T_{\vec{u}} \; : \; \mathcal{P} \longrightarrow \mathcal{P}, \quad A \in \mathcal{P} \longmapsto T_{\vec{u}}(A) = B, \quad \text{if by definition} \quad \overrightarrow{AB} = \vec{u}.$$

See **Figure 2.2**. This mapping is well-defined because if $\overrightarrow{AB} = \vec{u} = \overrightarrow{AC}$, then $B = C$.

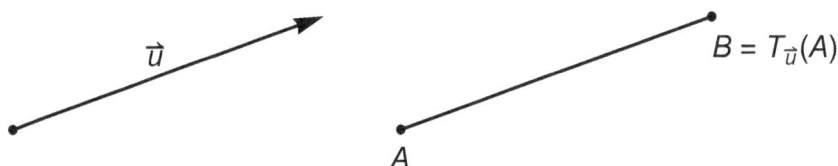

Figure 2.2: **Parallel translation $T_{\vec{u}}$ by a free vector ü**

This mapping is an isometry. See **Figure 2.3**. Indeed, if $T_{\vec{u}}(A) = B$ and $T_{\vec{u}}(C) = D$, then $\overrightarrow{AB} = \vec{u} = \overrightarrow{CD}$. Thus, $AB = [\mu(AB) = \mu(CD)] = CD$ and $AB \parallel CD$. Therefore $ABDC$ is a parallelogram and so $AC = BD$. So, we obtain the relation

$$d[T_{\vec{u}}(A), T_{\vec{u}}(C)] = d(B, D) = d(A, C),$$

which proves that $T_{\vec{u}}$ is an isometry, by the definition of isometry.

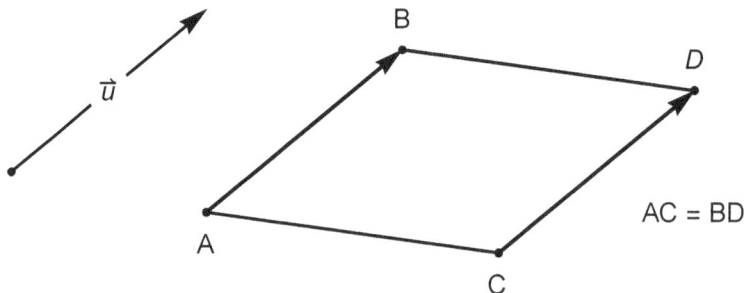

Figure 2.3: **Parallel translation is an isometry**

Notice that if $\vec{u} = \vec{0}$, then $T_{\vec{0}} = I$ the identity.

(2) Rotation Around a Point by an Oriented Angle

We consider any point $O \in \mathcal{P}$ and any oriented angle $\angle\theta$. The mapping of rotation around O by the oriented angle $\angle\theta$ (kept fixed) is defined by:

$$R_{[O,\theta]} \; : \; \mathcal{P} \longrightarrow \mathcal{P}, \quad A \in \mathcal{P} \longmapsto R_{[O,\theta]}(A) = B,$$

such that, $OA = OB$ and oriented angle $\angle(OA, OB) = \theta$. See **Figure 2.4**.

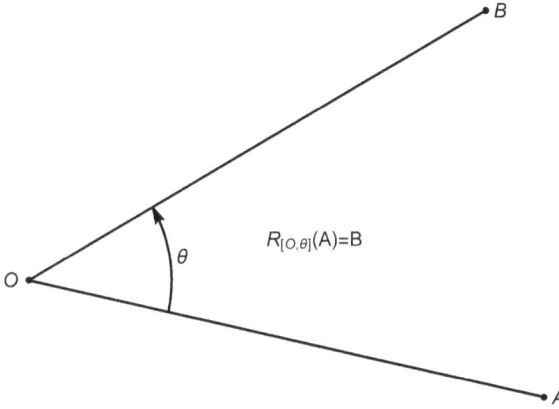

Figure 2.4: Rotation $R_{[O,\theta]}$ with center O and rotation angle θ

By basic geometry, the point B, defined in this way, is proven to be unique. The angle θ is called the **angle of the rotation**, or **rotation angle** and the fixed point O is called the **center of rotation**, or **rotation center**.

This mapping $R_{[O,\theta]}$ is an isometry. See **Figure 2.5**. Indeed:
If $R_{[O,\theta]}(A) = B$ and $R_{[O,\theta]}(C) = D$, then $OA = OB$ and $OC = OD$. Also,

$$\angle(OA, OC) = \angle(OB, OD) = \theta - \angle(OC, OB).$$

Therefore, the triangles OAC and OBD are equal and so $AC = BD$. That is,

$$d[R_{[O,\theta]}(A), R_{[O,\theta]}(C)] = d(B, D) = d(A, C).$$

A special rotation is the **half-turn**. This is a rotation $R_{[O,\pi]}$, with center any point O and rotation angle π. A half-turn is also called **symmetry with respect to a point**, namely the center O of the rotation.

A point O is called **center of symmetry of a figure** Ω, if $R_{[O,\pi]}(\Omega) = \Omega$. In such a case, we say that Ω is set-wise invariant under the half-turn $R_{[O,\pi]}$.

For example, any line is invariant under a half-turn about any of its points. The common midpoint of the diagonals of a rectangle is its center of symmetry. Notice that a half-turn maps a vector to the opposite vector and any half line to a half line of opposite direction. (Make figures for these situations.)

Notice that if $\theta = 0$, (mod 2π), then $R_{[O,0]} = I$ the identity, for any $O \in \mathcal{P}$.

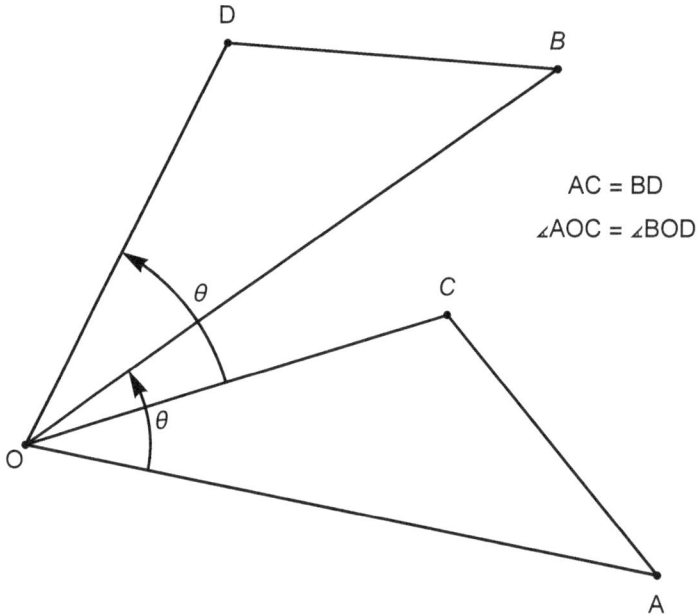

Figure 2.5: A rotation is an isometry

(3) Reflection or Symmetry in a Straight Line

We consider a straight line l in the plane $\mathcal{P}$. The reflection or symmetry in the straight line l is defined by

$$R_l \; : \; \mathcal{P} \longrightarrow \mathcal{P}, \quad A \in \mathcal{P} \longmapsto R_l(A) = B,$$

such that l is perpendicular to the straight segment AB at its midpoint M. ($l \perp AB$ and $AM = MB$.) The point B, thus defined, is uniquely determined. See **Figure 2.6**.

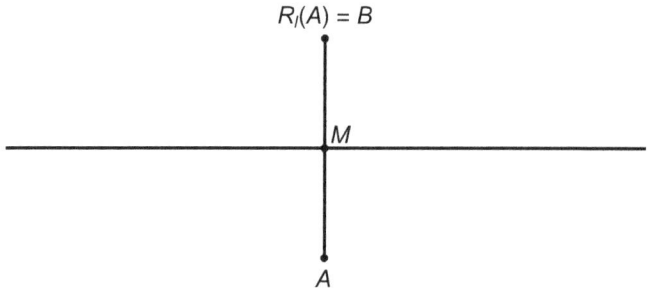

Figure 2.6: Reflection or symmetry R_l in a straight line l

This mapping is an isometry. Indeed, if

$$R_l(A) = B \quad \text{and} \quad R_l(C) = D,$$

then l is perpendicular to the segments AB and CD at their midpoints. See **Figure 2.7**. Therefore, $AB \parallel CD$ and $ACDB$ is an isosceles trapezium, which is not a skew parallelogram; so its two lateral sides are equal (and its two diagonals are also equal). Hence, $AC = BD$. That is,

$$d[R_l(A), R_l(C)] = d(B, D) = d(A, C).$$

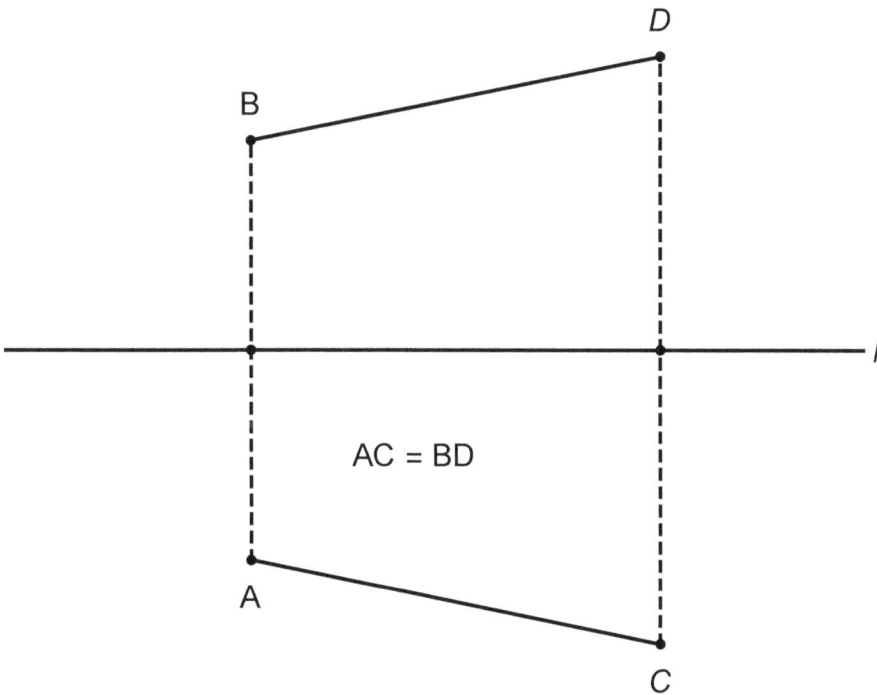

Figure 2.7: Reflection R_l, or symmetry in a straight line l, is an isometry

The straight line l is called the **axis of the reflection in l, or the symmetry in l, R_l**. We observe that for any point $A \in l$, $R_l(A) = A$ and therefore $R_l(l) = l$. We say that l is invariant under R_l point-wise (i.e., point by point). Also, for a straight line $k \neq l$, $R_l(k) = k$ iff $k \perp l$. Such a straight line is invariant by R_l set-wise (i.e., as a whole set). In **Figure 2.7**, the lines AB and CD are invariant set-wise.

We note that the identity is not a reflection. That is, for any line l in the plane $\mathcal{P}$, $R_l \neq I$. So, *the set of reflections does not contain the identity and therefore is not an algebraic group.*

2.2   General Properties of Plane Isometries

The set of isometries of the plane has the following properties.

1. Every isometry is continuous at every point of the plane and it is uniformly continuous on the whole plane.

2. Every isometry is one to one.

3. Every isometry is onto.

4. The inverse of an isometry is an isometry.

5. The composition of two isometries is an isometry.

6. The identity mapping is an isometry.

7. The set of the plane isometries is an algebraic group under the operation of composition, which is non-commutative.

8. An isometry maps a straight line to a straight line (a straight segment to a straight segment and the interior of a straight segment or a vector to the interior of a straight segment or a vector), a circle to an equal circle, a triangle to a directly or oppositely equal triangle, an oriented angle to an equal or opposite oriented angle.

Proofs: We Consider $F,\ G\ :\ \mathcal{P} \longrightarrow \mathcal{P}$ two isometries of the plane.

(1) We consider any $A \in \mathcal{P}$ and we allow the point B to approach A. Then as $B \longrightarrow A$, we have $d(B, A) \longrightarrow 0$. But then,

$$d[F(B), F(A)] = d(B, A) \longrightarrow 0, \quad \text{as} \quad B \longrightarrow A.$$

Thus, by **Definition 1.1.1, Property 2** (non-degeneracy),

$$F(B) \longrightarrow F(A), \quad \text{as} \quad B \longrightarrow A.$$

By the definition of continuity, F is continuous at A.

Now the uniform continuity follows from the fact that $\forall\ A,\ B$ points of $\mathcal{P}$, it holds that $d[F(A), F(B)] = d(A, B)$. Therefore, for any given $\epsilon > 0$ and any $A,\ B$ points of $\mathcal{P}$, if $d(A, B) < \delta := \epsilon$, then $d[F(A), F(B)] < \epsilon$, which is the definition of uniform continuity.

(2) Suppose that for $A,\ B \in \mathcal{P}$, we have $F(A) = F(B)$. Then,

$$d(A, B) = d[F(A), F(B)] = 0.$$

Hence, by **Definition 1.1.1, property 2** (non-degeneracy), $A = B$, which proves that F is one to one.

(3) To prove that F is onto we must know some point-set topology. We pick any point $B \in \mathcal{P}$ and must prove that there is a point $A \in \mathcal{P}$ such that

$$F(A) = B.$$

For a point $O \in \mathcal{P}$, let $Q = F(O)$. If $B = Q$, then $A = O$. Otherwise, we let $r = d(Q, B)$ and consider the circles $\mathcal{C}_1 = C[O, r]$ and $\mathcal{C}_2 = C[Q, r]$ with equal radii r and centers O and Q respectively. Since F is an isometry, we have $F(\mathcal{C}_1) \subseteq \mathcal{C}_2$. See **Figure 2.8**.

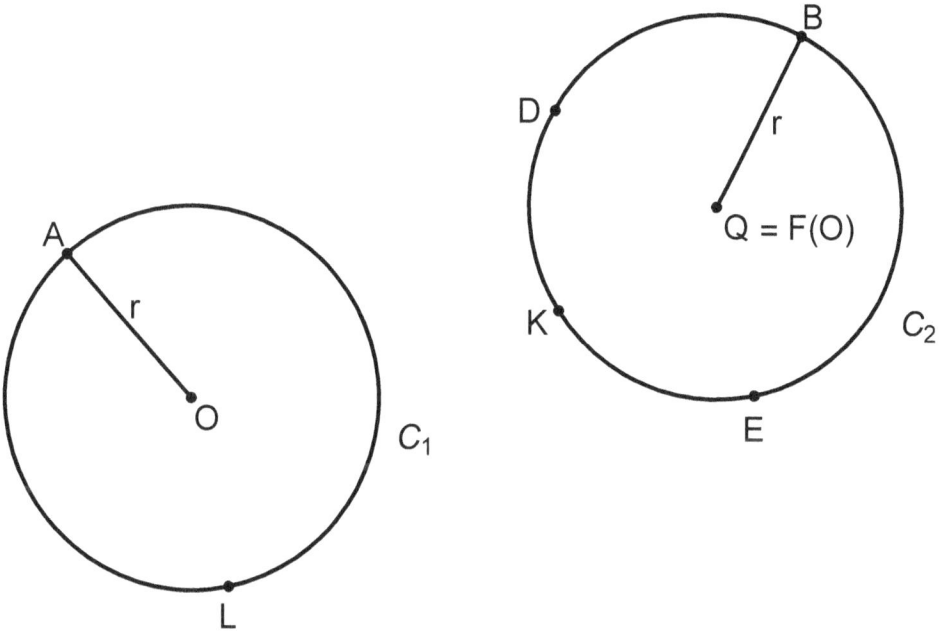

Figure 2.8: Isometries are onto mappings

Now, $\mathcal{C}_1$ is a compact and connected set and F is continuous. So, $F(\mathcal{C}_1)$ is a compact and connected subset of $\mathcal{C}_2$. Since F is one to one, $F(\mathcal{C}_1)$ cannot be just one point. Therefore, $F(\mathcal{C}_1)$ is either the whole $\mathcal{C}_2$ or a closed connected arc $\widehat{DE} \subset \mathcal{C}_2$ of it with non-empty interior. In the latter case, we pick any interior point K of $F(\mathcal{C}_1)$ and let $L \in \mathcal{C}_1$ be the unique point such that $F(L) = K$. Then, the set $F(\mathcal{C}_1) - \{K\} = F(\mathcal{C}_1 - \{L\})$ has two connected components, the half-closed and half-open arcs $\widehat{KD}$ and $\widehat{KE}$, with K not included, whereas $\mathcal{C}_1 - \{L\}$ has one connected component. This is impossible because F is continuous. So, $F(\mathcal{C}_1) = \mathcal{C}_2$. Since $B \in \mathcal{C}_2 \subset \mathcal{P}$, there is an $A \in \mathcal{C}_1 \subset \mathcal{P}$ such that $F(A) = B$ proving that $F : \mathcal{P} \longrightarrow \mathcal{P}$ is onto.

(4) Since F is one to one and onto the set-theoretic inverse F^{-1} is a well-defined function. Suppose $F^{-1}(Y) = X$ and $F^{-1}(Z) = W$. Then $F(X) = Y$

and $F(W) = Z$. Since F is an isometry, we get

$$d[F(X), F(W)] = d(X, W),$$

and so

$$d(Y, Z) = d[F^{-1}(Y), F^{-1}(Z)]$$

proving that F^{-1} is an isometry.

(5) For any two points A, $B \in \mathcal{P}$ we have:

$$d[G \circ F(A), G \circ F(B)] = d\{G[F(A)], G[F(B)]\} = d[F(A), F(B)] = d(A, B)$$

proving that $G \circ F$ is an isometry.

(6) Since for the identity mapping we have $I(A) = A$, for all $A \in \mathcal{P}$, the result follows immediately.

(7) Let $\mathcal{I}$ be the set of the plane-isometries. Properties (4), (5), (6) prove that $(\mathcal{I}, \circ)$ is an algebraic group, since the composition of relations is always associative. But, composition is not commutative, in general, and so $(\mathcal{I}, \circ)$ is not a commutative group.

(8) Consider any straight line k and two different points A and B on k. We must prove that $F(k)$ is the straight line l through the points $F(A)$ and $F(B)$. That is, a straight line is mapped onto a straight line by isometries.

Of course, the points $F(A)$ and $F(B)$ belong to $F(k)$. We now pick any point $C \in k$ other than A and B. See **Figure 2.9**.

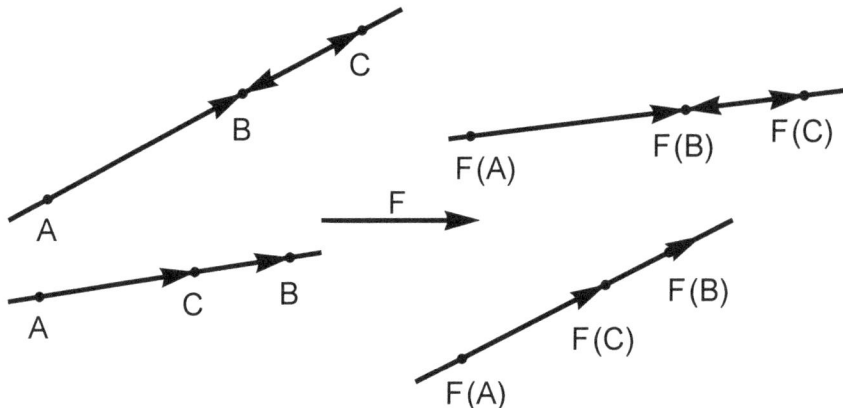

Figure 2.9: **Isometries map straight lines to straight lines**

Then, with A, B, and C on a straight, such as k, we have

$$\overrightarrow{AC} + \overrightarrow{CB} = \overrightarrow{AB} \qquad \text{and} \qquad \overrightarrow{F(A)F(C)} + \overrightarrow{F(C)F(B)} = \overrightarrow{F(A)F(B)}.$$

Also, since F is an isometry, we have that the distances satisfy

$$d(A, B) = d[F(A), F(B)],$$
$$d(A, C) = d[F(A), F(C)],$$
$$d(B, C) = d[F(B), F(C)].$$

Now, we use the triangle inequality, which says: *"the sum of any two sides of a triangle is greater than the third side which also implies that the absolute value of the difference of two sides is less than the third side"*. This implies that the point $F(C)$ cannot be outside the line l, defined by the points $F(A)$ and $F(B)$, for otherwise $F(A)F(B)F(C)$ would be an honest triangle and we would have a contradiction. That is, $F(C) \in l$. Also, we easily prove that if C is between A and B, then $F(C)$ is between $F(A)$ and $F(B)$ and if C is outside AB, then $F(C)$ is outside $F(A)F(B)$. (Write the argument in some detail.)

Thus, $F(k) \subseteq l$. Since F is one to one and onto, and preserves distances, it must be $F(k) = l$. (Complete the argument!) See **Figure 2.9**.

Next, by an argument similar to **(3)**, we have that a circle $C[K, r]$ with center K and radius r has image

$$F\{C[K, r]\} = C[F(K), r],$$

the circle with center $F(K)$ and radius r.

Finally, since F preserves distances, it maps any triangle ABC to another triangle $A'B'C'$ with corresponding sides equal. If these triangles are directly equal, then their corresponding oriented angles are equal. If they are oppositely equal, then their corresponding oriented angles are opposite.

∎

In particular, we have the following **basic results for the basic isometries**. [Their proofs follow from the properties of isometries exposed above, especially **(8)**, or are basic straightforward exercises left to the reader.]

(a) A **parallel translation** $T_{\vec{u}}$ with $u \neq \vec{0}$, so that $T_{\vec{u}}$ is not the identity, maps: See **Figures 2.10** and **2.11**.

(1) A straight line k to a parallel straight line $l = T_{\vec{u}}(k)$, i.e., $(l \parallel k)$. Also, $l \neq k \Longleftrightarrow \vec{u} \nparallel l$, and $k = l$ set-wise iff $l \parallel \vec{u}$.

(2) A circle $C[O, r]$ to an equal circle $C[O', r]$, such that $\overrightarrow{OO'} = \vec{u}$.

(3) A straight segment AB to straight segment $A'B' = T_{\vec{u}}(AB)$, such that, $A'B' = \parallel AB$ and $\overrightarrow{AA'} = \vec{u} = \overrightarrow{BB'}$.

(4) A vector $\overrightarrow{AB}$ to vector $\overrightarrow{A'B'} = T_{\vec{u}}(\overrightarrow{AB})$, such that, $\overrightarrow{A'B'} = \overrightarrow{AB}$ and $\overrightarrow{AA'} = \vec{u} = \overrightarrow{BB'}$.

(5) A half-line $\overrightarrow{Ox}$ to half line $\overrightarrow{O'x'} = T_{\vec{u}}\overrightarrow{Ox}$, such that, $\overrightarrow{O'x'} =\parallel \overrightarrow{Ox}$, the half-lines $\overrightarrow{O'x'}$ and $\overrightarrow{Ox}$ have the same direction and $\overrightarrow{OO'} = \vec{u}$.

(6) An oriented angle $\angle(Ox, Oy)$ to another equal oriented angle $(O'x', O'y')$. [Combine this with (5) above. $T_{\vec{u}}$ preserves orientation. Parallel translation is direct isometry.]

(7) A triangle ABC to a directly equal triangle $A'B'C'$.

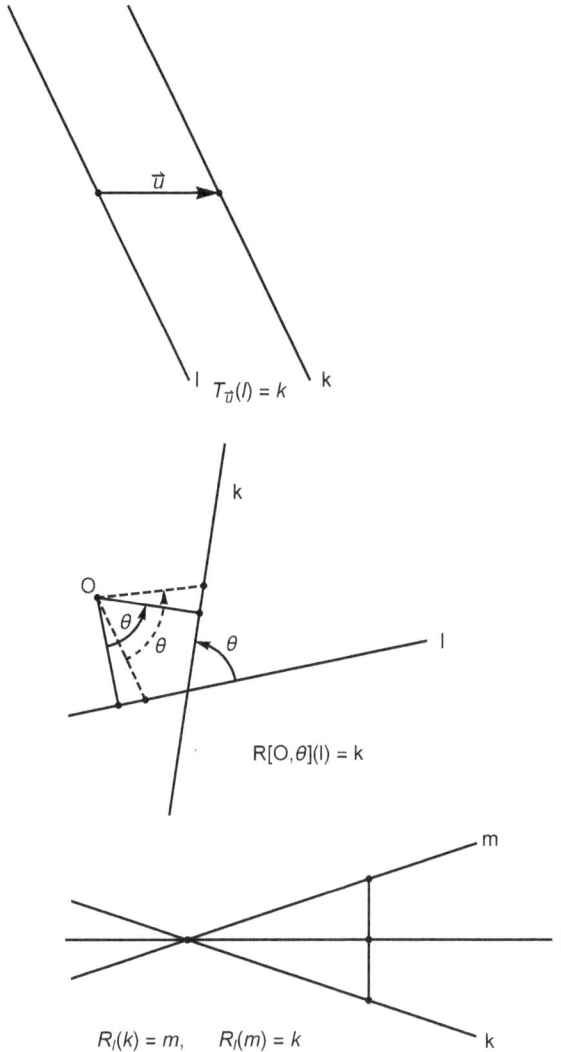

Figure 2.10: Straight lines are mapped to lines by the basic isometries

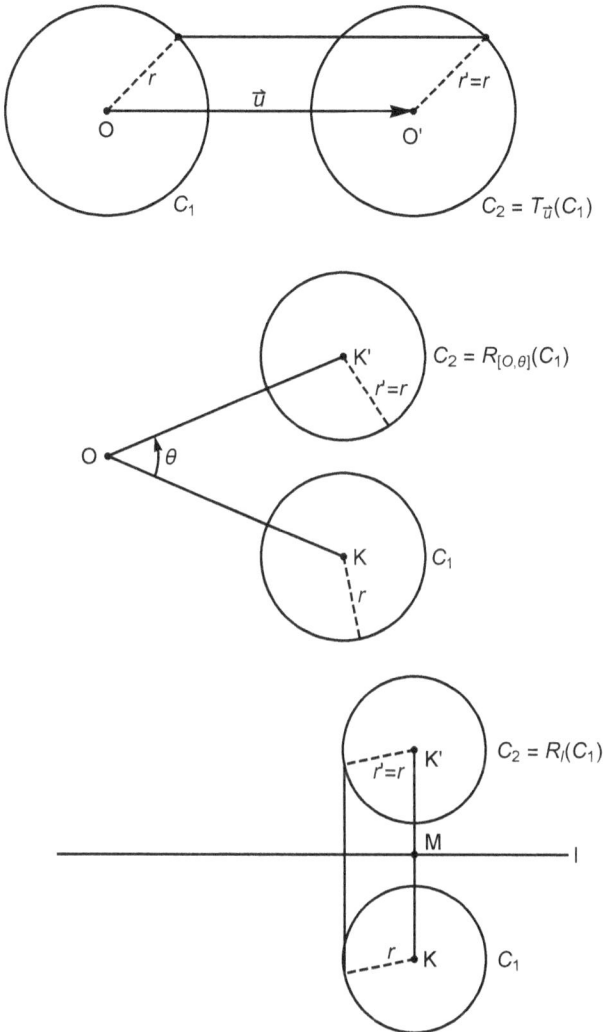

Figure 2.11: Circles mapped to circles by the basic isometries

(b) A **rotation** $R_{[O,\theta]}$ with $\theta \neq 0 \bmod 2\pi$, so that $R_{[O,\theta]}$ is not the identity, maps: (See **Figures 2.10** and **2.11**.)

(1) A straight line l to a straight line k, such that $\angle(l,k) = \theta$. Also, $k = R_{[O,\theta]}(l) = l$ set-wise iff $\theta = 0 \bmod \pi$.

(2) A circle $C[K,r]$ to an equal circle $C[K',r]$, such that $OK' = OK$ and $\angle(OK,OK') = \theta$.

(3) A straight segment AB to straight segment $A'B' = R_{[O,\theta]}(AB)$, such that, $A'B' = AB$ and $\angle(AB, A'B') = \theta$.

(4) A vector $\overrightarrow{AB}$ to vector $\overrightarrow{A'B'} = R_{[O,\theta]}(\overrightarrow{AB})$ of equal length, such that, the $\angle(\overrightarrow{AB}, \overrightarrow{A'B'}) = \theta$.

(5) An oriented angle $\angle(Kx, Ky)$ to another equal oriented angle $(K'x', K'y')$. [$R_{[O,\theta]}$ preserves orientation. Rotation is direct isometry.]

(6) A triangle ABC to a directly equal triangle $A'B'C'$, with angle between corresponding sides θ.

(c) A **half-turn** $R_{[O,\theta]}$ with $\theta = \pi \mod 2\pi$, O center of symmetry, maps: (Draw figure for each item below.)
(1) A straight line l to a straight line $k = R_{[O,\pi]}(l)$, such that $l \parallel k$ and the two lines are symmetrical about O. Also, $k = l$ set-wise iff $O \in l$.
(2) A circle $C[K, r]$ to an equal circle $C[K', r]$ and the two circles are symmetrical about O. Also, $OK' = OK$ and $\angle(OK, OK') = \pi$.
(3) A straight segment AB to straight segment $A'B' = R_{[O,\pi]}(AB)$, such that, $A'B' = AB$ and the two segments are symmetrical about O.
(4) A vector $\overrightarrow{AB}$ to vector $\overrightarrow{A'B'} = R_{[O,\pi]}(\overrightarrow{AB})$, such that, the two vectors are **opposite** and symmetrical about O.
(5) An oriented angle $\angle(Kx, Ky)$ to another equal oriented angle $(K'x', K'y')$ and the two angles are symmetrical about the point O. [$R_{[O,\pi]}$ preserves orientation. Rotation is direct isometry.]
(6) A triangle ABC to a directly equal triangle $A'B'C'$, and the two triangles are symmetrical about O.

Center of Symmetry.   (Draw figure for each item below.)
(1) The midpoint M of a straight segment AB is the center of symmetry of AB.
(2) Every point of a straight line l is a center of symmetry of l.
(3) The center K of a circle $C[K, r]$ is the center of symmetry of $C[K, r]$.
(4) The point of intersection of two straight lines k and l is the center of symmetry of their figure.
(5) Every point of the mid-parallel line of two parallel straight lines $k \parallel l$ is a center of symmetry of their figure.
(6) The center of a canonical polygon with **even** number of sides is the center of symmetry of the polygon.
(7) The common point of the diagonals of a parallelogram is the point of symmetry of the parallelogram.
(8) If two figures $\mathfrak{F}_1$ and $\mathfrak{F}_2$ are symmetrical about a point O, then so is their union and their intersection.
(9) If two figures $\mathfrak{F}_1$ and $\mathfrak{F}_2$ have a common center of symmetry, then this is the center of symmetry of their union and their intersection.

(d) A **reflection** R_l in a straight line l, axis of symmetry, maps: See **Figures 2.10** and **2.11**.

(1) A straight line k to a straight line m symmetrical to each other with respect to l. The axis of symmetry l is the bisector of the angle $\angle(k,l)$. Also, $R_l(l) = l$ point-wise, and $R_l(k) = k$ set-wise iff $k \perp l$.

(2) A circle $C[K,r]$ to an equal circle $C[K',r]$, symmetrical to each other with respect to l and so K' and K are points symmetrical with respect to l.

(3) A straight segment AB to an equal straight segment $A'B'$ of equal length, symmetrical to each other in l.

(4) A vector $\overrightarrow{AB}$ to a vector $\overrightarrow{A'B'}$, symmetrical to each other in l.

(5) An oriented angle to an **opposite** oriented angle, symmetrical to each other in l. (Reflection is opposite isometry.)

(6) A triangle ABC to an **oppositely equal** triangle $A'B'C'$.

Axis of symmetry.   (Draw figure for each item below.)

(1) A straight line has axes of symmetry itself and any straight line perpendicular to it.

(2) A straight segment has axes of symmetry its own straight line and the midpoint perpendicular to it.

(3) A circle has axes of symmetry every diameter of it. If a figure has an axis of symmetry in every direction, then it is a circle.

(4) An angle has, as axis of symmetry, its bisector.

(5) Two intersecting straight lines has axes of symmetry the bisectors of their two angles.

(6) Two parallel lines have axes of symmetry their mid-parallel line and any straight line perpendicular to them.

(7) An isosceles triangle has axis of symmetry the bisector of the angle between the two equal sides.

(8) A canonical polygon with n sides has n axes of symmetry. The midpoint perpendicular lines of its sides if n is odd, and the midpoint perpendicular lines of its sides and its diagonals defined by the pairs of opposite vertices, if n is even.

(9) If two figures $\mathfrak{F}_1$ and $\mathfrak{F}_2$ are symmetrical about a straight line l, then so is their union and their intersection.

(10) If two figures $\mathfrak{F}_1$ and $\mathfrak{F}_2$ have a common axis of symmetry l, then l is the axis of symmetry of their union and their intersection.

[See **Item (6)** in **Section 2.3** and state and justify the analogous basic results about **glides**. See also **Item (9)** and **Item (10) of Subsection 4.4.3** for the direct similarities between straight lines or circles (before the part of "Some Additional Results") and also "About opposite similarities between two straight lines or two circles" at the **end of Chapter 3** before **Section 4.6** of Applications. Among straight lines some of these similarities are isometries, but this is always the case among equal circles.]

2.2.1 Fixed Points and Determination of Isometries

Definition 2.2.1 *A point $x \in X$ is called a **fixed point** of a set-theoretic function $f : X \longrightarrow X$, if $f(x) = x$.*

We notice that:

1. The parallel translation $T_{\vec{u}} : \mathcal{P} \longrightarrow \mathcal{P}$ has no fixed point if $\vec{u} \neq \vec{0}$, since all points are moved by $\vec{u} \neq \vec{0}$. If $\vec{u} = \vec{0}$, then $T_{\vec{u}} = I$ (the identity) and all the points of the plane $\mathcal{P}$ are fixed points.

2. Any rotation $R_{[O,\theta]} : \mathcal{P} \longrightarrow \mathcal{P}$ in which $\theta \neq 0$, (mod 2π), has exactly one fixed point, namely the O. If $\theta = 0$, (mod 2π), then $R_{[O,\theta]} = I$ (the identity) and all the points of the plane $\mathcal{P}$ are fixed points.

3. The reflection $R_l : \mathcal{P} \longrightarrow \mathcal{P}$ in a straight line l has fixed points all the points of the straight line l, its axis, and no other.

We continue with some preliminary results in order to prove that an isometry is completely determined by three non-collinear points and their images and maps all oriented angles to equal oriented angles or reverses all angles to their opposite. We begin with the following Lemma.

Lemma 2.2.1 *Consider two triangles ABC and $A'B'C'$ with equal corresponding sides ($BC = B'C'$, $CA = C'A'$ and $AB = A'B'$) and a point P of the plane not on the straight line AB. Then:*
(1) The circles $C[A', AP]$ and $C[B', BP]$ intersect at two points P' and P'', such that $CP = C'P'$ and P'' is the symmetrical of P' in the straight line $A'B'$. The circle $C[C', CP]$ passes through P' (along with the circles $C[A', AP]$ and $C[B', BP]$).
(2) $C'P'' \neq C'P'$ (the two segments do not have equal lengths).
(3) If the triangles ABC and $A'B'C'$ are directly equal (they have the same orientation), then the triangles ABP and $A'B'P'$ are not degenerate and are directly equal. (Also, if the triangles BCP and $B'C'P'$ are not degenerate, then they are directly equal. Similar conclusion for the triangles ACP and $A'C'P'$.)
(4) If the triangles ABC and $A'B'C'$ are oppositely equal (they have opposite orientations), then the triangles ABP and $A'B'P'$ are not degenerate and are oppositely equal. (Also, if the triangles BCP and $B'C'P'$ are not degenerate, then they are oppositely equal. Similar conclusion for the triangles ACP and $A'C'P'$.)
(Analogous results, if in the above hypotheses we use the sides BC or CA, instead of AB.)

Proof In **Figure 2.12** the given triangles ABC and $A'B'C'$ are directly equal and in **Figure 2.13** are oppositely equal.

We will prove the results when ABC and $A'B'C'$ are directly equal and we have analogous work when they are oppositely equal. We consider P not to belong to any of the straight lines of the sides of the triangle in order to avoid the degenerate cases of triangles (that is, they become straight segments). But, in the end, we can see that the results are also valid in the degenerate cases as well. Also, there are more subcases depending on the relative positions of C and P. For keeping the exposition short, we are not going to examine all of the subcases but what we prove gives sufficient information to the reader to easily check the remaining ones.

(1) Since $AB = A'B'$ and ABP is a non-degenerate triangle, the circles $C[A', AP]$ and $C[B', BP]$ intersect at two points P' and P'' which are symmetrical with respect to the line $A'B'$. We give ABP the same orientation as ABC. The lines AB and $A'B'$ separate the plane into two half planes. If P and C are in the same half plane with respect to the straight line AB, then we choose P' to be on the same half plane as C' with respect to the straight line $A'B'$. Otherwise, we choose P' in the other half plane.

With **Figures 2.12** and **2.13**, we treat the subcase with the points P and C in the same half plane and also the points P' and C' in the same half plane. (Analogous work otherwise.) In **Figure 2.12**, the triangles $A'B'P'$ and ABP are directly equal by construction. Therefore, $\measuredangle(BP, BA) = \measuredangle(B'P', B'A')$.

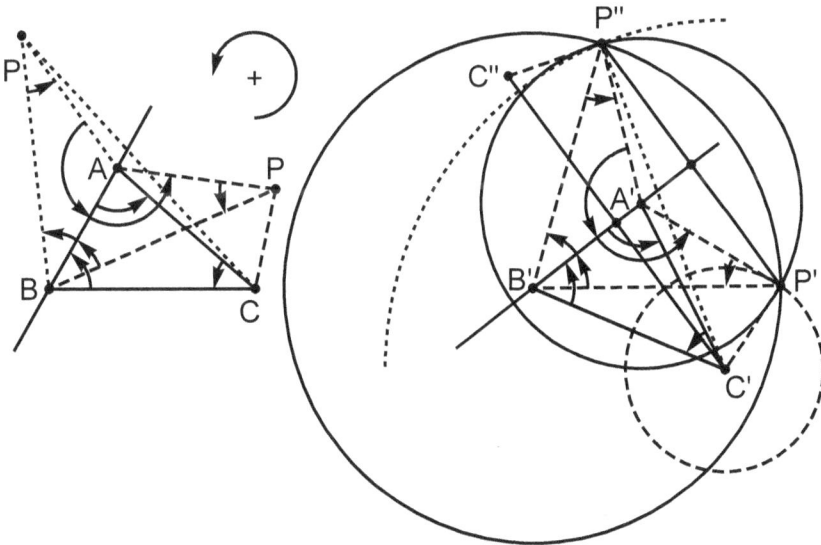

Figure 2.12: For Lemma 2.2.1 with the triangles ABC and $A'B'C'$
directly equal

Then, the triangles BCP and $B'C'P'$ are directly equal since $BC = B'C'$, $BP = B'P'$ and

$$\measuredangle(BC, BP) = \measuredangle(BC, BA) - \measuredangle(BP, BA)$$
$$= \measuredangle(B'C', B'A') - \measuredangle(B'P', B'A') = \measuredangle(B'C', B'P').$$

So, $CP = C'P'$ and therefore the circle $C[C', CP]$ passes through P'.

(2) Let C'' be the point symmetrical to C' in the straight line $A'B'$ as P'' is also symmetrical to C'. Then, $C'P'P''C''$ is an isosceles trapezium, which is not a skew parallelogram. (Why?) In such a trapezium the two diagonals are equal and greater than each of the two equal sides. (Check this!) So, in the case examined here, $C'P'$ is one of the equal sides of this trapezium and $C'P''$ is a diagonal. Therefore, $C'P' < C'P''$.

(Check that if P' and C' lie in different half planes with respect to the straight line $A'B'$, then we have the opposite inequality $C'P' > C'P''$.)

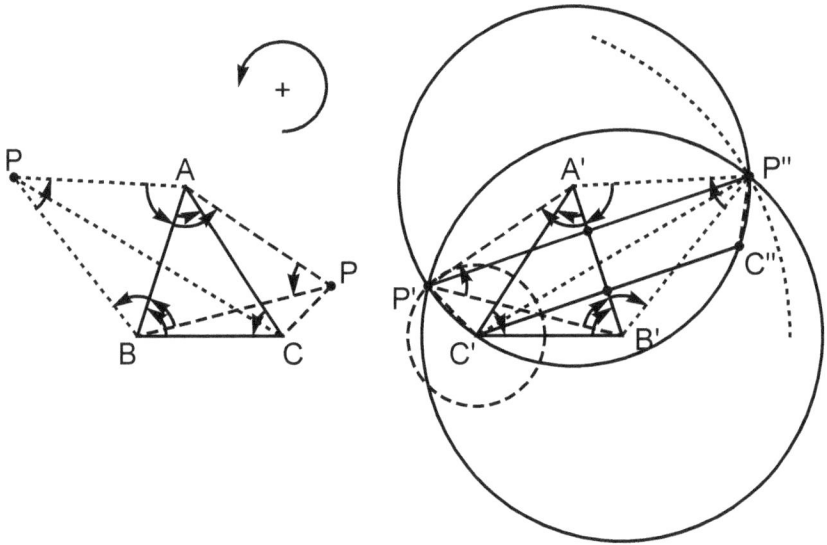

Figure 2.13: For Lemma 2.2.1 with the triangles ABC and $A'B'C'$
oppositely equal

(3) The triangles ABC and ABP have the same orientation and so do the triangles $A'B'C'$ and $A'B'P'$. Since the triangles ABC and $A'B'C'$ have the same orientation by hypothesis, we conclude that the triangles ABP and $A'B'P'$ have the same orientation. Since they also have equal corresponding side the triangles ABP and $A'B'P'$ are directly equal.

(4) The triangles ABC and ABP have the same orientation and so do the triangles $A'B'C'$ and $A'B'P'$. Since the triangles ABC and $A'B'C'$ have opposite orientation by hypothesis, we conclude that the triangles ABP and $A'B'P'$ have opposite orientation. Since they also have equal corresponding side the triangles ABP and $A'B'P'$ are oppositely equal.

∎

Theorem 2.2.1 (Determination of an isometry by 3 pairs of points)
An isometry $F : \mathcal{P} \longrightarrow \mathcal{P}$ is completely determined by the images A', B', and C' of three given points A, B, and C of the plane that do not lie on a straight line (3 non-collinear points).

Proof We have $F(A) = A'$, $F(B) = B'$, and $F(C) = C'$. The three given points not lying on the same straight line determine a triangle ABC. Since F is an isometry, it must be $AB = A'B'$, $BC = B'C'$, and $CA = C'A'$. So the triangle $A'B'C'$ is directly or oppositely equal to ABC. See **Figures 2.14** and **2.15**.

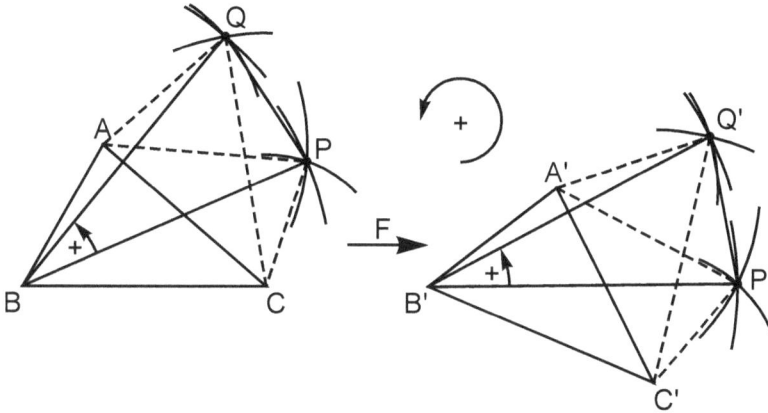

Figure 2.14: **Determination of an isometry by a triangle and its image. Here, the triangles ABC and $A'B'C'$ are directly equal**

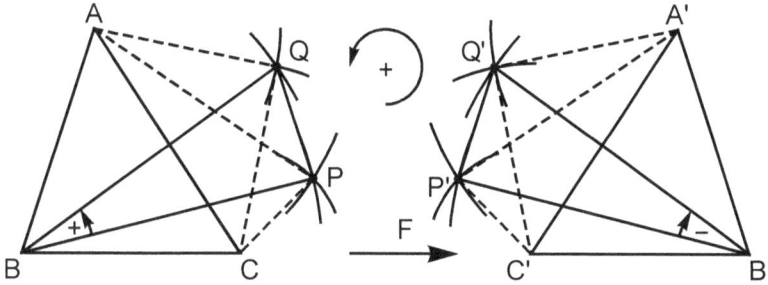

Figure 2.15: **Determination of an isometry by a triangle and its image. Here, the triangles ABC and $A'B'C'$ are oppositely equal**

Now given any point $P \in \mathcal{P}$, the segments AP, BP, and PC are known. The three circles $C[A, AP]$, $C[B, BP]$, and $C[C, CP]$ intersect just at the point P. Then, since F is an isometry, the image of P, $P' = F(P)$, must satisfy the conditions $AP = A'P'$, $BP = B'P'$, and $CP = C'P'$. But by **Lemma 2.2.1**, **(1)** and **(2)**, the three circles $C[A', AP]$, $C[B', BP]$, and $C[C', CP]$ intersect exactly at one point and therefore this must be the point $P' = F(P)$.

Also the mapping F defined by $F(P) = P'$, with P' constructed as the intersection of the three circles $C[A', AP]$, $C[B', BP]$, and $C[C', CP]$, is an isometry. Indeed, if we pick another point Q in the plane and we find the Q' in the same way, the we have that $BP = B'P'$, $BQ = B'Q'$ and $\angle(BP, BQ) = \pm\angle(B'P', B'Q')$. So, the triangles PBQ and $P'B'Q'$ are either directly or oppositely equal. Therefore $PQ = P'Q'$. That is, $d[F(P), F(Q)] = d(P, Q)$ and so F is an isometry.

$\blacksquare$

Remarks: (1) If in the **previous Theorem**, the three points A, B, and C are on the same straight line l (collinear), then the image of l under an isometry F is a straight line l'. If a point P is not on l, then the intersection of the three circles, as in the proof of the **Theorem**, is two points P' and P'', which are symmetrical in the line l'. If a point Q is in l, then this intersection is one point $Q' \in l'$. In the first case there are two choices, P' and P'', for the image of P under F and so F is not uniquely determined. See **Figure 2.16**.

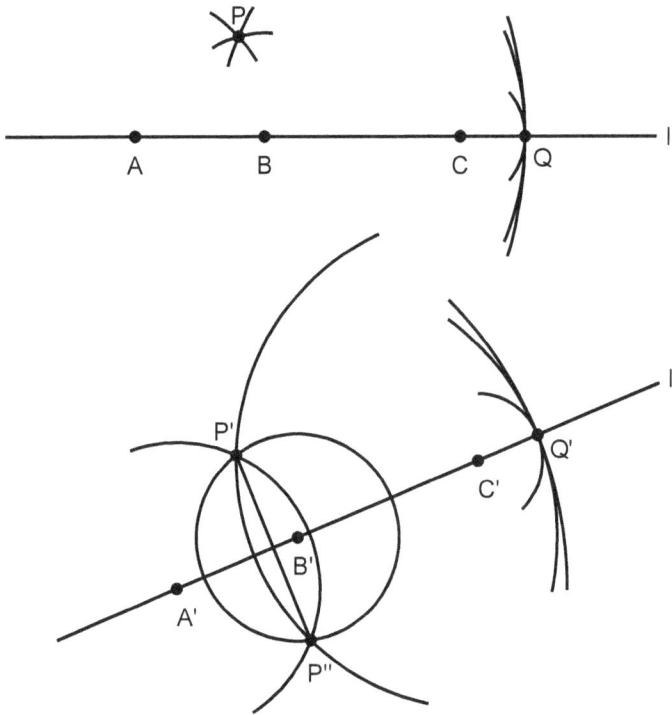

Figure 2.16: Three collinear points or two points only and their images may not determine a unique isometry

(2) If we assign the images A' and B' of two points A and B, such that $AB = A'B'$, then there are exactly two isometries, one direct and one opposite, that map A to A' and B to B'. This conclusion is established through this Chapter. (See, **General Conclusion Drawn from Chapter 2**, at the end of this Chapter, before the **Section 2.6**.)

Since an isometry preserves distances (lengths), maps a triangle to either a directly equal or an oppositely equal triangle. By comparing distances and using the **Lemma 2.2.1** and its proof, and **Theorem 2.2.1**, we can prove the following corollary:

Corollary 2.2.1 *If an isometry F maps **one** triangle to a directly equal triangle or to an oppositely equal triangle, then F maps any triangle to a directly equal triangle or to an oppositely equal triangle, respectively. (No intermixed case.)*

Proof See **Figures 2.17** and **2.18**.

In **Figure 2.17** the isometry F maps the triangle ABC to $A'B'C'$ and the two triangles are directly equal.

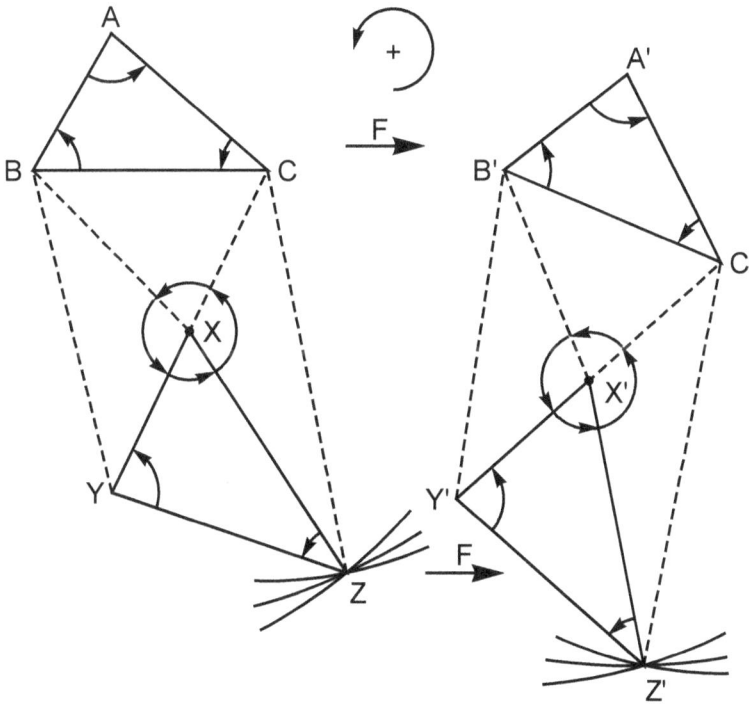

Figure 2.17: If an isometry maps one triangle to a directly equal triangle, then it does the same thing to any other triangle

Now, consider any triangle XYZ with the same orientation as ABC and let $X'Y'Z'$ be its image under the isometry F. We must show that XYZ is directly equal to $X'Y'Z'$. Since they are equal, we only need to show that they have the same orientation by showing that their corresponding oriented angles are equal (not opposite).

We give the triangles XCB, BYX and XZC the same orientation as ABC. In **Lemma 2.2.1**, we proved that the triangle XCB has the same orientation as the triangle $X'C'B'$ and so the two triangles are directly equal. Then, for the same reason, the triangle YXB has the same orientation as $Y'X'B'$ and so they are directly equal and so are the triangles ZCX and $Z'C'X'$. Therefore, using equality of angles we get $\measuredangle(XY, XZ) = \measuredangle(X'Y', X'Z')$. Similarly we prove that $\measuredangle(ZX, ZY) = \measuredangle(Z'X', Z'Y')$ and $\measuredangle(YZ, YX) = \measuredangle(Y'Z', Y'Z')$. Hence, the triangles XYZ and $X'Y'Z'$ have the same orientation and are directly equal.

In **Figure 2.18** the isometry F maps the triangle ABC to $A'B'C'$ and the two triangles are oppositely equal.

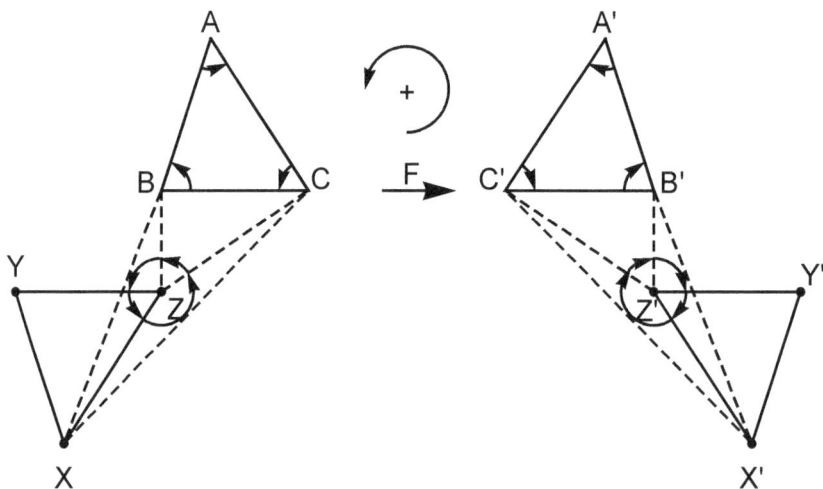

Figure 2.18: If an isometry maps one triangle to an oppositely equal triangle, then it does the same thing to any other triangle

Then, if for a triangle XYZ, its image under this isometry is the triangle $X'Y'Z'$, these two triangles are oppositely equal. The proof is analogous to the previous case.

∎

Since for any angle in $(0, \pi)$ we can consider a triangle containing it, and for any angle in $(\pi, 2\pi)$ we can consider its vertical angle, and in degenerate cases we use **Property (8)** of **Section 2.2**, we obtain the following corollary.

Corollary 2.2.2 *An isometry of the plane either maps any oriented angle to an equal angle or any oriented angle to an opposite angle. (No intermixed case.)*

So, we give the definition:

Definition 2.2.2 *An isometry of the plane that maps oriented angles to equal angles is called* **direct isometry** *and an isometry that maps oriented angles to opposite angles is called* **opposite isometry**.

We also have the following corollary of **Theorem 2.2.1**.

Corollary 2.2.3 *If an isometry* $F : \mathcal{P} \longrightarrow \mathcal{P}$ *has three different fixed points that do not lie on the same straight line, then* $F = I$, *the identity.*

Proof This follows immediately from the proof of the **Theorem**, since $A' = F(A) = A$, $B' = F(B) = B$, and $C' = F(C) = C$.

Theorem 2.2.2 *Suppose an isometry* $F : \mathcal{P} \longrightarrow \mathcal{P}$ *is not the identity and has two different fixed points* A *and* B. *Then, all the points of the straight line* $l := AB$ *are also fixed, and* $F = R_l$, *the reflection in* l.

Proof We have $F(A) = A$ and $F(B) = B$. We now consider any point C on the line $l := AB$, other than A and B. (Whenever we write $l := AB$, we mean the whole straight line l that passes through the points A and B.) The image of C, $F(C) = C'$, satisfies

$$d(A, C') = d[F(A), F(C)] = d(A, C),$$
$$d(B, C') = d[F(B), F(C)] = d(B, C).$$

Therefore, C' must lie on the two circles with centers A and B and radii $d(A, C) = d(A, C')$ and $d(B, C) = d(B, C')$, respectively. But, if, e.g., C is between A and B, as in **Figure 2.19**, then

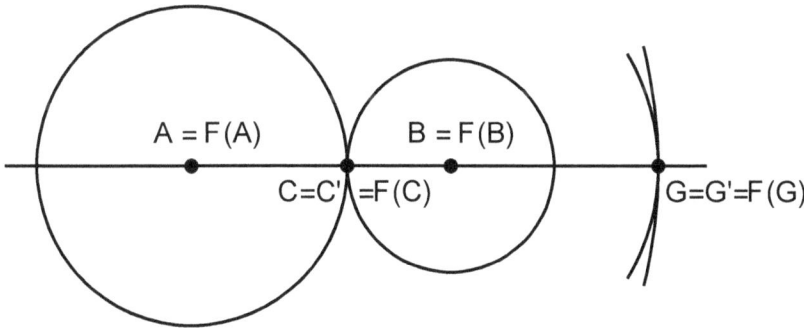

Figure 2.19: An isometry different than the identity and
with two fixed points is a reflection

$$d(A, C) + d(C, B) = d(A, B),$$

since A, B, C are on the same line. So,

$$d(A, C') + d(C', B) = d(A, B).$$

Hence, the two circles are externally tangent to each other at the point C and C'. Therefore, $C = C' = F(C)$, i.e., C is a fixed point.

Analogous work is done for a point G on the line $l := AB$ and outside the segment AB In this case, the two circles are internally tangent. Finally, all the points of the line AB are fixed points.

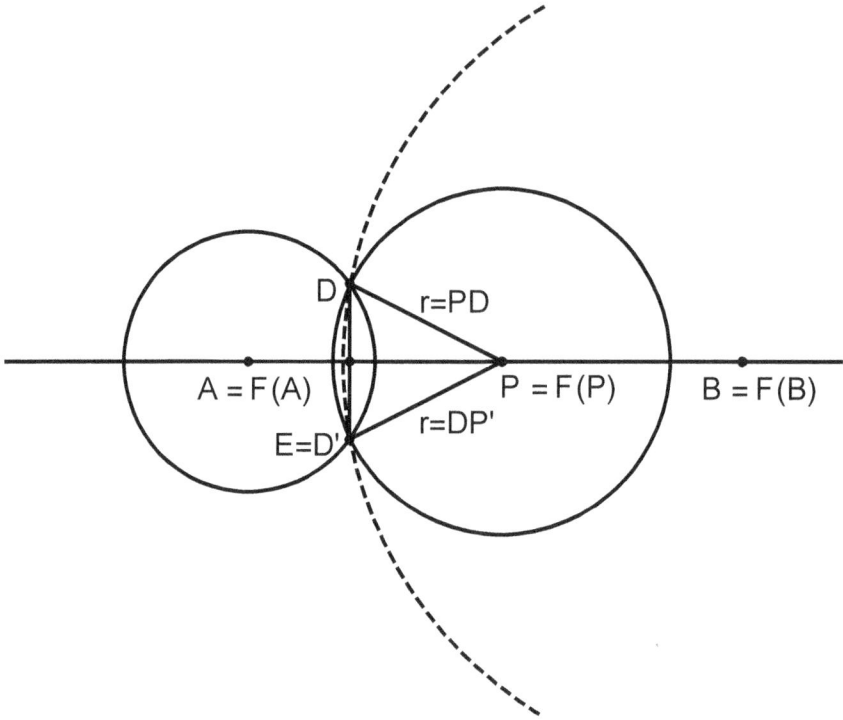

**Figure 2.20: An isometry different than the identity and
with two fixed points is a reflection**

We now consider any point D not on the line AB, as in **Figure 2.20**. By
the previous **Corollary**, the point D cannot be fixed, since F is assumed to
be other than the identity. That is, $D' := F(D) \neq D$. We consider the circles
with centers the points P of the line $l := AB$, and radii the segments PD. For
example, $P = A$, B, any P on l, as in **Figure 2.20**. Besides D, these circles
have another point of intersection E, because D is not on their center line AB.
So, $d(P, E) = d(P, D)$ for every such circle and E is symmetrical to D about l.

Also, for the image of D, $F(D) = D'$, we have that for every point $P \in l$

$$d(P, D) = d[F(P), F(D)] = d(P, D'),$$

because F is an isometry and P is a fixed point. So, the image of D, $F(D) = D'$,
lies on these circles too. Then, by basic geometry, it must be that $E = D'$ and
the center-line of the above circles, $l = AB$, is the perpendicular bisector of the
segment DD'. So, $D' = F(D)$ is symmetrical to D with respect to the line l,
and so it coincides with E. ($D' = E$.) Therefore, $F = R_l$ is the reflection in the
line $l := AB$.

2.2.2 Fixed Sets

Definition 2.2.3 *A subset $S \subseteq X$ is called a **fixed** or **invariant set** of a set-theoretic function $f : X \longrightarrow X$, if $f(S) = S$.*

   This does not necessarily mean that $\forall\ u \in S$, $f(u) = u$. It simply means that the two sets S and $f(S)$ are equal. That is, $f(S)$ is a shuffling of S. Of course, it may happen that $f(u) = u$, for every $u \in S$, i.e., $f_{|S} = I_{|S}$, which implies that $f(S) = S$. In this case, we say that the set S is **fixed by** f **point-wise**. Otherwise, we say that S is **fixed by** f **set-wise**. Obviously, if S is fixed by f and g, then S is also fixed by $g \circ f$ and any composition $f \circ f \circ \ldots \circ f$.

   We can geometrically prove as easy exercises the following **examples**, as depicted in **Figure 2.21**.

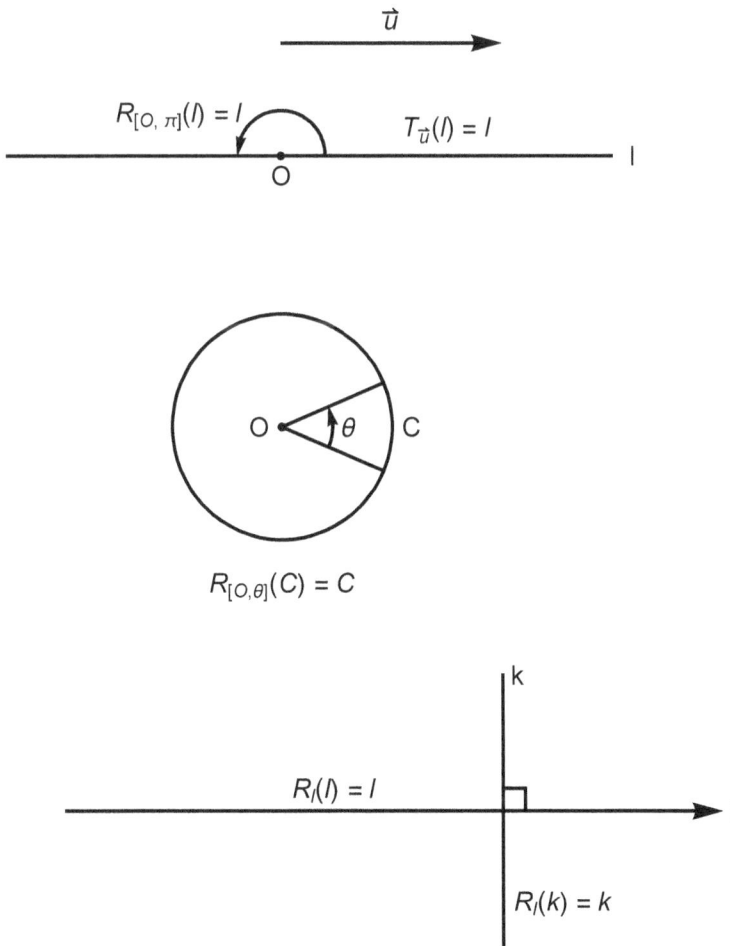

Figure 2.21: Some fixed sets of the three basic isometries

1. A straight line l is a set-wise fixed set of a parallel translation $T_{\vec{u}}$ with $\vec{u} \neq \vec{0}$ if and only if $l \parallel \vec{u}$.

   A straight line l is also a set-wise fixed set of a half-turn about any of its points.

   If $\vec{u} = \vec{0}$, then $T_{\vec{u}} = I$ (the identity) and all the points of $\mathcal{P}$ are fixed points. So, if $\vec{u} = \vec{0}$, any subset $S \subseteq \mathcal{P}$ is point-wise fixed for $T_{\vec{0}} = I$.

2. A circle is a set-wise fixed set of a rotation $R_{[O,\theta]}$ with $\theta \neq 0$, (mod 2π) if and only if the center of the circle is the center O of the rotation.

   The singleton $\{O\}$ is fixed for any rotation $R_{[O,\theta]}$.

   If $\theta = 0$, (mod 2π), then $R_{[O,\theta]} = I$ (the identity) and all the points of $\mathcal{P}$ are fixed points. So, if $\theta = 0$, (mod 2π), any subset $S \subseteq \mathcal{P}$ is point-wise fixed.

3. A straight line k is a set-wise fixed set of a reflection R_l if and only if $k = l$ or $k \perp l$. But, the axis of the reflection l is a point-wise fixed set.

Remark: Notice that a union of fixed sets is a fixed set. In this way, we can find more fixed sets of functions. [See **case 1 of paragraph (3)** below and **Section 2.5**. For example, any set symmetrical with respect to a line l is a set-wise fixed set of the reflection R_l, etc. Give more examples of fixed sets.]

2.3   Compositions of Basic Isometries

We face the question what isometries we obtain, if we compose a number of basic isometries. We deal with this question in this section, in the order they appear below.

(1) Compositions of Parallel Translations

Given any two parallel translations $T_{\vec{u}}$ and $T_{\vec{v}}$ (i.e., the vectors $\vec{u}$ and $\vec{v}$ are known), we have that
$$T_{\vec{v}} \circ T_{\vec{u}} = T_{\vec{u}+\vec{v}}.$$
That is, the composition of two translations is a new translation by the vector equal to the sum of the vectors of the two given translations. See **Figure 2.22**.

This follows easily from the following observation. If $A \in \mathcal{P}$, then let $T_{\vec{u}}(A) = B$ and $T_{\vec{v}}(B) = C$. We have $\overrightarrow{AB} = \vec{u}$ and $\overrightarrow{BC} = \vec{v}$. Hence, $\overrightarrow{AC} = \overrightarrow{AB} + \overrightarrow{BC} = \vec{u} + \vec{v}$.

Similarly,
$$T_{\overrightarrow{u_1}} \circ T_{\overrightarrow{u_2}} \circ \ldots \circ T_{\overrightarrow{u_n}} = T_{\overrightarrow{u_1}+\overrightarrow{u_2}+\ldots+\overrightarrow{u_n}}, \quad \forall\, n \in \mathbb{N}.$$

Since $\vec{u} + \vec{v} = \vec{v} + \vec{u}$, we obtain
$$T_{\vec{v}} \circ T_{\vec{u}} = T_{\vec{u}+\vec{v}} = T_{\vec{v}+\vec{u}} = T_{\vec{u}} \circ T_{\vec{v}}.$$

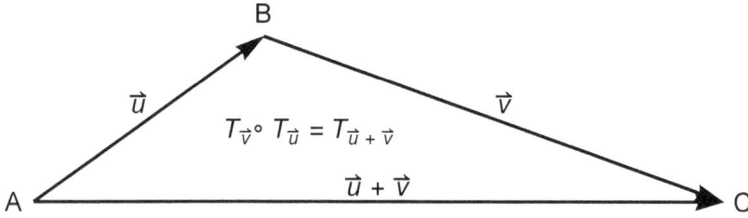

Figure 2.22: Compositions of parallel translations

That is, *the composition of parallel translations is commutative.*
We also have that
$$T_{\vec{0}} = I, \quad \text{the identity.}$$

Then, $T_{-\vec{u}} \circ T_{\vec{u}} = T_{\vec{u}} \circ T_{-\vec{u}} = T_{-\vec{u}+\vec{u}} = T_{\vec{0}} = I$. Therefore, for any free vector $\vec{u}$, the inverse of $T_{\vec{u}}$ is

$$T_{\vec{u}}^{-1} = T_{-\vec{u}}, \quad \text{the parallel translation by the opposite vector.}$$

The above facts prove the **result**: *The set T, of the parallel translations in P, is a commutative (Abelian) subgroup of the isometries $(I, \circ)$, under the operation of composition. Also, any parallel translation can be decomposed into a composition of parallel translations, in infinitely many ways.*
The composition of a parallel translation with itself satisfies the rule

$$\forall \quad n \in \mathbb{Z}, \quad T_{\vec{u}}^{(\circ, n)} = T_{n\vec{u}}.$$

(Here, we have introduced the **notation** $f \circ f \circ \ldots \circ f = f^{(\circ, n)}$ for the composition of a function f with itself n times, $n \in \mathbb{N}$. When $n = 0$, the composition gives the identity. When $n < 0$, then we compose T^{-1} with itself $-n > 0$ times.)

(2) Compositions of Rotations of the Same Center

Given any two rotations of the same center $R_{[O,\theta]}$ and $R_{[O,\phi]}$ (i.e., the point O and the oriented angles θ and ϕ are given), we have that

$$R_{[O,\phi]} \circ R_{[O,\theta]} = R_{[O,\theta+\phi]}.$$

That is, the composition of two rotations of the same center is a new rotation of the same center by the angle equal to the sum of the angles of the two given rotations. See **Figure 2.23**.
This follows easily from the following observation. If $A \in P$, then let $R_{[O,\theta]}(A) = B$ and $R_{[O,\phi]}(B) = C$. We have $OA = OB = OC$, $\angle(OA, OB) = \theta$ and $\angle(OB, OC) = \phi$. Hence, $\angle(OA, OC) = \angle(OA, OB) + \angle(OB, OC) = \theta + \phi$.
Similarly,

$$R_{[O,\theta_1]} \circ R_{[O,\theta_2]} + \ldots + R_{[O,\theta_n]} = R_{[O,\theta_1+\theta_2+\ldots+\theta_n]}, \quad \forall n \in \mathbb{N}.$$

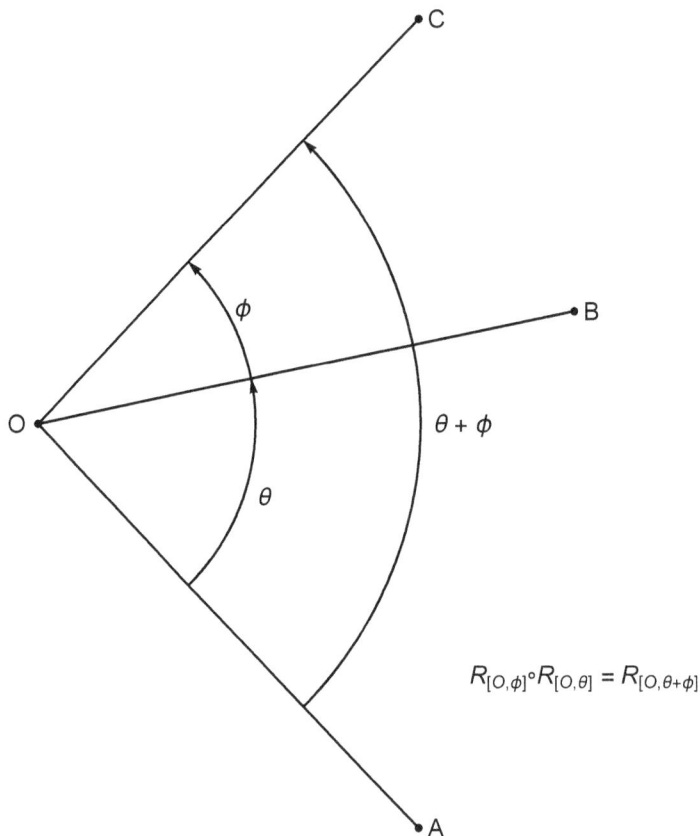

Figure 2.23: Composition of two rotations of the same center is a rotation with the same center

Since $\theta + \phi = \phi + \theta$, we obtain

$$R_{[O,\phi]} \circ R_{[O,\theta]} = R_{[O,\theta+\phi]} = R_{[O,\phi+\theta]} = R_{[O,\theta]} \circ R_{[O,\phi]}.$$

That is, the composition of rotations of the same center is commutative.

We also have that

$$R_{[O,0]} = I, \text{ the identity,}$$

or more general $\forall\ \phi = 0$, (mod 2π), i.e., $\phi = 2k\pi$ with k an integer, we have

$$R_{[O,\phi]} = R_{[O,2k\pi]} = I.$$

Then we obtain:

(1) $R_{[O,-\theta]} \circ R_{[O,\theta]} = R_{[O,-\theta+\theta]} = R_{[O,0]} = I$. Therefore, any rotation $R_{[O,\theta]}$, has inverse the rotation of the same center and opposite rotation angle, i.e.,

$$R_{[O,\theta]}^{-1} = R_{[O,-\theta]}.$$

(2) If $\theta - \phi = 0$, (mod 2π), i.e., $\theta = \phi + 2k\pi$ with k an integer number, then

$$R_{[O,\theta]} = R_{[O,\phi+2k\pi]} = R_{[O,\phi]} \circ R_{[O,2k\pi]} = R_{[O,\phi]} \circ I = R_{[O,\phi]}.$$

The above facts prove the **result**: *The set* $\mathcal{R}_O$, *of rotations of the same center a point* $O \in \mathcal{P}$, *is a commutative (Abelian) subgroup of the isometries* $(\mathcal{I}, \circ)$, *under the operation of composition. Also, any rotation can be decomposed into a composition of rotations with the same center, in infinitely many ways.*

The composition of a rotation with itself satisfies the rule

$$\forall \quad n \in \mathbb{Z}, \qquad R_{[O,\theta]}^{(o,n)} = R_{[O,n\theta]}.$$

(3) Compositions of Two Reflections

We are given two straight lines k and l in $\mathcal{P}$ and consider the reflections R_k and R_l. In composing these two reflections, we have the following three cases.

Case 1: The lines coincide, $k = l$. Then, by the definition of a reflection, we have that

$$R_l \circ R_l = I, \quad \text{identity.}$$

See **Figure 2.24**. $(R_l \circ R_l)(A) = R_l[R_l(A)] = R_l(B) = A[= I(A)]$.

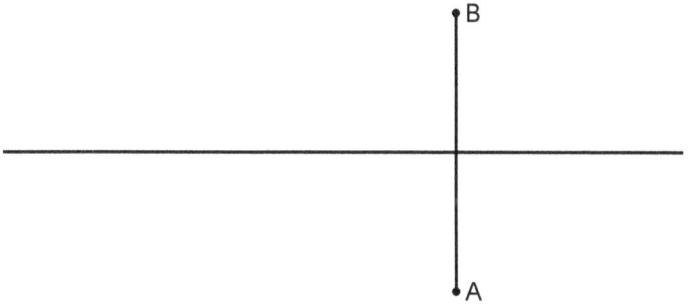

Figure 2.24: **A reflection is an involutory transformation, $R_1 \circ R_1 = I$**

Definition 2.3.1 *Any transformation* $G \ : \ \mathcal{P} \longrightarrow \mathcal{P}$ *that satisfies* $G \circ G = I$ *is called **involutory transformation** or simply **involution**.*

Besides the symmetry R_l in a line l, a half-turn $R_{[O,\pi]}$ is also an involutory transformation, since

$$R_{[O,\pi]} \circ R_{[O,\pi]} = R_{[O,\pi+\pi]} = R_{[O,2\pi]} = I.$$

Some properties of involutory transformations.

Any involutory transformation $G \ : \ \mathcal{P} \longrightarrow \mathcal{P}$ satisfies the following properties.

(1) If $G(P) = G(Q)$, then $G \circ G(P) = G \circ G(Q)$ and so $I(P) = I(Q)$, that is $P = Q$, proving that G is **one to one**.

(2) Now, for a $Q \in \mathcal{P}$, we let $P = G(Q)$. Then $G(P) = G \circ G(Q) = I(Q) = Q$, proving that G is **onto**.

(3) Thus, the inverse G^{-1} of an involutory transformation exists and is $G^{-1} = G$. This follows from

$$G^{-1} = G^{-1} \circ I = G^{-1} \circ (G \circ G) = (G^{-1} \circ G) \circ G = I \circ G = G.$$

(4) So, the composition of an involutory transformation G with itself satisfies the rule

$$\forall \quad n \in \mathbb{Z}, \qquad G^{(\circ, n)} = \begin{cases} I, & \text{if} \quad n \quad \text{is even,} \\[2mm] G = G^{-1} & \text{if} \quad n \quad \text{is odd.} \end{cases}$$

(5) If G_1 and G_2 are two involutory transformations, then

$$(G_1 \circ G_2)^{-1} = G_2^{-1} \circ G_1^{-1} = G_2 \circ G_1.$$

(6) For any involutory transformation G and any subset $V \subseteq \mathcal{P}$, the set $V \cup G(V)$ is a set-wise fixed set of G. This follows from

$$G[V \cup G(V)] = G(V) \cup (G \circ G)(V) = G(V) \cup I(V) = G(V) \cup V = V \cup G(V).$$

So, in accordance with **(6)** above, we have that any reflection $G := R_l$ and any half-turn $G := R_{[O,\pi]}$ have many set-wise fixed sets of the type $V \cup G(V)$. They are inverses of themselves, i.e., $R_l^{-1} = R_l$ and $R_{[O,\pi]}^{-1} = R_{[O,\pi]}$, and they satisfy all the six properties listed above.

A figure Ω invariant under a reflection R_l is called **symmetrical with respect to the line** l. The line l is called **the axis of symmetry** of Ω. For example, any circle with center on l is an invariant set of the reflection R_l. A figure Ω invariant under a half-turn $R_{[O,\pi]}$ is called **symmetrical with respect to the point** O. The point O is called **the center of symmetry** of Ω. For example, any rectangle has center of symmetry the common point of its diagonals, which is their midpoint.

Case 2: The lines are different, $k \neq l$, and parallel, $k \parallel l$.

We want to find $R_l \circ R_k$. In this case, the non-zero constant free vector $\overrightarrow{u} \neq \overrightarrow{0}$ with initial point on k, terminal point on l and perpendicular to these lines is well determined. (If the two lines coincide, $\overrightarrow{u} = \overrightarrow{0}$ and we are in the previous case.) In such situations the following definition is convenient:

Definition 2.3.2 *We call **free vector determined by two parallel lines** $k \parallel l$ **in the order** (k, l), and we denote it by $\overrightarrow{kl}$, the vector with initial point on k, terminal point on l and perpendicular to both k and l. (That is, $\overrightarrow{kl}$ is normal vector to both lines, its length is equal to the distance of the two parallel lines and its direction is from k to l).*

Notice that $\overrightarrow{lk} = -\overrightarrow{kl}$. In the sequel, we will use this convenient notation.

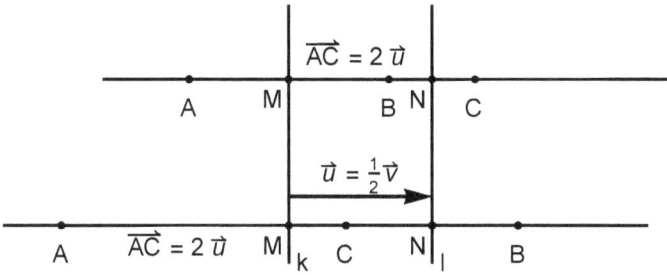

Figure 2.25: The composition of two reflections in two parallel lines is a parallel translation

In this case, according to this notation, we have the rule:

$$R_l \circ R_k = T_{2\overrightarrow{kl}}, \quad \text{the parallel translation by the vector } 2\overrightarrow{kl}.$$

See **Figure 2.25**.

Indeed: Let $A \in \mathcal{P}$, $R_k(A) = B$ and $R_l(B) = C$ and so $(R_l \circ R_k)(A) = C$. Then, by the definition of reflection, $k \perp \overrightarrow{AB}$ at its midpoint M and $l \perp \overrightarrow{BC}$ at its midpoint N. So, $\overrightarrow{AM} = \overrightarrow{MB}$ and $\overrightarrow{BN} = \overrightarrow{NC}$. Since $k \parallel l$, the points A, B and C are on the same straight line which is perpendicular to both k and l. Then, $\overrightarrow{MB} + \overrightarrow{BN} = \overrightarrow{kl}$ and $\overrightarrow{AC} = \overrightarrow{AM} + \overrightarrow{MB} + \overrightarrow{BN} + \overrightarrow{NC} = 2(\overrightarrow{MB} + \overrightarrow{BN}) = 2\overrightarrow{kl}$.

Notice that if $\overrightarrow{lk} \neq \overrightarrow{0}$, then

$$R_k \circ R_l = T_{2\overrightarrow{lk}} = T_{-2\overrightarrow{kl}} \neq T_{2\overrightarrow{kl}} = R_l \circ R_k.$$

So, *the composition of reflections is not commutative.*

Conversely: Any parallel translation $T_{\overrightarrow{v}}$ can be considered as the composition of two reflections $R_l \circ R_k$ in any two parallel lines k and l such that $\overrightarrow{kl} = \frac{1}{2}\overrightarrow{v}$. The distance of the parallel lines k and l is $\frac{1}{2} \times$ (length of $\overrightarrow{v}$). See **Figure 2.25**. So, we have the following **Result**:

Any parallel translation can be decomposed into a composition of two reflections in parallel lines, in infinitely many ways.

Case 3: The lines are non-parallel, $k \nparallel l$ (and therefore different).

We want to find $R_l \circ R_k$. In this case, the two lines intersect at a point $O \in \mathcal{P}$. We let $\theta := \angle(k, l)$, the oriented angle with vertex O and direction from k to l. See **Figure 2.26**. Then,

$R_l \circ R_k = R_{[O, 2\theta]}$, the rotation with center O by the oriented angle 2θ.

Indeed: Let $A \in \mathcal{P}$, $R_k(A) = B$ and $R_l(B) = C$ and so $(R_l \circ R_k)(A) = C$. Then, by the definition of reflection, $k \perp \overrightarrow{AB}$ at its midpoint M and $l \perp \overrightarrow{BC}$ at its midpoint N. So, $\overrightarrow{AM} = \overrightarrow{MB}$ and $\overrightarrow{BN} = \overrightarrow{NC}$. Then, $OA = OB = OC$ and

$\angle AOM = \angle MOB$ and $\angle BON = \angle NOC$. Also, $\angle MOB + \angle BON = \angle\theta$, and $\angle AOC = \angle AOM + \angle MOB + \angle BON + \angle NOC = 2(\angle MOB + \angle BON) = 2\angle\theta$.

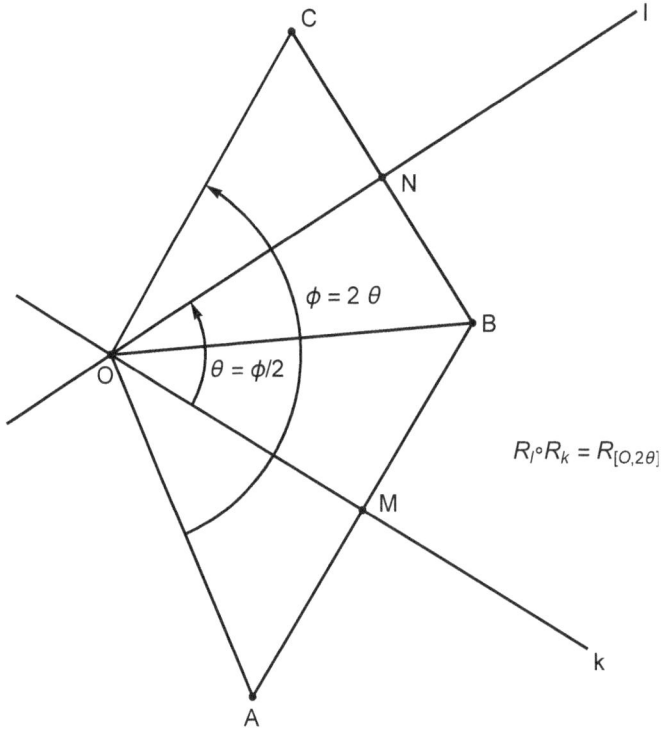

Figure 2.26: Composition of two reflections in two non-parallel lines is a rotation

We see that if $l \perp k$, then $R_l \circ R_k = R_{[O,\pi]}$; **this rotation is the half-turn or symmetry about the center** O.

Notice that if $\theta \neq 0$, (mod π), then

$$R_k \circ R_l = R_{[O,-2\theta]} \neq R_{[O,2\theta]} = R_l \circ R_k.$$

So, *the composition of reflections is not commutative.*

Conversely: Any rotation $R_{[O,\phi]}$ can be considered as the composition of two reflections $R_l \circ R_k$ on any two lines k and l intersecting each other at O and making oriented angle $\angle(k,l) = \dfrac{1}{2}\angle\phi$. See **Figure 2.26**. So, we have the following **Result**:

Any rotation can be decomposed into a composition of two reflections in lines through its center, in infinitely many ways.

(This freedom of decomposition will be used a lot in the sequel.)

(4) Compositions of Two Rotations of Different Centers

Given any two rotations $R_{[O_1,\theta_1]}$ and $R_{[O_2,\theta_2]}$ of different centers $O_1 \neq O_2$, we would like to find

$$R_{[O_2,\theta_2]} \circ R_{[O_1,\theta_1]}.$$

So, we know the two different center points O_1 and O_2 and the two oriented angles θ_1 and θ_2. In the sequel we will use the following **readjusting Remark:**

Since for any rotation it holds $R_{[O,\alpha]} = R_{[O,\alpha+2k\pi]}$, $k \in \mathbb{Z}$, **without loss of generality, we can adjust** θ_1 **and** θ_2 **and assume** $0 \leq \theta_1, \theta_2 < 2\pi$.

We consider the line $l = O_1O_2$ and we decompose $R_{[O_1,\theta_1]}$ and $R_{[O_2,\theta_2]}$ in the following way. We consider the line l_1 through O_1 making oriented angle $\sphericalangle(l_1,l) = \sphericalangle\dfrac{\theta_1}{2}$ and the line l_2 through O_2 making oriented angle $\sphericalangle(l,l_2) = \sphericalangle\dfrac{\theta_2}{2}$. Then, by the converse of the previous paragraph **(3)**, we have:

$$R_{[O_1,\theta_1]} = R_l \circ R_{l_1} \quad \text{and} \quad R_{[O_2,\theta_2]} = R_{l_2} \circ R_l, \quad \text{and so,}$$
$$R_{[O_2,\theta_2]} \circ R_{[O_1,\theta_1]} = (R_{l_2} \circ R_l) \circ (R_l \circ R_{l_1}) = R_{l_2} \circ (R_l \circ R_l) \circ R_{l_1}$$
$$= R_{l_2} \circ I \circ R_{l_1} = R_{l_2} \circ R_{l_1}.$$

Hence: *The composition of two rotations with different centers simplifies to a composition of two reflections initially, and so it is either a parallel translation or a rotation.* We are going to examine these cases.

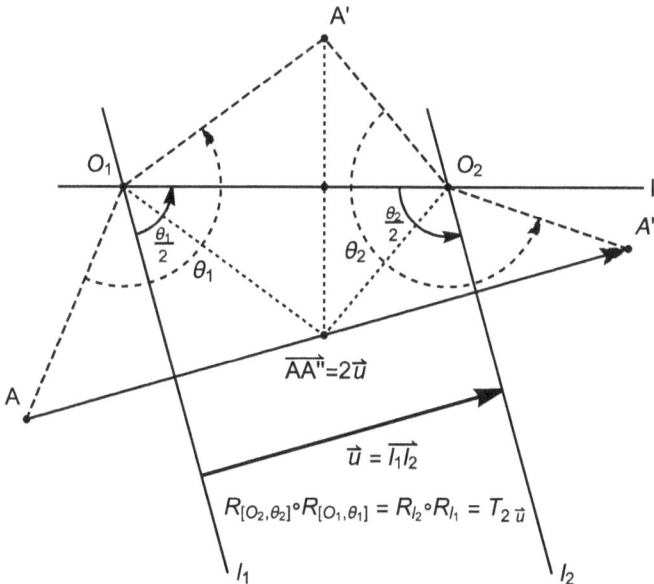

Figure 2.27: The composition of two rotations of different centers is a parallel translation iff $l_1 \parallel l_2 \iff \dfrac{\theta_1 + \theta_2}{2} = 0, \pmod{\pi}$.

First, we examine the **case of parallel translation** which happens iff

$$l_1 \parallel l_2 \iff \frac{\theta_1 + \theta_2}{2} = 0, \quad (\bmod \pi).$$

Then, see **Figure 2.27**,

$$R_{[O_2,\theta_2]} \circ R_{[O_1,\theta_1]} = R_{l_2} \circ R_{l_1} = T_{2\vec{u}}$$

is the **parallel translation** by the vector $2\vec{u}$, **where for** $0 \le \theta_1,\ \theta_2 < 2\pi$,

$$\vec{u} = \overrightarrow{l_1 l_2} = \overrightarrow{O_1 O_2} \cdot \sin\left(\frac{\theta_1}{2}\right) = \overrightarrow{O_1 O_2} \cdot \sin\left(\frac{\theta_2}{2}\right)$$

is the vector perpendicular to both l_1 and l_2, with initial point on l_1 and terminal point on l_2 (see **Definition 2.3.2**).

For example, the **composition of two half-turns**

$$R_{[O_2,\pi]} \circ R_{[O_1,\pi]} = T_{2\overrightarrow{O_2 O_1}},$$

as $l_1 \parallel l_2 \perp O_2 O_1$ and $\sin\left(\frac{\pi}{2}\right) = 1$. See **Figure 2.28**.

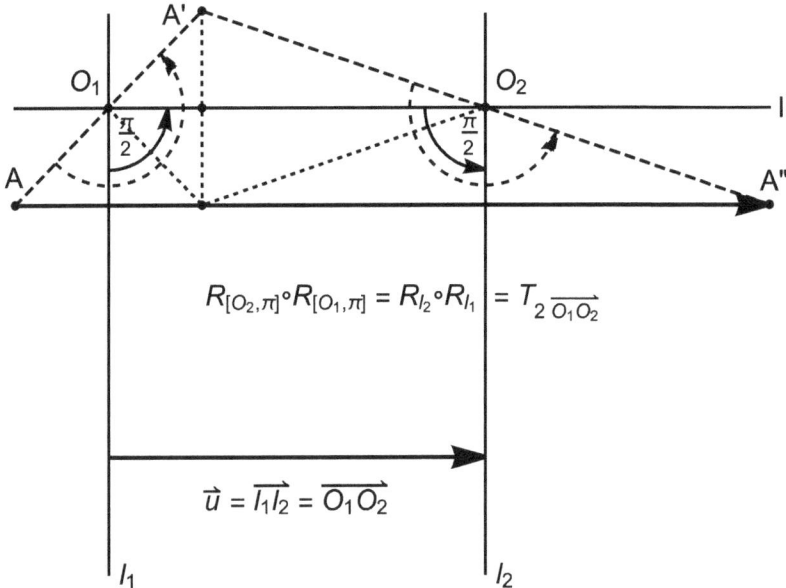

$$R_{[O_2,\pi]} \circ R_{[O_1,\pi]} = R_{l_2} \circ R_{l_1} = T_{2\overrightarrow{O_1 O_2}}$$

$$\vec{u} = \overrightarrow{l_1 l_2} = \overrightarrow{O_1 O_2}$$

Figure 2.28: The composition of two half-turns of different centers is a parallel translation

Second, we examine the **general case** in which we obtain **a rotation of some center**. This happens iff

$$l_1 \nparallel l_2 \iff \frac{\theta_1 + \theta_2}{2} \neq \pi, \quad (\mathrm{mod}\ 2\pi).$$

In this case l_1 and l_2 intersect at a point $O \in \mathcal{P}$ and the oriented angle at O from line l_1 to line l_2 is

$$\angle(l_1, l_2) = \frac{\theta_1 + \theta_2}{2}.$$

Then, we have

$$R_{[O_2,\theta_2]} \circ R_{[O_1,\theta_1]} = R_{l_2} \circ R_{l_1} = R_{[O,\theta_1+\theta_2]},$$

which is a **new rotation**, as it is illustrated in the self-explained **Figure 2.29**. (Provide the details.)

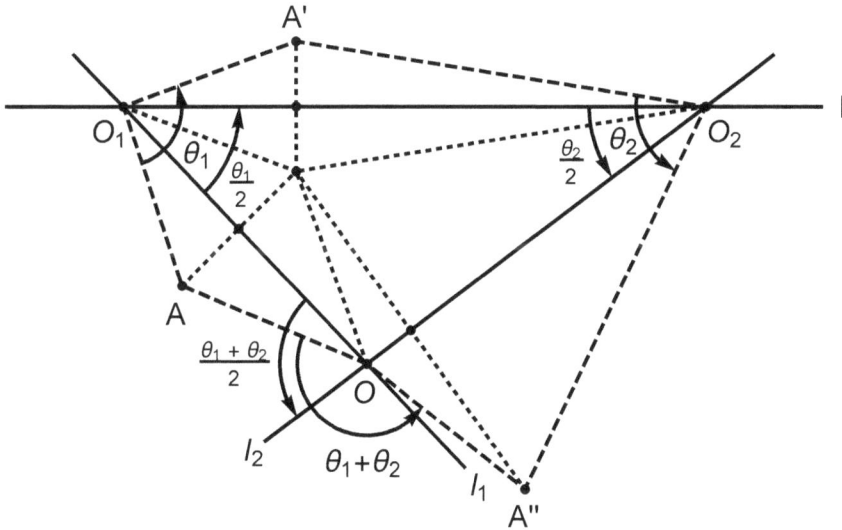

$$R_{[O_2,\theta_2]} \circ R_{[O_1,\theta_1]}\ (A) = R_{l_2} \circ R_{l_1}\ (A) = R_{[O,\theta_1+\theta_2]}(A) = A''$$

Figure 2.29: The composition of two rotations of different centers is a rotation with a new center, in general

So, by extrapolating to compositions of several rotations, by the cases presented here and in paragraph **(3)**, we obtain the following **Result:**

The composition of any number of rotations is: (1) either a new rotation of some center and by an angle equal to the sum of the angles of all of the rotations, or (2) a parallel translation. It is a parallel translation (including the case of the identity) iff the sum of all of the rotation angles is zero (mod π), i.e., it is an integer multiple of π. This translation is the identity if it has a fixed point!

When $O_1 \neq O_2$, the composition, $R_{[O_2,\theta_2]} \circ R_{[O_1,\theta_1]}$, **is not commutative.** Working as before, we find

$$R_{[O_1,\theta_1]} \circ R_{[O_2,\theta_2]} = R_{[O',\theta_1+\theta_2]},$$

where $O' = R_{O_1O_2}(O)$, the reflection of O in the line O_1O_2 (and so the line O_1O_2 is the perpendicular bisector of segment OO'). See **Figure 2.30.**

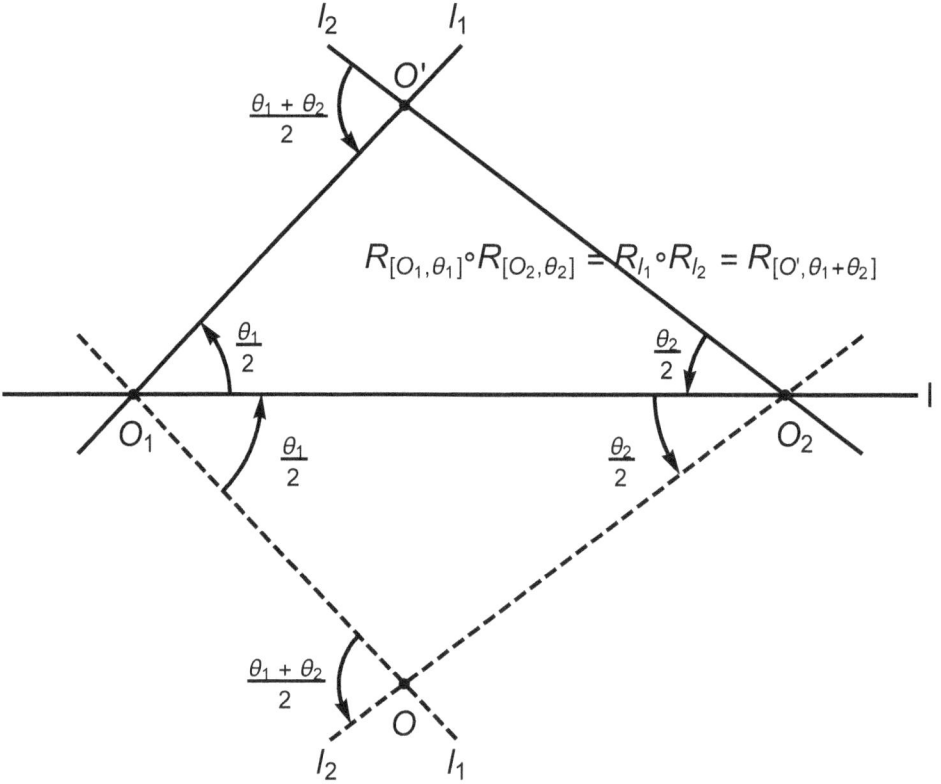

Figure 2.30: The composition of two rotations of different centers
is not commutative

(5) Compositions of a Rotation with a Parallel Translation

Given a rotation $R_{[O,\theta]}$, with $\angle\theta \neq 0$, (mod 2π), and a parallel translation $T_{\vec{u}}$, with $\vec{u} \neq \vec{0}$, so that either one is not the identity, we would like to find

$$R_{[O,\theta]} \circ T_{\vec{u}} \quad \text{and} \quad T_{\vec{u}} \circ R_{[O,\theta]}.$$

So, we are given the point O, the angle θ (oriented) and the vector $\vec{u} \neq \vec{0}$. Without loss of generality, we consider $0 < \theta < 2\pi$.

Case $0 < \theta < \pi$.

(i) The subcase $R_{[O,\theta]} \circ T_{\vec{u}}$.

We consider the line l through O and perpendicular to the vector $\vec{u}$. We also consider the line $l_1 \parallel l$ such that $\overrightarrow{l_1 l} = \dfrac{\vec{u}}{2}$ and the line l_2 through O such that $\angle(l, l_2) = \angle\dfrac{\theta}{2}$. Then, by paragraphs **(2)** and **(3)**, we have

$$R_{[O,\theta]} = R_{l_2} \circ R_l \quad \text{and} \quad T_{\vec{u}} = R_l \circ R_{l_1}.$$

The lines l_1 and l_2 intersect at a point Q and $\angle(l_1, l_2) = \angle\dfrac{\theta}{2}$. See **Figure 2.31**.

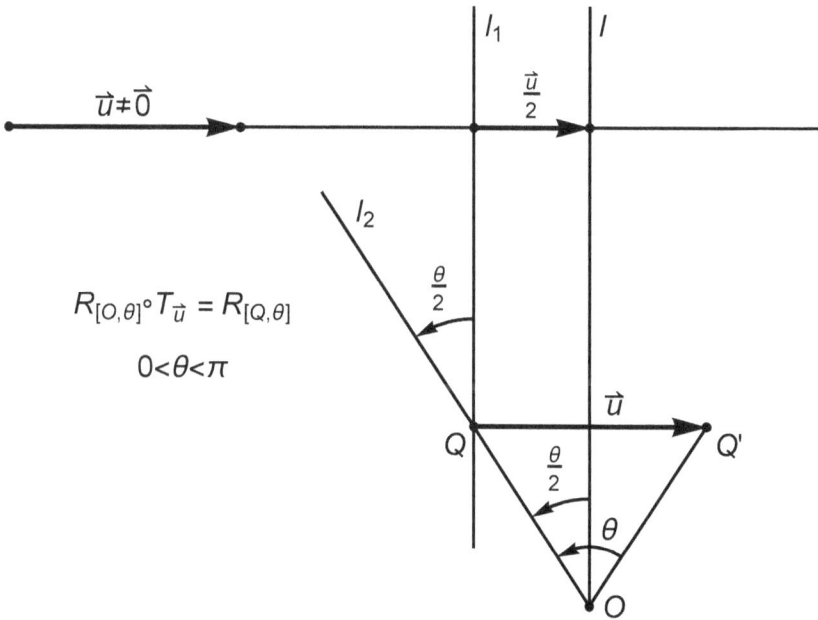

Figure 2.31: The composition of a parallel translation followed by a rotation is a new rotation

Therefore, $R_{[O,\theta]} \circ T_{\vec{u}} = (R_{l_2} \circ R_l) \circ (R_l \circ R_{l_1}) = R_{l_2} \circ (R_l \circ R_l) \circ R_{l_1}$

$$= R_{l_2} \circ I \circ R_{l_1} = R_{l_2} \circ R_{l_1} = R_{[Q,\theta]}.$$

The point Q is on l_2, as in the **Figure 2.31**, and

$$\text{length}(OQ) = \frac{\text{length}(\vec{u})}{2} \times \csc\left(\left|\frac{\theta}{2}\right|\right).$$

[The point Q, by construction, is directly proven to be fixed under $R_{[O,\theta]} \circ T_{\vec{u}}$. As we see in **Figure 2.31**, Q is mapped to Q' by $T_{\vec{u}}$ and then Q' is mapped back to Q by $R_{[O,\theta]}$. So, $\left(R_{[O,\theta]} \circ T_{\vec{u}}\right)(Q) = R_{[O,\theta]}\{T_{\vec{u}}(Q)\} = R_{[O,\theta]}(Q') = Q.$]

(ii) The subcase $T_{\vec{u}} \circ R_{[O,\theta]}$.

We consider the line l through O and perpendicular to the vector $\vec{u}$. We also consider the line $l_1 \parallel l$ such that $\overrightarrow{ll_1} = \dfrac{\vec{u}}{2}$ and the line l_2 through O such that $\angle(l_2, l) = \angle\dfrac{\theta}{2}$. Then, by paragraphs **(2)** and **(3)**, we have

$$R_{[O,\theta]} = R_l \circ R_{l_2} \quad \text{and} \quad T_{\vec{u}} = R_{l_1} \circ R_l.$$

The lines l_1 and l_2 intersect at a point Q [different than the center in **subcase (i)** above] and $\angle(l_2, l_1) = \angle\dfrac{\theta}{2}$. See **Figure 2.32**.

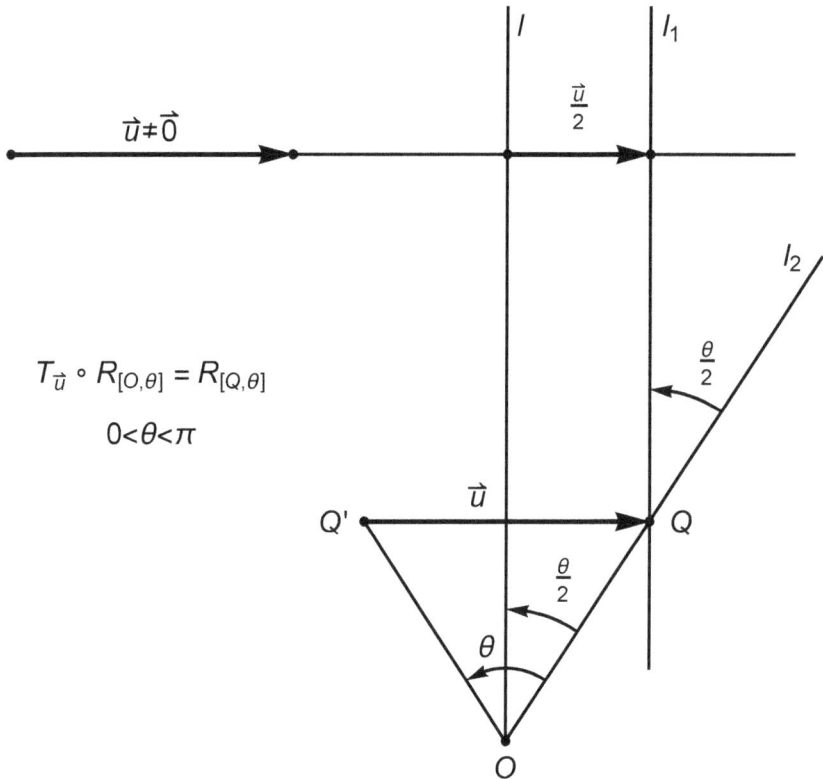

**Figure 2.32: The composition of a rotation followed a
parallel translation is a new rotation**

Therefore, $\quad T_{\vec{u}} \circ R_{[O,\theta]} = (R_{l_1} \circ R_l) \circ (R_l \circ R_{l_2}) = R_{l_1} \circ (R_l \circ R_l) \circ R_{l_2}$
$$= R_{l_1} \circ I \circ R_{l_2} = R_{l_1} \circ R_{l_2} = R_{[Q,\theta]}.$$

Again, the point Q is on l_2, as in the **Figure 2.31**, and

$$\text{length}(OQ) = \frac{\text{length}(\vec{u})}{2} \times \csc\left(\left|\frac{\theta}{2}\right|\right).$$

This is a different rotation than the one found in **subcase (i)**. [Again, the point Q, by construction, is directly proven to be fixed under $T_{\vec{u}} \circ R_{[O,\theta]}$. As we see in **Figure 2.32**, Q is mapped to Q' by $R_{[O,\theta]}$ and then Q' is mapped back to Q by $T_{\vec{u}}$. So, $(T_{\vec{u}} \circ R_{[O,\theta]})(Q) = T_{\vec{u}}\{R_{[O,\theta]}(Q)\} = T_{\vec{u}}(Q') = Q.$]

So, **in the two subcases** of

$$R_{[O,\theta]} \circ T_{\vec{u}} = R_{[Q_1,\theta]} \quad \text{and} \quad T_{\vec{u}} \circ R_{[O,\theta]} = R_{[Q_2,\theta]} \quad \textbf{of the case} \quad 0 < \theta < \pi,$$

the fixed points Q_1 and Q_2 can be simply found by constructing an isosceles triangle with vertex O, vertex angle θ (so the base angles are equal to $\frac{\pi - \theta}{2}$) and base parallel to $\vec{u}$ and equal to the length of $\vec{u}$. See **Figure 2.33**. (The construction of such a triangle is basic.) The base is $Q_1 Q_2$.

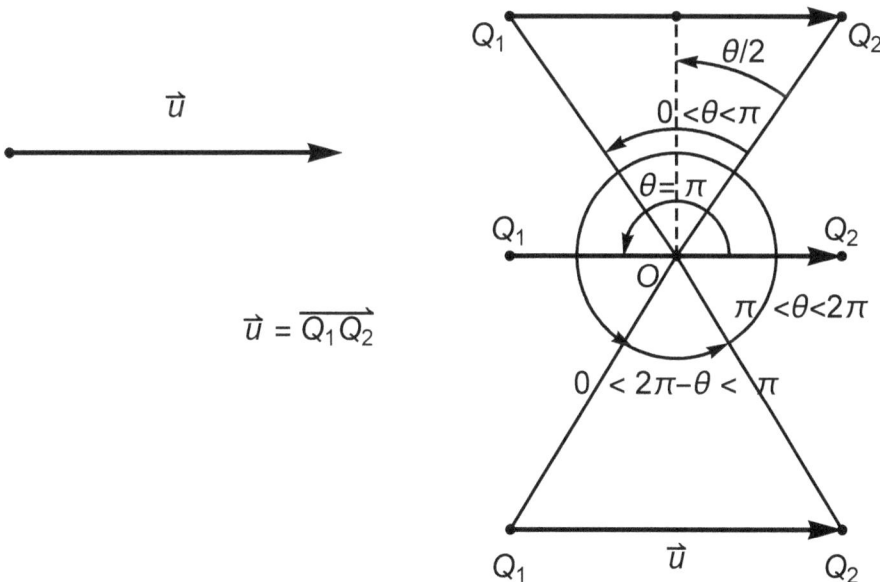

Figure 2.33: **Fixed point(s) of the composition of a parallel translation and a rotation**

Case $\theta = \pi$.
Here the rotation is a half-turn. Understanding the construction of the previous case, we directly see that in this case, the fixed points Q_1 and Q_2 of

the compositions

$$R_{[O,\pi]} \circ T_{\vec{u}} = R_{[Q_1,\pi]} \quad \text{and} \quad T_{\vec{u}} \circ R_{[O,\pi]} = R_{[Q_2,\pi]}$$

are the endpoints of the straight segment Q_1Q_2, such that, O is the midpoint of it and $\overrightarrow{Q_1Q_2} = \vec{u}$. See **Figure 2.33.**

Case $\pi < \theta < 2\pi$.

In this case, to find the fixed points the fixed points Q_1 and Q_2 of the compositions

$$R_{[O,\pi]} \circ T_{\vec{u}} = R_{[Q_1,\pi]} \quad \text{and} \quad T_{\vec{u}} \circ R_{[O,\pi]} = R_{[Q_2,\pi]},$$

we construct the isosceles triangle Q_1OQ_2 with vertex angle at O equal to $2\pi - \theta$ and $\overrightarrow{Q_1Q_2} = \vec{u}$. But in this case, the position of the triangle is up-side-down with respect to the triangle of case $0 < \theta < \pi$. See **Figure 2.33**.

In conclusion: *Given a rotation $R_{[O,\theta]}$, with $\angle\theta \neq 0$, (mod 2π), and a parallel translation $T_{\vec{u}}$, with $\vec{u} \neq \vec{0}$, the compositions*

$$R_{[O,\theta]} \circ T_{\vec{u}} = R_{[Q_1,\theta]} \quad \text{and} \quad T_{\vec{u}} \circ R_{[O,\theta]} = R_{[Q_2,\theta]}$$

yield new rotations with centers Q_1 and Q_2, constructed above and such that $O \neq Q_1 \neq Q_2 \neq O$, and the same rotation angle θ for all three of them.

By the **Results** of this paragraph and paragraphs **(1)**, **(2)** and **(4)**, before, **we conclude that:**

The union of all parallel translations $\mathcal{T}$ with all rotations $\mathcal{R}o$ of the plane $\mathcal{P}$, i.e., $\mathcal{T} \cup \mathcal{R}o$, is a non-commutative subgroup of the group $(\mathcal{I}, \circ)$, of isometries of the plane, under composition.

(6) Compositions of a Parallel Translation with a Reflection

Given a parallel translation $T_{\vec{u}}$, with $\vec{u} \neq \vec{0}$, and a reflection R_l, we would like to find

$$T_{\vec{u}} \circ R_l \quad \text{and} \quad R_l \circ T_{\vec{u}},$$

in the simplest possible way. So, we are given the vector $\vec{u} \neq \vec{0}$ and the straight line l in $\mathcal{P}$.

(i) The case $T_{\vec{u}} \circ R_l$.
We decomposed the vector as $\vec{u} = \vec{w} + \vec{v}$ with $\vec{w} \parallel l$ and $\vec{v} \perp l$. So,

$$T_{\vec{u}} \circ R_l = T_{\vec{w}} \circ T_{\vec{v}} \circ R_l.$$

Next, we decompose $T_{\vec{v}} = R_m \circ R_l$ where $m \parallel l$ and $\overrightarrow{lm} = \dfrac{\vec{v}}{2}$. So the line m is well determined. See **Figure 2.34.**

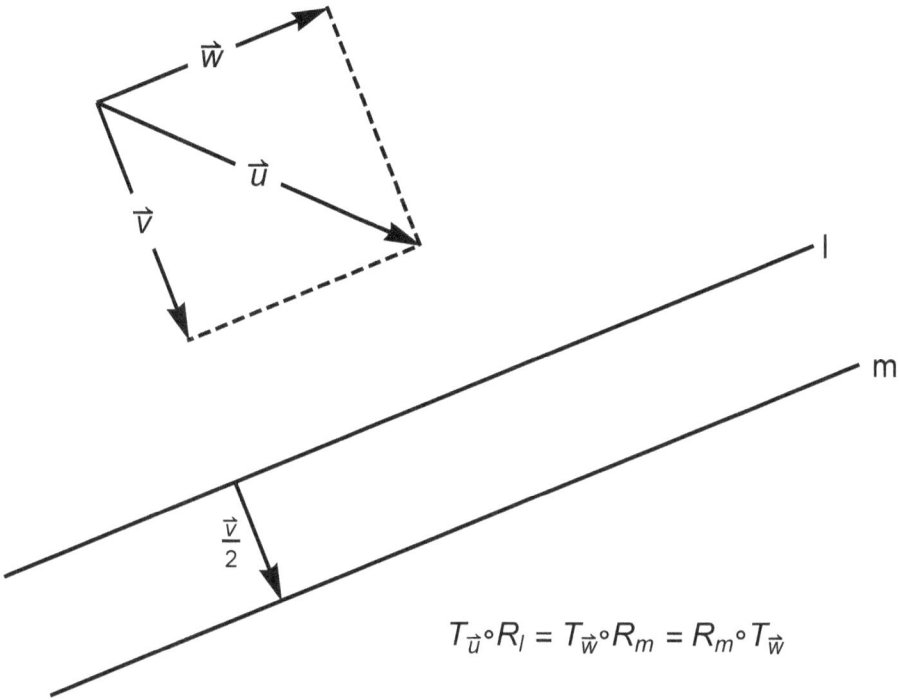

$$T_{\vec{u}} \circ R_l = T_{\vec{w}} \circ R_m = R_m \circ T_{\vec{w}}$$

Figure 2.34: The composition of a reflection followed by a parallel translation is a glide, in general

Then,

$$T_{\vec{u}} \circ R_l = T_{\vec{w}} \circ R_m \circ R_l \circ R_l = T_{\vec{w}} \circ R_m \circ I = T_{\vec{w}} \circ R_m.$$

(1) Notice that, $\vec{w} = \vec{0} \iff \vec{u} = \vec{v} \perp l$. In such a case, when $\vec{w} = \vec{0}$, we have

$$T_{\vec{u}} \circ R_l = T_{\vec{0}} \circ R_m = I \circ R_m = R_m \quad \text{is a new reflection in the line } m \parallel l.$$

(2) If $\vec{0} \neq \vec{w} \parallel l \parallel m$, then

$$T_{\vec{u}} \circ R_l = T_{\vec{w}} \circ R_m$$

is a special composition of a reflection R_m and a parallel translation $T_{\vec{w}}$ with $\vec{0} \neq \vec{w} \parallel m$. In this case, we have the following standard definition:

Definition 2.3.3 *An isometry F is called a **glide** if $F = T_{\vec{w}} \circ R_m$, where $\vec{w} \neq \vec{0}$ and $\vec{w} \parallel m$.*

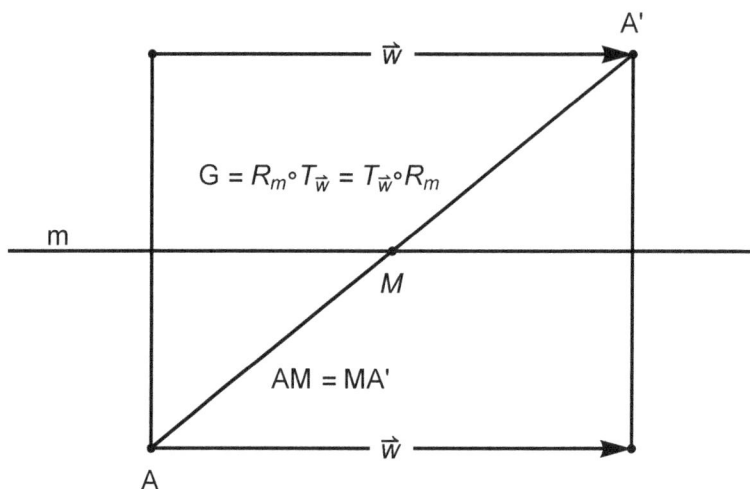

Figure 2.35: A glide and its properties

We state **the important properties of glides**, all of which can be easily verified geometrically and make the glide a special and convenient isometry. See **Figure 2.35**.

(1) A glide $T_{\vec{w}} \circ R_m$ is an isometry, since it is a composition of two isometries. It does not have any fixed point, as $\vec{w} \neq \vec{0}$. The line m is a set-wise fixed set of the glide. Also, the union of any two straight lines parallel to l, one in either side of l, and equally distanced from l is a set-wise fixed set of the glide.

(2) A glide $T_{\vec{w}} \circ R_m$ has the special and convenient geometrical property that $T_{\vec{w}} \circ R_m = R_m \circ T_{\vec{w}}$. That is, in any glide $T_{\vec{w}} \circ R_m$, $T_{\vec{w}}$ and R_m commute.

(3) If $(T_{\vec{w}} \circ R_m)(A) = A'$, then the midpoint M of the segment AA' lies on the axis m.

(4) A glide maps a straight line to a straight line, a circle to an equal circle, which can be easily geometrically determined. It maps an oriented angle to its opposite and so a glide is an opposite isometry.

(5) The inverse of a glide is

$$(T_{\vec{w}} \circ R_m)^{-1} = T_{-\vec{w}} \circ R_m = R_m \circ T_{-\vec{w}},$$

which is a new glide (since $-\vec{w} \neq \vec{w} \neq \vec{0}$).

This follows by the commutativity of $T_{\vec{w}}$ and R_m in this case, $R_m^{-1} = R_m$ and $T_{\vec{w}}^{-1} = T_{-\vec{w}}$.

(6) Since $T_{\vec{w}}$ and R_m commute, we get

$$\forall \quad n \in \mathbb{Z}, \quad (T_{\vec{w}} \circ R_m)^{(\circ, n)} = \begin{cases} T_{n\vec{w}}, & \text{if} \quad n \text{ is even,} \\ \\ T_{n\vec{w}} \circ R_m = R_m \circ T_{n\vec{w}}, & \text{if} \quad n \text{ is odd.} \end{cases}$$

(7) A glide $T_{\vec{w}} \circ R_m$ or just a reflection R_m (in which case $\vec{w} = \vec{0}$) can be determined by two pairs of corresponding points (A, A') and (B, B'). These four points must satisfy the necessary for an isometry condition $AB = A'B'$. Then we have that:

(a) The axis m of the glide (or the reflection) is the straight line MN where M and N are the midpoints of the segments AA' and BB' [by **property (3)**].

(b) The vector $\vec{w} \parallel m$ is found as explained later in **Lemmata 2.4.2** and **2.4.3**, their **Corollaries** and **Figures 2.50-2.54** and **2.78-2.80**, where we discuss more geometrical properties.

(8) A glide $T_{\vec{w}} \circ R_m = R_m \circ T_{\vec{w}}$ can be written as a special composition of **three** reflections

$$T_{\vec{w}} \circ R_m = (R_{l_2} \circ R_{l_1}) \circ R_m = R_m \circ T_{\vec{w}} = R_m \circ (R_{l_2} \circ R_{l_1}),$$

where l_1 and l_2 are any two straight lines perpendicular to the axis of the glide m (and to $\vec{w}$, as $\vec{w} \parallel l$) and such that $\overrightarrow{l_1 l_2} = \dfrac{\vec{w}}{2}$. See **Figure 2.36**.

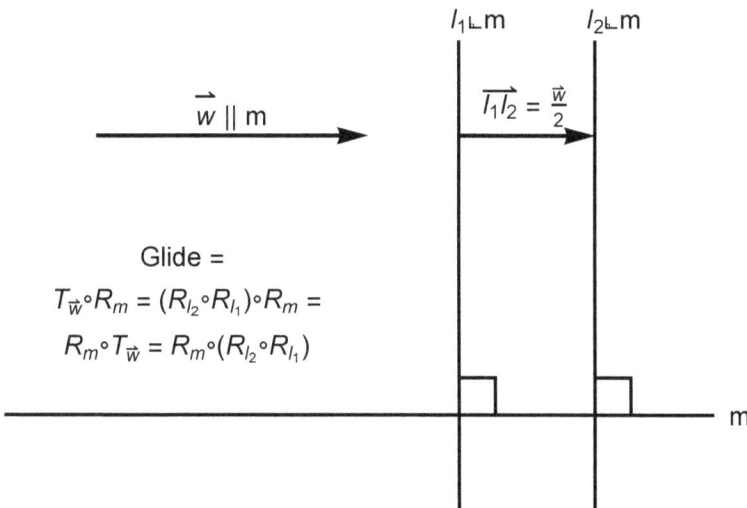

Figure 2.36: A glide is a special composition of three reflections

This follows from the fact that $\vec{w} \parallel m$ and using **paragraph (3), case (2)**, of **this Section**.

(ii) The case $R_l \circ T_{\vec{u}}$.

We decomposed the vector as $\vec{u} = \vec{w} + \vec{v} = \vec{v} + \vec{w}$ with $\vec{w} \parallel l$ and $\vec{v} \perp l$.
So,

$$R_l \circ T_{\vec{u}} = R_l \circ T_{\vec{v}} \circ T_{\vec{w}}.$$

Next, we decompose $T_{\vec{v}} = R_l \circ R_m$, where $m \parallel l$ and $\overrightarrow{ml} = \dfrac{\vec{v}}{2}$. So, the line m is well determined. See **Figure 2.37**.

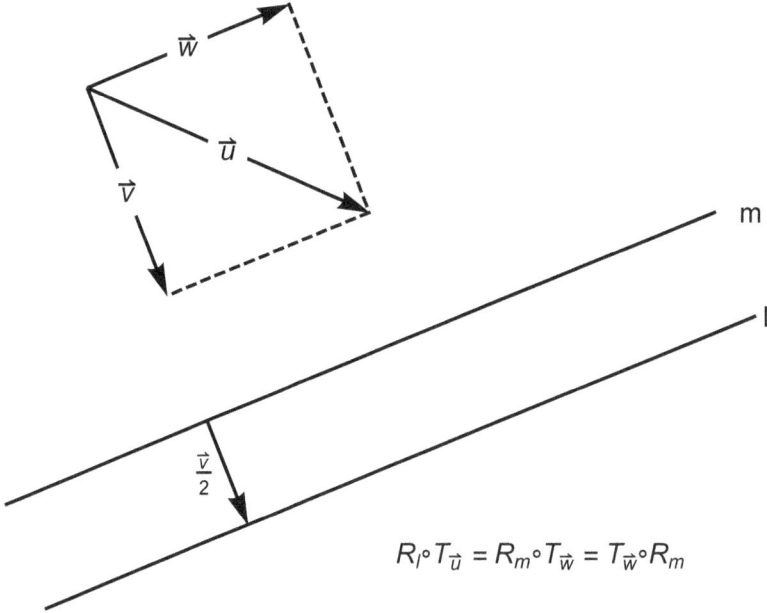

$$R_l \circ T_{\vec{u}} = R_m \circ T_{\vec{w}} = T_{\vec{w}} \circ R_m$$

Figure 2.37: **The composition of a parallel translation followed by a reflection is a glide, in general**

Then,

$$R_l \circ T_{\vec{u}} = R_l \circ R_l \circ R_m \circ T_{\vec{w}} = I \circ R_m \circ T_{\vec{w}} = R_m \circ T_{\vec{w}} = T_{\vec{w}} \circ R_m.$$

As before, this is equal to the reflection R_m, if $\vec{w} = \vec{0}$ ($\Longleftrightarrow \vec{u} = \vec{v} \perp l$), or is the glide $T_{\vec{w}} \circ R_m$ [different than the one found in **case (i)**], if $\vec{w} \neq \vec{0}$.

(7) Compositions of a Rotation with a Reflection

Given a rotation $R_{[O,\theta]}$ and a reflection R_l, we would like to find

$$R_{[O,\theta]} \circ R_l \quad \text{and} \quad R_l \circ R_{[O,\theta]},$$

in the simplest possible way. So, we are given the angle $\angle \theta$ [$\neq 0$, (mod 2π), so that the rotation is not the identity], the point O and the straight line l.

(i) The case $R_{[O,\theta]} \circ R_l$.

We consider the lines l_1 and l_2 through O and such that $l_1 \parallel l$ and the angle $\angle(l_1, l_2) = \dfrac{\theta}{2}$. Then, $R_{[O,\theta]} = R_{l_2} \circ R_{l_1}$ and so

$$R_{[O,\theta]} \circ R_l = R_{l_2} \circ R_{l_1} \circ R_l.$$

Since, $l_1 \parallel l$, we let $\vec{u} = \overrightarrow{ll_1}$ and have that $R_{l_1} \circ R_l = T_{2\vec{u}}$. We decompose $\vec{u} = \vec{v} + \vec{w}$ such that $\vec{v} \perp l_2$ and $\vec{w} \parallel l_2$. See **Figure 2.38**. Hence,

$$R_{[O,\theta]} \circ R_l = R_{l_2} \circ T_{2\vec{u}} = R_{l_2} \circ T_{2\vec{v}} \circ T_{2\vec{w}}.$$

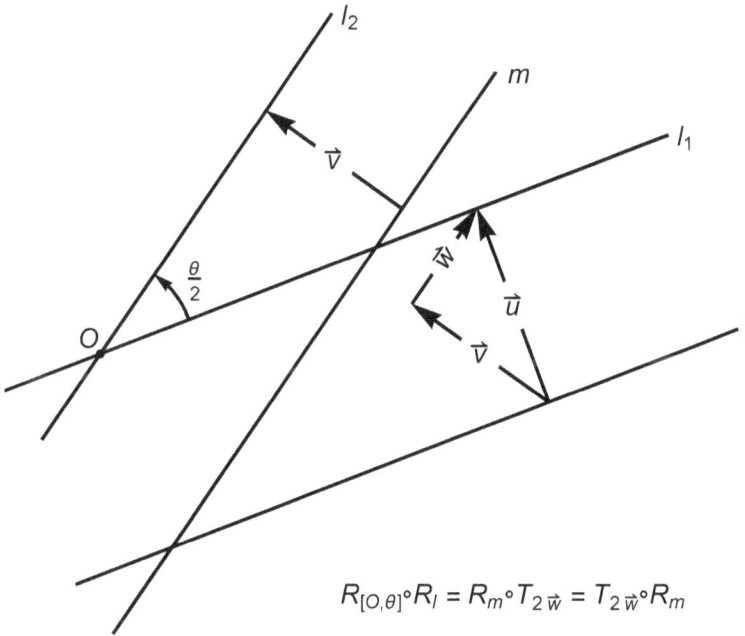

$$R_{[O,\theta]} \circ R_l = R_m \circ T_{2\vec{w}} = T_{2\vec{w}} \circ R_m$$

Figure 2.38: The composition of a reflection followed by a rotation is a glide

We now consider the line $m \parallel l_2$ such that $\overrightarrow{ml_2} = \vec{v}$. Then, $T_{2\vec{v}} = R_{l_2} \circ R_m$. So,

$$R_{[O,\theta]} \circ R_l = R_{l_2} \circ R_{l_2} \circ R_m \circ T_{2\vec{w}} = I \circ R_m \circ T_{2\vec{w}} = R_m \circ T_{2\vec{w}}.$$

Since $\theta \neq 0$, $(\mathrm{mod}\ 2\pi)$, $\vec{0} \neq \vec{w} \parallel l_2 \parallel m$. Therefore, the final outcome is a glide, completely determined.

Note: If the given line l passes through the given point O, then $l_1 = l$ and $\vec{u} = \vec{0}$. In this situation we have

$$R_{[O,\theta]} \circ R_l = R_{l_2} \circ R_{l_1} \circ R_l = R_{l_2} \circ R_l \circ R_l = R_{l_2} \circ I = R_{l_2}.$$

(ii) The case $R_l \circ R_{[O,\theta]}$.

We consider the lines l_1 and l_2 through O and such that $l_2 \parallel l$ and the angle $\angle(l_1, l_2) = \dfrac{\theta}{2}$. Then, $R_{[O,\theta]} = R_{l_2} \circ R_{l_1}$ and so

$$R_l \circ R_{[O,\theta]} = R_l \circ R_{l_2} \circ R_{l_1}.$$

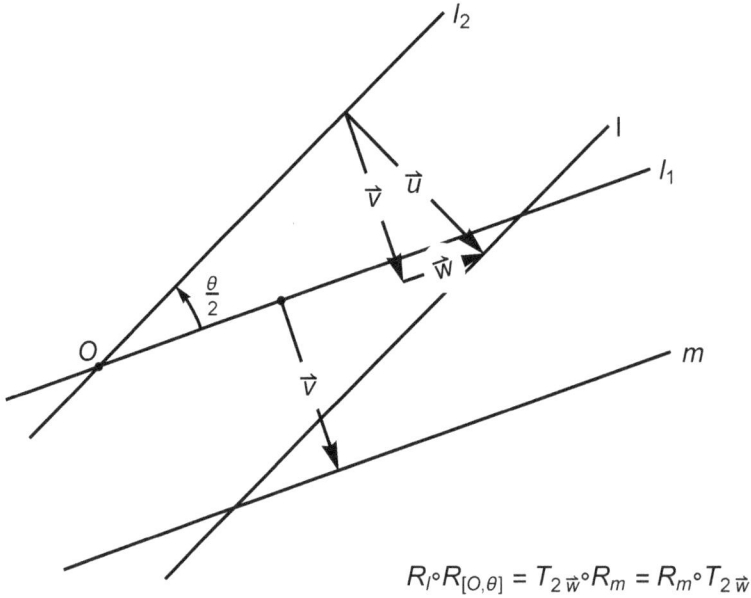

$$R_l \circ R_{[O,\theta]} = T_{2\vec{w}} \circ R_m = R_m \circ T_{2\vec{w}}$$

Figure 2.39: The composition of a rotation followed by a reflection is a glide

Since, $l_2 \parallel l$, we let $\vec{u} = \overrightarrow{l_2 l}$ and have that $R_l \circ R_{l_2} = T_{2\vec{u}}$. We decompose $\vec{u} = \vec{v} + \vec{w}$ such that $\vec{v} \perp l_1$ and $\vec{w} \parallel l_1$. See **Figure 2.39**. Hence,

$$R_l \circ R_{[O,\theta]} = T_{2\vec{u}} \circ R_{l_1} = T_{2\vec{w}} \circ T_{2\vec{v}} \circ R_{l_1}.$$

We now consider the line $m \parallel l_1$ such that $\overrightarrow{l_1 m} = \vec{v}$. Then, $T_{2\vec{v}} = R_m \circ R_{l_1}$. So,

$$R_l \circ R_{[O,\theta]} = T_{2\vec{w}} \circ R_m \circ R_{l_1} \circ R_{l_1} = T_{2\vec{w}} \circ R_m \circ I = T_{2\vec{w}} \circ R_m.$$

Since $\theta \neq 0$, (mod 2π), $\vec{0} \neq \vec{w} \parallel l_1 \parallel m$. Therefore, the final outcome is a glide, completely determined. [This glide is different from the glide in **case (i)** above.]

Note: If the given line l passes through the given point O, then $l_2 = l$ and $\vec{u} = \vec{0}$. In this situation we have

$$R_l \circ R_{[O,\theta]} = R_l \circ R_{l_2} \circ R_{l_1} = R_l \circ R_l \circ R_{l_1} = I \circ R_{l_1} = R_{l_1}.$$

(8) Compositions of Three Reflections

We are given three different lines l_1, l_2 and l_3 in the plane $\mathcal{P}$ and consider the composition $R_{l_3} \circ R_{l_2} \circ R_{l_1}$ of the three related reflections. We would like to simplify this composition as much as possible. To this end, we examine the following five cases:

(i) **Case:** $l_1 \parallel l_2 \parallel l_3$.

We let $\vec{u} = \overrightarrow{l_1 l_2}$ and $\vec{v} = \overrightarrow{l_2 l_3}$ and we consider the line $m \parallel l_1 \parallel l_2 \parallel l_3$ such that $\overrightarrow{l_2 m} = \vec{v} - \vec{u}$. Then,

$$R_{l_3} \circ R_{l_2} \circ R_{l_1} = R_m.$$

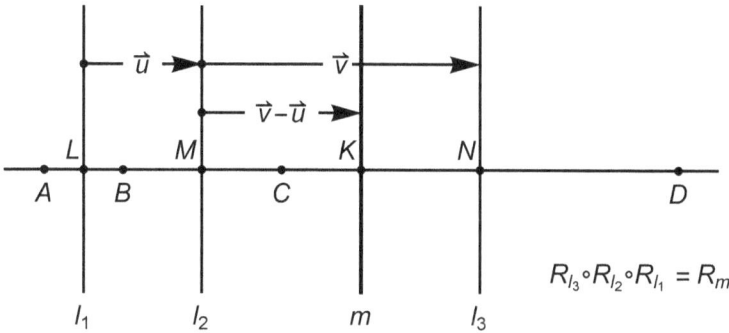

Figure 2.40: Composition of three reflections in three parallel lines

We prove this in the following way. See **Figure 2.40**. Let $A \in \mathcal{P}$ and $B = R_{l_1}(A)$, $C = R_{l_2}(B)$ and $D = R_{l_3}(C)$. Then, $l_1 \perp AB$ at its midpoint L, $l_2 \perp BC$ at its midpoint M and $l_3 \perp CD$ at its midpoint N. In this case, A, L, B, M, C, N, D are on the same straight line, which is perpendicular to the lines $l_1 \parallel l_2 \parallel l_3$. Let K be the intersection of the line AD and the line m.

Then, $\overrightarrow{AL} = \overrightarrow{LB} := \vec{a}$, $\overrightarrow{BM} = \overrightarrow{MC} := \vec{b}$, $\overrightarrow{CN} = \overrightarrow{ND} := \vec{c}$, $\vec{u} = \overrightarrow{LB} + \overrightarrow{BM} = \vec{a} + \vec{b}$ and $\vec{v} = \overrightarrow{MC} + \overrightarrow{CN} = \vec{b} + \vec{c}$.

Then, $\vec{v} - \vec{u} = \vec{c} - \vec{a}$ and

$$\overrightarrow{AK} = \vec{a} + \vec{a} + \vec{b} + \vec{v} - \vec{u} = \vec{a} + \vec{a} + \vec{b} + \vec{b} + \vec{c} - \vec{a} - \vec{b}$$
$$= \vec{a} + \vec{b} + \vec{c}.$$

and

$$\overrightarrow{KD} = \vec{b} - (\vec{v} - \vec{u}) + \vec{c} + \vec{c} = \vec{b} - \vec{b} - \vec{c} + \vec{a} + \vec{b} + \vec{c} + \vec{c}$$
$$= \vec{a} + \vec{b} + \vec{c}.$$

Thus, $\overrightarrow{AK} = \overrightarrow{KD}$ and this equality proves the result claimed.

(ii) Case: $l_1 \cap l_2 \cap l_3 = \{O\}$.

In this case the three different lines intersect at the same point O. We let $\theta = \sphericalangle(l_1, l_2)$ and $\phi = \sphericalangle(l_2, l_3)$ and we consider the line m through O such that $\sphericalangle(l_2, m) = \phi - \theta$. Then

$$R_{l_3} \circ R_{l_2} \circ R_{l_1} = R_m.$$

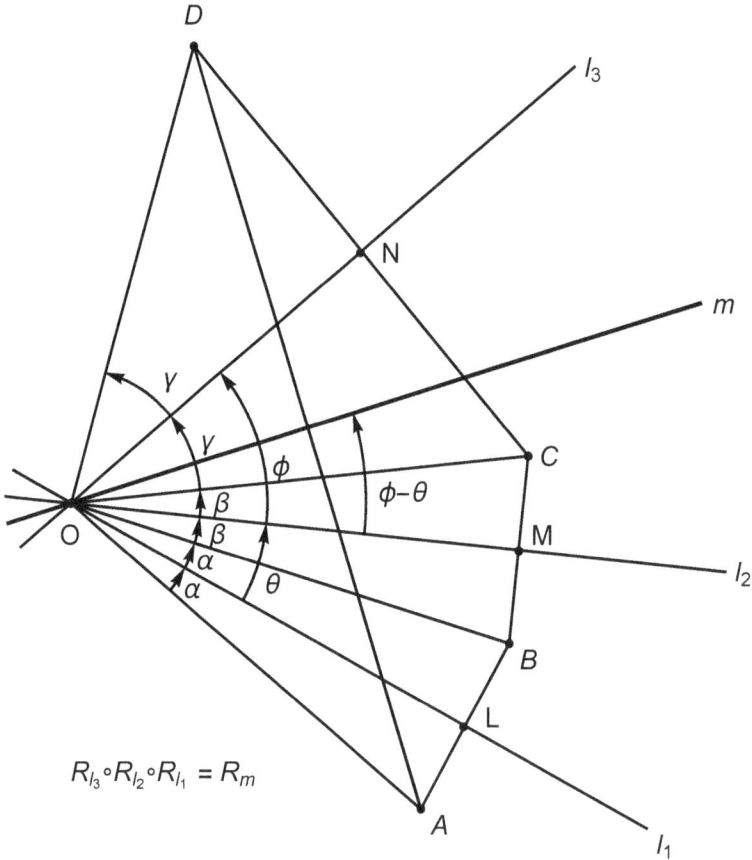

Figure 2.41: The composition of three refections in three concurrent lines reduces to one reflection

We prove this in the following way. See **Figure 2.41**. Let $A \in \mathcal{P}$ and $B = R_{l_1}(A)$, $C = R_{l_2}(B)$ and $D = R_{l_3}(C)$. Then $l_1 \perp AB$ at its midpoint L, $l_2 \perp BC$ at its midpoint M and $l_3 \perp CD$ at its midpoint N. In this situation, we have:

$OA = OB = OC = OD,$ $\qquad\qquad \sphericalangle(OA, OL) = \sphericalangle(OL, OB) := \sphericalangle\alpha,$

$\sphericalangle(OB, OM) = \sphericalangle(OM, OC) := \sphericalangle\beta,$ $\qquad \sphericalangle(OC, ON) = \sphericalangle(ON, OD) := \sphericalangle\gamma,$

$\sphericalangle\theta = \sphericalangle(OL, OB) + \sphericalangle(OB, OM) = \sphericalangle\alpha + \sphericalangle\beta,$

$\sphericalangle\phi = \sphericalangle(OM, OC) + \sphericalangle(OC, ON) = \sphericalangle\beta + \sphericalangle\gamma.$

Then,

$$(1) \qquad \gamma - \alpha = \phi - \theta,$$

$$(2) \quad \measuredangle(OA, m) = \measuredangle\alpha + \measuredangle\alpha + \measuredangle\beta + \measuredangle\phi - \measuredangle\theta$$
$$= \measuredangle\alpha + \measuredangle\alpha + \measuredangle\beta + \measuredangle\beta + \measuredangle\gamma - \measuredangle\alpha - \measuredangle\beta = \measuredangle\alpha + \measuredangle\beta + \measuredangle\gamma,$$

and (3) $\measuredangle(m, OD) = \measuredangle\beta - (\measuredangle\phi - \measuredangle\theta) + \measuredangle\gamma + \measuredangle\gamma$
$$= \measuredangle\beta - \measuredangle\beta - \measuredangle\gamma + \measuredangle\alpha + \measuredangle\beta + \measuredangle\gamma + \measuredangle\gamma = \measuredangle\alpha + \measuredangle\beta + \measuredangle\gamma.$$

Thus, $\measuredangle(OA, m) = \measuredangle(m, OD)$ and since $OA = OD$, these two equalities prove the result claimed.

(iii) Case: $l_1 \nparallel l_2 \parallel l_3$. See **Figure 2.42**.
We let $\vec{u} = \overrightarrow{l_2 l_3}$, $\{O\} = l_1 \cap l_2$ and $\theta = \measuredangle(l_1, l_2)$. Then, this case can be

considered as $\quad R_{l_3} \circ R_{l_2} \circ R_{l_1} = T_{2\vec{u}} \circ R_{l_1} \quad$ or $\quad R_{l_3} \circ R_{l_2} \circ R_{l_1} = R_{l_3} \circ R_{[O,2\theta]}.$

Both of these compositions were studied in paragraphs **(6)** and **(7)** and were found to be certain glides.

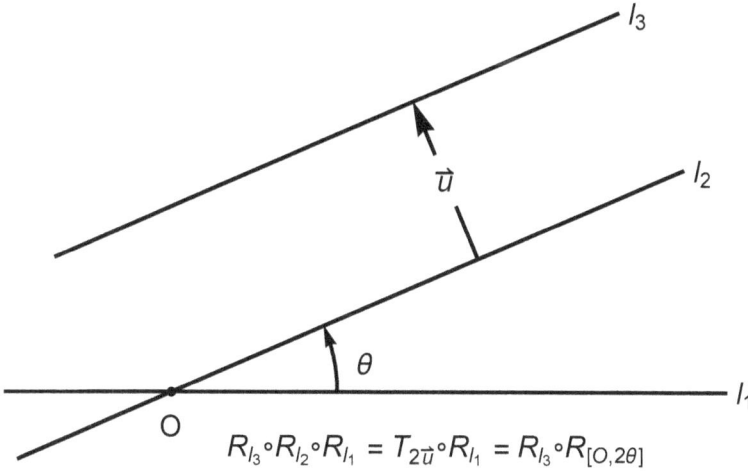

$$R_{l_3} \circ R_{l_2} \circ R_{l_1} = T_{2\vec{u}} \circ R_{l_1} = R_{l_3} \circ R_{[O,2\theta]}$$

Figure 2.42: **Composition of three refections in** $l_1 \nparallel l_2 \parallel l_3$

(iv) Case: $l_1 \parallel l_2 \nparallel l_3$. See **Figure 2.43**.
We let $\vec{u} = \overrightarrow{l_1 l_2}$, $\{O\} = l_2 \cap l_3$ and $\theta = \measuredangle(l_2, l_3)$. Then, this case is reduced

to either $\quad R_{l_3} \circ R_{l_2} \circ R_{l_1} = R_{l_3} \circ T_{2\vec{u}}, \quad$ or $\quad R_{l_3} \circ R_{l_2} \circ R_{l_1} = R_{[O,2\theta]} \circ R_{l_1}.$

Both of these compositions were studied in paragraphs **(6)** and **(7)** and were found to be certain glides.

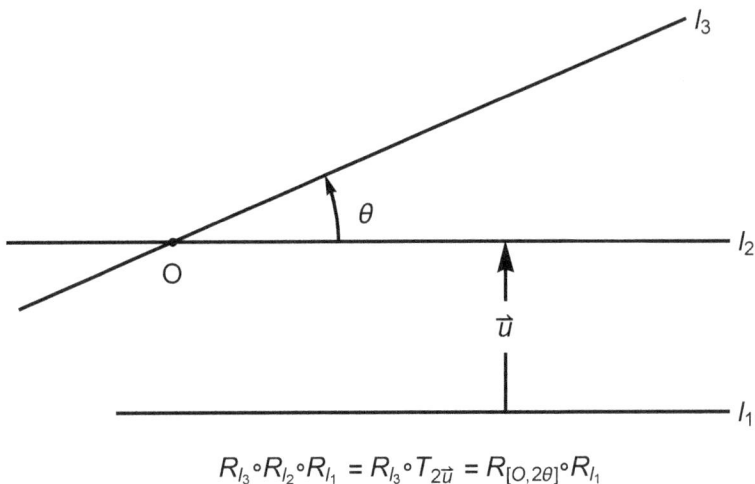

$$R_{l_3} \circ R_{l_2} \circ R_{l_1} = R_{l_3} \circ T_{2\vec{u}} = R_{[O,2\theta]} \circ R_{l_1}$$

Figure 2.43: Composition of three reflections in $l_1 \parallel l_2 \nparallel l_3$

(v) **Case:** $l_1 \nparallel l_2$ and $l_2 \nparallel l_3$. See **Figure 2.44**.
   We let $\theta = \angle(l_1, l_2)$, $\{O\} = l_1 \cap l_2$, $\phi = \angle(l_2, l_3)$ and $\{Q\} = \angle(l_2, l_3)$. Then, this case reduces to

either $R_{l_3} \circ R_{l_2} \circ R_{l_1} = R_{l_3} \circ R_{[O,2\theta]}$,   or $R_{l_3} \circ R_{l_2} \circ R_{l_1} = R_{[Q,2\phi]} \circ R_{l_1}$.

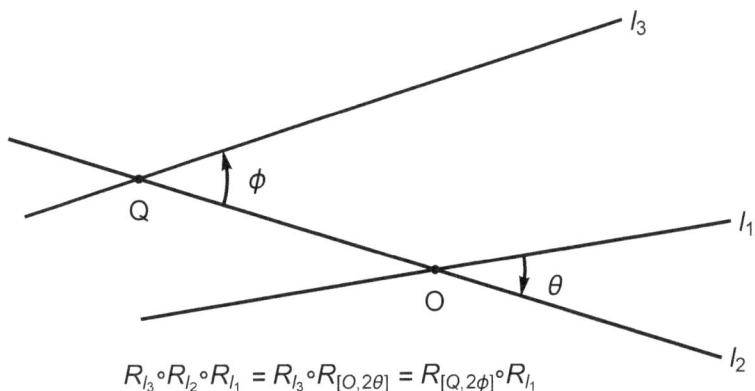

$$R_{l_3} \circ R_{l_2} \circ R_{l_1} = R_{l_3} \circ R_{[O,2\theta]} = R_{[Q,2\phi]} \circ R_{l_1}$$

Figure 2.44: Composition of three reflections in $l_1 \nparallel l_2$ and $l_2 \nparallel l_3$

Both of these compositions were studied in paragraph **(7)** and were found to be certain glides. (Elaborate!)
   Note: The **composition of a glide with another glide or one of the other basic isometries** reduces to one of the cases already studied above. See also some exercises on reflections and glides, **Section 3.3**.

2.3.1 General Result

Manipulating the results proven in paragraphs **(1)-(8)** and the properties of glides, we easily obtain that any finite string of compositions of basic isometries of the Euclidean plane can be reduced to the composition of at most three (3) reflections. Namely, we have seen that:

(a) The identity is $I = R_l \circ R_l$, for any straight line l of the plane.

(b) One reflection satisfies the result, already.

(c) A parallel translation and or a rotation is a composition of two reflections (in infinitely many ways) and vice versa.

(d) Three reflections may be reduced to one reflection or a glide, which in turn can be written as a special composition of three reflections, and vice versa.

(e) All other compositions reduce to one of the above four cases.

Remarks: (1) *The identity, the parallel translations and the rotations are direct isometries and the reflections and the glides are opposite isometries.* (This follows from the fact that a reflection is an opposite isometry.) Therefore, *the composition of an even number of reflections is direct and the composition of an odd number of reflections is opposite.*

(2) If the composition of basic isometries has exactly one fixed point, then it simplifies to a composition of at most two (2) reflections.

(3) Another method of proving this general result, by mapping a triangle to another directly or oppositely equal triangle, is presented in **Subsection 2.4.1**.

2.4 The Set of All of Isometries

The question we face here is, if there are isometries of the plane other than those studied thus far in this chapter. The answer is no! That is, all isometries of the plane are exactly those studied so far. This claim is proven by examining the different cases that arise by the number and the relative positions of the fixed points that an isometry may have.

In **Theorems 2.2.1, 2.2.2** and **Corollary 2.2.3**, we have proved that an isometry with **3** non-collinear fixed points is the identity and if it is not the identity and has **2** fixed points then it is the reflection with axis the line through these two fixed points. In the latter case all the points of the axis of the reflection are fixed. So, we must examine the cases in which the number of fixed points is either **1**, or **0**.

The Case of One Fixed Point: We have seen that a rotation that is not the identity has exactly one fixed point, namely the center of the rotation. Now, we must prove the converse. For this, we first need the following lemma:

Lemma 2.4.1 *An isometry F with exactly **1** fixed point (and so not identity) cannot map a non-zero oriented angle with vertex the fixed point to its opposite.*

Proof Let O be the fixed point of F $[F(O) = O]$ and suppose there is an oriented angle $\angle xOy \neq 0$ which is mapped to $\angle x'Oy' = -\angle xOy \neq 0$, so that, the ray Ox is mapped to the ray Ox' and the ray Oy to the ray Oy'. See **Figure 2.45**.

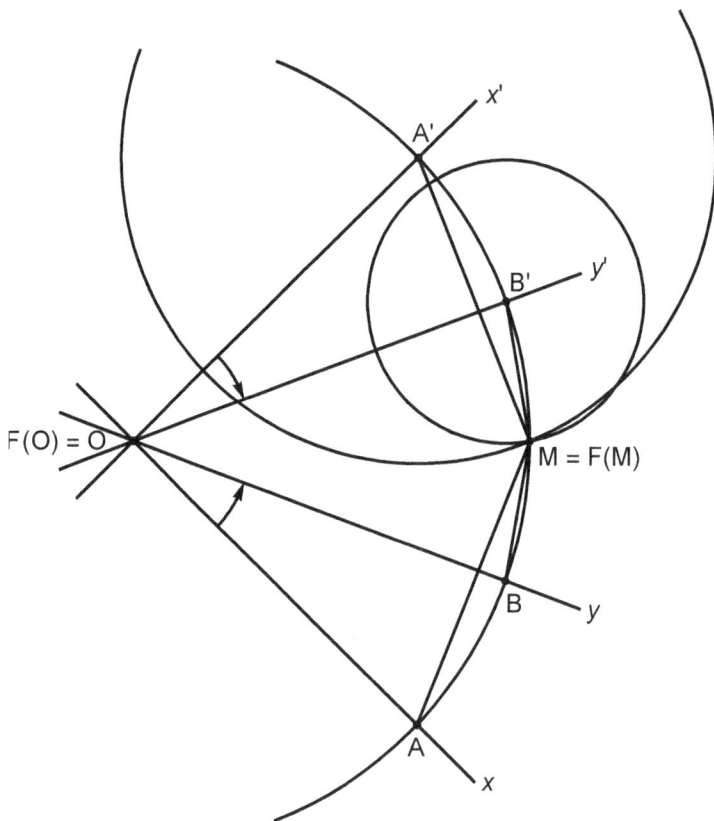

Figure 2.45: About an isometry with exactly one fixed point

Since F is an isometry and O is the only fixed point of it, we note that:

(1) $Ox \neq Ox'$ and $Oy \neq Oy'$. [By **Section 2.2, item (8)**, the whole line Ox (not just the ray), is mapped to the whole line Ox' and similarly, line Oy is mapped to line Oy'.]

(2) Points P and P' on the rays Ox and Ox' respectively, such that $OP = OP'$, satisfy $F(P) = P'$, and similarly for analogous points on the rays Oy and Oy'.

Now, we consider a circle $C[O, r]$ (of center O and a radius $r > 0$). This intersects the rays Ox, Oy, Ox' and Oy' at the points point A, B, A' and B', respectively. On this circle, consider the arc $\overset{\frown}{AA'} \neq \overset{\frown}{0}$, oriented with the sense of the angle $\angle xOy$. This contains the points B and B'. Let M be the midpoint of this arc. Then, $M \neq O$ and $AM = A'M$. Also, since $\angle x'Oy' = -\angle xOy$, we also obtain $BM = B'M$.

Since $F\{C[O,r]\} = C[O,r]$, as F is an isometry with fixed point O, we conclude that the image point $M' = F(M)$ must lie on the three circles, $C[O,r]$, $C[A', AM = A'M']$ and $C[B', BM = B'M']$. Under the assumed conditions, the points O, A' and B' are not collinear. Therefore, the three circles can have at most one point in common. By construction, this point is the point M itself. That is, $F(M) = M$, or M is a fixed point of F other than O. This is contradiction to the hypothesis that the isometry F has exactly **1** fixed point. Therefore, the claim of this Lemma is proved, by contradiction.

■

Remark: By **Corollary 2.2.2**, the isometry of the **above Lemma** is a direct isometry.

Theorem 2.4.1 *An isometry with exactly* **1** *fixed point (and so not identity) is a rotation by some oriented angle* $\neq 0$, *(mod* 2π*), and with center the fixed point.*

Proof We consider two different points A and B and different from the unique fixed point O in the plane and their images $F(A) = A' \neq A$ and $F(B) = B' \neq B$.

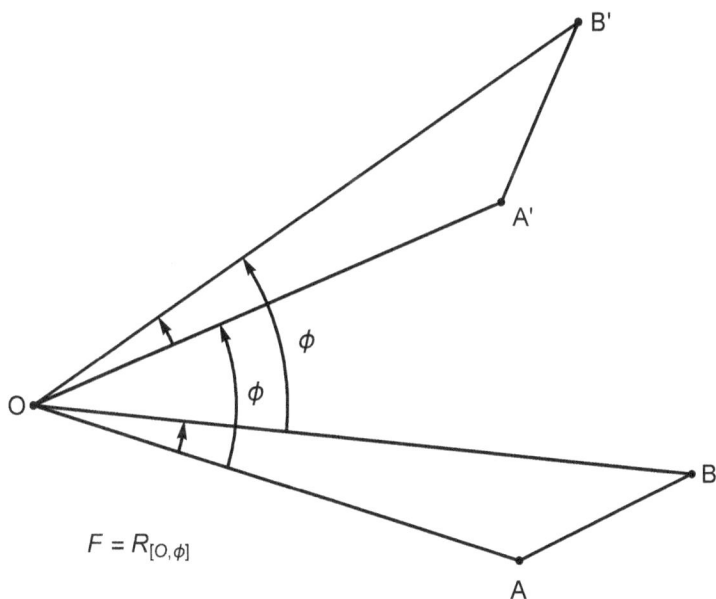

Figure 2.46: An isometry with exactly one fixed point is a rotation

The points A' and B' are different than $O = F(O)$ and different from each other, since F is one-one. Also,

$$OA = F(O)F(A) = OA', \quad OB = F(O)F(B) = OB', \quad AB = F(A)F(B) = A'B'$$

and, by the **previous Lemma**, $\angle(OA, OB) = \angle(OA', OB')$. [This angle could be 0 (mod π), if O, A and B are collinear.] See **Figure 2.46**.

To prove the Theorem, we must show

$$\angle(OA, OA') = \angle(OB, OB').$$

We have,

$$\angle(OA, OA') = \angle(OA, OB) + \angle(OB, OA')$$
$$= \angle(OB, OA') + \angle(OA', OB') = \angle(OB, OB').$$

Now, we keep A constant (and so A' is also constant) and let B vary. Then, for any $B \in \mathcal{P}$, we find $\angle(OB, OB') = \angle(OA, OA') := \phi$ fixed. Therefore, $F = R_{[O,\phi]}$. Since $F \neq I$, we have $\phi \neq 0$, (mod 2π).

∎

The Case of No Fixed Point: We have seen that a parallel translation which is not the identity, and a glide have no fixed points. Now, we must prove the converse. We let F be an isometry of the plane with no fixed points, so it cannot be a rotation or a reflection. Here, we need to give the definition:

Definition 2.4.1 *Given two different parallel lines $l_1 \parallel l_2$, we call the straight line m the **mid-parallel line** (or simply **mid-parallel**) of l_1 and l_2 if $m \parallel l_1 \parallel l_2$ and $\overrightarrow{l_1 m} = \overrightarrow{m l_2}$.*

Next, we must examine the following cases:

Case 1: We suppose that F maps two different points A and B to $F(A) = A' \neq F(B) = B'$, such that the lines AB and $A'B'$ are parallel, and the segments AA' and BB' are perpendicular to $AB \parallel A'B'$. See **Figure 2.47**.

Then, $AA'B'B$ is a rectangle (since, $AB =\parallel A'B'$ and AA', $BB' \perp AB \parallel A'B'$) and so $\overrightarrow{AA'} = \overrightarrow{BB'}$. Hence, A' and B' may be obtained as images of A and B respectively, by either the parallel translation $T_{\overrightarrow{AA'}}$, or by the reflection R_l, where l is the mid-parallel line of the lines $AB \parallel A'B'$, or by "something else". The second option is excluded, since we have assumed that F has no fixed points. We must also exclude the "something else", so that the parallel translation $T_{\overrightarrow{AA'}}$ is the only possibility.

By **Section 2.2, item (8)**, the whole line AB is mapped onto the line $A'B'$. Next, consider any point C of the plane $\mathcal{P}$ not on the line AB. Since F is an isometry, its image must lie on the two circles $C[A', AC]$ and $C[B', BC]$. Since ABC is an "honest" triangle, these circles intersect at two distinct points C' and C'', which are easily proven to be collinear with C. Also, the line $C'C''$ is perpendicular to the lines $AB \parallel A'B'$. Then, for one of these points, the C' let us say, we have $\overrightarrow{CC'} = \overrightarrow{AA'} = \overrightarrow{BB'}$ (prove this). If we let $F(C) = C'$, i.e., C' is also obtained by the parallel translation by the free vector $\overrightarrow{AA'}$, then, by **Theorem 2.2.1**, $F = T_{\overrightarrow{AA'}}$ as the isometry is determined by the three points A, B, and C and their images A', B', and C'.

If we let $F(C) = C''$, the other point of the intersection of the above two circles, then we easily prove that $R_l(C) = C''$. We also have $R_l(A) = A'$ and $R_l(B) = B'$. Thus, by **Theorem 2.2.1**, these three pairs of points would prove

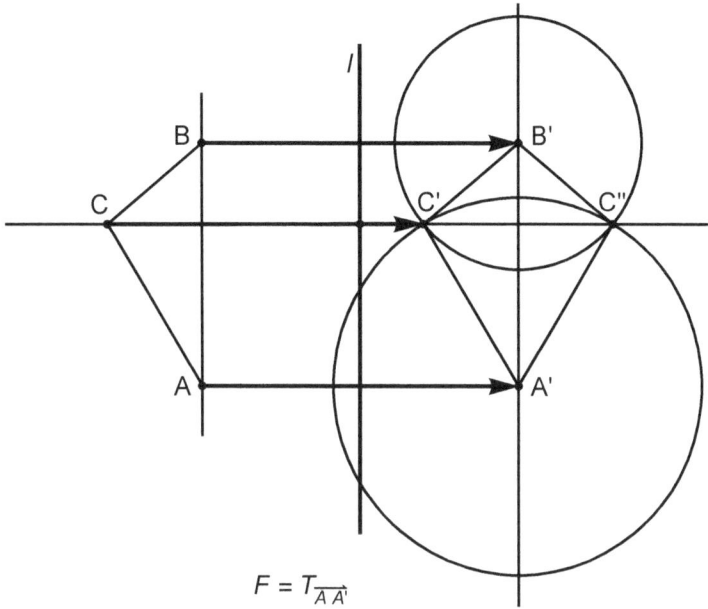

$$F = T_{\overrightarrow{AA'}}$$

**Figure 2.47: In case 1, the isometry with no fixed point
is a parallel translation**

that $F = R_l$. But, as we have said before, this cannot be the case in the absence of fixed points.

Finally, in the case considered here, $F = T_{\overrightarrow{AA'}}$, where $\overrightarrow{AA'} \neq \overrightarrow{0}$.

Case 2: We suppose that the isometry F has no fixed points and maps two different points A and B to $F(A) = A' \neq F(B) = B'$, such that the lines $l_1 = AB$ and $l_2 = A'B'$ are parallel, and the segments AA' and BB' are **not** perpendicular to $AB \parallel A'B'$. (Then, $AA'B'B$ is a parallelogram, since $AB =\parallel A'B'$. – The notation $=\parallel$ means equal and parallel.)

Subcase 1: The segments AA' and BB' are the diagonals of the parallelogram $AA'B'B$. See **Figure 2.48**.

Then, the segments AA' and $B'B$ intersect at their midpoint O and the figure of the lines l_1 and l_2 is a set-wise fixed set of the rotation $R_{[O,\pi]}$ with $R_{[O,\pi]}(A) = A'$ and $R_{[O,\pi]}(B) = B'$. But in the absence of fixed points, $F \neq R_{[O,\pi]}$, since $R_{[O,\pi]}$ has a fixed point, namely O.

We now let $O' = F(O) \neq O$. This must be the point of intersection, other than O, of the two circles $C[A', OA]$ and $C[B', OB]$. Then, by using elementary geometry, we prove that the straight line $k := OO'$ is perpendicular to $l_2 \parallel l_1$ and $\overrightarrow{l_1 l_2} = \overrightarrow{OO'} = \overrightarrow{A_1 A'} = \overrightarrow{B_1 B'}$, where A_1 and B_1 are the reflections of A and B in the line $k := OO'$.

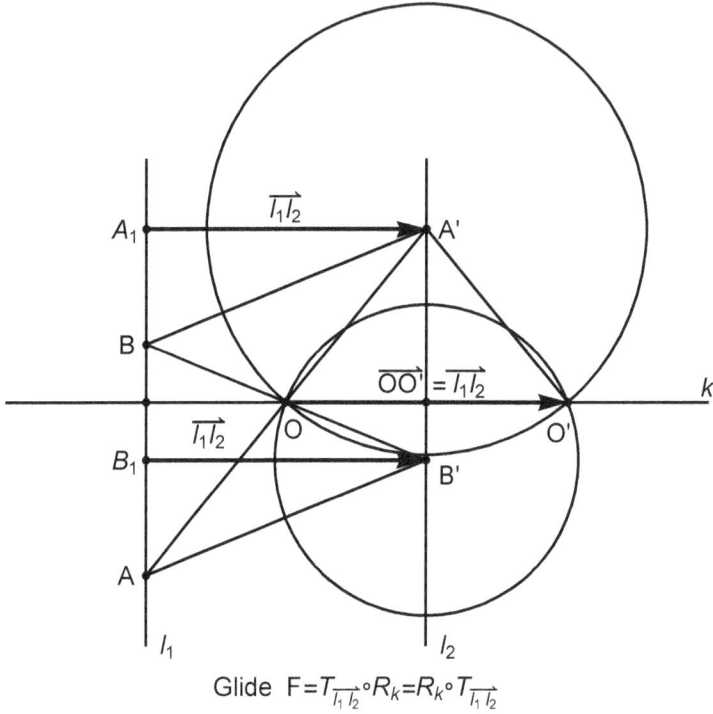

Glide $F = T_{\overrightarrow{l_1 l_2}} \circ R_k = R_k \circ T_{\overrightarrow{l_1 l_2}}$

Figure 2.48: In case 2 subcase 1, the isometry with no fixed point is a glide

Notice now that the glide $R_k \circ T_{\overrightarrow{l_1 l_2}} = T_{\overrightarrow{l_1 l_2}} \circ R_k$, maps A, B, and O to maps A', B', and O', respectively. Since the three points A, B and O are not on a straight line and $F(A) = A'$, $F(B) = B'$ and $F(O) = O'$ are known, by **Theorem 2.2.1**, we have that the isometry $F = R_k \circ T_{\overrightarrow{l_1 l_2}} = T_{\overrightarrow{l_1 l_2}} \circ R_k$, (a glide, with no fixed point).

Subcase 2: The segments AA' and BB' are the two parallel sides of the parallelogram $AA'B'B$. Then, $\overrightarrow{AA'} = \overrightarrow{BB'}$. See **Figure 2.49**. Then we have:

(a) We see that the line $l_1 = AB$ is mapped to $l_2 = A'B'(\| l_1 = AB)$ by the parallel translation $T_{\overrightarrow{AA'}}$. So, F could be the parallel translation $T_{\overrightarrow{AA'}}$, which satisfies the hypotheses that F has no fixed point and $F(A) = A'$ and $F(B) = B'$.

(b) There is one more option. For the parallel lines $l_1 = AB$ and $l_2 = A'B'$, we let $\overrightarrow{u} = \overrightarrow{l_1 l_2}$. We let m be the mid-parallel line of l_1 and l_2, and decompose $\overrightarrow{AA'} = \overrightarrow{u} + \overrightarrow{v} = \overrightarrow{BB'}$, where $\overrightarrow{v} \parallel l_1 \parallel l_2 \parallel m$.

We see that the figure of the straight lines l_1 and l_2 is a set-wise fixed set of the glide $F := T_{\overrightarrow{v}} \circ R_m = R_m \circ T_{\overrightarrow{v}}$. This option satisfies the hypotheses that

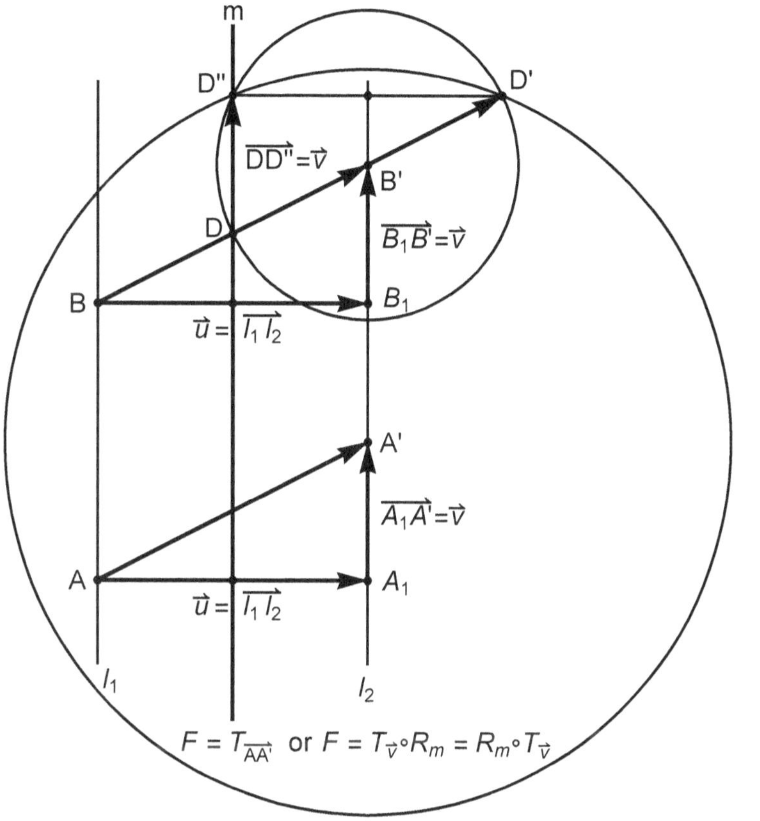

Figure 2.49: In case 2 subcase 2, the isometry with no fixed point is either a parallel translation or a glide

F has no fixed point, $F(A) = A'$ and $F(B) = B'$.

(c) There is no other option. To prove this, we consider, e.g., $D = m \cap BB'$ which is the midpoint of segment BB'. Its image $F(D)$ must lie on the two circles $C[B', BD]$ and $C[A', AD]$. We use elementary geometry to prove the following: These circles intersect at two points: (1) D' the point diametrically opposite to D on the circle $C[B', BD]$. Then $D' = F(D)$ is obtained by the parallel translation $F := T_{\overrightarrow{AA'}}$, applied to D. (2) D'' on the line m and such that $\overrightarrow{DD''} = \vec{v}$. Then, $D'' = F(D)$ is also obtained by the glide

$$F := T_{\vec{v}} \circ R_m = R_m \circ T_{\vec{v}}$$

applied to D.

Since the three points A, B, and D are not on the same straight line and A', B' and D' or D'' are known, **Theorem 2.2.1** proves that there are no other options for the isometry F, but $F = T_{\overrightarrow{AA'}}$ or $F = T_{\vec{v}} \circ R_m = R_m \circ T_{\vec{v}}$.

Case 3: We suppose that the isometry F has no fixed points and maps two different points A and B to $F(A) = A' \neq F(B) = B'$, such that the straight lines $l_1 = AB$ and $l_2 = A'B'$ are **not** parallel and so they intersect at a point W. So, $AB = A'B'$, but $\overrightarrow{AB} \neq \overrightarrow{A'B'}$, as they are not parallel.

Subcase 1: Assume that the points W, A, B on line l_1 and W, A', B' on line l_2 satisfy $WA = WA'$, $WB = WB'$ and the necessary condition $AB = A'B'$. (These conditions imply $AA' \parallel BB'$.)

Under these assumptions, none of these four segments must be zero, for otherwise either $A = W = A'$ or $B = W = B'$ and so there would be a fixed point of F. Also, given that $AB = A'B'$, we must have that W is either outside both AB and $A'B'$ or inside both AB and $A'B'$. (Check this!) See **Figures 2.50** and **2.51**.

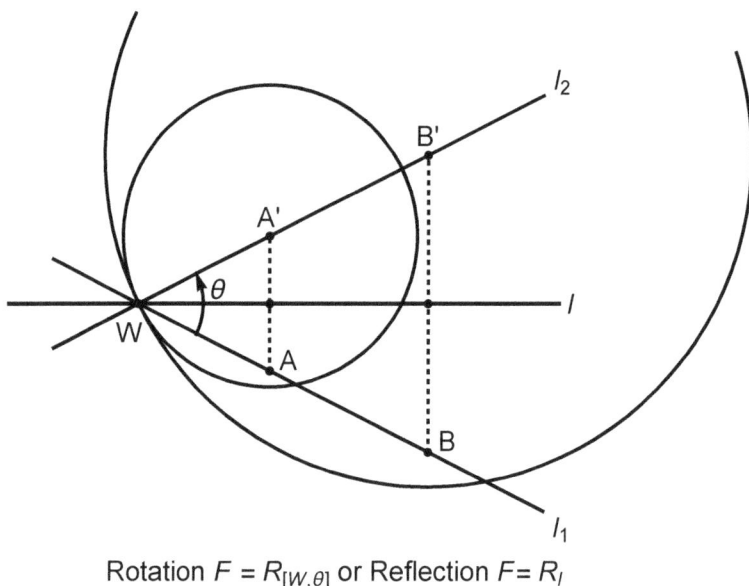

Rotation $F = R_{[W,\theta]}$ or Reflection $F = R_l$

Figure 2.50: About an isometry with no fixed point.
This situation is excluded from Case 3.

The point $W' := F(W)$ must lie on the two circles $C[A', AW]$ and $C[B', BW]$. In **Figure 2.50**, we have $A'B' = WB' - WA' = WB - WA$ and so these circles are internally tangent at W.
In **Figure 2.51**, we have $A'B' = WB' + WA' = WB + WA$ and so these circles are externally tangent at W.

In either case, the circles have W as the only common point. Therefore, $W' = F(W) = W$ is a fixed point. So, this subcase is excluded from **Case 3** because F is assumed to have no fixed point.

{But there are two obvious isometries F such that $F(A) = A' \neq F(B) = B'$.

(1) The line l_1 is mapped to l_2 by the rotation $R_{[W,\theta]}$, where $\theta = \angle AWA'$, and this rotation satisfies $R_{[W,\theta]}(A) = A'$ and $R_{[W,\theta]}(B) = B'$.

(2) The figure of the lines l_1 and l_2 is a set-wise invariant set under the reflection R_l, where l is the angle bisector of $\angle AWA'$.

In either case, W is a fixed point. Hence, these options are excluded from **Case 3**.}

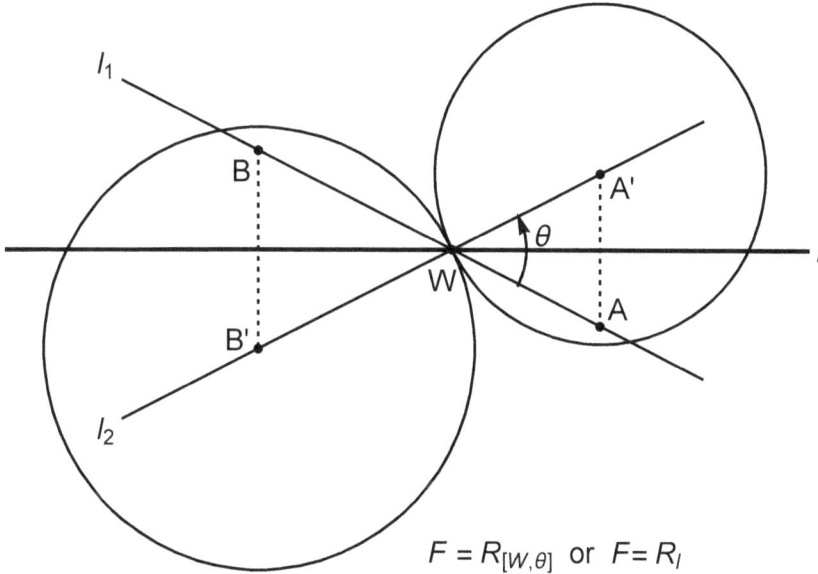

Figure 2.51: About an isometry with no fixed point.
This situation is excluded form Case 3.

Remark: Under the condition $AB = A'B'$ the circles $C[A', AW]$ and $C[B', BW]$ are always tangent, no matter how the points A and B are located in l_1 and A' and B' are located on l_2. (Check this! See also **Figures 2.58** and **2.59**.)

Before we resolve **Case (3)** when $WA \neq WA'$ or $WB \neq WB'$, we need to prove the following **Lemmata** and their **Corollaries**.

Lemma 2.4.2 *We consider two straight lines $l_1 \nparallel l_2$ intersecting at the point W, segment AB on l_1 and segment $A'B'$ on l_2, such that, $AB = A'B'$ and the lines AA' and BB' do not intersect in the interiors of the angles $\angle AWA'$ and its equal vertical angle (at W). Let M be the midpoint of AA' and N the midpoint BB'. Then, the line $m := MN$ is parallel to the bisector l of the angle $\angle BWB'$ (i.e., $m \parallel l$ or $m \equiv l$) and so perpendicular to l' the bisector of the supplementary angle of $\angle BWB'$ at W ($m \perp l'$).*

Proof Figure 2.52 shows a configuration of this situation. There are more configurations, depending on the positions of the points A, B, A' and B' in relation to the point W, but the proofs are analogous in each case.

We have assumed $AB = A'B'$. We consider the points A_1 and B_1 on l_2 such that $WA_1 = WA$ and $WB_1 = WB$ and we draw the bisector l of the angle $\angle BWB'$ and l' the bisector of its supplementary angle at W. Let S be the midpoint of AA_1 and T the midpoint of BB_1.

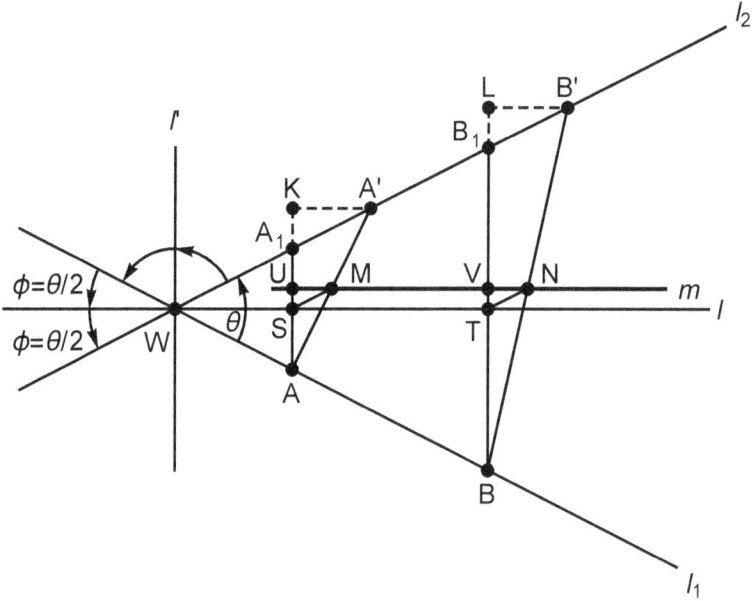

Figure 2.52: A configuration for Lemma 2.4.2

Then, we get: $A_1B_1 = AB = A'B'$ and so $A_1A' = B_1B'$. Hence,
$$SM = \frac{A_1A'}{2} = \frac{B_1B'}{2} = TN \text{ and } SM \parallel l_2 \parallel TN. \text{ Therefore, } MNTS$$
is a parallelogram. So, $m := MN \parallel ST := l \perp l'$. (Fill in the details.) ∎

From **this Lemma**, we observe the following:

Corollary 2.4.1 *(a) We keep the points B and B' of **the Lemma** fixed and we let points A and A' move along l_1 and l_2, respectively and in either direction, under the conditions:*
(1) $AB = A'B'$, and
(2) the lines AA' and BB' do not intersect in the interiors of the angles $\angle AWA'$ and its equal vertical angle (at W).
*Then, the midpoint M of AA' traces the line m of **the Lemma**.*

*(b) We keep the segments AB and $A'B'$ of **the Lemma** fixed and we consider points C and C' move along l_1 and l_2, respectively and in either direction, under the conditions:*

(1) $CA = C'A'$ and $CB = C'B'$, and
(2) the lines CC', AA' and BB' do not intersect in the interiors of the angles $\angle AWA'$ and its equal vertical angle (at W).
Then, the midpoint K of CC' traces the line m of **the Lemma**.

(c) The projections of all vectors $\overrightarrow{CC'}$, in **(b)**, on the line m (or the bisector l) are equal (they have the same length).

(d) The right triangles KA_1A' and LB_1B' are equal and similar to the equal the triangles USM and VTN with ratio of similarity $\dfrac{1}{2}$.

Proof (a) Since the segment BB' is fixed, its midpoint N is also fixed. The line l is fixed and $MN \parallel l$. It follows that the line $MN := m$ is always the same.

(b) Follows from **(a)**.

(c) See **Figure 2.52**. Let K be the reflection of A on m and L the reflection of B on m. Since M and N are midpoints of AA' and BB', respectively, then KA' and LB' are parallel to $m \parallel l$. So, $\angle AKA' = \angle BLB' = \dfrac{\pi}{2}$. The right triangles A_1KA' and B_1LB' have equal angles and $A_1A' = B_1B'$ and therefore they are equal. So, $KA' = LB'$ and these are equal to the projections of the vectors $\overrightarrow{AA'}$ and $\overrightarrow{BB'}$ on the line m (or the bisector l). Hence, any vector $\overrightarrow{CC'}$, as in **(b)**, has the same projection on the line m (or the bisector l) and so all of these projections are equal (they have the same length).

(d) In **(c)**, we proved that the triangles KA_1A' and LB_1B' are equal. Similarly the triangles USM and VTN are equal because they have equal angles and $SM = TN = \dfrac{A_1A'}{2} = \dfrac{B_1B'}{2}$. Also the latter triangles are similar to the former because their corresponding angles are equal. The ratio of similarity is $\dfrac{1}{2}$. ■

Corollary 2.4.2 *If in **the previous Lemma** and / or **the previous Corollary** the two different lines l_1 and l_2 are parallel ($l_1 \parallel l_2$), then, under the remaining conditions, $l \equiv m$ is the mid-parallel line of the lines l_1 and l_2.*

The Corollary follows easily. See **Figure 2.53**.

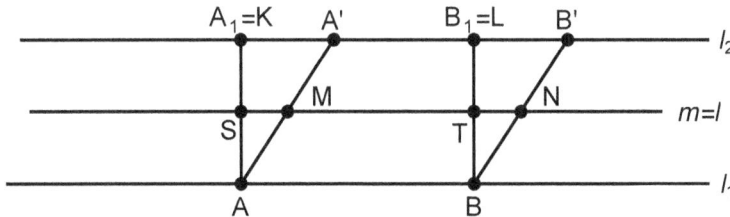

Figure 2.53: For Lemma 2.4.2 and its Corollaries, when $l_1 \parallel l_2$

Lemma 2.4.3 *We consider two straight lines $l_1 \nparallel l_2$ intersecting at the point W, segment AB on l_1 and segment $A'B'$ on l_2, such that, $AB = A'B'$, and the lines AA' and BB' do not intersect in the exteriors of the angles AWA' and its equal vertical angle (at W). Let M be the midpoint of AA' and N the midpoint BB'. Then, the line $m' := MN$ is perpendicular to the bisector l of the angle $\angle BWB'$, ($m' \perp l$), and so it is parallel to l' the bisector of the supplementary angle of $\angle BWB'$ at W. ($m' \parallel l'$).*

Proof: **Figure 2.54** shows a configuration of this situation. There are more configurations, depending on the positions of the points A, B, A' and B' in relation to the point W, but the proofs are analogous in each case.

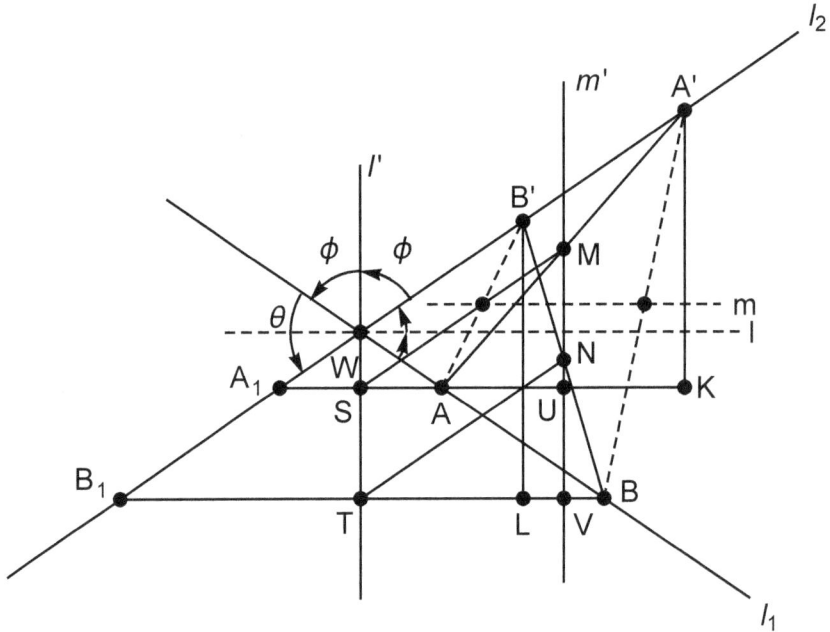

Figure 2.54: For Lemma 2.4.3

We have $AB = A'B'$, we take points A_1 and B_1 on l_2 such that $WA_1 = WA$ and $WB_1 = WB$ and we draw the bisector l of the angle $\angle BWB'$ and l' the bisector of its complementary. Let S be the midpoint of AA_1 and T the midpoint of BB_1.

Then, we get: $A_1B_1 = AB = A'B'$ and so $A_1A' = B_1B'$. Hence,

$$SM = \frac{A_1A'}{2} = \frac{B_1B'}{2} = TN \quad \text{and} \quad SM \parallel l_2 \parallel TN.$$

Therefore, $MNTS$ is a parallelogram. So, $m' := MN \parallel ST := l'$ and so $m' \perp l$. (Fill in the details.)

∎

From **this Lemma**, we also observe the following:

Corollary 2.4.3 *(a) We keep the points B and B′ of **the Lemma** fixed and we let points A and A′ move along l_1 and l_2, respectively and in either direction, under the following two conditions:*
(1) $AB = A′B′$.
(2) The lines AA′ and BB′ do not intersect in the exteriors of the angles AWA′ and its equal vertical angle (at W).
*Then, the midpoint M of AA′ traces the line m′ of **the Lemma**.*

  *(b) We keep the segments AB and A′B′ of **the Lemma** fixed and we consider points C and C′ move along l_1 and l_2, respectively and in either direction, under the following two conditions:*
(1) $CA = C′A′$ and $CB = C′B′$.
(2) The lines CC′, AA′ and BB′ do not intersect in the exteriors of the angles $\angle AWA′$ and its equal vertical angle (at W).
*Then, the midpoint K of CC′ traces the line m′ of **the Lemma**.*

  (c) The projections of all vectors $\overrightarrow{CC′}$ on the line m′ (or the bisector l′) are equal and so they have the same length.

  (d) The right triangles $KA_1A′$ and $LB_1B′$ are equal and similar to the equal the triangles USM and VTN with ratio of similarity $\dfrac{1}{2}$.

The proof is similar to the proof of **Corollary 2.4.1**. See **Figure 2.54**.
[For (c) prove analogously that $KA′ = LB′$.]

∎

Corollary 2.4.4 *If in **the above Lemma** and / or **the previous Corollary** the two different lines l_1 and l_2 are parallel ($l_1 \parallel l_2$), then $M = N$ and this is the midpoint of any segment AA′, BB′, etc. (All the other segments CC′, have midpoint the point $M = N$ which is the center of symmetry of the whole figure.)*

See **Figure 2.55**.

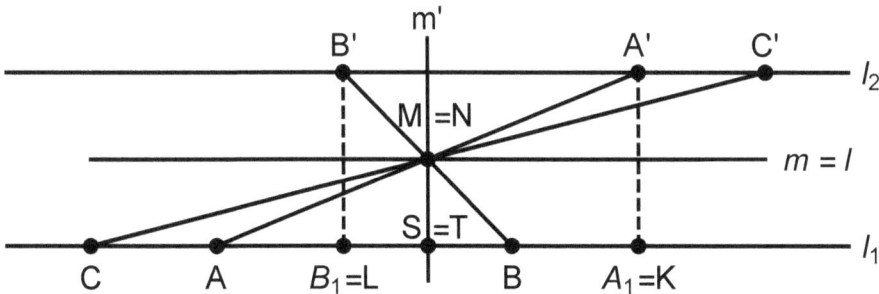

Figure 2.55: For Lemma 2.4.3 and its Corollaries, when $l_1 \parallel l_2$

Next we continue with another subcase of the case we deal with here.

Subcase 2 of **Case (3)**. Along the necessary for an isometry condition $AB = A'B'$, in order to have a glide and not just a reflection, we must assume

$$WA \neq WA' \quad \Longleftrightarrow \quad WB \neq WB'.$$

Compare **Figures 2.52 and 2.54** with **Figures 2.56 and 2.57** and let $\overrightarrow{AA'} = \overrightarrow{AK} + \overrightarrow{KA'}$ with $\overrightarrow{AK} \perp \overrightarrow{MN}$ and $\overrightarrow{KA'} \parallel \overrightarrow{MN}$, and similarly, $\overrightarrow{BB'} = \overrightarrow{BL} + \overrightarrow{LB'}$ with $\overrightarrow{BL} \perp \overrightarrow{MN}$ and $\overrightarrow{LB'} \parallel \overrightarrow{MN}$.

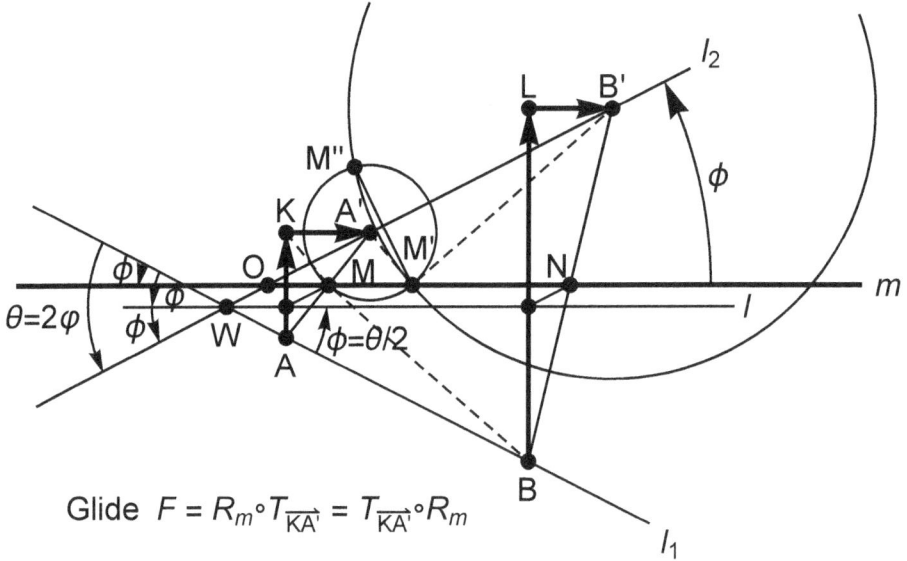

Glide $F = R_m \circ T_{\overrightarrow{KA'}} = T_{\overrightarrow{KA'}} \circ R_m$

Figure 2.56: An isometry with no fixed point, that is not a parallel translation, is a glide

See **Figure 2.56** and use **Lemma 2.4.2** to prove that $\overrightarrow{KA'} = \overrightarrow{LB'}$. Then, $F = R_m \circ T_{\overrightarrow{KA'}}$ is a glide (no fixed point) that maps $F(A) = A'$ and $F(B) = B'$, as we readily see.

Now, $F(M)$ must be on the circles $C[A', MA]$ and $C[B', MB]$, which intersect at two points, $M' \in m$ and $M'' = R_{l_2}(M')$.

For $M' \in m$, we have $\overrightarrow{MM'} = \overrightarrow{KA'}$ (prove!). Therefore, $F = R_m \circ T_{\overrightarrow{KA'}}$ is the above glide, since F also satisfies $F(M) = M'$.

For $M'' = R_{l_2}(M')$, since $F = R_m \circ T_{\overrightarrow{KA'}} = T_{\overrightarrow{KA'}} \circ R_m$, we have that

$$M'' = R_{l_2} \circ \left(R_m \circ T_{\overrightarrow{KA'}} \right)(M) = \left(R_{l_2} \circ T_{\overrightarrow{KA'}} \right) \circ R_m(M).$$

Then, the isometry $R_{l_2} \circ R_m \circ T_{\overrightarrow{KA'}}$ maps the three non-collinear points A, B and M to A', B' and M'', respectively. But, $R_{l_2} \circ R_m = R[O, 2\phi]$, where $\{O\} = l_2 \cap m$ and $\phi = \angle(m, l_2)$. Then, $R[O, 2\phi] \circ T_{\overrightarrow{KA'}}$ is a new rotation [by paragraph (5)]. So, this isometry cannot be F, as F has no fixed point.

Now, we check the **Figure 2.57**. The straight line $m' := MN$ is perpendicular to l, the bisector of the angle $\angle AWA'$, and therefore m' is parallel to the bisector l' of the angle $\angle B'WD$, the supplementary angle of $\angle AWA'$. Also, using the **Lemma 2.4.3**, we prove that

$$\overrightarrow{KA'} = \overrightarrow{LB'}.$$

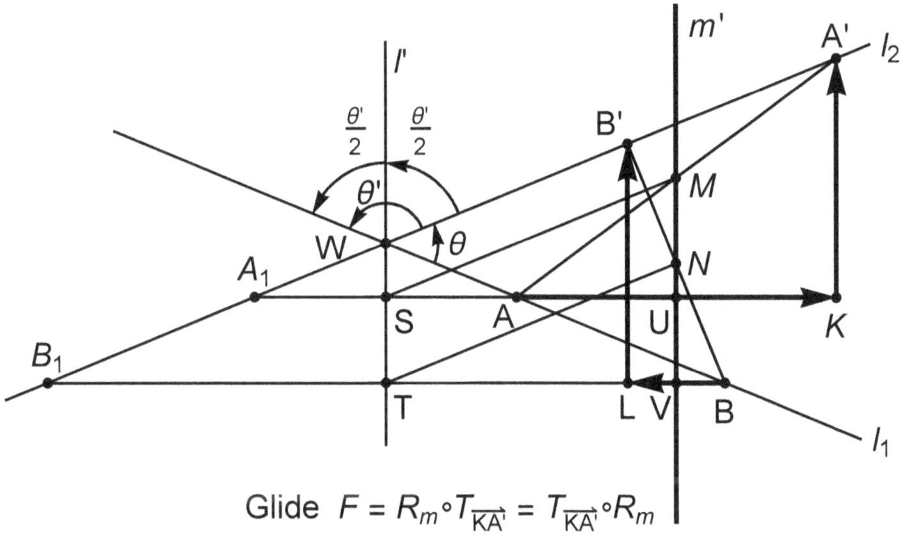

$$\text{Glide } F = R_m \circ T_{\overrightarrow{KA'}} = T_{\overrightarrow{KA'}} \circ R_m$$

Figure 2.57: **An isometry with no fixed point, that is not a parallel translation, is a glide**

Now, the proof is analogous to the proof of the previous case along with the necessary adjustments.

General Conclusions:

(1) All isometries of the plane are the basic isometries and their compositions. Their set is a (non-commutative) group under composition.

(2) All compositions of isometries can be reduced to at most 3 reflections.

*(3) If an isometry is a composition that contains an even number of reflections, then it preserves oriented angles and is a **direct isometry**. If it contains an odd number of reflections, then it maps any oriented angle to its opposite and is an **opposite isometry**.*

Other configurations of the glides analogous to those presented in **Figures 2.56 and 2.57** are in **Figures 2.58 and 2.59**.

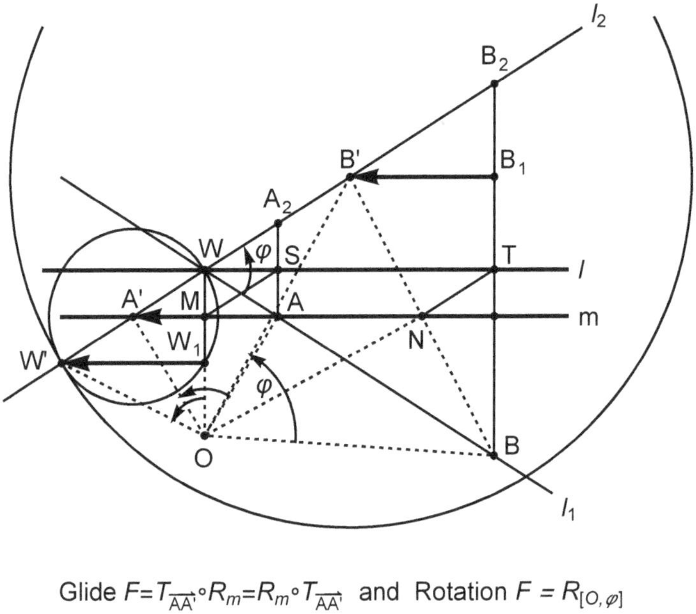

Glide $F=T_{\overrightarrow{AA'}}\circ R_m=R_m\circ T_{\overrightarrow{AA'}}$ and Rotation $F = R_{[O,\varphi]}$

Figure 2.58: An isometry with no fixed point, that is not a
parallel translation, is a glide

Glide $F=T_{\overrightarrow{AA'}}\circ R_m=R_m\circ T_{\overrightarrow{AA'}}$

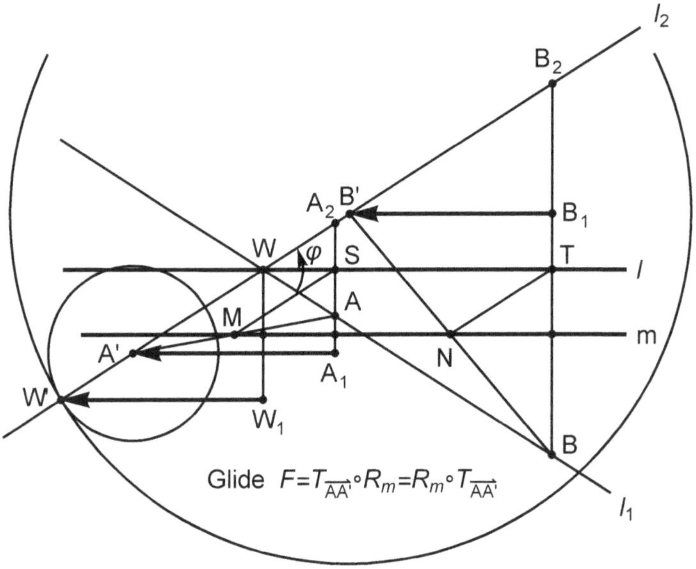

Figure 2.59: An isometry with no fixed point, that is not a
parallel translation, is a glide

2.4.1 General Result Revisited

The fact that any finite composition of isometries can be reduced to at most three (3) reflections, summarized in **Subsection 2.3.1**, can be proven faster by using **Theorem 2.2.1** and **Corollary 2.2.3** in the following way:

Suppose the isometry F maps the triangle ABC to triangle $A'B'C'$, so that $F(A) = A'$, $F(B) = B'$ and $F(C) = C'$. Then $AB = A'B'$, $BC = B'C'$ and $CA = C'A'$. Therefore the two triangles, having the corresponding sides equal, are either directly equal or oppositely equal.

Case (1): If $A = A'$, $B = B'$, and $C = C'$, then $F = I$, the identity, and so $F = R_l \circ R_l$, where R_l is any reflection.

Case (2): If $A = A'$, $B = B'$, and $C \neq C'$, then $AB = A'B'$ is the perpendicular bisector of the segment CC'. Hence, $F = R_{AB}$, is the reflection with axis $AB = A'B'$ and maps the triangle ABC to $A'B'C'$. See **Figure 2.60**.

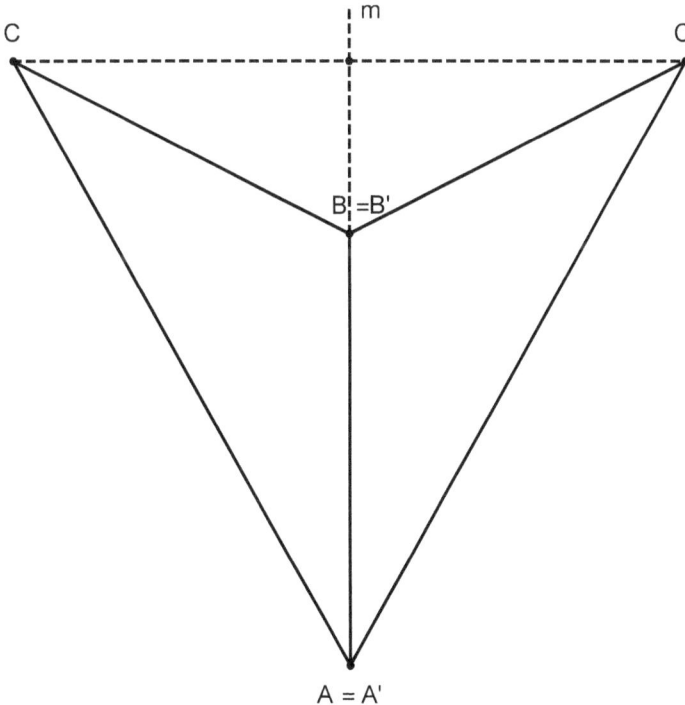

Figure 2.60: The isometry determined by a triangle

Case (3): If $A = A'$, $B \neq B'$, and $C \neq C'$, then A lies on m, the perpendicular bisector of BB'. Then, $R_m(A) = A = A'$ and $R_m(B) = B'$ and thus this case is reduced to one of the two previous cases. That is, F is either the reflection R_m, or the composition of R_m with another reflection. See **Figure 2.61**.

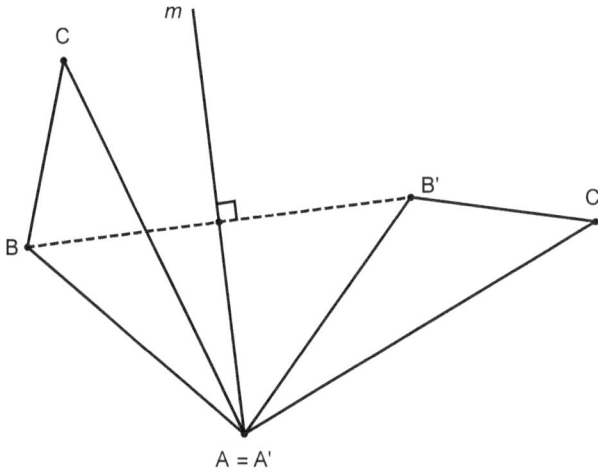

Figure 2.61: The isometry determined by a triangle

Case (4): If $A \neq A'$, $B \neq B'$, and $C \neq C'$, then we consider m the perpendicular bisector of AA'. Thus, $R_m(A) = A'$, and this case is reduced to one of the three previous cases. That is, F is either the reflection R_m, or the composition of R_m with one or two additional reflections. See **Figure 2.62**.

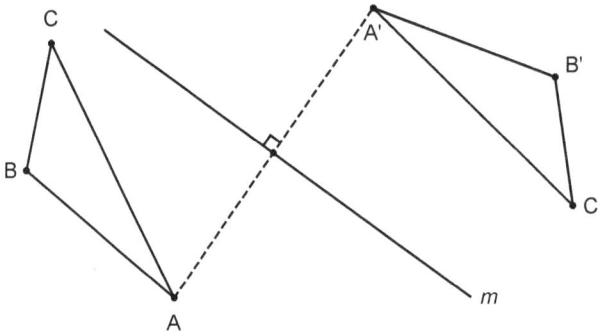

Figure 2.62: The isometry determined by a triangle

Remarks: (1) This is a concise proof of the existence of the three (3) reflections of the stated claim. But, the lengthy proof developed in **Subsection 2.3.1**, through a number of results and various cases of compositions, derives many partial results and the terminal reductions of various compositions of isometries. These facts have their own merit and are useful for proofs of other results and solutions of problems.

(2) If the isometry has exactly one fixed point, then it can be written as the composition of at most two (2) reflections.

2.5 Geometric Determination of Basic Isometries

2.5.1 Parallel Translations and their Characterization

As we saw in **Section 2.2, item (8)**, a parallel translation $T_{\vec{u}}$ maps, a line k to a parallel line l, ($\| k$), a circle $C[O, r]$ to an equal circle $C[O', r]$, such that $OO' = \vec{u}$, and preserves the equality of oriented angles. See **Figures 2.10** and **2.11**.

In fact: If $T_{\vec{u}}(A) = A'$ and $T_{\vec{u}}(B) = B'$, then $\overrightarrow{AA'} = \vec{u}$, $\overrightarrow{BB'} = \vec{u}$. So, we get,

$$\overrightarrow{A'B'} = \overrightarrow{A'A} + \overrightarrow{AB} + \overrightarrow{BB'} = -\vec{u} + \overrightarrow{AB} + \vec{u} = \overrightarrow{AB}.$$

Therefore, $\overrightarrow{AB} = \overrightarrow{A'B'}$. That is, the image of the whole vector $\overrightarrow{AB}$ is $\overrightarrow{A'B'} = T_{\vec{u}}(AB)$. So:

1. Given any two points A and A' in the plane $\mathcal{P}$, the transformation that corresponds to any point $B \in \mathcal{P}$ the point $B' \in \mathcal{P}$, such that

$$\overrightarrow{A'B'} = \overrightarrow{AB},$$

is the parallel translation $T_{\vec{u}}$, with $\vec{u} = \overrightarrow{AA'}$. (Here, the quadrilateral $ABB'A'$ is a parallelogram. See **Figure 2.63**.)

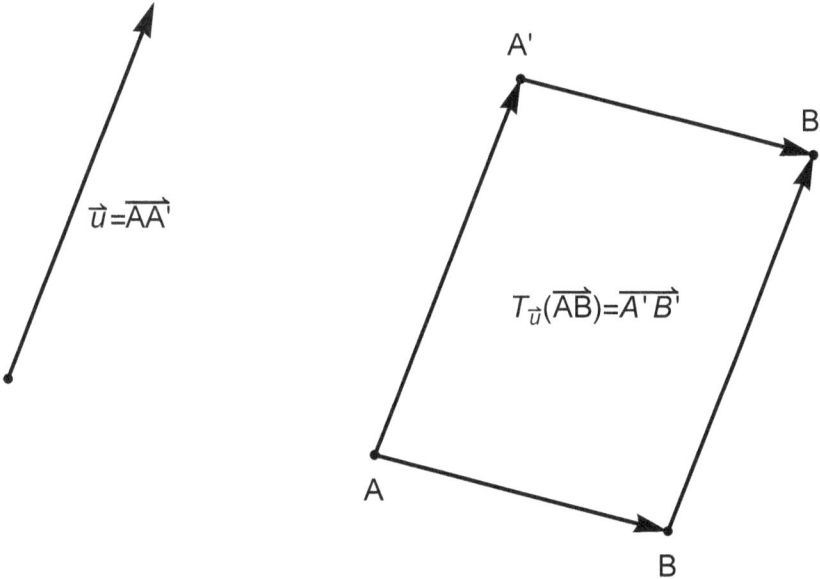

Figure 2.63: Parallel translation

2. If two vectors are equal, $\overrightarrow{AB} = \overrightarrow{A'B'}$, then, there exists one and only one parallel translation $T_{\vec{u}}$, such that $T_{\vec{u}}(\overrightarrow{AB}) = T_{\vec{u}}(\overrightarrow{A'B'})$. This is the parallel translation with $\vec{u} = \overrightarrow{AA'} = \overrightarrow{BB'}$. See **Figure 2.63**.

So, we have the following Theorem:

Theorem 2.5.1 (Characterization of Parallel Translations) *A function* $S : \mathcal{P} \longrightarrow \mathcal{P}$, *such that for any vector* $\overrightarrow{AB}$, *it holds* $\overrightarrow{S(A)S(B)} := \overrightarrow{A'B'} = \overrightarrow{AB}$ *[where for any point* P, *we denote* $P' := S(P)$*], is the parallel translation* $T_{\vec{u}}$, *with* $\vec{u} = \overrightarrow{AA'}$.

Proof Choose any free vector $\overrightarrow{AB}$, in $\mathcal{P}$ (and so the points A and B are any). Let $A' = S(A)$ and $B' = S(B)$. To prove the Theorem, we must show that $\overrightarrow{AA'} = \overrightarrow{BB'}$. See **Figure 2.63**.

By hypothesis $\overrightarrow{A'B'} = \overrightarrow{AB}$. Therefore, the quadrilateral $ABB'A'$ is a parallelogram. So, $AA' \parallel = BB'$ and they are directed in the same direction. So, $\overrightarrow{AA'} = \overrightarrow{BB'}$. Keeping A fixed (and so A' is also fixed), we prove that $\overrightarrow{AA'} = \overrightarrow{BB'}$, for all $B \in \mathcal{P}$. Hence, $S = T_{\vec{u}}$, with $\vec{u} = \overrightarrow{AA'}$.

∎

3. A parallel translation preserves free vectors. Therefore, it maps a ray (half-line) Ax to a parallel ray $A'x'$ in the same direction and $A' = T_{\vec{u}}(A)$. So, a parallel translation is a direct isometry, i.e., it preserves oriented angles. See **Figure 2.64**, in which $\vec{u} = \overrightarrow{AA'}$ and the claim follows easily.

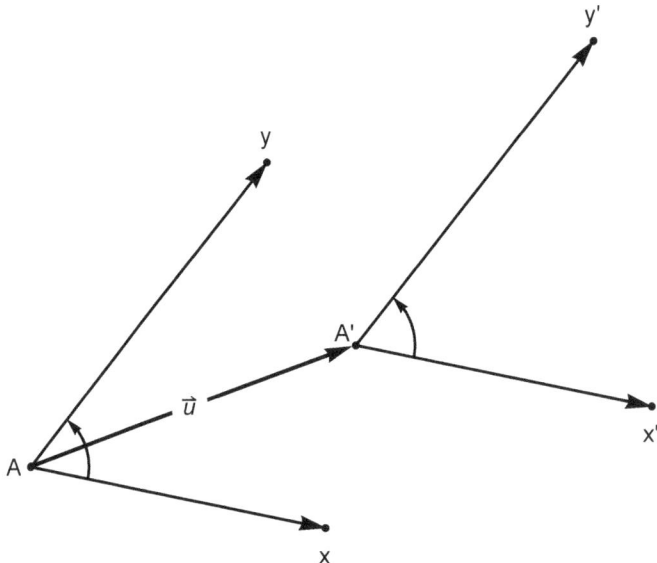

Figure 2.64: A parallel translation preserves oriented angles

2.5.2 Rotations

As we saw in **Section 2.2, item (8)**, a rotation $R_{[O,\theta]}$ maps: (1) a line k to a line l, such that $\angle(k,l) = \theta$, (2) a circle $C[K,r]$ to an equal circle $C[K',r]$, such that $OK' = OK$ and $\angle(OK,OK') = \theta$, and (3) preserves the equality of oriented angles (direct isometry). See **Figures 2.10** and **2.11**. (See also **Lemma 2.4.1**, **Corollary 2.2.1** and **Theorem 2.4.1**.) We now continue with the theorem:

Theorem 2.5.2 *Consider a rotation* $R_{[O,\theta]}$ *and a non-zero vector* $\overrightarrow{AB} \neq \overrightarrow{0}$ *such that the straight line AB does not contain O.*
Let $A' = R_{[O,\theta]}(A)$ *and* $B' = R_{[O,\theta]}(B)$*, and so* $R_{[O,\theta]}(\overrightarrow{AB}) = \overrightarrow{A'B'}$*. Then:*
(1) The triangles OAB and OA'B' are directly equal.
(2) AB=A'B' (the straight segments are equal).
(3) The oriented angle $\angle(\overrightarrow{AB}, \overrightarrow{A'B'}) = \theta$*.*

Proof As we have seen in the **proof of Theorem 2.4.1** the triangles OAB and $OA'B'$ are directly equal. Therefore, $AB = A'B'$ (the straight segments are equal). Also $\angle(\overrightarrow{AB}, \overrightarrow{AO}) = \angle(\overrightarrow{A'B'}, \overrightarrow{A'O})$. See **Figure 2.65**.

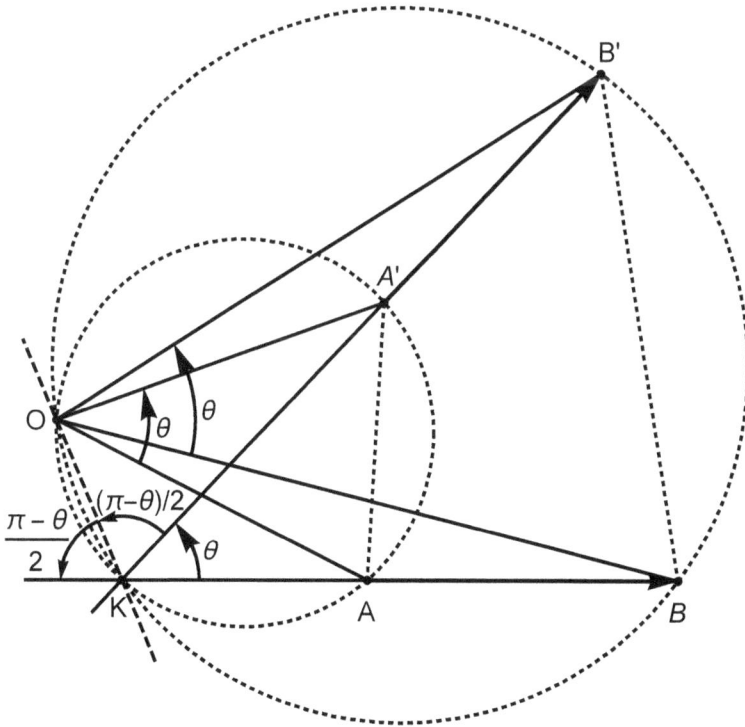

Figure 2.65: Rotation related to two vectors and the oriented angles

Now, the angle of the straight lines AB and $A'B'$, at their common point K as in **Figure 2.65**, is the rotation angle θ, since

$$\angle(\overrightarrow{AB}, \overrightarrow{A'B'}) = \angle(\overrightarrow{AB}, \overrightarrow{AO}) + \angle(\overrightarrow{AO}, \overrightarrow{A'O}) + \angle(\overrightarrow{A'O}, \overrightarrow{A'B'})$$

$$= \angle(\overrightarrow{AB}, \overrightarrow{AO}) + \angle(\overrightarrow{OA}, \overrightarrow{OA'}) - \angle(\overrightarrow{A'B'}, \overrightarrow{A'O}) = \angle(\overrightarrow{OA}, \overrightarrow{OA'}) = \theta.$$

(This can also be proven by comparing triangles and angles in **Figure 2.65**.)
∎

Remark. Examine the above **Theorem** when $\overrightarrow{AB} = \overrightarrow{0}$ or the straight line AB contains the point O, the center of the rotation, etc. State and prove the resulting degenerate results.

Corollary 2.5.1 *Under the conditions of the **Theorem** and if $AB \nparallel A'B'$, [that is, $\theta = \angle(\overrightarrow{AB}, \overrightarrow{A'B'}) \neq 0$, (mod π)], we let K be the point of intersection of the straight lines AB and $A'B'$. Then, we have:*

1. *The triangles OAA' and OBB' are similar isosceles triangles.*

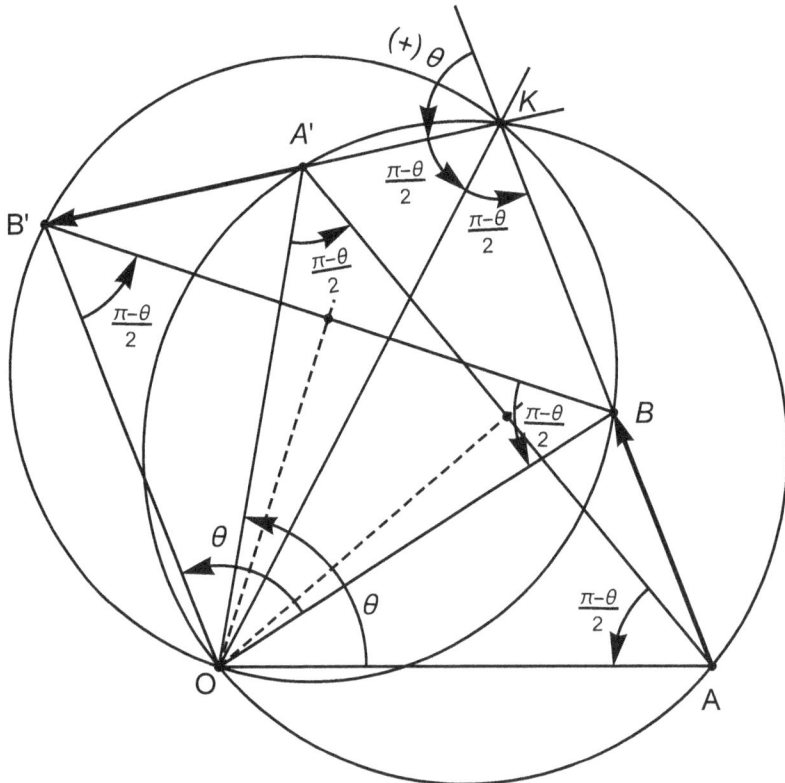

Figure 2.66: On the properties of rotations

2. At the point K, we have

$$\measuredangle(\overrightarrow{A'B'}, -\overrightarrow{AB}) = \begin{cases} \pi - \theta, & \text{if} \quad \theta > 0, \\ \\ -\pi - \theta, & \text{if} \quad \theta < 0. \end{cases}$$

3. The quadrilaterals $OAKA'$ and $OBKB'$ are inscribable.

4. The line OK is the bisector of the angle $\measuredangle(\overrightarrow{A'B'}, -\overrightarrow{AB})$ at the point K. [See also **Corollary 4.4.3, (b)**.]

5. If $AA' \nparallel BB'$, the rotation center O is the point of intersection of the perpendicular bisectors of the segments AA' and BB'. (**Otherwise**, see **Figures 2.68** and **2.71**.)

6. The rotation center O is the point of intersection other than K of the circles AKA' and BKB' [**unless** $O \equiv K$ (see **Figure 2.73**), or these circles cannot be determined (see **Figure 2.74** where $A = B' = K$), etc.].

Proof See **Figures 2.66, 2.67** and **2.65**.

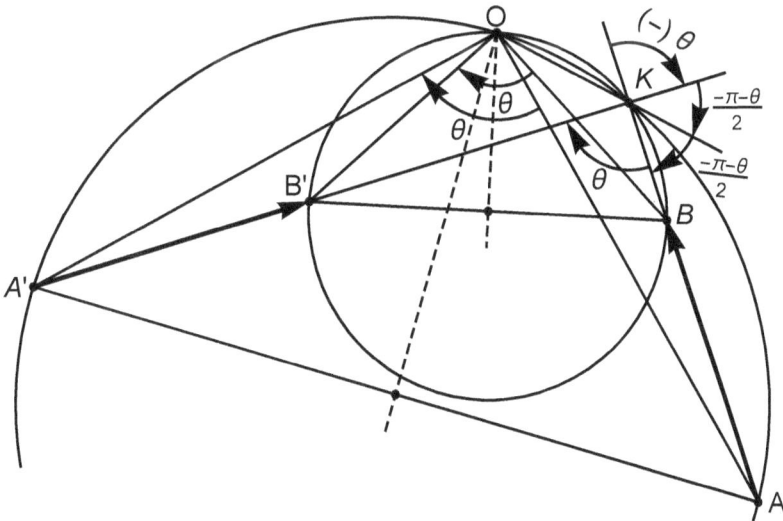

Figure 2.67: On the properties of rotations

(1) The two triangles OAA' and OBB' are isosceles since $OA = OA'$ and $OB = OB'$. Both have vertex angle (at O) the rotation angle θ. So, their base angles are equal to $\dfrac{\pi - \theta}{2}$. Therefore they are similar.

(2) Follows immediately form the relation $\angle(\overrightarrow{AB}, \overrightarrow{A'B'}) = \theta$ and the angle-relations between vectors as explained in **Chapter 1**.

(3) Follows from **(2)**.

(a) If in $OAKA'$ and / or $OBKB'$, the vertex K is opposite to vertex O, then the absolute values of the angles at its vertices O and K add up to $\pi[=|\theta| + (\pi - |\theta|)]$ (**Figure 2.66**).

(b) If O and K lie on the same side, then they subtend angles with the opposite side equal to θ (**Figure 2.67**).

(4) By **(3)** and **(1)**,

$$\angle AKO = \frac{\pi - \theta}{2} = \angle OKA'.$$

[See also **Corollary 4.4.3, (b)**.]

(5) Follows from **(1)**.

(6) Follows from **(3)**.
                                                                         ∎

Corollary 2.5.2 *Under the conditions of the* **Theorem**, *we obtain:*

1. *If $AB =\parallel A'B'$ and $\overrightarrow{AB} = -\overrightarrow{A'B'} \neq \vec{0}$, then the rotation is the half-turn*

$$R_{[O,\pi]}$$

with O the point of intersection of AA' and BB'.

2. *If $AA' \parallel BB'$ and $AB =\nparallel A'B'$ (so, $\overrightarrow{AB} \neq \overrightarrow{A'B'}$), the rotation center O of the rotation $R_{[O,\theta]}$ is the point of intersection of the lines AB and $A'B'$ and the rotation angle*
$$\theta = \angle(AB, A'B').$$

3. *If $\overrightarrow{AB} = \overrightarrow{A'B'}$ (equal as free vectors), and $A \neq A'$ (so, $B \neq B'$), then there is no Euclidean rotation mapping A to A' and B to B', but the parallel translation*
$$T_{\vec{u}}, \quad with \quad \vec{u} = \overrightarrow{AA'} = \overrightarrow{BB'} \neq \vec{0},$$
maps A to A' and B to B', in this case.

4. *If $\overrightarrow{AB} = \overrightarrow{A'B'}$ (equal as free vectors), and $A = A'$ (so, $B = B'$), then the Euclidean rotation mapping A to A' and B to B' is the rotation by angle zero and so the identity, which is also the parallel translation $T_{\vec{0}}$. (In this case, $\overrightarrow{AA'} = \overrightarrow{BB'} = \vec{0}$.)*

Proof Draw figures for each of the above parts of the Corollary and follow the hints below.

(1) The given hypotheses imply that the quadrilateral $ABA'B'$ is a parallelogram. See **Figure 2.70**. The two diagonals AA' and BB' intersect at their midpoint O. The figure is symmetrical about O. So, the rotation is $R_{[O,\pi]}$.

(2) The given hypotheses imply that the quadrilateral $AA'B'B$ is an isosceles trapezium, which is not a skew parallelogram. The two sides AB and $A'B'$ intersect at a point O. See **Figures 2.73, 2.78** and **2.79**. So, the rotation is $R_{[O,\theta]}$ and this maps A to A' and B to B'. (Also, the figure is symmetrical on the bisector of the angle $\widehat{AOA'} = \widehat{BOB'}$.)

(3) Under the given hypotheses, by **Section 2.5.1, item 2**, by letting vector $\vec{u} = \overrightarrow{AA'} = \overrightarrow{BB'}$ we have that $T_{\vec{u}}(\overrightarrow{AB}) = T_{\vec{u}}(\overrightarrow{A'B'})$. So, the parallel translation $T_{\vec{u}}$ maps A to A' and B to B'. See **Figures 2.68** and **2.63**.

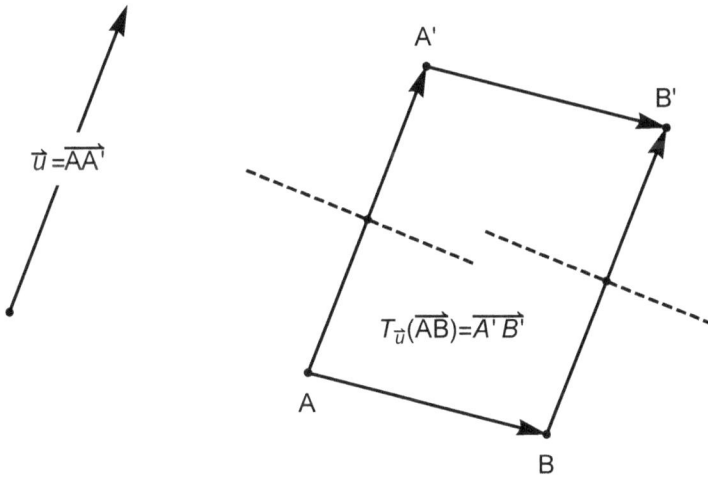

Figure 2.68: **Parallel translation instead of rotation**

Since $AB =\parallel A'B'$, the quadrilateral $AA'B'B$ is a parallelogram. The perpendicular bisectors of the sides AA' and BB' either coincide [when $AA'B'B$ is an orthogonal parallelogram (rectangle)] or they are different parallel (otherwise). So, the rotation O cannot be defined unless we consider it to be a point at infinity (the infinity point of the two parallel perpendicular bisectors) and the rotation angle to be zero, in which case the rotation is a parallel translation.

(4) This claim is immediate.

2.5.3   Determination of the Center of a Rotation

(I) We want to find the center O of the rotation that maps a given point $A \in \mathcal{P}$ to a given point $A' \in \mathcal{P}$ and has rotation-angle a given angle θ.

(1) If $\theta = 0$, (mod 2π), the rotation is the identity. In this case, it must be that $A = A'$ and any center $O \in \mathcal{P}$ can be any point of the plane.

(2) If $A = A'$ and $\theta \neq 0$, (mod 2π), then the rotation is not the identity. Since, a rotation has one fixed point, then the center $O = A = A'$. Hence, the rotation is $R_{[A,\theta]}$.

(3) If $\theta = \pi$, (mod 2π), then the rotation is a half-turn. The center O is the midpoint of the segment AA'. [If $A = A'$, then $O = A = A'$, as we have just seen in **(2)**.]

So, in general, we assume that $A \neq A'$ and $0 < \theta < 2\pi$, (mod 2π). We distinguish the following cases with respect to the size of the angle θ:

Suppose $0 < \theta < \dfrac{\pi}{2}$, as in **Figure 2.69**. According to **Corollary 2.5.1**, the sought center O is the intersection of the straight line m, the perpendicular bisector of the segment AA', and the circle with cord AA' and such that any point P of one of its arc $\overset{\frown}{AA'}$ subtends angle $\measuredangle(PA, PA') = \theta$ with the segment AA'. (In **Figure 2.69**, O is such a point on the respected arc.)

This circle is constructed in the following way. Let M be the midpoint of AA'. Then, $m \perp AA'$ at the point M, and so $\widehat{mMA} = \dfrac{\pi}{2}$. We construct the angle $\measuredangle(AM, Ax) = \dfrac{\pi}{2} - \theta$ and let C be the point of intersection of Ax with m. Then, C is the center of the sought circle and the radius is $r := CA = CA'$. So the circle is $C[C, r]$, the center of the rotation is the point $O = m \cap \overset{\frown}{AA'}$ and the rotation is $R_{[O,\theta]}$.

If $\theta = \dfrac{\pi}{2}$, then Ax coincides with AA'. The center of the new circle is $C = M$ and the radius is $r = MA = MA'$. Now the segment AA' is a diameter of the circle. (Make a new figure!)

If $\dfrac{\pi}{2} < \theta < \pi$, then, $-\dfrac{\pi}{2} < \dfrac{\pi}{2} - \theta < 0$ and so we construct the new triangle MAC as the reflection of the old triangle AMC in the line AA'. So, the new center C is the reflection of the old center C in the line AA', etc. (Make a new figure!)

If $\theta = \pi$, then as we have explained in **(3)** above, the rotation is a half-turn. The center O of this half-turn is the midpoint M of the segment AA'.

Finally, if $\pi < \phi < 2\pi$, we let $\theta = \phi - \pi \Leftrightarrow \phi = \pi + \theta$. Then $0 < \theta < \pi$ and we make the previous construction as in **Figure 2.69** with $\theta = \phi - \pi$. Then, the center of the rotation is the point K in the figure and the rotation angle is $\phi = \pi + \theta$. So we get the rotation $R_{[K,\phi]}$.

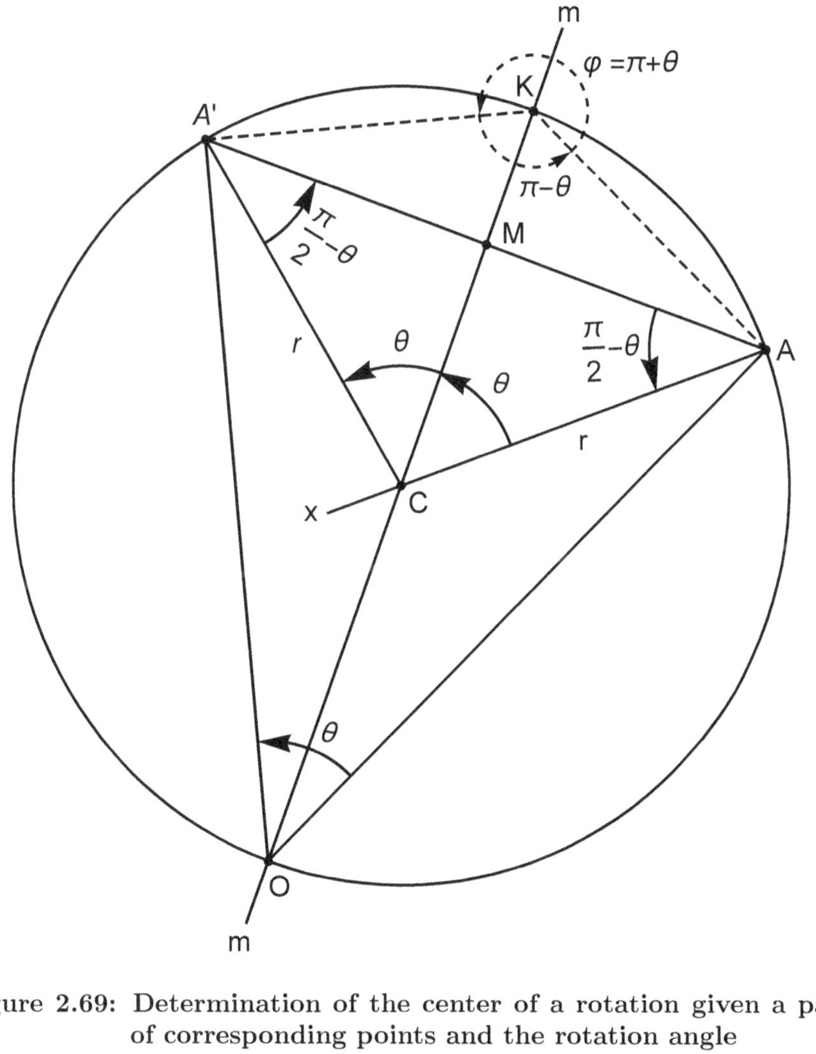

Figure 2.69: **Determination of the center of a rotation given a pair of corresponding points and the rotation angle**

From the above, we can also conclude:

(II) *(1) If we are given two corresponding points A and $A' \neq A$ of a rotation and the rotation angle $\theta \neq 0, \pi, \pmod{2\pi}$, then the rotation center O is the point of intersection of the perpendicular bisector of the segment AA' and the arc of the circle that passes through A and A' and subtends angle θ with AA'.*

(2) If $A = A'$ and θ is any angle, then $O = A = A'$ and the rotation is the $R_{[A,\theta]}$.

(3) If $\theta = \pi$, (mod 2π), then the rotation is the half-turn $R_{[O,\pi]}$, with O the midpoint of AA'.

(4) If $\theta = 0$, (mod 2π), then the rotation is the identity.

(III) *We want to find the center O and the rotation-angle θ of the rotation that maps a given vector $\overrightarrow{AB} \neq \overrightarrow{0}$ to a given vector $\overrightarrow{A'B'} \neq \overrightarrow{0}$, such that the lengths of the segments $AB = A'B'$ are equal, but $\overrightarrow{AB} \neq \overrightarrow{A'B'}$.*

(1) If $AB \parallel A'B'$, since we have assumed $\overrightarrow{AB} \neq \overrightarrow{A'B'}$ and $AB = A'B'$, then it must be

$$\overrightarrow{AB} = -\overrightarrow{A'B'}.$$

See **Figure 2.70**.

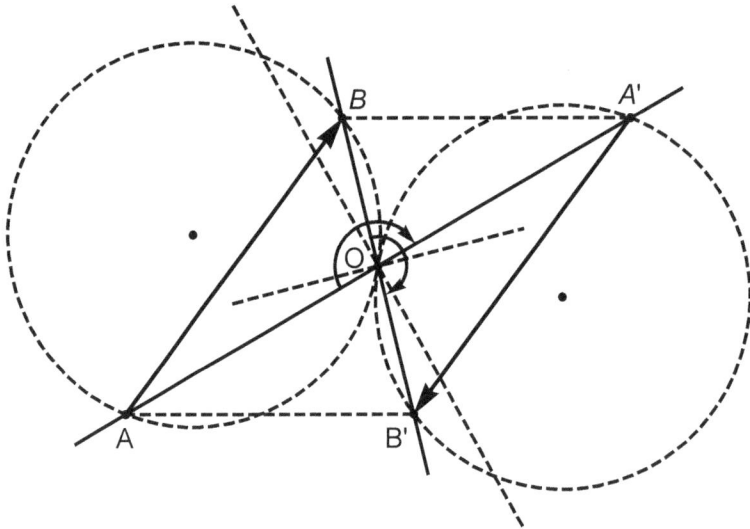

Figure 2.70: Determination of the center of a half-turn

In this case, the rotation center O is the common midpoint of the segments AA' and BB' (as $ABB'A'$ is a parallelogram with diagonals AA' and BB'), which is also the common point of the perpendicular bisectors of AA' and BB'. The rotation angle is $\theta = \pi$, (mod 2π). That is, the rotation is the half-turn $R_{[O,\pi]}$. [Since the straight lines AB and $A'B'$ meet at an infinity point ($AB \parallel A'B'$), the circles (ABO) and $(A'B'O)$ have nothing to do with the determination of the center of the half-turn, but prove that they are externally tangent at O.]

(2) If $\overrightarrow{AB} = \overrightarrow{A'B'} \neq \overrightarrow{0}$, and $A = A'$ and $B = B'$, then the rotation has more than one fixed points, and so the rotation is the identity.

(3) If $\overrightarrow{AB} = \overrightarrow{A'B'} \neq \overrightarrow{0}$, and $A \neq A'$ and $B \neq B'$, then $ABA'B'$ is a parallelogram. The perpendicular bisectors of the parallel straight segments $AA' \neq 0$ and $BB' \neq 0$ are parallel. These perpendicular bisectors are different, if $ABA'B'$ is a skew parallelogram, or they coincide, if $ABA'B'$ is an orthogonal parallelogram (rectangle). So, **no center** of rotation can be determined in this way. In this case, $\overrightarrow{AA'} =\parallel \overrightarrow{BB'} \neq \overrightarrow{0}$ and the parallel translation $T_{\overrightarrow{AA'}}$ maps $\overrightarrow{AB}$ to $\overrightarrow{A'B'}$. See **Figure 2.71**.

(In this limiting case, the parallel translation may be considered like a rotation by angle zero and center the point at infinity of the parallel perpendicular bisectors of the parallel straight segments $AA' \neq 0$ and $BB' \neq 0$.)

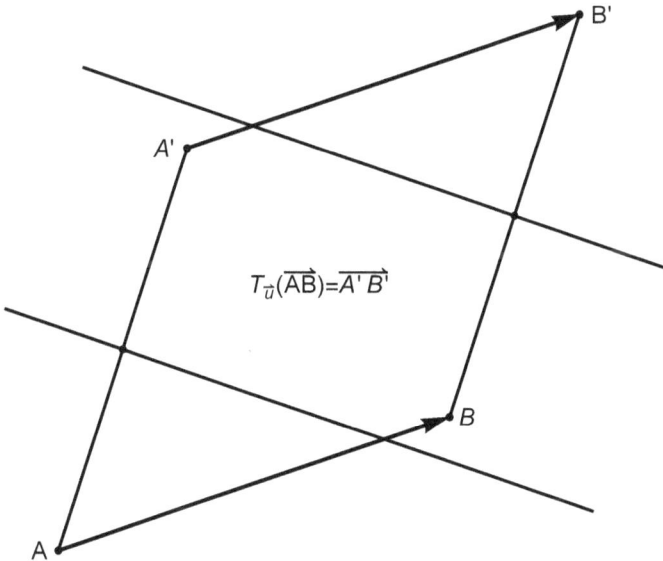

Figure 2.71: Parallel translation instead of rotation

(4) So, in general, we assume that $AB =\nparallel A'B' \neq 0$ (equal $\neq 0$ but not parallel). (See also **Corollary 2.5.2, item 3**.)

Then, we have that the rotation is $R_{[O,\theta]}$, where:

(1) **The rotation center** O is the point of intersection of the perpendicular bisectors of the segments AA' and BB', as long as they intersect. See **Figure 2.72**. (If the perpendicular bisectors coincide, then **Figure 2.73** explains how to find the rotation center in two ways.)

(2) **The rotation-angle is**

$$\theta = \measuredangle(\overrightarrow{AB}, \overrightarrow{A'B'}) = \measuredangle(OA, OA') = \measuredangle(OB, OB').$$

Now, let K be the point of intersection of the lines AB and $A'B'$. See **Figure 2.72**. Then, we also have:

(1) **The rotation center** O *is the other point of intersection of the two circles circumscribed to the triangles* AKA' *and* BKB' *(unless* O *and* K *coincide, as it happens in the special case presented below).*

(2) **The rotation angle** *is* $\theta = \measuredangle(\overrightarrow{AB}, \overrightarrow{A'B'}) = \measuredangle(OA, OA') = \measuredangle(OB, OB').$

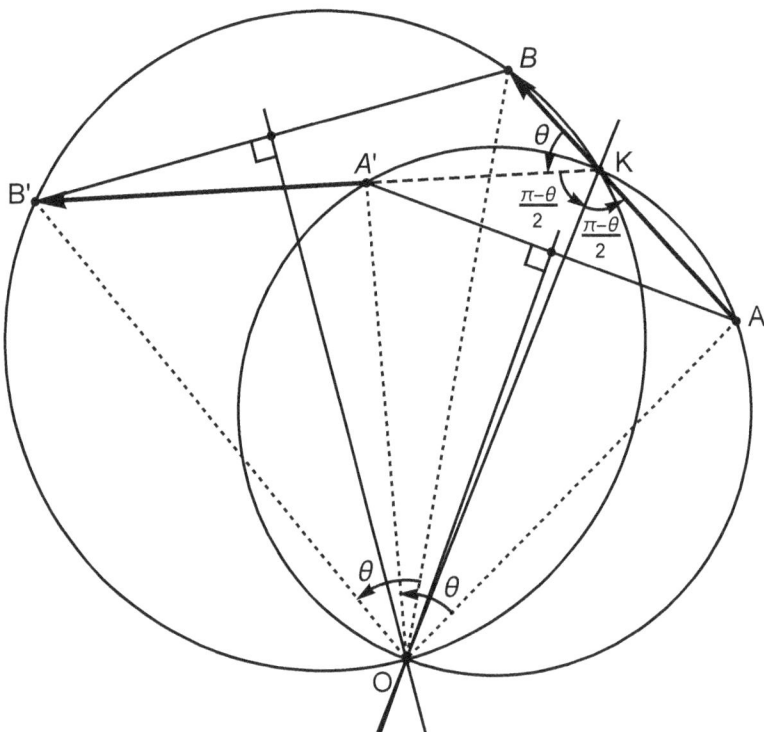

Figure 2.72: Determination of the center of a rotation

Remark: As we have seen in **Figures 2.66** and **2.67**, the rotation center O lies also on the bisector of one of the angles between the lines AB and $A'B'$. In **Figure 2.72**, KO is the bisector of $\pi - \measuredangle(\overrightarrow{AB}, \overrightarrow{A'B'}) = \pi - \theta$. [See also **Corollary 4.4.3, (b)**, for another determination of the center of a rotation as an intersection of two special straight lines.]

A special case is to determine the center of rotation that maps $\overrightarrow{AB}$ to $\overrightarrow{A'B'}$, when $AB =\!\!\!\not\parallel A'B'$ and $AA' \parallel BB'$. Then, $ABB'A'$ is an isosceles trapezium (which is not a skew parallelogram) and the perpendicular bisectors of AA' and BB' coincide. The center of rotation is $O = K$, the point of intersection of the lines AB and $A'B'$, which also coincides with the point of intersection of the circles AKA' and BKB' (unique in this case). The rotation angle is $\theta = \measuredangle(OA, OA') = \measuredangle(OB, OB')$. See **Figure 2.73**.

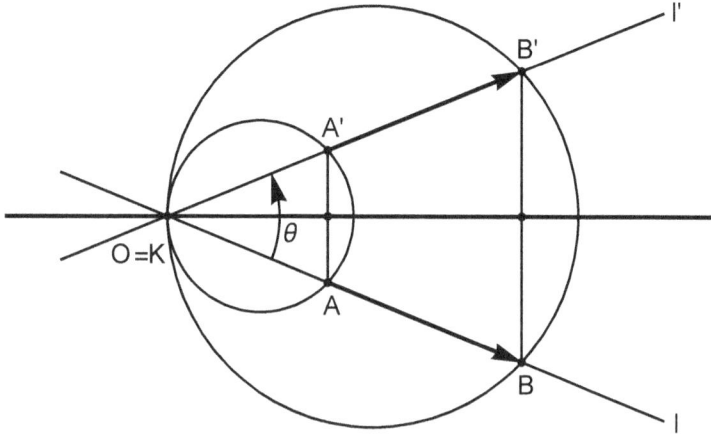

Figure 2.73: Special case of determination of a rotation by two pairs of corresponding points (A, A') and (B, B')

Another special case in which the circle method does not work is the following: $\overrightarrow{AB} \nparallel \overrightarrow{A'B'}$, $AB = A'B' \neq 0$ and $A = B' = K$. See **Figure 2.74**.

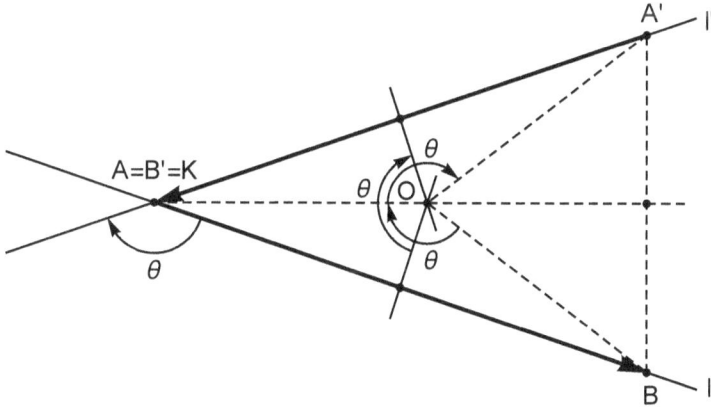

Figure 2.74: Special case of determination of a rotation by two pairs of corresponding points (A, A') and (B, B')

The circles AKA' and BKB' are not defined. But, the perpendicular bisectors of AA' and BB' intersect at the circumcenter O of the isosceles triangle ABA' and this is the center of the rotation that maps $\overrightarrow{AB}$ to $\overrightarrow{A'B'}$. The rotation angle is $\theta = \angle(\overrightarrow{AB}, \overrightarrow{A'B'})$, which is equal to the angle from the perpendicular bisector of AB to the perpendicular bisector of $A'B'$.

Now study, interpret and compare with **Figures 2.75** and **2.76** below.

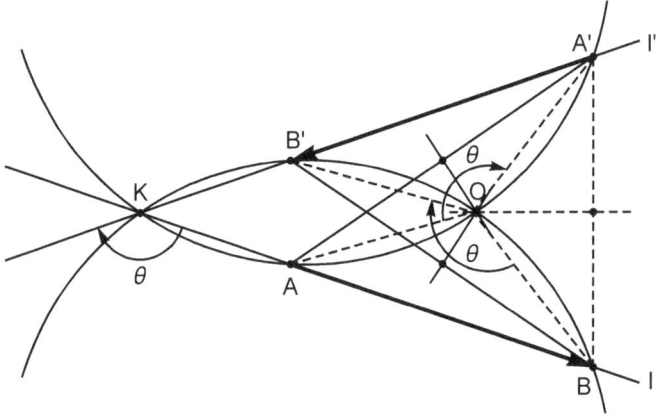

Figure 2.75: **Special case of determination of a rotation by two pairs of corresponding points** (A, A') **and** (B, B'). **Here** $KA = KB'$

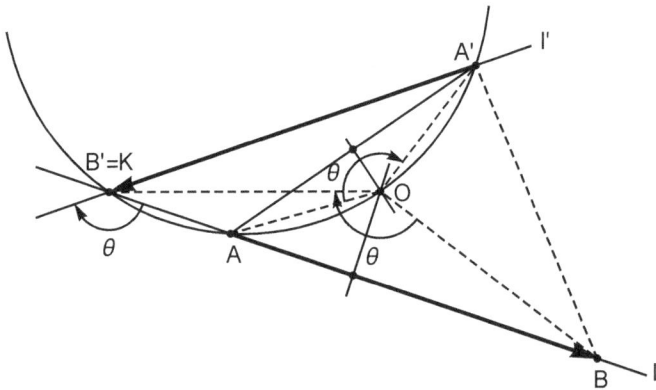

Figure 2.76: **Special case of determination of a rotation by two pairs of corresponding points** (A, A') **and** (B, B')

Combining all of the results developed in the current **Subsection 2.5.3** and the **Subsection 2.5.2**, we obtain the following theorem that characterizes the rotations in the plane.

Theorem 2.5.3 (Characterization of Rotations) *(a) Given a rotation $R_{[O,\theta]}$ and a vector $\overrightarrow{AB}$, let*

$$R_{[O,\theta]}(\overrightarrow{AB}) = \overrightarrow{A'B'}.$$

Then,

$$AB = A'B' \quad \text{(the segments are equal)}$$

and when $\overrightarrow{AB} \neq \vec{0}$ the oriented angle of the vectors $\angle(\overrightarrow{AB}, \overrightarrow{A'B'}) = \theta$.

*(b) **Conversely**: Given two different non-zero vectors $\overrightarrow{AB} \neq \overrightarrow{A'B'}$ of equal length $AB = A'B'$, or $\overrightarrow{AB} = \overrightarrow{A'B'}$ but with $A = A'$ (so, $B = B'$), then there exists unique rotation $R_{[O,\theta]}$ mapping A to A' and B to B' (and so $\overrightarrow{AB}$ to $\overrightarrow{A'B'}$), where its center O has been determined above (in this **Subsection 2.5.3**) and*

$$\theta = \angle(\overrightarrow{AB}, \overrightarrow{A'B'}), \quad (mod 2\pi).$$

If $\overrightarrow{AB} = \overrightarrow{A'B'}$ and $A \neq A'$ (and so $B \neq B'$), then $\overrightarrow{AA'} = \overrightarrow{BB'}$ and the rotation becomes the parallel translation

$$T_{\overrightarrow{AA'}} = T_{\overrightarrow{BB'}}.$$

Or:

*(c) Given a rotation $R_{[O,\theta]}$ and two corresponding points A and $A' = R_{[O,\theta]}(A)$: If $B' = R_{[O,\theta]}(B)$, for any point $B \in \mathcal{P}$, then the following **two conditions** are satisfied:*

$$A'B' = AB, \quad \text{and when } A \neq B, \quad \text{then } \angle(\overrightarrow{AB}, \overrightarrow{A'B'}) = \theta.$$

Also, the triangles OAB and $OA'B'$ are directly equal.
*(d) **Conversely**: Given any two points A and A' in $\mathcal{P}$, the transformation that maps any point $B \in \mathcal{P}$ to the point $B' \in \mathcal{P}$ determined by the above two conditions in **(c)**, where $\theta \neq 0$, (mod 2π), is a rotation $R_{[O,\theta]}$, with angle θ and center O as determined in **this Subsection 2.5.3**.*

If $\theta = 0$, (mod 2π), then:
(1) If $A = A'$, the transformation is the identity and so a rotation by the zero angle, or a translation by the zero vector.
(2) If $A \neq A'$, the transformation is the parallel translation $T_{\vec{u}}$, with $\vec{u} = \overrightarrow{AA'}$.

2.5.4   Reflections, Opposite Isometries and Glides

As we saw in **Section 2.2, item (8)**, a reflection R_l, maps: A line k to a line m symmetrical to k with respect to l; a circle $C[K, r]$ to an equal circle $C[K', r]$ symmetrical to $C[K, r]$ with respect to l and so K' is symmetrical to K with respect to l; and oriented angles to opposite angles. Also, **any reflection is involutory**, i.e., $R_l \circ R_l = I$ (identity) and so, $R_l^{-1} = R_l$.

We continue with the following observation: If we consider any two lines l_1 and l_2 in the plane $\mathcal{P}$, as $R_{l_2} \circ R_{l_2} = I$, we have

$$R_{l_1} = R_{l_1} \circ (R_{l_2} \circ R_{l_2}) = (R_{l_1} \circ R_{l_2}) \circ R_{l_2} = F \circ R_{l_2},$$

where $F = R_{l_1} \circ R_{l_2}$ is either a parallel translation, when $l_1 \parallel l_2$, or a rotation, when $l_1 \nparallel l_2$, as explained in **Section 2.3, (3)**.

This kind of decomposition of R_{l_1} can also be done as

$$R_{l_1} = (R_{l_2} \circ R_{l_2}) \circ R_{l_1} = R_{l_2} \circ (R_{l_2} \circ R_{l_1}) = R_{l_2} \circ G,$$

where $G = R_{l_2} \circ R_{l_1} \left[= R_{l_2}^{-1} \circ R_{l_1}^{-1} = (R_{l_1} \circ R_{l_2})^{-1} = F^{-1} \right]$ is either a parallel translation, when $l_1 \parallel l_2$, or a rotation, when $l_1 \nparallel l_2$. So, we have $R_{l_1} = F \circ R_{l_2} = R_{l_2} \circ F^{-1}$ and in general we state:

Any reflection R_{l_1} can be analysed in a composition of another reflection R_{l_2} and a parallel translation or a rotation in either order of composition, in infinitely many ways.

We have called an **isometry opposite**, if it maps oriented angles to their opposite. As an isometry, the opposite isometry preserves the lengths of straight segments. The reflections in straight lines are opposite isometries. By the theory developed in **Section 2.4**, we have that any composition of isometries can be reduced to either one reflection or to a composition of two or three reflections. The composition of the three reflections in the final reduction yields a glide, which can be rewritten as the composition of one reflections and a parallel translation) Eventually we have:

Any opposite isometry can be reduced to one reflection or to the composition of one reflection with a direct isometry which is either a parallel translation or a rotation. (The latter two cases can be reduced to glides.)

Now, we consider an opposite isometry G that maps the vector $\overrightarrow{AB} \neq \overrightarrow{0}$ to the vector $\overrightarrow{A'B'}$. That is $G(\overrightarrow{AB}) = \overrightarrow{A'B'}$, $G(A) = A'$ and $B' = G(B)$. Then:

(1) $AB = A'B'$ (the segments are equal).

(2) Any point C is mapped to the point $C' = G(C)$ determined by the following **two conditions**:

(1) $AC = A'C'$ and (2) $\angle(\overrightarrow{A'B'}, \overrightarrow{A'C'}) = -\angle(\overrightarrow{AB}, \overrightarrow{AC}) = \angle(\overrightarrow{AC}, \overrightarrow{AB})$ (✠).

[Or, (1) $BC = B'C'$ and (2) $\angle(\overrightarrow{B'A'}, \overrightarrow{B'C'}) = -\angle(\overrightarrow{BA}, \overrightarrow{BC}) = \angle(\overrightarrow{BC}, \overrightarrow{BA})$.]

So: *An opposite isometry G that corresponds the non-zero vector, $\overrightarrow{AB} \neq \overrightarrow{0}$ to $\overrightarrow{A'B'} \neq \overrightarrow{0}$, i.e., $G(\overrightarrow{AB}) = \overrightarrow{A'B'}$, satisfies the above two conditions in (✠).* See **Figure 2.77**.

Next, we prove the converse in the following:

Theorem 2.5.4 (Characterization of Opposite Isometries) *Given two equal non-zero segments $AB = A'B'$, the transformation G that maps a point $C \in \mathcal{P}$ to a point $C' \in \mathcal{P}$, by satisfying the two conditions*

$$AC = A'C' \quad and \quad \measuredangle(\overrightarrow{A'B'}, \overrightarrow{A'C'}) = -\measuredangle(\overrightarrow{AB}, \overrightarrow{AC}) = \measuredangle(\overrightarrow{AC}, \overrightarrow{AB}),$$

*is an opposite isometry G, which satisfies $G(\overrightarrow{AB}) = \overrightarrow{A'B'}$. (See **Figure 2.77**.)*

Proof We pick any straight line l and consider the reflection R_l with respect to it. Let $R_l(\overrightarrow{AB}) = \overrightarrow{A''B''}$. Then, under the first hypothesis, $A''B'' = AB = A'B'$.

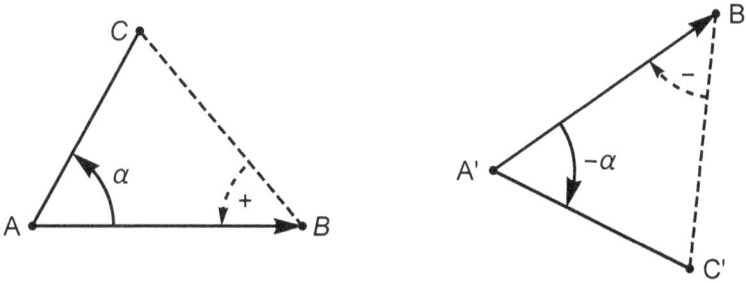

Figure 2.77: Characterization of an opposite isometry

By the theory developed in the **previous two Subsections**, there is a direct isometry F such that $F(\overrightarrow{A''B''}) = \overrightarrow{A'B'}$.

We let $\mathcal{G} := F \circ R_l$. This is an opposite isometry for which we have $\mathcal{G}(\overrightarrow{AB}) = F[R_l(\overrightarrow{AB})] = F(\overrightarrow{A''B''}) = \overrightarrow{A'B'}$, that is:

$$\mathcal{G}(\overrightarrow{AB}) = \overrightarrow{A'B'}.$$

If we let $\mathcal{G}(C) = C''$, then for the point $C \in \mathcal{P}$, by the two conditions in ($\maltese$) before this Theorem, we have: $A'C'' = AC$ and $\measuredangle(\overrightarrow{A'B'}, \overrightarrow{A'C''}) = -\measuredangle(\overrightarrow{AB}, \overrightarrow{AC})$. But, by the hypotheses, we also have

$$\measuredangle(\overrightarrow{A'B'}, \overrightarrow{A'C'}) = -\measuredangle(\overrightarrow{AB}, \overrightarrow{AC}) \quad and \quad AC = A'C'.$$

Then, we obtain

$$A'C' = A'C'' \quad and \quad \measuredangle(\overrightarrow{A'B'}, \overrightarrow{A'C''}) = \measuredangle(\overrightarrow{A'B'}, \overrightarrow{A'C'}),$$

which proves that $C' = C''$, i.e., $\mathcal{G}(C) = C'' = C' = G(C)$, for all $C \in \mathcal{P}$.

Finally $G = \mathcal{G} = F \circ R_l$ is an opposite isometry, as a composition of a direct isometry and a reflection.

The relation $G(\overrightarrow{AB}) = \overrightarrow{A'B'}$ follows from the definition of G, the hypothesis $AB = A'B'$ and the obvious relation $\measuredangle(\overrightarrow{A'B'}, \overrightarrow{A'B'}) = 0 = -\measuredangle(\overrightarrow{AB}, \overrightarrow{AB})$. ∎

Corollary 2.5.3 *By the Theorem and the paragraph before it (its converse), we have:*

(1) A transformation that preserves the lengths of straight segments and reverses oriented angles is an opposite isometry.

(2) Two opposite isometries that have two pairs of corresponding points, (A, A') and (B, B'), in common, are equal.

(3) An opposite isometry can be decomposed into a composition of a reflection of arbitrary axis and a direct isometry, in infinitely many ways.

(4) An opposite isometry that has a fixed point A is a reflection in an axis passing through A.

Proof (1) This result follows from the **Theorem** immediately.

(2) We would have $AB = A'B'$, as isometries preserve the lengths of straight segments, and then both isometries map a point $C \in \mathcal{P}$ to C' according to the conditions $AC = A'C'$ and $\angle(\overrightarrow{A'B'}, \overrightarrow{A'C'}) = -\angle(\overrightarrow{AB}, \overrightarrow{AC})$. So, for both isometries the image point C' is the same for any point C and therefore, they are equal.

(3) This was proven in the proof of the **Theorem**, since the line l was arbitrarily picked as the axis of the reflection R_l.

(4) Consider any point $B \neq A$ whose image under the isometry is B'. If $B = B'$, then B is another fixed point, then, by the **above Theorem** or by **Theorem 2.2.2**, the given opposite isometry is the reflection in the line AB.

Otherwise, we let N be the midpoint of BB' and so $BN = NB'$. If $N = A$, then we choose a different B so that $A \neq N$ (isometries are one to one functions). Since $A = A'$ is fixed, we have $AB = A'B' = AB'$. Therefore the triangle ABB' is isosceles and AN is median on its base and so it is also the bisector of its vertex angle $\widehat{BAB'}$. See **Figure 2.78**. That is, $\angle(\overrightarrow{AB'}, \overrightarrow{AN}) = -\angle(\overrightarrow{AB}, \overrightarrow{AN})$.

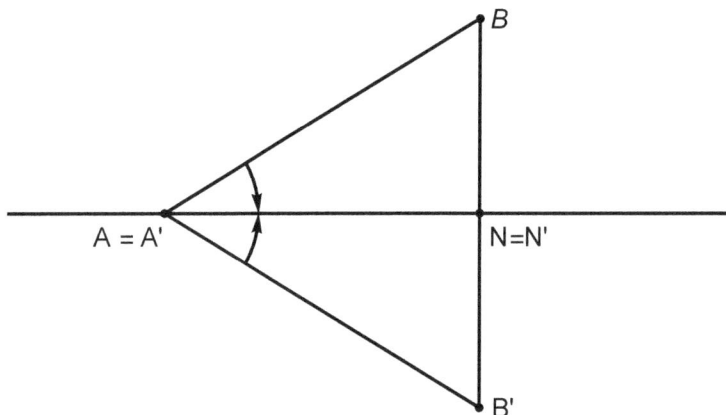

Figure 2.78: Determination of a reflection by a fixed point and
a pair of corresponding points

Let N' is the image of N. Since $A = A'$, we have $AN = A'N' = AN' \neq 0$. By the **Theorem** and the previous relation, $\angle(\overrightarrow{AB'}, \overrightarrow{AN}) = -\angle(\overrightarrow{AB}, \overrightarrow{AN})$, we get $\angle(\overrightarrow{AB'}, \overrightarrow{AN'}) = \angle(\overrightarrow{A'B'}, \overrightarrow{A'N'}) = -\angle(\overrightarrow{AB}, \overrightarrow{AN}) = \angle(\overrightarrow{AB'}, \overrightarrow{AN})$.

Therefore, we have obtained the two conditions

$$AN = AN' \neq 0 \qquad \text{and} \qquad \angle(\overrightarrow{AB'}, \overrightarrow{AN}) = \angle(\overrightarrow{AB'}, \overrightarrow{AN'}).$$

These imply $\overrightarrow{AN} = \overrightarrow{AN'}$ and so $N = N' \neq A$. Hence, this opposite isometry has two different fixed points A and N and so it is a reflection with axis AN. ∎

In **Section 2.3, paragraphs (6), (7)** and **(8)**, we have proved that the composition of three reflections, or a reflection and a rotation, or a reflection and a parallel translation can be simplified to a new reflection or to a glide. The geometrical constructions of the new reflections and the glides are fully presented there. As we have explained, the glides have several convenient properties which help to solve many geometrical problems. We have proved:

An opposite isometry G (and so not identity) such that $G(AB) = A'B'$, with $AB \neq 0$, is either a reflection or a glide or can be reduced to one of these.

We summarize **the important results** for an opposite isometry G:

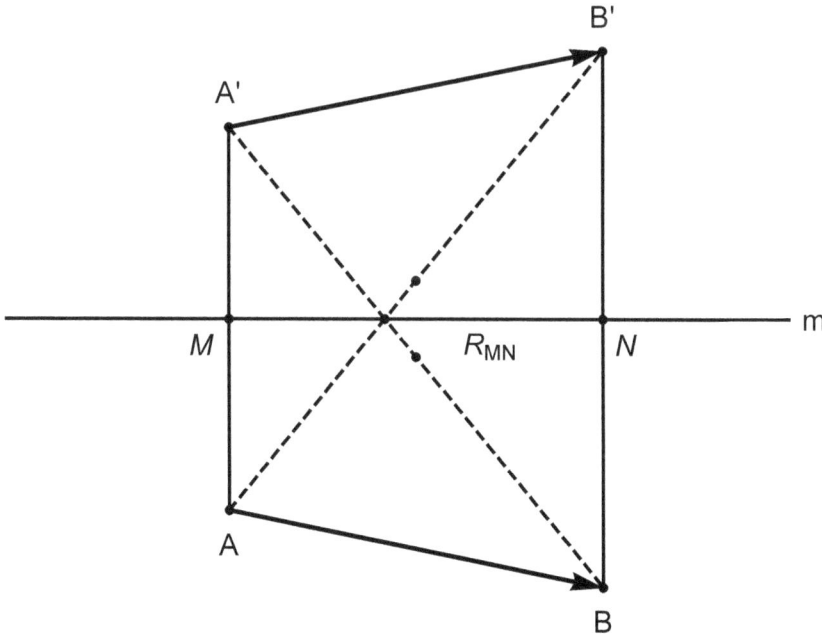

Figure 2.79: Determination of a reflection R_{MN} by two pairs of corresponding points

Consider two different points A and B in the plane, the opposite plane isometry G and the images of the points under G, $G(A) = A'$ and $G(B) = B'$. Then:

(a) If $A = A'$ **or** $B = B'$, i.e., at least one of these points is fixed, then G is a reflection. Its axis is:

(a1) The line AB, if both points A and B are fixed points ($A = A'$ and $B = B'$).

(a2) The line AN, if $B \neq B'$, where N is the midpoint of BB'. (**Figure 2.78**.)

(a3) The line MB, if $A \neq A'$, where M is the midpoint of AA'. (Draw figure.)

(b) If $A \neq G(A) = A'$ **and** $B \neq G(B) = B'$ (so $AB = A'B'$) **and** the quadrilateral $ABB'A'$ **is** an isosceles trapezium (so $0 \neq AA' \parallel BB' \neq 0$), which is not a skew parallelogram (but it can be an orthogonal parallelogram), and M and N are the midpoints of the segments AA' and BB', respectively, then the line $m := MN$ is perpendicular to both AA' and BB' and the isometry G is the reflection R_{MN}. That is, $G = R_{MN} := R_m$. See **Figures 2.79** and **2.80**.

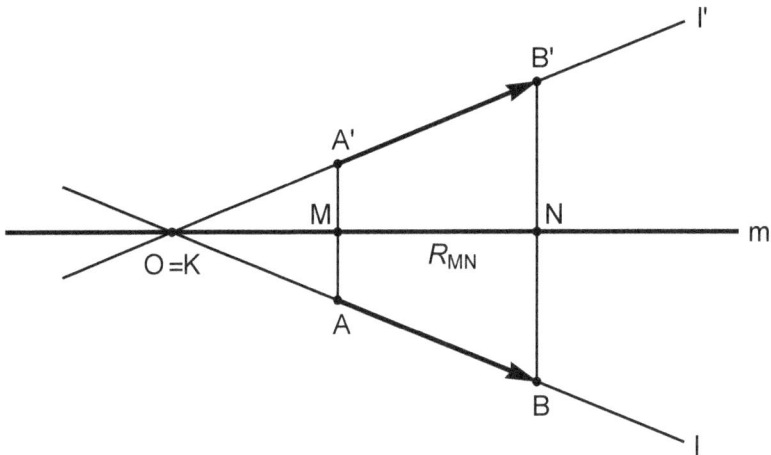

Figure 2.80: Determination of a reflection R_{MN} by two pairs of corresponding points

(c) If $A \neq G(A) = A'$ and $B \neq G(B) = B'$ (so $AB = A'B'$) and the quadrilateral $ABB'A'$ **is not** an isosceles trapezium (so, $0 \neq AA' \nparallel BB' \neq 0$), or $ABB'A'$ **is** a skew parallelogram (so, $0 \neq AA' \parallel = BB' \neq 0$ and no angle is a right angle), then we find the midpoints M and N of the segments AA' and BB', respectively, and the vector $\overrightarrow{A''A'}$, where $A'' = R_{MN}(A) \neq A'$. In such a case, the opposite isometry G is the **glide** with **axis** the line $m := MN$ and **translation vector** $\overrightarrow{A''A'} \neq \overrightarrow{0}$ (parallel to MN), i.e.,

$$G = R_m \circ T_{\overrightarrow{[A''A']}}.$$

[Here, $A'' = R_{MN}(A) \neq A'$, for otherwise, $ABB'A'$ would be an isosceles trapezium or an orthogonal parallelogram. See also **Section 2.4, cases (2)** and **(3)**.]

Thus, study **Figures 2.81, 2.82**, and **2.83** to see how the glide

$$G = R_m \circ T_{[\overrightarrow{A''A'}]}.$$

is formed. (Review **Lemmata 2.4.2, 2.4.3** and their proofs and then prove that $\overrightarrow{A''A'} = \overrightarrow{B''B'} \parallel MN$, etc.)

(a) In **Figures 2.81** and **2.82**, the straight lines AB and $A'B'$ intersect at a point W.

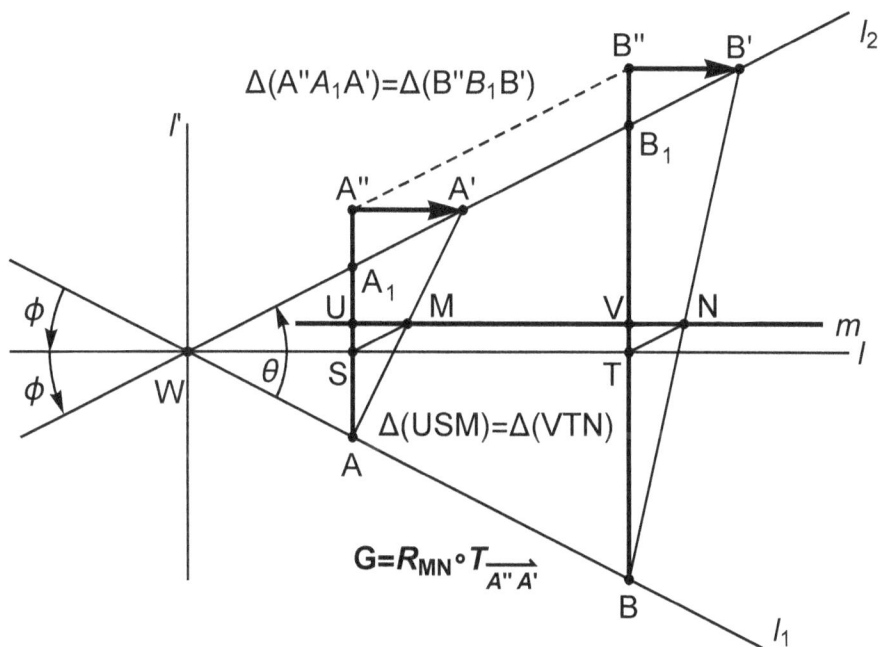

Figure 2.81: Determination of a glide G by two pairs
of corresponding points

(b) When $\overrightarrow{AB} = -\overrightarrow{A'B'}$, **Figure 2.82** reduces to **Figure 2.83** ($ABA'B'$ is a parallelogram, skew or orthogonal). Then, we have $AB = \parallel A'B'$, the segments AA' and BB' intersect at their midpoint $M = N$, and the straight line m, the axis of the glide, is the perpendicular to AB and $A'B'$ through $M = N$. (See also **Figure 2.56**.)

(c) If $\overrightarrow{AB} = \overrightarrow{A'B'}$, then see **Figure 2.53** and observe, in this case, how the glide

$$G = R_m \circ T_{[\overrightarrow{A_1A'}]}.$$

is formed. If in **Figure 2.53**, $\overrightarrow{A_1A'} = \overrightarrow{0}$, then G is the reflection R_m in the line m. That is, $G = R_m$.

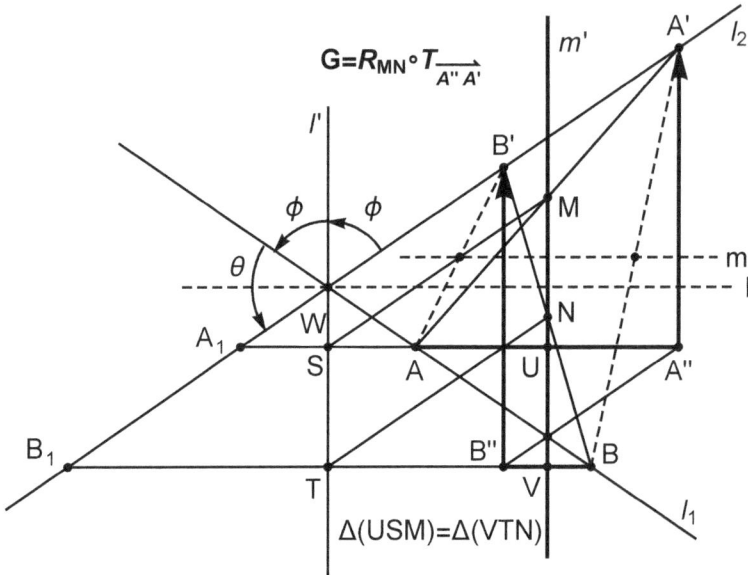

Figure 2.82: Determination of a glide G by two pairs of corresponding points

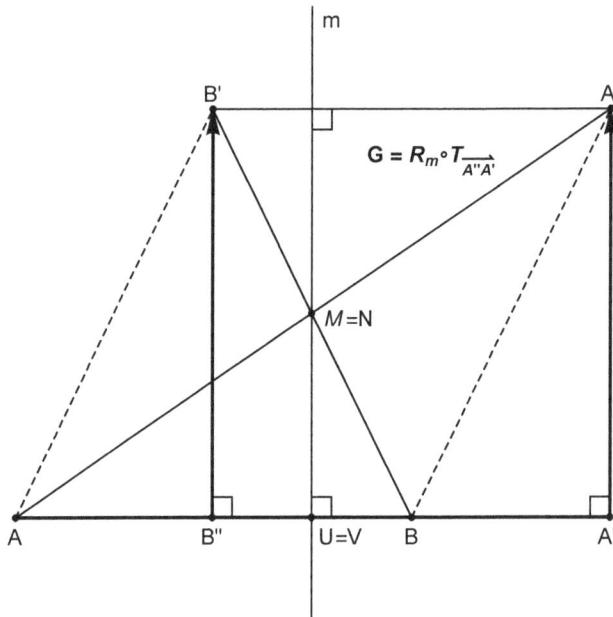

Figure 2.83: Determination of a glide G by two pairs of corresponding points

General Conclusion Drawn from Chapter 2.

(a) Given any two non-zero vectors $\overrightarrow{AB} \neq \vec{0}$ and $\overrightarrow{A'B'} \neq \vec{0}$ in the plane $\mathcal{P}$, such that $AB = A'B'$, there is a unique direct isometry and a unique opposite isometry of the plane that maps $\overrightarrow{AB}$ onto $\overrightarrow{A'B'}$.

(b) If the isometry is direct, then its direct-isometry-angle is $\theta = \measuredangle(\overrightarrow{AB}, \overrightarrow{A'B'})$.

(c) A direct isometry preserves oriented angles (orientation).

(d) An opposite isometry reverses oriented angles (orientation).

(e) The image of a triangle by a direct isometry is a triangle directly equal to the original triangle.

(f) The image of a triangle by an opposite isometry is a triangle oppositely equal to the original triangle.

The following table states some basic pieces of information about isometries.

Isometry	Number of Reflections	Preserves Orientation?	Number of Fixed Points	Involution?
I, Identity	0 2	Yes	∞, all points of the plane, $\mathcal{P}$	Yes
$T_{\vec{u}}$, Parallel Translation by $\vec{u} \neq \vec{0}$	2	Yes	0	No
$R_O(\theta)$, Rotation by $\theta \neq 0 \bmod \pi$	2	Yes	1, point O	No
$R_O(\theta = \pi)$, Half Turn, $\theta = \pi \bmod 2\pi$	2	Yes	1, point O	Yes
R_l, Reflection in a str. line l	1	No	∞, all points of the axis l	Yes
$T_{\vec{u}} \circ R_l$, Glide with $l \parallel \vec{u} \neq \vec{0}$	3=2+1	No	0	No

Isometric Equality and Coincidence

Two planar figures are **isometrically equal**, if there exists an isometry of their plane that maps one of the figures onto the other. If the above isometry is direct, we say the two figures are **directly equal**. If the isometry is opposite, we say the two figures are **oppositely equal**. Some figures, such as two equal: circles, isosceles triangles, etc., admit both direct and opposite isometries.

Two directly equal figures can be moved *without leaving the plane* by a finite sequence of parallel translation(s) and rotation(s) to make them coincide. (This is the **Principle of superposition**, used by Euclid and many after him,

but without using or mentioning the direct isometries, to prove equality as
coincidence.

But, if two figures are only oppositely equal, (e.g., two non-isosceles triangles
symmetrical in a straight line) *they cannot be moved without leaving the plane
to make them coincide.* To make two such figures coincide we must leave the
plane and use the space. (Draw two non-isosceles triangles symmetrical in a
straight line and try to make them coincide.)

Hilbert[1] in his book *"The Foundations of Geometry"*, noticed the problem
and introduced an appropriate axiom in order to remove the proofs that were
based on superposition.

2.6   Applications of Isometries

In this section, we present a few examples of applications of the plane isometries.

2.6.1   Examples of Applications of Parallel Translation

Example 2.6.1 Application 1.

Consider two straight lines in the plane l and l' and a straight segment
AB. Find points $P \in l$ and $P' \in l'$ such that the quadrilateral $PABP'$ is a
parallelogram.

Solution: See **Figure 2.84**.

The quadrilateral $PABP'$ is a parallelogram if and only if $\overrightarrow{PP'} = \overrightarrow{AB}$.

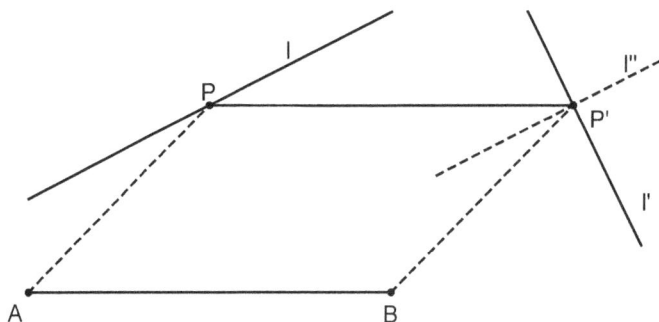

Figure 2.84: Example, A.P.T. 1

This indicates that P' is the image of P under the parallel translation $T_{\overrightarrow{AB}}$.
At the same time P' must be on l' and P on l.

So, we consider $l'' = T_{\overrightarrow{AB}}(l)$ the parallel translation of l by $T_{\overrightarrow{AB}}$ and let
$P' = l' \cap l''$. Having thus determined the point P', we find the point

---

[1]David Hilbert, great German mathematician, January 23, 1862, February 14, 1943.
Among many mathematical achievements, he developed a Categorical Axiomatic Foundation
of Geometry.

$$P = T_{-\overrightarrow{AB}}(P') = T_{\overrightarrow{BA}}(P').$$

Then, $\overrightarrow{PP'} = \overrightarrow{AB}$ and so the quadrilateral $PABP'$ is a parallelogram.

If $l' \parallel \neq l''(\parallel l)$ the points P' is not determined and the construction is not possible. Draw a figure for such a situation. Can there be more than one solutions? How many?

▲

Example 2.6.2 Application 2.

We consider a circle and a vector $\overrightarrow{U}$ in the plane. We want to find a chord of the circle AB such that $\overrightarrow{AB} = \overrightarrow{U}$.

Solution: See **Figure 2.85**.

The point B is the image of A under the parallel translation $T_{\overrightarrow{U}}$. Therefore, B belongs to circle which is the image of the given circle under the same parallel translation. If the center of the given circle is O then $O' = T_{\overrightarrow{U}}(O)$ (so, $\overrightarrow{OO'} = \overrightarrow{U}$) and the radii are the same. In this way we find B and then $A = T_{-\overrightarrow{U}}$.

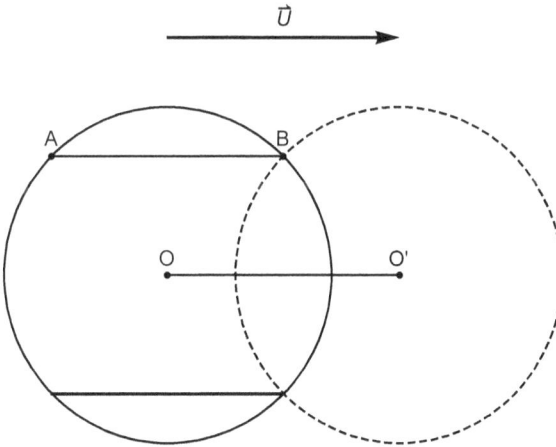

Figure 2.85: Example, A.P.T. 2

Note that depending on the sizes of $\overrightarrow{U}$ and the given circle, we may have 0 or 1 or 2 solutions. In **Figure 2.85** we have 2 solutions. Namely:

1. We have no solution if the length of $\overrightarrow{U}$ is greater than the diameter of the circle.

2. We have one solution if the length of $\overrightarrow{U}$ is equal to the diameter of the circle.

3. We have two solutions if the length of $\overrightarrow{U}$ is less than the diameter of the circle. (Provide figures for the first two situations.)

Note: Instead of a circle we could have another figure and ask the same question. Again, the solution is similar.

▲

Example 2.6.3 Application 3.

Consider two circles and a straight line l in the plane. We want to find a straight line parallel to l and intersecting both circles such that the two chords thus determined are equal.

Solution: See **Figure 2.86**.

In the figure, we have the two circles with centers O and O' and the line l.

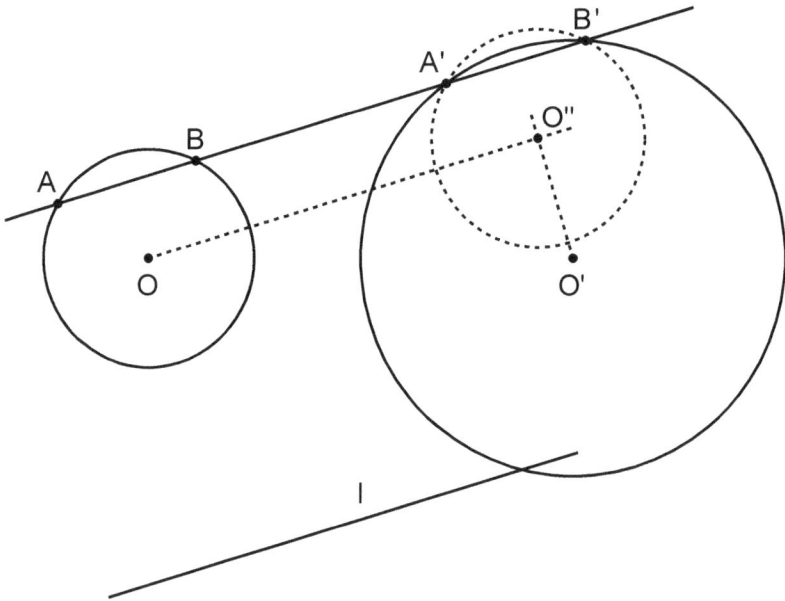

Figure 2.86: Example, A.P.T. 3

If we assume that we have the solution in **Figure 2.86**, we see that the parallel translation $T_{\overrightarrow{AA'}}$ maps the chord AB to $A'B'$ and the circle (O) to a circle (O''). The circles (O') and (O'') intersect at two points A' and B', and $OO'' \parallel l$.

So, to find O'', we draw the line from O parallel to l and the line through O' perpendicular to it, since this line is perpendicular to $A'B'$ and $A'B' \parallel l \parallel OO''$. Then, we draw the circle with center O'' and equal to circle (O) and thus determine the points of intersection of (O'') with (O'), A' and B'. The line $A'B'$ extended long enough will intersect the circle (O) at A and B. It will be $AB = A'B'$, since (O) and (O'') are equal circles and their center line OO'' is parallel to AB'. (Easy to prove.)

Note: Besides the solution in **Figure 2.86**, depending on the sizes of the two given circles and the direction of the given line l, there may be no solution [if (O') and (O'') do not intersect], or a trivial solution [if (O') and (O'') touch at one point]. Investigate these situations and provide figures.

▲

Example 2.6.4 Application 4.

Consider a fixed circle with center O and radius R and a fixed straight line l. We consider a circle with center K and radius a given r rolling on the given circle (O) externally. We want to find the geometrical locus of the points of tangency of all tangents to (K) which are parallel to l as it rolls on (O) externally.

Solution: See **Figure 2.87**.

We have the two circles with centers O and K and the line l.

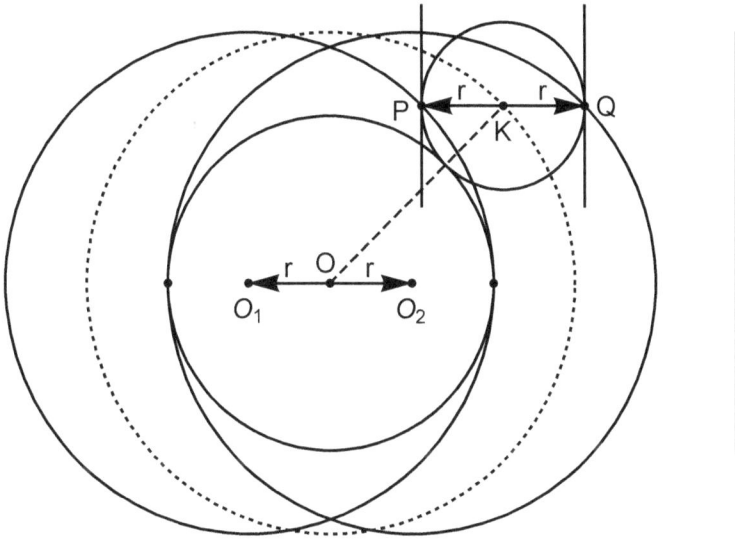

Figure 2.87: Example, A.P.T. 4

Though the rolling, the center K traces a circle with center O and radius $R + r$. As (K) rolls, at every position, there are two diametrical points of tangency P and Q on tangents parallel to l. These are images of K by the parallel translations $T_{\overrightarrow{KP}}$ and $T_{\overrightarrow{KQ}}$. The vectors $\overrightarrow{KP}$ and $\overrightarrow{KQ}$ are opposite, they have length r, they are perpendicular to l and their initial point moves along the circle with center O and radius $R + r$. So, they are known vectors at every moment of the rolling. We call them $\overrightarrow{u}$ and $-\overrightarrow{u}$, respectively.

Hence, the points P and Q trace the circles obtained from the circle with center O and radius $R + r$ under the parallel translations $T_{\overrightarrow{u}}$ and $T_{-\overrightarrow{u}}$. Therefore, the sought geometrical locus is the two circles $C[O_1, R+r]$ and $C[O_2, R+r]$, where $\overrightarrow{OO_1} = \overrightarrow{u}$ and $\overrightarrow{OO_2} = -\overrightarrow{u}$.

Note: If the circle $C[K, r]$ rolls on $C[O, R]$ internally (we need $R > r$, in this case), then K traces the circle $C[O, R-r]$ and the sought geometrical locus is the two circles $C[O_1, R - r]$ and $C[O_2, R - r]$. Make a figure!

▲

2.6.2   Examples of Applications of Rotation

Example 2.6.5 Application 1.

Consider two equal circles (O) and (O') intersecting at two points A and B. We consider the rotation $R_{[A,\theta]}$, such that the image of (O) is (O'). So, $\theta = \angle(AO, AO')$. Prove that any two corresponding points P and P' under this rotation on the two circles define a straight line passing through B.

Solution: See **Figure 2.88**.

We have the two circles with centers O and O' intersecting at A and B and we have two corresponding points P and P' under the rotation $R_{[A,\theta]}$, that is, $R_{[A,\theta]}(P) = P'$.

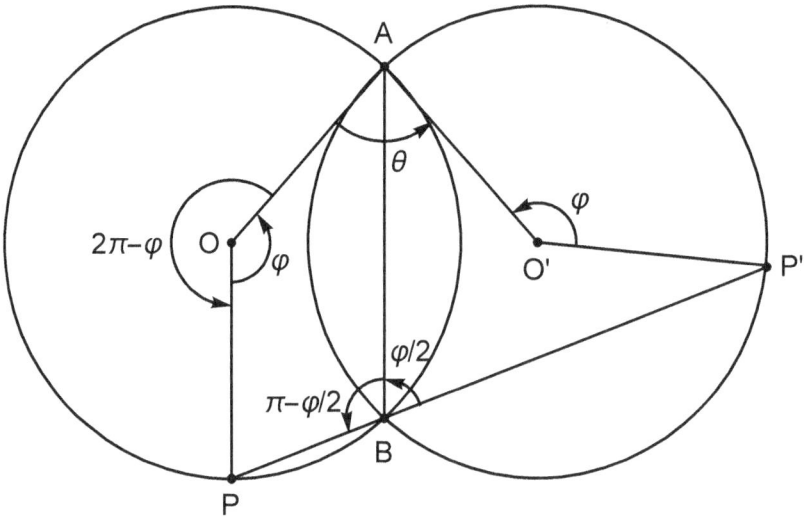

Figure 2.88: Example, A.RO. 1

We also have $R_{[A,\theta]}(O) = O'$ and so (as in the figure)

$$\angle(\overrightarrow{OP}, \overrightarrow{OA}) = \angle(\overrightarrow{O'P'}, \overrightarrow{O'A}) := \phi.$$

Then, from elementary geometry, we have

$$\angle(\overrightarrow{BP'}, \overrightarrow{BA}) = \frac{\phi}{2} \quad \text{and} \quad \angle(\overrightarrow{BA}, \overrightarrow{BP}) = \frac{2\pi - \phi}{2}.$$

Therefore,

$$\angle(\overrightarrow{BP'}, \overrightarrow{BP}) = \angle(\overrightarrow{BP'}, \overrightarrow{BA}) + \angle(\overrightarrow{BA}, \overrightarrow{BP}) = \frac{\phi}{2} + \frac{2\pi - \phi}{2} = \pi.$$

Hence, P, B, P' are on the same straight line.

▲

Example 2.6.6 Application 2.

Consider two equal circles (O) and (O') and two fixed points A and A' on them, respectively. We choose corresponding points P and P' on (O) and (O'), such that the corresponding arcs on the circles are similarly oriented and equal in length, i.e.,

$$\overset{\frown}{AP} = \overset{\frown}{A'P'}.$$

Prove that the perpendicular bisectors of all segments PP' pass through a fixed point Q.

Solution: See **Figure 2.89**.

(a) Assume that $\angle(\overrightarrow{OA}, \overrightarrow{O'A'}) \neq 0$, (mod 2π). Since, $OA = O'A'$, there is a rotation that maps $\overrightarrow{OA}$ to $\overrightarrow{O'A'}$. [See **Theorem 2.5.3, (b)**.]

Since two circles are equal, this rotation maps circle (O) onto (O'), and under the condition of the correspondence of points, the point P to point P'.

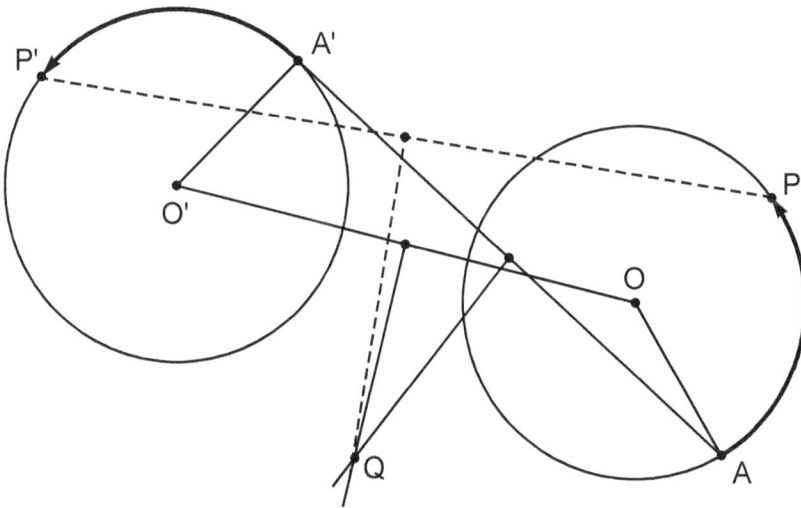

Figure 2.89: Example, A.RO. 2

Then, the perpendicular bisectors of all segments PP' pass through the center of this rotation, which is a fixed point Q and can be predetermined as the common point of the perpendicular bisectors of segments OO' and AA', which are fixed.

(b) If $\angle(\overrightarrow{OA}, \overrightarrow{O'A'}) = 0$, (mod 2π), then $\overrightarrow{OA} = \overrightarrow{O'A'}$. Hence, by **Theorem 2.5.3, (b)**, the vectors $\overrightarrow{OA}$ and $\overrightarrow{O'A'}$ correspond to each by the parallel translation $T_{\overrightarrow{OO'}}$. In this case, the perpendicular bisectors of all segments PP' are perpendicular OO' and so they are parallel. Conventionally, they pass through a point at infinity.

▲

Example 2.6.7 Application 3.

Consider two concentric circles $\mathcal{F}$ and $\mathcal{F}'$ with center O and a point A on $\mathcal{F}$, the smaller circle. Construct a square $ABCD$, such that B is on $\mathcal{F}$ and C and D are on $\mathcal{F}'$.

Solution: See **Figures 2.90** and **2.91**.

If we assume that the square $ABCD$ has been already constructed. Then we observe:

$$AD = AB \quad \text{and} \quad \measuredangle(\overrightarrow{AB}, \overrightarrow{AD}) = \frac{\pi}{2}$$

or

$$AD = AB \quad \text{and} \quad \measuredangle(\overrightarrow{AB}, \overrightarrow{AD}) = \frac{-\pi}{2}.$$

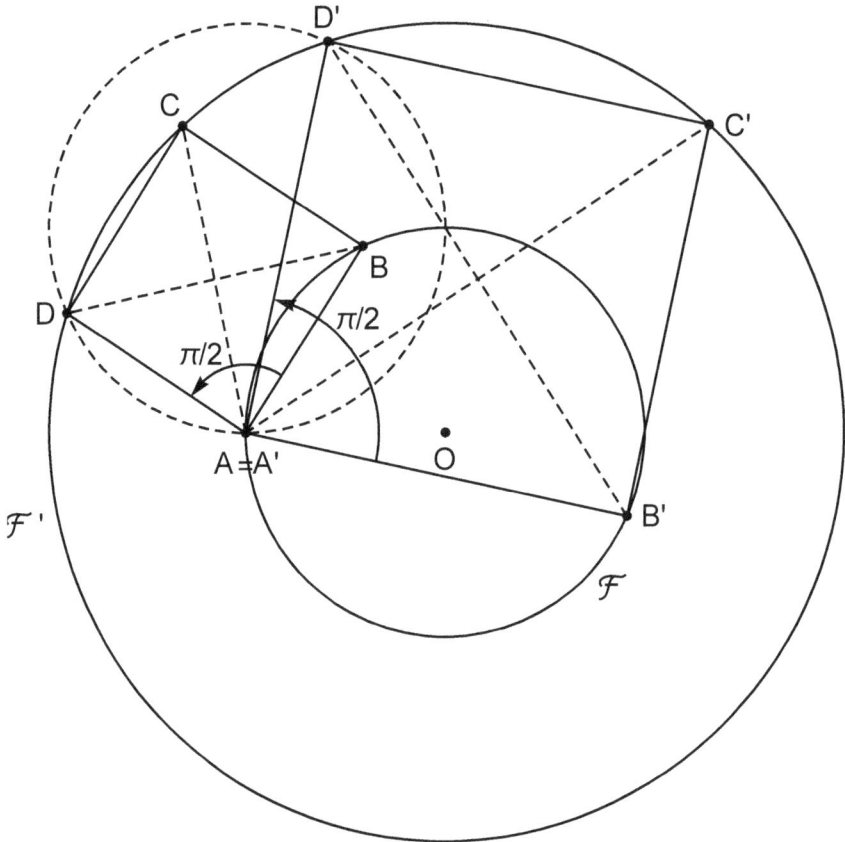

Figure 2.90: Example, A.RO. 3

In the first case, $D \in \mathcal{F}'$ is the image of B by the rotation $R_{[A, \frac{\pi}{2}]}$. So, we find the intersection of the circles $\mathcal{F}'$ and $R_{[A, \frac{\pi}{2}]}(\mathcal{F})$. There are 0 or 1 or 2 points of intersection. (In **Figure 2.90**, we have two points D and D'.)

Having found D, we find C as the symmetrical of A on the perpendicular bisector of BD, or as the intersection of the circle $\mathcal{F}'$ and the line from B and parallel to AD. (Prove that $ABCD$ thus constructed is a square.)

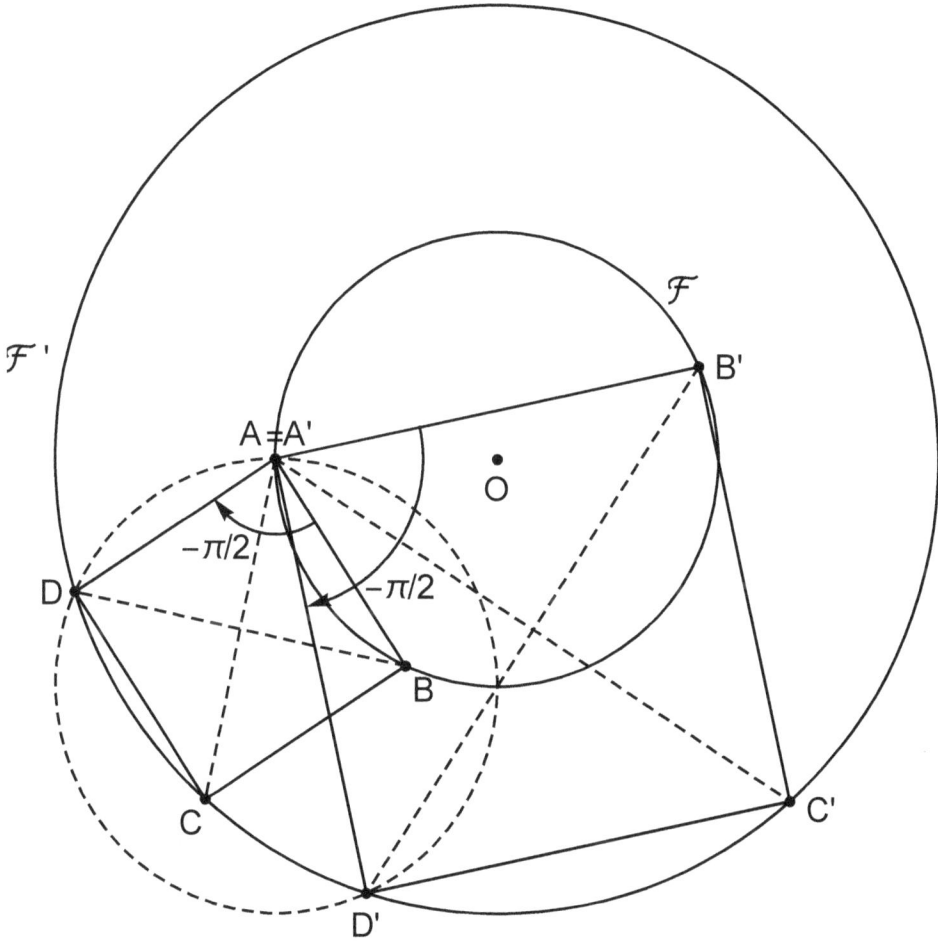

Figure 2.91: Example, A.RO. 3

In the second case, $D \in \mathcal{F}'$ is the image of B by the rotation $R_{[A,\frac{-\pi}{2}]}$. So, we find the intersection of the circles $\mathcal{F}'$ and $R_{[A,\frac{-\pi}{2}]}(\mathcal{F})$. There are 0 or 1 or 2 points of intersection. (In **Figure 2.91**, we have two points D and D'.) Then, the construction is done as in the previous case.

Depending on the sizes of the two circles, the number of solutions is 0 or 2 or 4. Between the two figures, we have provided here, the solutions are 4. Investigate the cases of 0 or 2 solutions and provide figures.

▲

Example 2.6.8 Application 4.

Consider two straight lines l and l', angle $0 < \phi < \pi$, and a point A. Construct an isosceles triangle ABC such that, $AB = AC$, $B \in l$ and $C \in l'$ and $\widehat{BAC} = \phi$.

Solution: See **Figure 2.92**.

The point $C \in l'$ is also a point of the line l'', the image of the line l under the rotations $R_{[A,\pm\phi]}$. Having found the two points C and C', in this way, we find

$$B = R_{[A,-\phi]}(C) \qquad \text{and} \qquad B' = R_{[A,\phi]}(C').$$

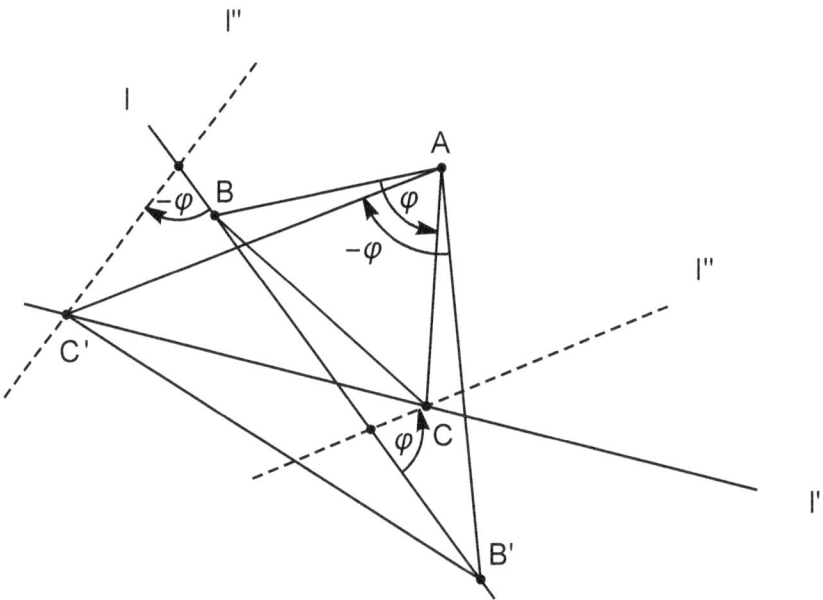

Figure 2.92: Example, A.RO. 4

Depending on the positions of the lines and l and l', the position of the point A, and the size of the angle ϕ, we may have 0, 1, or 2 solutions. Provide figures in the cases of 0 or 1 solution.

▲

Example 2.6.9 Application 5.

We consider a circle with center O and passing through a point A. We choose any points B and C on this circle, such that the angle $\widehat{BAC} = \theta < \pi$ is given fixed.

(1) Find the geometrical locus, of the point P symmetrical of B on the line AC, (as B and C move on the given circle such that the angle $\widehat{BAC} = \theta < \pi$ given fixed).

(2) Prove that the line BP passes through a fixed point.

Solution: See **Figure 2.93**.

(1) For any point P on the geometrical locus, it holds

$$AP = AB \quad \text{and} \quad \angle(\overrightarrow{AB}, \overrightarrow{AP}) = 2(AB, AC) = 2\theta.$$

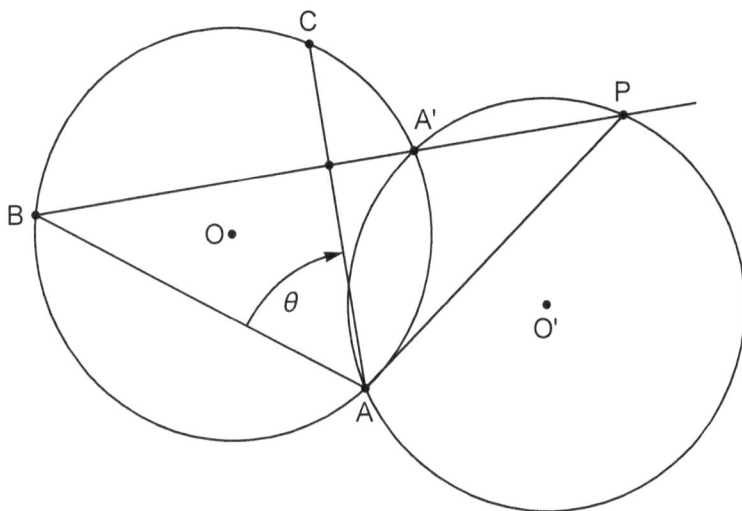

Figure 2.93: Example, A.RO. 5

So, P is the image of B under the rotation $R_{[A,2\theta]}$. Therefore the set of P's is the image circle of the given circle under this rotation.

(2) If A' is the other point of intersection of these two circles as we know from **Example 2.6.5**, the line BP passes through A', which is a fixed point.

▲

Example 2.6.10 Application 6.

Consider a triangle ABC. We construct equilateral triangles ABC', BCA', and CAB' in the same orientation. Prove:

(1) The segments AA', BB' and CC' are equal.

(2) The segments AA', BB' and CC' are concurrent, i.e., they pass through the same point S, called the **Fermat[2]-Torricelli[3]-Steiner[4] point**.

(3) $\widehat{A'SC} = \widehat{CSB'} = \ldots = \widehat{BSA'} = \dfrac{\pi}{3}$.

(4) S is the common intersection of the circles ABC', BCA', and CAB'.

---

[2]Pierre de Fermat, French lawyer and mathematician, 1607-1665.
[3]Evangelista Torricelli, Italian physicist and mathematician, 1608-1647.
[4]Jakob Steiner, Swiss mathematician, great geometer, 1796-1863.

Solution: See **Figure 2.94**.

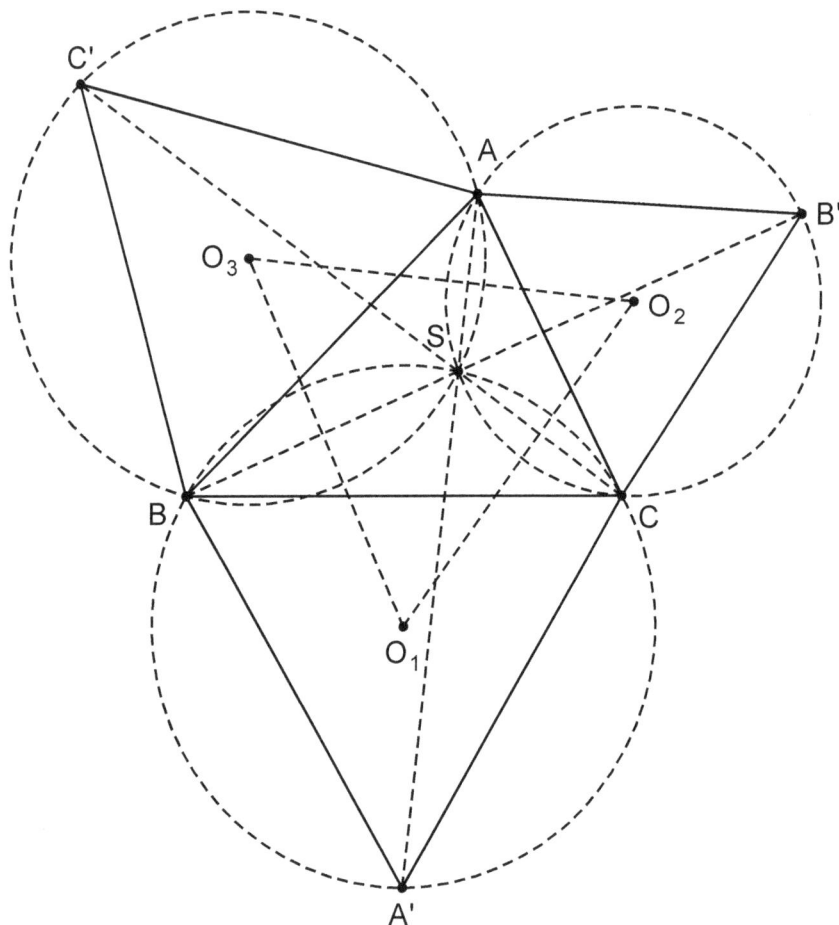

Figure 2.94: Example, A.RO. 6

The rotation $R_{[A,\frac{\pi}{3}]}$ maps C' to B and C to B'. As rotations are isometries, we get

$$CC' = BB' \quad \text{and} \quad \angle(\overrightarrow{C'C}, \overrightarrow{BB'}) = \frac{\pi}{3} = \angle BAC'.$$

Let S be the common point of CC' and BB'. Then, by the equality of the above angles, the points C', B, A, and S are concyclic, and so are the points C, B', A, and S.

Similarly, using the rotation $R_{[B,\frac{\pi}{3}]}$, we obtain $AA' = CC'$ and if AA' and CC' intersect at S', then the points A', B, C, and S' are concyclic, and so are the points A, B, C', and S'. But then, the circle ABC' intersects the segment $C'C$ at the points C' and S' and S. So, it must hold $S = S'$.

If each angle of the given triangle ABC is smaller than $\dfrac{2\pi}{3}$, then S is an interior point of ABC. In this case, this point minimizes the sum $XA + XB + XC$ of the distances from a point X of the plane to the vertices of the triangle. This minimum is

$$\min = SA + SB + SC = AA' = BB' = CC'.$$

Notes: (a) If an angle is greater or equal to $\dfrac{2\pi}{3}$, then the minimizing point is the vertex of that angle. (See bibliography, items 6, 10, and 17.)

(b) Notice that all the angles between two successive straight segments around the point S, as in **Figure 2.94**, are $\dfrac{\pi}{3}$. (Justify this!)

(c) The centers O_1, O_2, O_3 of the above three circles (or the equilateral triangles) are vertices of an equilateral triangle.
{Hint: Let the angle $\angle(\overrightarrow{AB}, \overrightarrow{AC})$ be positive. Consider the three rotations

$$R_{\left[O_i, \frac{-2\pi}{3}\right]}, \quad i = 1, 2, 3.$$

Since $\dfrac{-2\pi}{3} \times 3 = -2\pi$, by **Section 2.3, (4)**, the composition

$$R_{\left[O_3, \frac{-2\pi}{3}\right]} \circ R_{\left[O_2, \frac{-2\pi}{3}\right]} \circ R_{\left[O_1, \frac{-2\pi}{3}\right]}$$

is a parallel translation (in **Figure 2.94**). But, the vertex B of the triangle is a fixed point for this composition. Hence, the composition is the identity and so

$$R_{\left[O_3, \frac{-2\pi}{3}\right]} \circ R_{\left[O_2, \frac{-2\pi}{3}\right]} = R_{\left[O_1, \frac{-2\pi}{3}\right]}^{-1} = R_{\left[O_1, \frac{2\pi}{3}\right]}.$$

Use this to show that the triangle $O_1O_2O_3$ is equilateral.}

(d) For the lengths of SA, SB and SC see item [17] of the bibliography. The point S is very important to applications. For instance, it is used in constructing minimizing graphs and in civil engineering.

▲

Example 2.6.11 Application 7.
Consider an equilateral triangle ABC and a point A'. We construct the equilateral triangles $CA'B'$ and $BA'C'$, similarly oriented with respect to ABC. Prove that the quadrilateral $A'B'AC'$ is a parallelogram.
Solution: See **Figure 2.95**.
The rotation $R_{\left[B, \frac{-\pi}{3}\right]}$ maps A to C and C' to A', and so $\overrightarrow{AC'}$ to $\overrightarrow{CA'}$. The rotation $R_{\left[A', \frac{\pi}{3}\right]}$ maps $\overrightarrow{CA'}$ to $\overrightarrow{B'A'}$. So, the composition of these two rotations maps $\overrightarrow{AC'}$ to $\overrightarrow{B'A'}$. But, by **Section 2.3, (4)**, this composition is a parallel translation and so

$$\overrightarrow{AC'} = \overrightarrow{B'A'}.$$

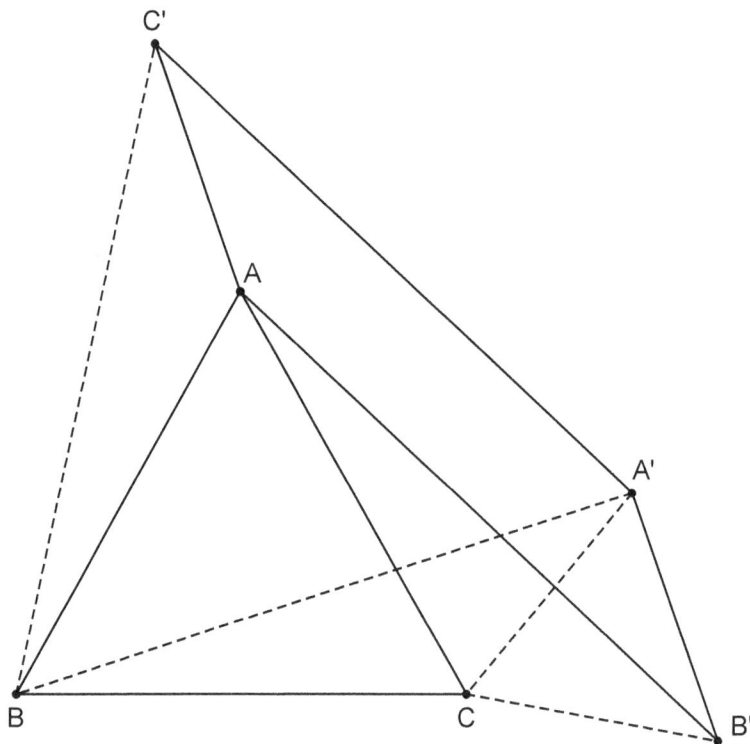

Figure 2.95: Example, A.RO. 7

Hence, $A'B'AC'$ is a parallelogram.

▲

Example 2.6.12 Application 8.

Consider a triangle ABC and a point P. Let A_1, B_1 and C_1 be the symmetrical points of P with respect to the midpoints A', B' and C' of the sides BC, CA, and AB of the triangle, respectively. Prove that the triangle $A_1B_1C_1$ is symmetrical to ABC with respect to the point O, which is the intersection AA_1, BB_1 and CC_1.

Solution: See **Figure 2.96**.

The composition

$$R_{[A',\pi]} \circ R_{[C',\pi]}$$

(of half-turns) is the parallel translation by the vector $2\overrightarrow{C'A'}$ [by **Section 2.3, (4)**]. Hence,

$$\overrightarrow{C_1A_1} = 2\overrightarrow{C'A'} = -\overrightarrow{CA},$$

and similarly,

$$\overrightarrow{A_1B_1} = -\overrightarrow{AB} \quad \text{and} \quad \overrightarrow{B_1C_1} = -\overrightarrow{BC}.$$

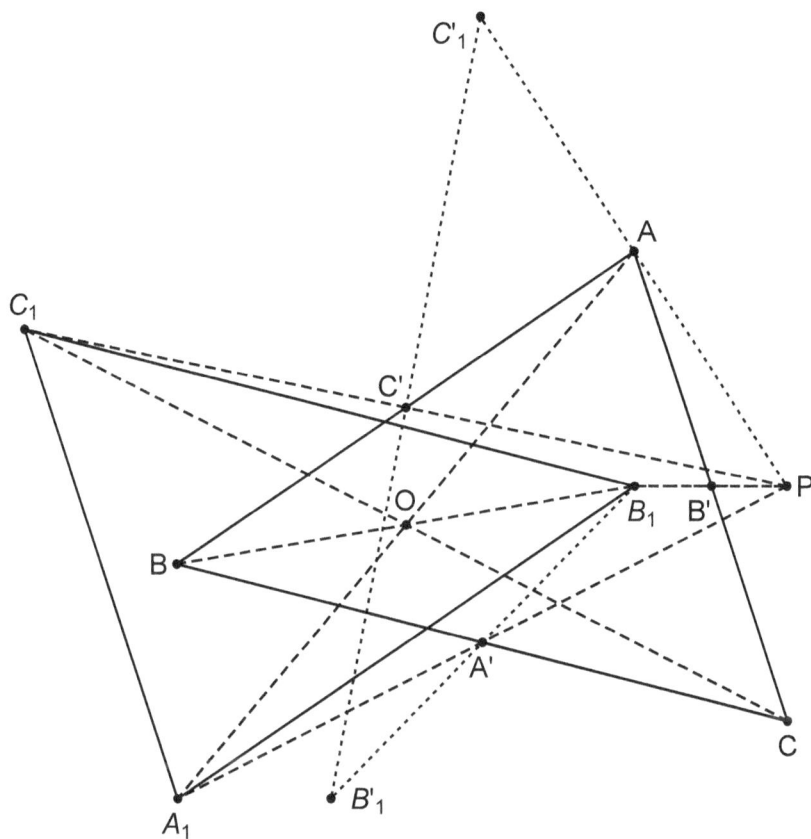

Figure 2.96: Example, A.RO. 8

Therefore, the triangles ABC and $A_1B_1C_1$ are symmetrical with respect to the common midpoint of the segments AA_1, BB_1 and CC_1. (How many parallelograms can you draw in the above figure?)

Prove also: If $C_1' = R_{[C',\pi]} \circ R_{[A',\pi]} \circ R_{[B',\pi]}(P)$, then the midpoint of the straight segment PC_1' is A. Thus, we can construct the triangle ABC if we are given the midpoints A', B', C' of its sides, by finding the vertex A, etc.

This construction yields a unique answer and generalizes to the construction of any polygon with odd number of sides given the midpoints of its sides. (As above, we apply the corresponding half-turns, composed in some order, to any point of the plane P. One vertex is the midpoint of the segment defined by P and the final image of P. Then the polygon is easily constructed.) But, explain why this method yields infinitely many answers, if the number of sides is even. For a triangle or a pentagon there are also other methods of construction. Find them.

▲

2.6.3   Examples of Applications of Reflection

Example 2.6.13 Application 1.

Consider a circle with center O and a straight line l. Find a point A on the circle and a point B on the line, so that the straight segment AB has perpendicular bisector a given line m.

Solution: See **Figure 2.97**.

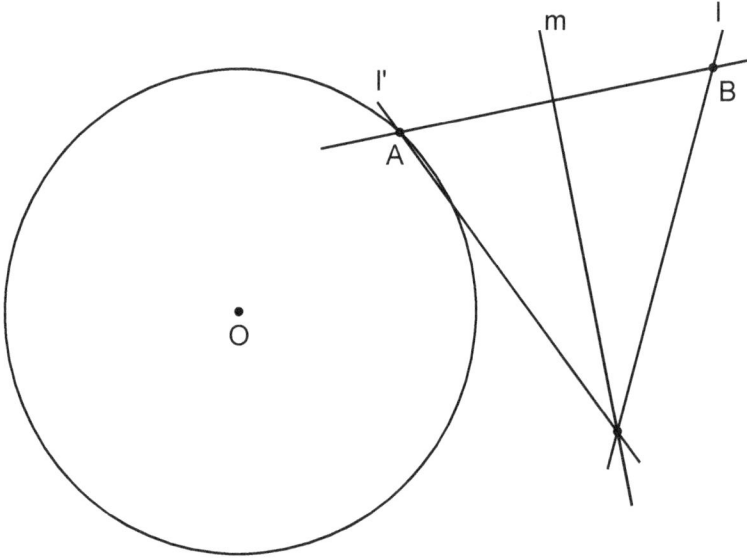

Figure 2.97: Example, A.RE. 1

The point A has B as symmetrical point with respect to m. Therefore, A must also belong to the line l', symmetrical of l with respect to m. Hence, we construct the point A as the point of intersection of the given circle and l'.

Depending on the positions of the lines, l and m, and the circle the number of solutions is 0, or 1, or 2. Make figures!

▲

Example 2.6.14 Application 2.

Prove that in any triangle the distance of a vertex from the orthocenter is twice the distance of the circumcenter from the opposite side.

Solution: See **Figure 2.98**.

We consider a triangle ABC and the vertex A. We know that H_1 (otherwise, provide the proof), the symmetrical of the orthocenter H on the line BC, belongs to the circumscribed circle of the triangle. Then, the symmetrical of H_1 on the line $x'Ox$ parallel of AB through the center O of the circle, is the vertex A.

So, H is mapped to A by the composition of reflections $R_{x'x} \circ R_{BC}$, which is a parallel translation by the vector $2\overrightarrow{O_1O}$. [See **Section 2.3, (3), Case 2.**]

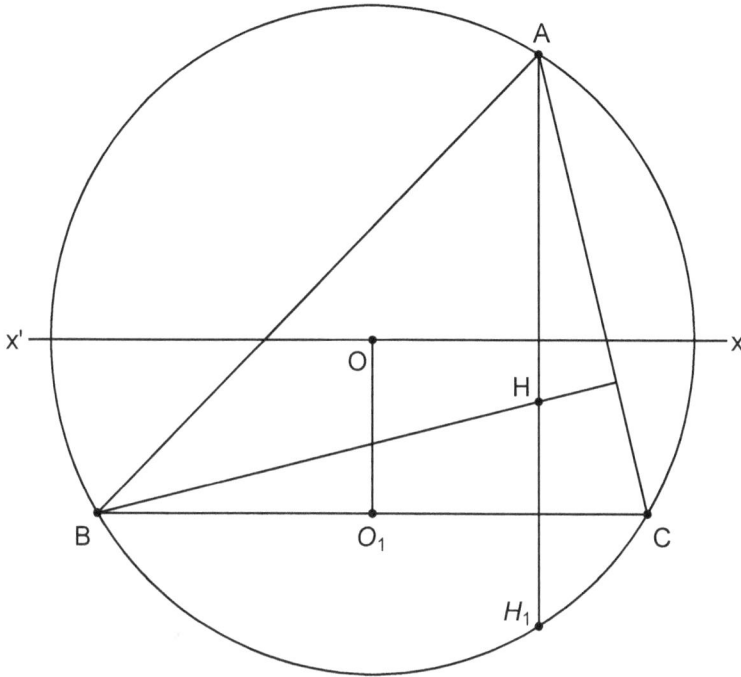

Figure 2.98: Example, A.RE. 2

Thus,

$$\overrightarrow{HA} = 2\overrightarrow{O_1O}.$$

▲

Example 2.6.15 Application 3.

Consider any triangle ABC and a point P on its circumscribed circle with center O.

(1) The points P_1, P_2 and P_3 symmetrical of P on the sides of the triangle BC, CA and AB, respectively, lie on the same line (**Steiner line**), which passes through the orthocenter H of the triangle.

(2) The projections A', B' and C' of P on the sides of the triangle BC, CA and AB, respectively, lie on the same line (**Simson**[5] **line**), which passes through the midpoint of the segment PH.

[The line (**2**) was known to Simson, but Wallace[6] published it first in 1797 (after Simson). So, the line is also called **Wallace line**. Wallace also extended the result from the orthogonal projections of the point P on the sides of the triangle to projections with projection-angles all equal and acute and arranged in the same direction.]

---

[5]Robert Simson, Scottish mathematician, 1687-1768.
[6]William Wallace, Scottish mathematician and astronomer, 1768-1843.

Solution: See **Figure 2.99**.

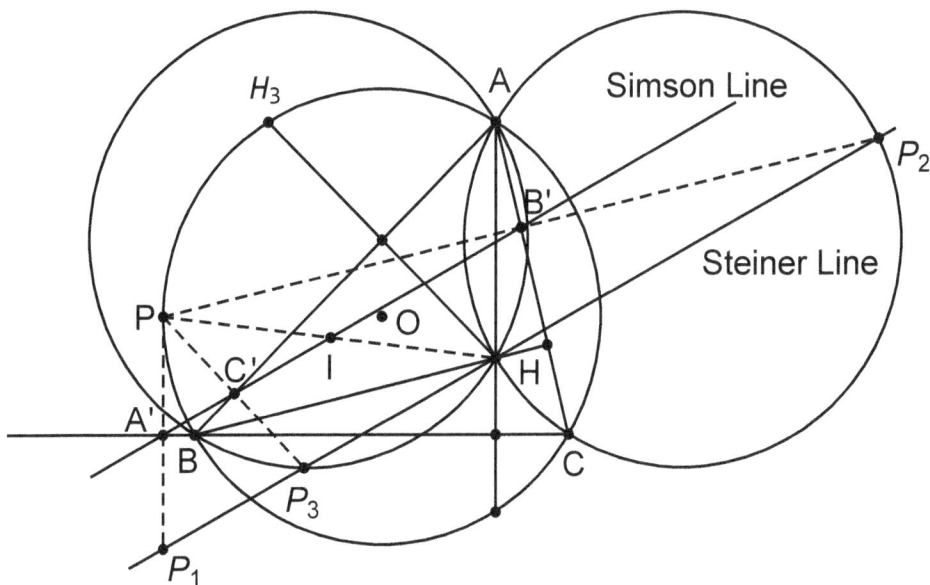

Figure 2.99: Example, A.RE. 3

(1) We know, (otherwise provide the proof), that the symmetrical points H_1, H_2 and H_3, of the orthocenter H on the sides BC, CA and AB, belong to the circumscribed circle of the triangle. The reflection R_{AB} maps P_3 to P and the circle (ABH) to the circle (ABH_3), which coincides with the circumscribed circle (ABC) of the triangle. Similarly, the reflection R_{AC} maps P to P_2 and the circumscribed circle (ABC) of the triangle to the equal circle (ACH).

The composition of these reflections is the rotation

$$R_{AC} \circ R_{AB} = R_{[A, 2\angle(AB,AC)]}.$$

[See, **Section 2.3, (3), Case 3**.] This rotation maps P_3 to P_2 and the circle (ABH) to the circle (ACH).

Therefore, by the result of **Example 2.6.5**, the line P_2P_3 passes through the common point H of these circles. Hence the points P_2, P_3 and H lie on the same straight line.

Analogously, we prove that the points P_1, P_2, and H are on a line. Thus the points P_1, P_2, P_3, and H are on the same line, called the **Steiner line**.

(2) The projections A', B' and C' of P on the sides of the triangle BC, CA and AB, respectively, are the midpoints of the segments PP_1, PP_2 and PP_3, respectively. Then, by **(1)**, these points and the midpoint I of the segment PH are on the same line, which is parallel to the Steiner line. This new line is called the **Simson line**.

Notes: (a) The circles (ABH), (BCH), and (CBH), are symmetrical to the circle (ABC) with respect to the lines AB, BC, and CA, respectively, and so are equal to it. (This was observed by **Carnot**[7], among other several facts.)

(b) The point I is on the nine-point circle, according to the result in **Application 3 of Example 4.2.3, property (a)**.

▲

Example 2.6.16 Application 4.

Consider three different straight lines k, l and m. Construct a straight line n perpendicular to m and such that, if A, B, C are its common points with the lines m, l and k, respectively, then it holds $AB = AC$.

Solution: See **Figure 2.100**.

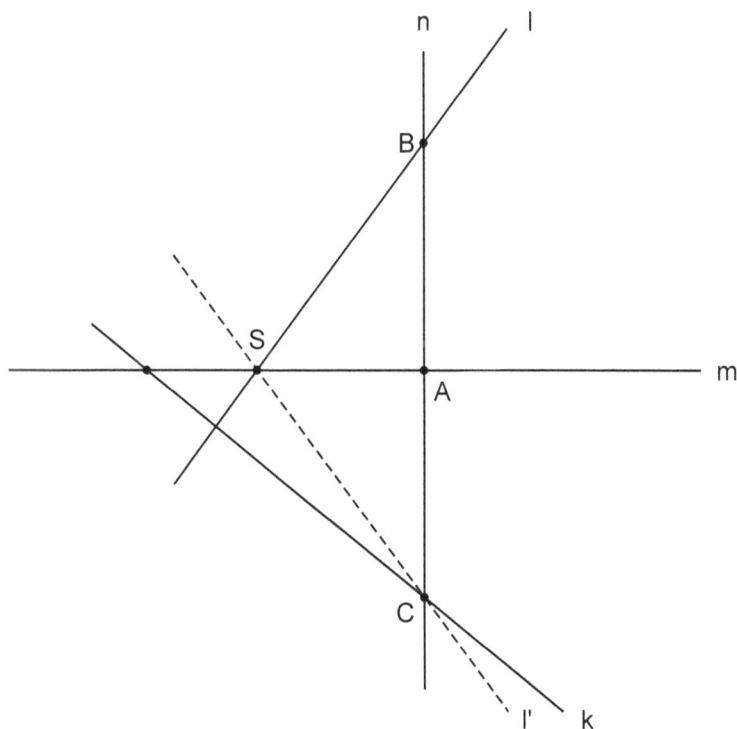

Figure 2.100: Example, A.RE. 4

Suppose we have found a solution line n. Then, the point C is symmetrical to B, with regard to line m. But, since $B \in l$, a known line, the point C is also a point of the line l', the reflection of l with regard to m. So, l' is also known.

Therefore, to solve the problem, we construct the line l', the reflection of l on m. Let C be the intersection point of l' and k. Then, the line n from C and perpendicular to m is the solution.

---

[7]Lazare Nicolas Marguérite Carnot, French mathematician and engineer, 1753-1823.

Indeed, the common point B of $n(\perp m)$ and l is the symmetrical of C with respect to m, because m is the axis of reflection of the figure made by l and l'. So, $AB = AC$.

We notice that in order for the problem to have solution: l' must not be parallel to and different than k. Also, n must intersect both l and k, and so not both of them can be perpendicular to m. Then, the problem has one solution. If $l \perp m$ and C exists as the common point of l and k, then $n = l$ and B is the symmetrical of C with respect to the line m.

Otherwise, the problem has no solution, unless $l' \equiv k$, in which case, it has infinitely many solutions. In the latter case, every line perpendicular to m is a solution. (Make figures for all of these cases.)

▲

Example 2.6.17 Application 5. Fagnano's[8] Problem.

Consider a scalene triangle ABC (all angles are acute) and points D, E, and F on the sides BC, CA, and AB, respectively. Prove that:

(1) Keeping D fixed, the perimeter of the variable triangle DEF is the least when EF passes through the symmetrical points M and N of D on the sides AB and AC, respectively.

(2) The perimeter of the triangle DEF is minimum, when the points D, E, and F are the feet of the altitudes of the given fixed triangle ABC.

Solution: See **Figure 2.101**.

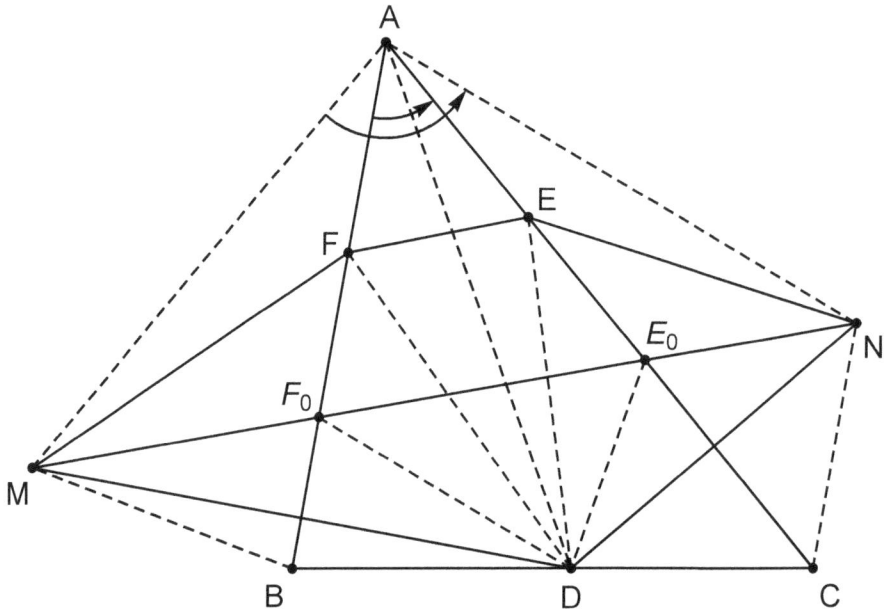

Figure 2.101: Example, A.RE. 5

---

[8]Giovanni Francesco Fagnano dei Toschi, Italian mathematician, 1715-1797.

(1) The line MN intersects the sides AB and AC of ABC, at F_0 and E_0, respectively. Then, we get

$$\text{perimeter of } DEF = DE + EF + FD = NE + EF + FM > MN$$
$$= MF_0 + F_0E_0 + E_0N = DF_0 + F_0E_0 + E_0D = \text{ perimeter of } DE_0F_0.$$

That is, the triangle DE_0F_0 has the least perimeter, when D is fixed in BC.

(2) When D is fixed, the least perimeter is the segment MN. So, we must make MN minimum with respect to the position of D. Since D is symmetrical to M with respect to the line AB and to N with respect to line AC, N is the image of M under the composition

$$R_{AC} \circ R_{AB} = R_{[A,2\angle(AB,AC)]}.$$

Hence, by the reflections of D on the sides AB and AC, we have

$$AM = AD = AN.$$

Thus, the triangle AMN is isosceles with basis MN and vertex angle $2\widehat{A}$ fixed. The basis MN becomes least (simultaneously with the equal sides $AM = AN$), when $AM \perp BM$. On account of the reflection, this happens when $AD \perp BC$ (notice that $\angle BMA = -\angle BDA$). Therefore, the point D that yields the least basis MN (and perimeter) for the triangle AMN is the foot of the altitude of ABC, drawn from A.

Finally, to obtain the triangle DEF with the least perimeter and inscribed in the given scalene triangle ABC, as we just proved about the vertex D, so the vertices E and F must be the feet of the other two altitudes of ABC.

The triangle DEF whose vertices are the feet of the heights of the triangle ABC on its sides is called **the orthic triangle of the triangle** ABC.

Another justification of this result and three properties of the orthic triangle are suggested in the **Remark** at the end of **Example 2.6.18 of Application 6** next. (The orthic triangle has many properties and has been studied a lot.)

▲

Example 2.6.18 Application 6. Heron's[9] Principle.

Consider a straight line m in the plane. This separates the plane into two half-planes. In one of the half plane we pick two points A and B. If D is any point of the line m, we would like to find the special point that minimizes the path from A to D to B, that is, the path $AD + DB$.

First of all, the parts AD and DB must be straight segments. See **Figure 2.102**. Then, the sought path would be minimum if the points A, D and B are on a straight line. This is obtained by the point C which is the common point of m and the segment $A'B$, where $A' = R_m(A)$ [that is, the intersection of the straight line m with the segment $B'A$, where $B' = R_m(B)$].

Indeed, for any point $D \neq C$ in m, the triangle BDA' is not degenerate and so, by the triangle inequality, we have

$$BC + CA' = BA' < BD + DA'.$$

---

[9]Heron of Alexandria, Greek mathematician, engineer and scientist, 10 -70 CE.

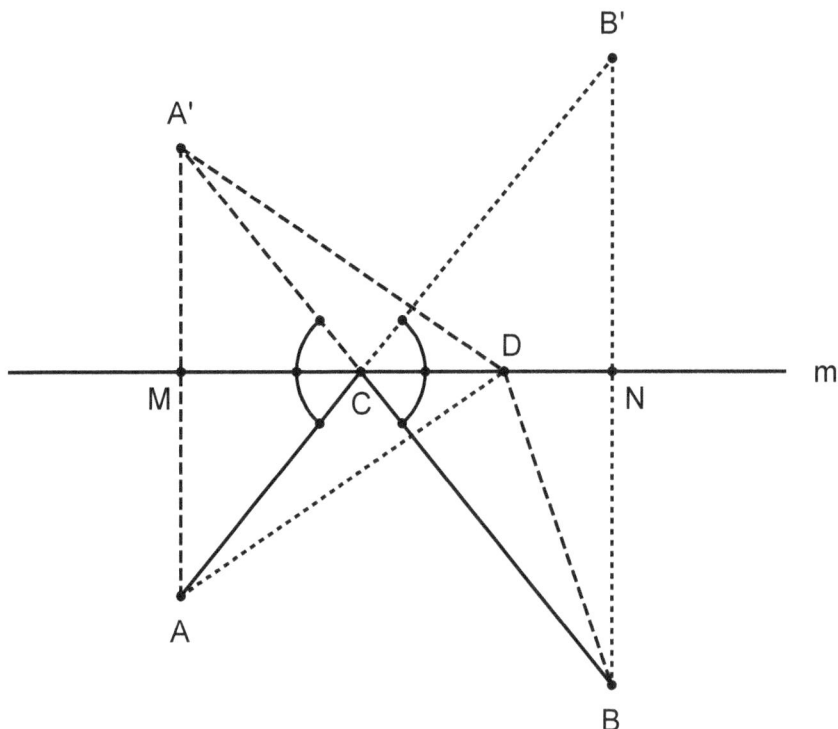

Figure 2.102: Example, A.RE. 6

But, by the reflection R_m in the line m, we have

$$CA' = CA \qquad \text{and} \qquad DA' = DA.$$

Therefore,

$$AC + CB = A'C + CB < AD + DB.$$

Hence, the point C minimizes the path from point A to a point of the line m to the point B (or in reverse order).

The solution of this minimization problem is called **Heron's principle**[10]. The found path is the path that a light beam will follow from A and been reflected in a straight reflector along the line m will hit the point B. From the geometrical solution presented here, we see that the angle of incidence $\widehat{ACM}$ is equal to the angle of the reflection $\widehat{NCB}$, as the principles of the geometrical optics require, etc.

---

[10]This principle is also the "physical law of the shortest path" that light obeys: "*If a ray of light is reflected in a (plane) mirror, then the angle of incidence of the ray is equal to the angle of its reflection.*" Heron exposed this law among many things in his book **Catoptrica**.

In **Figure 2.103**, we have an acute angle $\widehat{xOy}$ and two points A and B in it. We ask for the minimum path from A to a point on the side Ox, then to a point on the side Oy and finally to B.

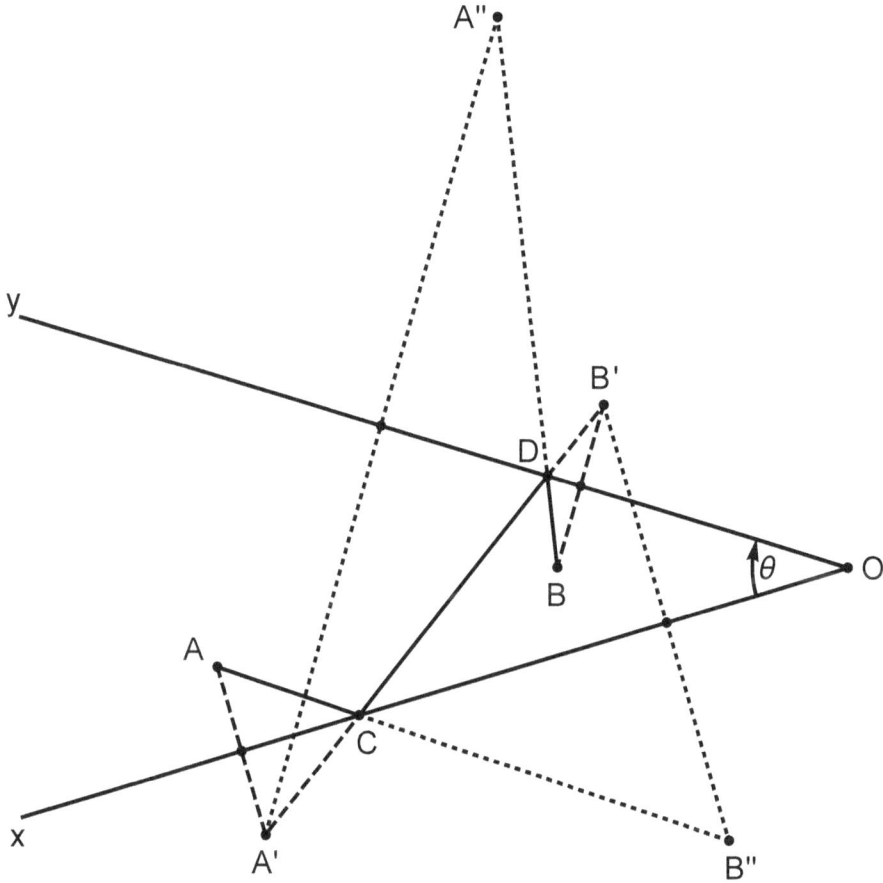

Figure 2.103: Example, A.RE. 6

Keeping in mind Heron's principle, the figure explains how to find the minimizing points $C \in Ox$ and $D \in Oy$, in several ways. (Describe these ways.) Here we have:

$$A' = R_{Ox}(A),$$
$$B' = R_{Oy}(B),$$
$$A'' = R_{Oy}(A') = R_{Oy} \circ R_{Ox}(A) = R_{[O,2\theta]}(A),$$
$$B'' = R_{Ox}(B') = R_{Ox} \circ R_{Oy}(B) = R_{[O,-2\theta]}(B),$$

where $\theta = \angle(Ox, Oy)$, etc. The polygonal path $ACDB$ is the solution.

In the next three figures, we may consider the depicted rectangle as a billiard-table and the points A and B as two billiard-balls.

In **Figure 2. 104**, we would like to hit the ball B by the ball A and making one reflection on one side of the table (one carom).

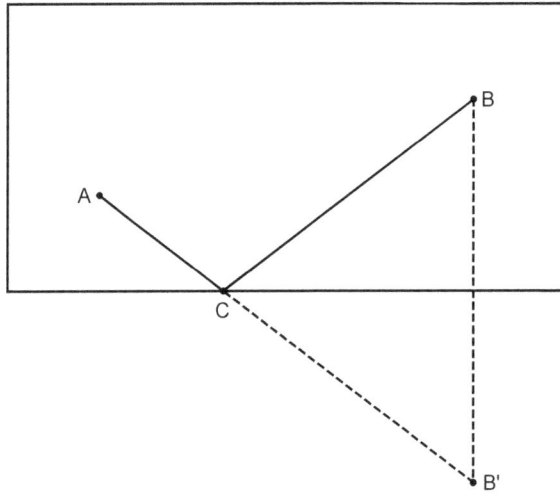

Figure 2.104: Example, A.RE. 6

In **Figure 2. 105**, we would like to hit the ball B by the ball A and making two reflections on two sides of the table (two caroms).

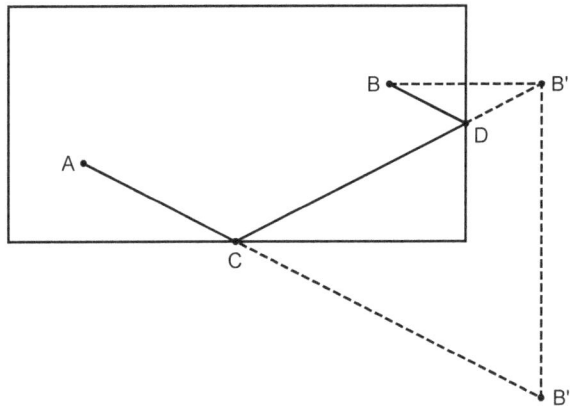

Figure 2.105: Example, A.RE. 6

In **Figure 2. 106**, we would like to hit the ball B by the ball A and making three reflections on three sides of the table (three caroms).

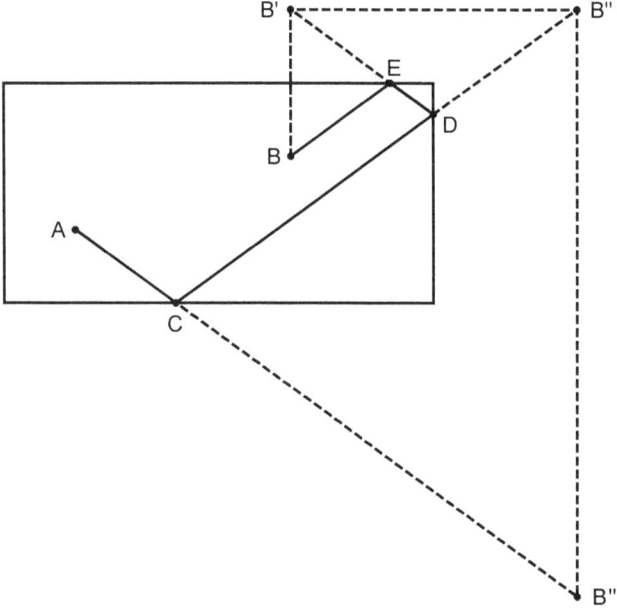

Figure 2.106: Example, A.RE. 6

Some solutions (not all) are depicted and can be explained by Heron's principle (the angle of incidence must be equal to the angle of reflection).

Next, we consider a circle with center O and two points A and B, such that there is a tangent of the circle that splits the plane into two closed half-planes, one of which contains the circle and the other half plane contains the two points. See **Figure 2. 107**.

The question is to find a point P on the circle such that the path from A to P and then B is minimum. To find this point is not easy, but we can easily find a necessary condition that must satisfy, based on Heron's principle. This condition is the following:

"*The tangent to the circle at the point P must make equal angles with the straight segments AP and BP.*"

That is, as in **Figure 2. 107**, the tangent line xy to the circle at P must satisfy the equality of (non-oriented) angles:

$$\widehat{xPA} = \widehat{yPB} \qquad \text{and so} \qquad \widehat{OPA} = \widehat{OPB},$$

and so the extension of the radius OP is the bisector of the angle $\widehat{AOB}$.

This is so, because if we pick another point S of the circle, we let T be the common point of AS and the tangent xy. Then under this condition, by Heron's

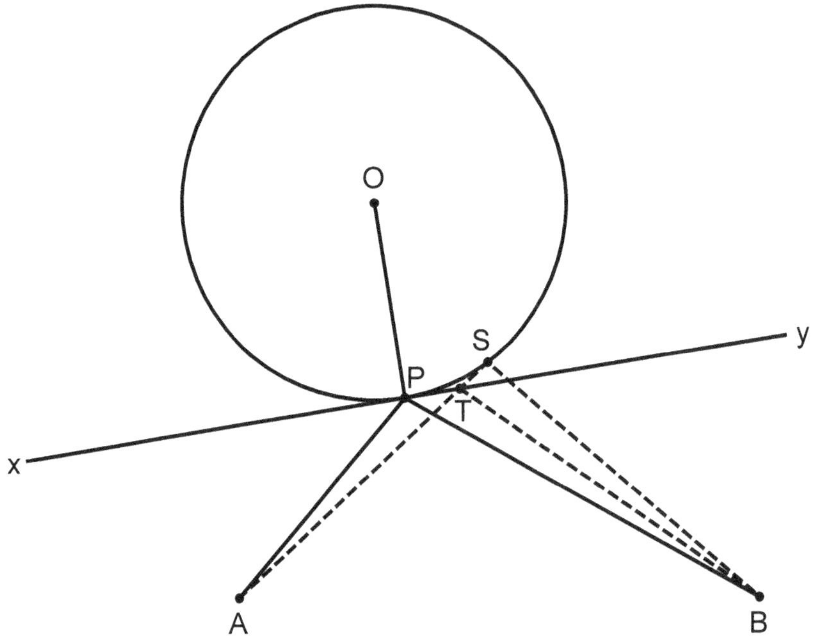

Figure 2.107: Example, A.RE. 6

principle, we have
$$AP + PB < AT + TB.$$

But, we also have
$$TB < TS + SB$$

by the triangle inequality. So,

$$AP + PB < AT + TB < AT + TS + SB = AS + SB.$$

[This principle was used by Fermat and Torricelli in order to find the point S in **Example 2.6.10, Application 6**. The solution presented there is due to Steiner.

Here, finding the point P analytically, leads to higher degree equation(s) that lead(s) to non-constructible numbers, in general. Only special cases are constructible, e.g., the two easy cases in which: (1) both points are on a tangent line of the given circle, or (2) they are equidistant from the center O. (Solve the problem in both cases!) But, the necessary condition we have exposed above, is a good step forward. Lagrange[11] multipliers can be an analytic method. We can also use trigonometry since we know the sides of the triangle OAB, the radius OP and the above necessary condition.]

---

[11] Joseph-Louis Lagrange, Italian-French mathematician, 1736-1813.

Remark: *Three properties of the orthic triangle*: (Decipher **Figure 2.108**.)

(1) Using the properties of the inscribable quadrilaterals, we prove that the heights of a scalene triangle are the angle bisectors of its orthic triangle. (Give a proof. Also, check what happens with a right or an obtuse triangle.)

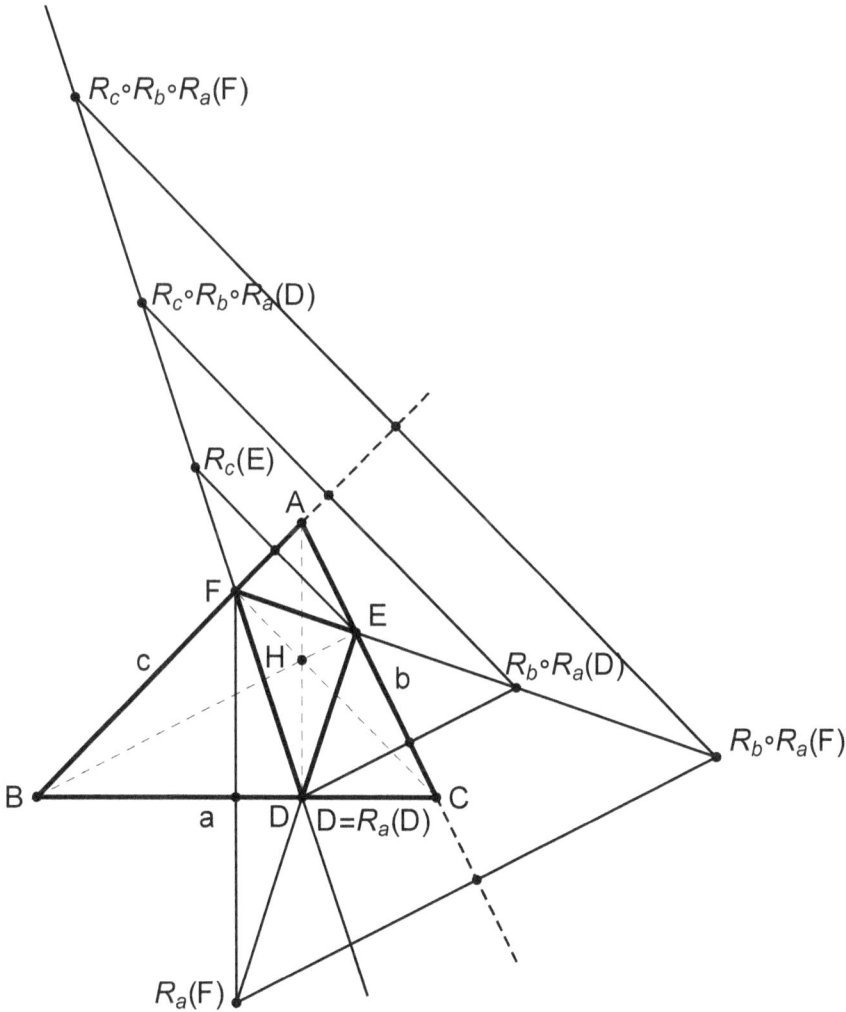

Figure 2.108: **Example, A.RE. 6**

(2) Using (1) and **Heron's principle**, we can give another proof of the result in **Example 2.6.17 of Application 5**. That is, if a triangle ABC is scalene, then its orthic triangle DEF, where $D \in BC := a$, $E \in CA := b$, and $F \in AB := c$, is the triangle inscribed in ABC with the smallest perimeter. In

this case, at first we keep a side fixed, e.g., EF, and we want to find D such that $ED + DF$ is minimum. By Heron's principle and the above property of the heights, this is obtained when D is the foot of the height AD on BC. Similarly for the other vertices E and F.

(3) Along these lines, we prove that the composition of the reflections in the sides a, b, and c of ABC, $R_c \circ R_b \circ R_a$, is the *glide* whose axis is the line DF and the length of its vector is $DE + EF + FD$, which is perimeter of the orthic triangle.

Notice that D, $R_c \circ R_b \circ R_a(D)$, F, and $R_c \circ R_b \circ R_a(F)$ are on the same straight line and so the line DF is fixed under $R_c \circ R_b \circ R_a$. Also,

$$D[R_c \circ R_b \circ R_a(D)] = DE + EF + FD = F[R_c \circ R_b \circ R_a(F)].$$

Now, the claims follow.

(Commute the reflections and prove analogous results.)

▲

For further results, applications, and examples study and solve the exercises that follow.

Chapter 3

Exercises on Isometries

We abbreviate "straight line", "-es" with "s.l.", "s.ls"; "geometrical locus", "-es" with "g.l.", "g.ls"; and "if and only if" with "iff".

3.1 Parallel Translations

3.1.1 General

1. Prove that a parallel translation maps a tangent to a circle to a tangent of the image circle.

2. Prove that two equal circles can be mapped onto each other by a parallel translation.

3. Prove that two parallel s.ls can be mapped onto each other by infinitely many parallel translations.

4. Find subsets of the plane that are set-wise fixed by a parallel translation (not listed in the text).

3.1.2 Geometrical Constructions

1. Construct a circle of a given radius, passing through a given point, and defining a chord of given length on a given line. Check the number of solutions.

2. Construct a quadrilateral if the three sides and the two angles adjacent to the fourth side are given.

3. Construct a straight segment equal to a given segment, parallel to a given s.l. and whose endpoints are on two given circles or two given s.ls, or one is on a given s.l. and the other on a given circle.

4. Construct a triangle ABC given the midpoints of the sides AB and AC and the vertices B and C are points of a given s.l and a given circle, respectively.

5. Construct a trapezium, given its diagonals, the angle of the diagonals and one of its sides.

6. Construct a s.l. of given direction and intersecting two circles in such a way that the two chords thus defined have a given sum, or a given difference.

7. We take two points A and B, one in each region outside the tape defined by two parallel lines l and l' ($l \parallel l'$). Find the points $P \in l$ and $P' \in l'$ such that the s.l. PP' has a given direction (that is, $PP' \parallel k$, with k a given s.l.) and one of the following conditions holds.

   (a) $AP = BP'$.

   (b) $AP + PP' + P'B$ is minimum.

8. Construct a s.l. parallel to a given s.l. and cutting on two given circles chords with given sum. Check the number of solutions.

3.1.3   Geometrical Loci

1. Find the g.l. of the points of tangency of the s.ls which are parallel to a given s.l. and are tangent to the circles of given radius and pass through a given point.

2. Consider a circle and two fixed points A and B. A point C moves on the circle. Find the g.l. of the point D if $ABCD$ is a parallelogram.

3. Consider all the isosceles triangles ABC whose vertex B is fixed, the vertex C is a point of a given s.l. Bx, and its circumscribed circle has a given radius. Find the g.l. of the vertex A.

4. Circle of constant radius but variable center passes through a given fixed point. At each position of the circle draw tangents parallel to a given s.l. Find the g.l. of the points of tangency.

5. We take a circle and a chord AB on it. Let P be a variable point on the circle. Find the g.ls of:

   (a) The orthocenter H of the triangle PAB.

   (b) The common points of the circles $C[P, PH]$ and $C[H, HP]$.

6. Consider two fixed points A and D and two s.ls l_1 and l_2 through A. A variable circle passing through A and D intersects l_1 and l_2 at the points B and C, respectively. Find the g.l. of the orthocenter of the triangle ABC.

3.2 Rotations

3.2.1 General

1. Prove that a rotation maps a tangent to a circle to a tangent of the image circle.

2. Prove that two equal circles can be mapped onto each other by infinitely many rotations. Find them all!

3. Prove that two s.ls can be mapped onto each other by infinitely many rotations of the same rotation angle. (Examine the case of non parallel lines and the case of parallel lines separately.)

4. Find subsets of the plane that are set-wise fixed by a rotation (not listed in the text).

5. Consider an isosceles trapezium $ABCD$, where $AB \parallel CD$. Find the rotation that maps $\overrightarrow{AD}$ to: (a) $\overrightarrow{BC}$.   (b) $\overrightarrow{CB}$.

6. Let xOx' and yOy' be two axes through O and having unit vectors of equal length. Take two points $A \in xOx'$ and $A' \in yOy'$ such that $OA \neq OA'$ fixed. Consider variable points $B \in xOx'$ and $B' \in yOy'$ such that $\overline{AB} = \overline{A'B'}$.
   Prove that all the circles (OBB') pass through a second fixed point lying on the bisector of one of the angles of the two axes.

7. On the sides AB and AC of a triangle ABC consider two points P and Q such that $BP = CQ$. Prove that the perpendicular bisectors of the straight segments PQ pass through a fixed point.

8. Consider two directly equal triangles ABC and $A'B'C'$ in the plane. $(AB = A'B', BC = B'C', CA = C'A'.)$ Prove the equality of the angles $\measuredangle(AB, A'B') = \measuredangle(BC, B'C') = \measuredangle(CA, C'A')$.

9. Consider a triangle such that its convex angle $\measuredangle(\overrightarrow{AB}, \overrightarrow{AC})$ has the positive orientation in the plane. On the sides BC, CA, AB construct equilateral triangles lying in the exterior of the triangle ABC. Let X, Y Z be the circumcenters of these equilateral triangles, respectively. Prove that the composition of the rotations $R_{[X,-\frac{2\pi}{3}]} \circ R_{[Y,-\frac{2\pi}{3}]} \circ R_{[Z,-\frac{2\pi}{3}]} = I$ is the identity.

10. Consider two equal circles $C[O,r]$ and $C[O',r]$ externally tangent at a point A. Point P traces the circle $C[O,r]$ and point P' traces the circle $C[O',r]$, in such a way that $\measuredangle(OP, O'P') = \dfrac{\pi}{2}$.

   (a) Prove that the perpendicular bisectors of the straight segments PP' pass through the vertex D of the right isosceles triangle $OO'D$ $\left(\widehat{D} = \dfrac{\pi}{2}\right)$, which is fixed point. Also, D lies on the common tangent s.l. at A.

(b) If E is the symmetrical of D in the s.l. OO' and $Q = R_{[E,-\frac{\pi}{2}]}(P)$, prove that the segments $P'Q$ are diameters of the circle $C[O',r]$ and so they pass through the fixed point O'.

11. Prove directly that the composition of two symmetries about two centers O and Q (half-turns about the points O and Q) is a parallel translation by the vector $2\overrightarrow{OQ}$.

12. Prove that a bounded plane figure cannot have two different centers of symmetry. Give examples of unbounded figures that admit more than one centers of symmetry.

13. Prove that the composition of the half-turns with centers three points A, B, C is a half-turn with center a point D such that $ABCD$ is a parallelogram.

14. (a) Consider a triangle ABC. Construct a square $ABC'K$ in the exterior of the triangle. Extend the height AD of the triangle from the side of A and pick on it the point S such that, $AS = BC$. Show that the triangle $C'BC$ is mapped onto BAS by the composition of the parallel translation $T_{\overrightarrow{BA}}$ followed by the rotation $R_{[A,\frac{\pi}{2}]}$. Then conclude that $BS \perp CC'$.

(b) We next construct a square $ACB'L$ in the exterior of the triangle. Prove that the s.l. from B and perpendicular to CC' and the s.l. from C and perpendicular to BB' intersect on the s.l. of the height AD of the triangle ABC.

(**Vecten's**[1] **Theorem**, published in 1817 in "les Annales de Gergonne"[2].)

15. Prove that the transformation $F = T_{-\overrightarrow{u}} \circ R_{[O,\phi]} \circ T_{\overrightarrow{u}}$ simplifies to $F = R_{[Q,\phi]}$, where Q is defined by $\overrightarrow{OQ} = -\overrightarrow{u}$.

16. Consider a triangle ABC and the three rotations $R_{[A,\frac{2\pi}{3}]}$, $R_{[B,\frac{2\pi}{3}]}$, and $R_{[C,\frac{2\pi}{3}]}$. Suppose a vector $\overrightarrow{DE} \neq \overrightarrow{0}$ is mapped to $\overrightarrow{D_1E_1}$ by the first rotation; the vector $\overrightarrow{D_1E_1}$ is mapped to $\overrightarrow{D_2E_2}$ by the second rotation; and the vector $\overrightarrow{D_2E_2}$ is mapped to $\overrightarrow{D_3E_3}$ by the third rotation. Prove:

(a) Oriented angle $\angle(\overrightarrow{DE}, \overrightarrow{D_3E_3}) = 2\pi \equiv 0$, (mod 2π).

(b) The composition $R_{[A,\frac{2\pi}{3}]} \circ R_{[B,\frac{2\pi}{3}]} \circ R_{[C,\frac{2\pi}{3}]}$ is equal to a parallel translation and find its vector.

(c) Prove that the composition in (**b**) is the identity iff ABC is equilateral.

17. (a) Prove that the composition of three half-turns is a half-turn.

(b) Consider a triangle and find the composition of the three half-turns with centers the midpoints of its sides.

---

[1] Vecten M., French mathematician (high school teacher), early 19th-century.
[2] Joseph Diez Gergonne, French mathematician, 1771-1859.

18. (a) Prove that the composition of four half-turns with distinct centers or otherwise, is a parallel translation and find its vector. When is this parallel translation the identity (i.e., its vector is the zero vector)?

(b) Consider a polygon $A_1 A_2 A_3 \ldots A_n$ with n even. Prove that the composition of the half-turns with centers the midpoints of its sides $A_1 A_2$, $A_2 A_3$, ..., $A_{n-1} A_n$ in this order (or any successive order) is equal to the identity.

(c) If in (b), n is odd, then the composition is the half-turn with center the vertex A_1.

3.2.2 Geometrical Constructions

1. Construct an isosceles triangle given its vertex A, the angle $\hat{A}$ and the vertices of the basis are points of:

(a) Two given circles.   (b) Two given s.ls.

(c) A given s.l. and a given circle.

2. Consider two equal circles $C[O, r]$ and $C[Q, r]$ and two fixed points $A \in C[O, r]$ and $B \in C[Q, r]$. Construct a straight segment ST of a given length such that $S \in C[O, r]$ and $T \in C[Q, r]$ and the oriented arcs $\overset{\frown}{AS}$ and $\overset{\frown}{BT}$ are directly equal.

3. Construct a square whose three of the vertices are points of three given parallel s.ls.

4. Construct an equilateral triangle whose vertices are points of:

(a) three concentric circles.   (b) three parallel s.ls.

5. Let O be the fixed center of a variable square $ABCD$ such that the s.l. AB is always tangent to a given circle. Find the four fixed circles that are tangent to the s.ls of the other sides of the square.

6. Construct a square $ABCD$ with center a given point, the s.l. AB being tangent to a given circle and the s.l. BC passing through a given point.

7. Consider a triangle ABC, such that $AB > AC$ and $\angle(AB, AC) = \dfrac{\pi}{3}$. Construct a point P on the line AB and a point Q on the line AC such that $BP = CQ$ and $2PQ = BC$.

8. Construct a square inscribed in a given parallelogram and whose vertices are on the sides of the parallelogram or their extensions.

9. Consider three parallel s.ls $l_1 \parallel l_2 \parallel l_3$ and a point $A \in l_1$. Construct an equilateral triangle ABC such that $B \in l_2$ and $C \in l_3$.

10. Construct a straight segment whose midpoint is given, and the endpoints are: (a) Points of two given s.ls.   (b) Points of two given circles.

(c) One is a point of a given circle and the other is a point of an given s.l.

11. Consider a s.l. l and a circle $\mathcal{C}$ and a point A. Construct an equilateral triangle ABC such that $B \in l$ and $C \in \mathcal{C}$.

12. In a given quadrilateral inscribe a parallelogram if its center is given.

13. Construct a pentagon given the midpoints of its sides. (Find two methods of construction.)

14. Construct a heptagon given the midpoints of its sides.

3.2.3   Geometrical Loci

1. Suppose that an equilateral triangle has a fixed vertex and a second vertex traces a given s.l., or a given circle. Find the g.l. of the third vertex in either case.

2. Consider two equal circles $C[O, r]$ and $C[O', r]$, fixed points $A \in C[O, r]$ and $A' \in C[O', r]$, and variable points $P \in C[O, r]$ and $P' \in C[O', r]$ such that the oriented arcs $\overset{\frown}{AP}$ and $\overset{\frown}{A'P'}$ are directly equal.

   (a) Find the center of the rotation that maps $\overset{\frown}{AP}$ to $\overset{\frown}{A'P'}$.

   (b) Decompose this rotation into a parallel translation and a rotation with center O'.

   (c) Find the g.l. of the midpoint of PP'.

   (d) Redo the previous questions if $\overset{\frown}{AP}$ and $\overset{\frown}{A'P'}$ are oppositely equal.

3. Consider a circle and a fixed point A. For any point B on the circle we construct an isosceles triangle ABC. Find the g.l. of the points C.

4. On the sides of an angle $\angle(Ox, Oy)$ we take two points P and P' such that $OP + OP' = c$, given length. Find the g.l. of the centers of the circumscribed circles of the triangles OPP'.

5. Consider two rotations $R_{[O_1, \phi_1]}$ and $R_{[O_2, \phi_2]}$. Given a point P, we let $R_{[O_1, \phi_1]}(P) = P_1$ and $R_{[O_2, \phi_2]}(P_1) = P_2$. Find the g.l. of the points P under the condition that $P_1 P_2 = \mu$, given length.

6. Find the g.l. of the centers of all rotations that map a given s.l. to another given s.l.

7. Consider a half-circle of diameter AB and a point C on it. We consider the points D on the half-line AC for which $AD = BC$. Find the g.l. of the points D.

8. Consider two fixed points P and Q. Find the g.l. of the points R, whose images R' under the rotation $R_{[P, \frac{\pi}{2}]}$ are on the line QR.

3.3 Reflections and Glides

3.3.1 General

1. Prove that a reflection in a line maps a tangent to a circle to a tangent of the image circle.

2. Prove that two equal circles can be mapped onto each other by a reflection.

3. Prove that two non-parallel s.ls can be mapped onto each other by two line reflections. If the s.ls are parallel and different, then only one line reflection maps one onto the other.

4. Find subsets of the plane that are set-wise fixed by a reflection or a glide (not listed in the text).

5. Prove that two parallel s.ls can be mapped onto each other by infinitely many glides.

6. Two figures are mapped onto each other by an opposite isometry which has a fixed point. Prove that they are reflections of each other in an axis through the fixed point.

7. Prove that if a figure has two different parallel axes of symmetry then:

   (a) The figure is unbounded.

   (b) The figure admits infinitely many axes of symmetry, parallel to the two given axes.

   (c) Give examples of such figures other than the plane, a half-plane, a s.l., or a tape (closed, open, or half closed and half-open) of parallel s.ls.

8. Suppose a figure has two axes of symmetry whose angle is $\frac{\pi}{n}$, for some $n = 2, 3, 4, \ldots$. (You may or may not assume that n is the maximum natural number with this property.) Prove:

   (a) There are n different rotations under which the given figure is invariant.

   (b) The figure admits n axes of symmetry all of which pass through the same point.

   (c) If $n = 2k$, even, then the axes in (b) appear in k perpendicular pair and their common point is center of symmetry of the given figure.

9. (a) Prove that if a figure different from the whole plane has an axis of symmetry in every direction, then this figure is a circle, or an open or a closed disc.

   (b)** Prove that if a figure has two axes of symmetry whose angle θ is a rational number, then this figure is a circle, or an open or a closed disc.

10. Consider two glides $G_1 = T_{\vec{u}} \circ R_l$ and $G_2 = T_{\vec{v}} \circ R_m$. Prove:

    (a) If $l \parallel m$ ($l = m$ is a subcase), then $G_2 \circ G_1 = T_{\vec{u}+\vec{v}+\vec{w}}$, where $w = 2\overrightarrow{lm}$. What is $G_1 \circ G_2$?

    (b) If $l \nparallel m$, then $G_2 \circ G_1$ is a rotation. Find its center and its angle. Do the same thing for $G_1 \circ G_2$.

11. Investigate and completely simplify the composition, in either order, of a glide with:

    (a) A parallel translation.   (b) A rotation.   (c) A reflection.
    (Indicate any special subcases in each of the above cases, if they exist.)

12. Investigate and simplify the composition of three glides.

13. Consider the parallel translation $T_{\vec{u}}$, $\vec{u} \neq \vec{0}$, the rotation $R_{[O,\phi]}$, $\phi \neq 0$, (mod 2π), and the reflection R_l in a s.l. l. Consider the six compositions of any two transformations (in either order) taken from these ones. Reduce these compositions to compositions of minimum number of reflections.

14. Consider a triangle ABC and a point P of its circumscribed circle.

    (a) We draw chords of the circumscribed circle $PA' \parallel BC$, $PB' \parallel CA$, and $PC' \parallel AB$. Prove that the triangles ABC and $A'B'C'$ are symmetrical in an axis.

    (b) Prove the same thing if the chords are drawn perpendicular to the sides of the triangle ABC.

15. Consider a s.l. l and two points A and B. Find a point P on l such that l bisects one of the angles $\angle PA, PB$. Prove that the sum $PA + PB$ is minimum if the points A and B are on the same side of l, otherwise the absolute value of the difference $|PA - PB|$ is maximum.

16. Consider a triangle ABC such that the oriented angle $\angle(\overrightarrow{AB}, \overrightarrow{AC})$ has the positive orientation in the plane. Construct two isosceles triangles AOB and $AO'C$ which are right at O and O' and lie in the exterior of the triangle.

    (a) Prove that the composition of the rotations $R_{[O,\frac{\pi}{2}]}$ and $R_{[O',\frac{\pi}{2}]}$ is a rotation that maps B to C. Conclude then that the center of this composition is the midpoint M of BC.

    (b) Prove that the triangle MOO' is isosceles and right at M.

17. (a) Consider a right triangle ABC ($\widehat{A} = 90^o$, and $a := BC$, $b := CA$ and $c := AB$), and let AD be its altitude on the hypotenuse a. Prove that $R_c \circ R_b \circ R_a = R_b \circ R_c \circ R_a$ is the glide with axis the s.l. AD and vector $\vec{u} = 2\overrightarrow{DA}$.

    (b) Commute the three reflections and prove analogous results in each case.

    (c) Draw and prove analogous conclusions, if $\widehat{A} > 90^o$.

18. Consider three points A, B, C in the plane. Let

$$\phi_1 = 2\angle(AC, AB), \quad \phi_2 = 2\angle(BA, BC), \quad \text{and} \quad \phi_3 = 2\angle(CB, CA).$$

Use decompositions with reflections to prove that the composition of the rotations

$$R_{[A,\phi_1]} \circ R_{[B,\phi_2]} \circ R_{[C,\phi_3]} = I, \quad \text{the identity.}$$

19. Consider an orthogonal parallelogram $ABCD$.

(a) Prove that

$$F := R_{CD} \circ R_{BC} \circ R_{AB} = T_{2\overrightarrow{BC}} \circ R_{BC}, \quad \text{(which is a glide).}$$

(b) Let $P' = F(P)$. Find the projection of $\overrightarrow{PP'}$ on BC.

20. The figure $\mathcal{F}_2$ is the image of a figure $\mathcal{F}_1$ under a rotation and the figure $\mathcal{F}_3$ is the image of $\mathcal{F}_2$ under a rotation. Prove that the three figures are the symmetrical figures of the same figure in three s.ls.

21. The image of the figure $\mathcal{F}_1$ under a parallel translation is the figure $\mathcal{F}_2$ and the image of the figure $\mathcal{F}_2$ under a rotation is the figure $\mathcal{F}_3$. Prove that the three figures are the symmetrical of the same figure in three s.ls.

22. Let T be an involutory transformation and f an invertible transformation of the plane. Prove that

$$f^{-1} \circ T \circ f \quad \text{and} \quad f \circ T \circ f^{-1}$$

are involutory transformations.

23. Let T_1 and T_2 be two involutory transformations. Prove that

$$T_1 \circ T_2 \quad \text{(or} \quad T_2 \circ T_1)$$

is involutory iff T_1 and T_2 commute.

24. Any involution G on the plane satisfies $G \circ G = I$ (identity) and so $G \circ G$ is continuous. Find an involution of the plane which is not continuous (and therefore it is not an isometry).

25. Consider a s.l. l and a point O in the plane. Prove that the composition $R_l \circ R_{[O,\frac{\pi}{2}]}$, or $R_{[O,\frac{\pi}{2}]} \circ R_l$, (reflection with half-turn) is involutory iff $O \in l$.

26. (a) Consider any two oppositely equal triangles ABC and $A'B'C'$ in the plane. Prove that the bisectors of the three angles of the pairs of the s.ls of the corresponding sides $\angle(BC, B'C')$, $\angle(CA, C'A')$ and $\angle(AB, A'B')$ are parallel.

(b) Prove the same result even if the two triangles are oppositely similar.

27. Consider an opposite isometry S and let $A' = S(A)$ for any point $A \in \mathcal{P}$. Prove:

(a) The midpoints of all segments AA' are on the same s.l. l.

(b) The projections of the segments AA' on the s.l. l are all equal.

(c) S is a glide with axis l and translation vector the vector of the projection of a vector $\overrightarrow{AA'}$ on l.

(d) S can be written as a composition of three reflections in three lines. It can also be written as the composition of a reflection in a line and a parallel translation.

28. Consider two equal and non-parallel straight segments AB and $A'B'$ in the plane. Let O be the common point of the s.ls AB and $A'B'$ and suppose $OA < OB$ and $OA' < OB'$. If $OA \neq OA'$, then prove that the opposite isometry that maps A to A' and B to B' has no fixed point and so it is a glide. Find its axis and its vector.

Otherwise, i.e., $OA = OA'$, prove that it is a symmetry and find its axis.

What happens if the equal segments AB and $A'B'$ are parallel?

29. Prove:

(a) Any canonical (regular) n-gon, $(n \geq 3)$ is set-wise invariant by n reflections in n axes of symmetry of the canonical n-gon and n rotations with common center the center of the canonical n-gon and angles

$$\frac{2\pi i}{n}, \quad i = 0,\ 1,\ 2,\ \ldots\ ,n-1.$$

These isometries make a non-commutative group of order $2n$ under composition. (**Dihedral group.**)

(b) Any circle is set-wise invariant by infinitively (uncountably) many reflections and infinitely (uncountably) many rotations (including the identity). These isometries make a group of infinite order under composition.

(c) Any orthogonal parallelogram that is not a square is set-wise invariant under two reflections and two rotations (including the identity). These isometries make a commutative group of order 4 under composition. (**Klein**[3] **four group.**)

30. Prove:

(a) Any subgroup of isometries of the plane that contains a non-trivial parallel translation or a glide must be of infinite order.

(b) A subgroup of isometries of the plane that contains a non-trivial rotation may be of finite or infinite order.

---

[3]Christian Felix Klein, German mathematician, 1849-1925.

(c) If a subgroup of isometries of the plane contains two non-trivial rotations that have different centers then:

(c1) If their angles are rational fractions of π, then the subgroup contains a non-trivial translation and so is of infinite order.

(c2) If at least one of the angles is not a rational fraction of π, then the subgroup is of infinite order.

(d) If a subgroup of isometries of the plane consists of rotations of the same center, give the conditions under which this subgroup has finite or infinite order.

(e) If a subgroup of isometries of the plane contains a non-trivial rotation and a reflection whose axis does not pass through the center of the rotation, then the subgroup contains a glide and so is of infinite order.

(f) In any finite subgroup of rotations of the plane, all rotations have the same center and the subgroup is cyclic. So, there is a rotation of the smallest positive angle about the common center that generates the whole subgroup.

(g) If a finite subgroup of isometries of the plane contains a rotation and a reflection, then:

(g1) All possible rotations have the same center.

(g2) The axes of all possible reflections pass through the common center of the rotations.

(g3) The number of all possible rotations is equal to the number of all possible reflections.

(g4) The subgroup is a dihedral group.

[The final two results (f) and (g) are called **Leonardo Da Vinci's**[4] **Theorem.**]

31. The points P of a straight line are mapped by a plane isometry onto the points P' of another straight line. Prove that the midpoints of the segments PP' either coincide, or the are distinct and collinear. Distinguish the cases in which these midpoints coincide. (**Hjelmslev's**[5] **Theorem.**)

    (Hint: Prove the claim for each basic isometry, half turn and glide. Only the rotation poses little difficulty. The others are easy.)

3.3.2 Geometrical Constructions

1. In each of the following two cases construct a straight segment whose perpendicular bisector is a given line and whose endpoints are on:

   (a) Two given s.ls.

   (b) Two given circles.

---

[4]Leonardo Da Vinci, Italian multi-talented sage, 1452-1519.
[5]Johannes Trolle Hjelmslev, Danish mathematician, 1873-1950.

2. Construct a triangle ABC if the sides AB and AC and the difference of the angles $\widehat{B}$ and $\widehat{C}$ are given.

3. Construct a triangle ABC if the vertices B and C, the foot of the altitude on the side BC, and the difference of the angles $\widehat{B}$ and $\widehat{C}$ are given.

4. Consider a s.l. l and two circles. Find point $P \in l$ such that l is the bisector of the angle formed by a tangent to one circle and a tangent to the other circle both drawn from P.

5. The given straight segments D_1E_1, D_2E_2, D_3E_3 of three different directions are the reflections of a non-given segment DE, in the lines of the sides of a triangle ABC. Construct the triangle.

6. Consider a s.l. l. In each of the three following cases, construct a square $ABCD$ such that B and D are on l and:

   (a) A is on a s.l. l_1 and C on another s.l. l_2.

   (b) A is on a s.l. and C is on a circle.

   (c) A is on a circle and C is on another circle.

7. Construct a triangle when the s.ls of the angle bisectors and a point of a side are given.

8. Consider two equal circles $\mathcal{C}$ and $\mathcal{D}$ and two points A and B on them, respectively. Construct a straight segment PQ with given length μ, with endpoints on the circles $P \in \mathcal{C}$ and $Q \in \mathcal{D}$ and the arcs $\overset{\frown}{BP}$ and $\overset{\frown}{BQ}$ are oppositely equal.

3.3.3 Geometrical Loci

1. Consider two fixed s.ls l_1 and l_2 intersecting at O, a point P, and a variable s.l. k through P. If the symmetrical s.ls of k in the lines l_1 and l_2 intersect at a point A, find the g.l. of A.

2. The vertices B and C of a triangle ABC lie on a given s.l. and its ortho-center is given.

   (a) Find the g.l. of the circumcenter.

   (b) Construct the triangle if the midpoint of the side BC is also given.

3. Consider two equal circles $C[O, r]$ and $C[O', r]$ and two fixed points A and A' on them, respectively. Let P and P' be variable points on these circles, respectively, such that $\measuredangle(AO, AP) = -\measuredangle(A'O', A'P')$.

   (a) Prove that the g.l. of the midpoints of the segments PP' is contained in a s.l. l and find this g.l.

   (b) Let Q and Q' be the points where the s.ls AP and $A'P'$ intersect the s.l. l. Prove that the vector $\overrightarrow{QQ'}$ is constant.

Chapter 4

Similarities

4.1 Definitions and Basic Similarities

Definition 4.1.1 *A function on the Euclidean plane*

$$S \; : \; \mathcal{P} \longrightarrow \mathcal{P}$$

*is called a (**plane**) **similarity** if it satisfies the property:*

$$\forall \; A, \; B \quad points \; of \; \; \mathcal{P}, \quad it \; holds \quad d[S(A), S(B)] = \lambda \cdot d(A, B),$$

*where $\lambda > 0$ fixed real number. This λ is called the **ratio of the similarity** or simply **similarity ratio**.*

See **Figure 4.1**.

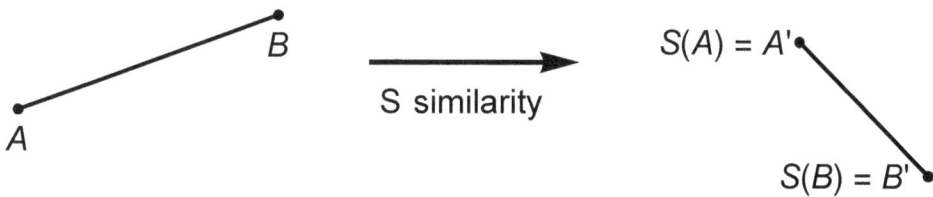

Figure 4.1: Similarity of ratio $\lambda = \dfrac{A'B'}{AB}$

In this context, instead of function, we may use the terms: map, mapping, and transformation.

4.1.1   Basic Similarities

To show that similarities exist and thus their definition is not vacuous, we first study the **basic similarities**. These are the following:

(1) Isometries. All plane isometries (studied in **Chapter 2**) are similarities with ratio $\lambda = 1$. This is obvious from the definitions of isometries and similarities.

(2) Homotheties. There are two kinds of a **homothety** (or **dilation** or **dilatation**) the positive and the negative, according to the following definitions:

Definition 4.1.2 *(a) Given a point $O \in \mathcal{P}$ and a positive real number $\lambda > 0$, we call **positive plane homothety** with **center** O and **positive ratio** $\lambda > 0$, the mapping*

$$H_{[O,\lambda]} \; : \; \mathcal{P} \longrightarrow \mathcal{P}$$

*defined as follows: $\forall \; A \in \mathcal{P}$, we define $H_{[O,\lambda]}(A) = A' \in \mathcal{P}$, such that, the point A' is on the ray OA (hence, O, A and A' are collinear and the points A and A' are on the same side of O) and $d[O, A'] = \lambda \cdot d(O, A)$, (or $OA' = \lambda \cdot OA$). See **Figure 4.2**.*

*(b) Given a point $O \in \mathcal{P}$ and a negative real number $\mu < 0$, we call $H_{[O,\mu]}$ **negative plane homothety** with **center** O and **negative ratio** $\mu < 0$ if $\forall \; A \in \mathcal{P}$, we define $H_{[O,\mu]}(A) = A' \in \mathcal{P}$, such that, the point A' lies on the ray opposite to OA (and so O, A and A' are collinear with O between A and A') and $d[O, A'] = |\mu| \cdot d(O, A)$, (or $OA' = |\mu| \cdot OA$). See **Figure 4.3**.*

*(c) For a homothety $H_{[O,\rho]}$, the two parts of this definition, **(a)** and **(b)**, can be joined in one by the condition*

$$\frac{\overline{OA'}}{\overline{OA}} = \rho \iff \overline{OA'} = \rho \cdot \overline{OA}, \quad \text{where} \quad \rho \neq 0 \quad \text{real number.}$$

Figure 4.2: **Positive homothety** $H_{[O,\lambda]}$, with ratio $\lambda = \dfrac{\overline{OA'}}{\overline{OA}}$

Figure 4.3: **Negative homothety** $H_{[O,\mu]}$, with ratio $\mu = \dfrac{\overline{OA'}}{\overline{OA}}$

Remarks: (1) It is an easy geometrical exercise to prove synthetically that any homothety $H_{[O,\rho]}$, $(\rho \neq 0)$ is a one to one and onto mapping of the plane. [Prove it now! A general proof will be given in **Section 4.3, Properties (2)** and **(3)**, since, as we shall prove in **item (e)** below, any homothety is a similarity. See also **item (h)**.]

(2) If $\rho = 0$, then every point $A \in \mathcal{P}$ is mapped to the point O, and so the mapping $H_{[O,0]}$ is constant $\{H_{[O,0]}(\mathcal{P}) = \{O\}\}$, and therefore not geometrically interesting.

Some Immediate Consequences of the Definition:

(a) If $\lambda = 1$, then $H_{[O,1]} = I$ is the **identity**.

(b) If $\mu = -1$, then $H_{[O,-1]} = R_{[O,\pi]}$ is the **half-turn or symmetry about** O. See **Figure 4.4**.

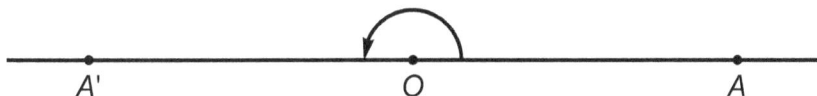

Figure 4.4: A negative homothety with ratio $\mu = -1$,
$$H_{[O,-1]} = R_{[O,\pi]}, \text{ is a half-turn}$$

(c) A negative homothety $H_{[O,\mu]}$, $(\mu < 0)$, is the composition of the positive homothety $H_{[O,\lambda=|\mu|=-\mu]}$ and the half-turn $R_{[O,\pi]}$. See **Figure 4.5**. This composition is commutative, that is,

$$H_{[O,\mu]} = R_{[O,\pi]} \circ H_{[O,\lambda=|\mu|=-\mu]} = H_{[O,\lambda=|\mu|=-\mu]} \circ R_{[O,\pi]}.$$

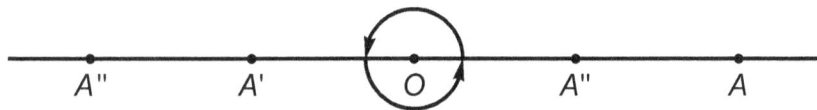

Figure 4.5: Negative homothety $\mu < 0$, $H_{[O,\mu]}(A) = (A')$,
$$H_{[O,\mu]} = R_{[O,\pi]} \circ H_{[O,\lambda=|\mu|=-\mu]} = H_{[O,\lambda=|\mu|=-\mu]} \circ R_{[O,\pi]}$$

In general, for any homothety (positive or negative) we have:

$$H_{[O,\mu]} = R_{[O,\pi]} \circ H_{[O,-\mu]} = H_{[O,-\mu]} \circ R_{[O,\pi]}.$$

(d) For any homothety $H_{[O,\rho]}$, the center O is a **fixed point**, that is, $H_{[O,\rho]}(O) = O$. All the straight lines through the center O are **set-wise fixed sets**.

If the ratio is positive ($\rho > 0$), set-wise **fixed sets** are also all the closed half-lines (rays), that start at the center O. (See previous **Figures 4.2, 4.3, 4.4** and **4.5**.)

If a homothety has two different fixed points, then $\rho = 1$ and the homothety is the identity.

(e) **Every homothety is a similarity with ratio the absolute value of the ratio of the homothety.**

To prove this, we consider a homothety $H_{[O,\rho]}$ and two points A and B in $\mathcal{P}$. We let A' and B' be their images under the homothety. (The image of O is $O' = O$, fixed point.)

Case 1: The points O, A and B are not on the same straight line. Then, this is also true for the points O, A' and B'. See **Figure 4.6**. Now, the triangles OAB and $OA'B'$ are similar because $\angle AOB = \angle A'OB'$ and $\dfrac{OA'}{OA} = \dfrac{OB'}{OB} = |\rho|$. Then we also have $\dfrac{A'B'}{AB} = |\rho|$ or $A'B' = |\rho| \cdot AB$, proving that the homothety is a similarity of ratio $|\rho| > 0$. Also, we get the necessary condition $AB \parallel A'B'$.

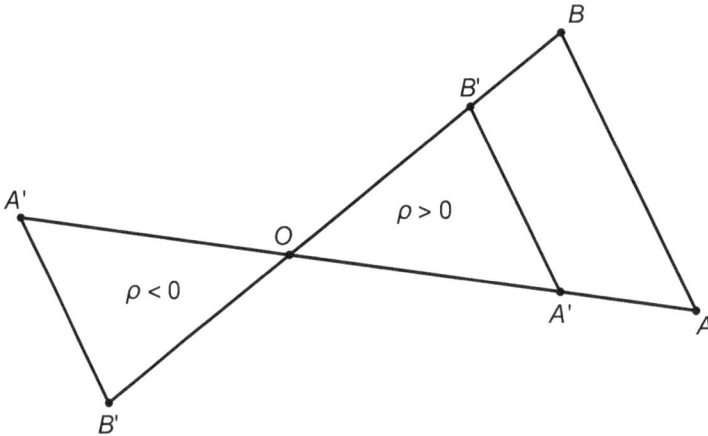

Figure 4.6: A homothety is a similarity

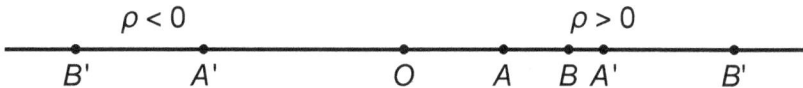

Figure 4.7: A homothety is a similarity

Case 2: The points O, A and B are on the same straight line l. See **Figure 4.7**. Hence, by the definition of a homothety, the points O, A' and B' are on l,

as well. Then, we have

$$|\rho| = \frac{OA'}{OA} = \frac{OB'}{OB} = \frac{-OA' + OB'}{-OA + OB} = \frac{A'B'}{AB}.$$

(f) Apart from the ratio of homothetic straight segments being $\dfrac{A'B'}{AB} = |\rho|$, [or $A'B' = |\rho| \cdot AB$], as found in (e), in the same way we also prove:

The homothetic of a vector $\overrightarrow{AB}$ is a vector $\overrightarrow{A'B'}$ such that

$$\overrightarrow{AB} \parallel \overrightarrow{A'B'} \quad \text{and} \quad \frac{\overrightarrow{A'B'}}{\overrightarrow{AB}} = \rho \ \left(\Longleftrightarrow \ \overrightarrow{A'B'} = \rho \cdot \overrightarrow{AB} \right).$$

Indeed: $\overrightarrow{OA'} = \rho \cdot \overrightarrow{OA}$ and $\overrightarrow{OB'} = \rho \cdot \overrightarrow{OB}$.   See **Figures 4.8** and **4.9**.

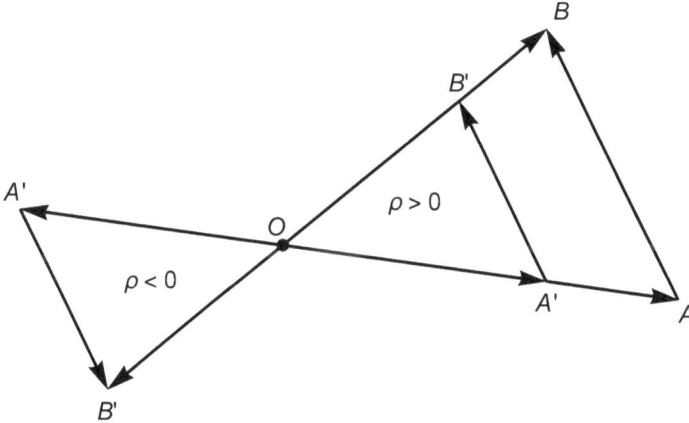

Figure 4.8: Homothetic vectors

Figure 4.9: Homothetic vectors

Then, $\overrightarrow{A'B'} = \overrightarrow{OB'} - \overrightarrow{OA'} = \rho \cdot \overrightarrow{OB} - \rho \cdot \overrightarrow{OA} = \rho \cdot (\overrightarrow{OB} - \overrightarrow{OA}) = \rho \cdot \overrightarrow{AB}.$

Reversing the arguments presented in (e) and (f), we conclude the following:

Result: *A homothety is determined by two non-zero, parallel (necessarily), corresponding vectors, $\overrightarrow{0} \neq \overrightarrow{AB} \parallel \overrightarrow{A'B'} \neq \overrightarrow{0}$. If $\overrightarrow{AB}$ is mapped to $\overrightarrow{A'B'}$, **the homothetic ratio is** $\rho = \dfrac{\overrightarrow{A'B'}}{\overrightarrow{AB}}$, and **the homothetic center O is the point** of intersection of the straight lines AA' and BB', in general. But:*

(1) If the parallel vectors are not equal $(\overrightarrow{AB} \neq \overrightarrow{A'B'})$ and on the same straight line, which is $AA' = BB'$ necessarily, then the center O is on this straight line and computed as shown above.

(2) If $\overrightarrow{AB} = \overrightarrow{A'B'}$, then $\rho = 1$ and the lines AA' and BB' are also parallel and the homothety is the parallel translation $T_{\overrightarrow{AA'}=\overrightarrow{BB'}}$, and so, its center O is the infinity point of the parallel lines AA' and BB'.

Now, we must state and prove the following useful **Lemma**:

Lemma 4.1.1 *If we know the ratio $\rho \neq 0$ of a homothety and a point A along with its image A' [so, (A, A') is a pair of corresponding points], then we can find the center O of the homothety.*

Proof (1) If $A = A'$ then the center $O = A = A'$ and so the homothety is $H_{[A,\rho]}$.

(2) If $\rho = 1$, then the homothety must be either the identity and so $A = A'$, or a parallel translation by the vector $\overrightarrow{AA'}$ (when $\rho = 1$ and $A \neq A'$) and so its center is the infinity point of the straight line AA'.

(3) Otherwise $A \neq A'$ and $0 \neq \rho \neq 1$. Then, the center O is on the straight line AA', the vector $\overrightarrow{AA'} \neq \overrightarrow{0}$ is known, and we have

$$\frac{\overrightarrow{OA'}}{\overrightarrow{OA}} = \rho = \frac{\rho}{1} \quad \text{and so,} \quad \frac{\overrightarrow{OA'} - \overrightarrow{OA}}{\overrightarrow{OA}} = \frac{\rho - 1}{1}.$$

$$\text{Hence,} \quad \frac{\overrightarrow{AA'}}{\overrightarrow{OA}} = \frac{\rho - 1}{1} \quad \text{and so,} \quad \overrightarrow{OA} = \frac{1}{\rho - 1}\overrightarrow{AA'}.$$

This relation determines O on the line AA', as in **Figures 4.8** and **4.9**.

Similarly, or more straightforward, we find

$$\overrightarrow{OA'} = \overrightarrow{OA} + \overrightarrow{AA'} = \left(\frac{1}{\rho - 1} + 1\right)\overrightarrow{AA'} = \frac{\rho}{\rho - 1}\overrightarrow{AA'}.$$

We can also use this relation to determine O on the line AA', as in **Figures 4.8** and **4.9**.

■

(g) *A homothety preserves oriented angles.* See **Figure 4.10**.

(1) Consider any positive homothety $H_{[O,\rho]}$, $(\rho > 0)$, and the angle $\angle(\overrightarrow{AB}, \overrightarrow{CD})$ of any two non-zero vectors $\overrightarrow{AB}$ and $\overrightarrow{CD}$. Let

$$\overrightarrow{A'B'} = H_{[O,\rho]}(\overrightarrow{AB}) \quad \text{and} \quad \overrightarrow{C'D'} = H_{[O,\rho]}(\overrightarrow{CD}).$$

By **(f)** above, we have $\overrightarrow{A'B'} = \rho\overrightarrow{AB}$ and $\overrightarrow{C'D'} = \rho\overrightarrow{CD}$. Since $\rho > 0$, the vectors $\overrightarrow{A'B'}$ and $\overrightarrow{C'D'}$ are of the same direction as the vectors $\overrightarrow{AB}$ and $\overrightarrow{CD}$, respectively. Therefore, $\angle(\overrightarrow{A'B'}, \overrightarrow{C'D'}) = \angle(\overrightarrow{AB}, \overrightarrow{CD})$.

(2) If now the homothety is negative, then, by **(c)** above, it is equal to a commutative composition of a half-turn and a positive homothety. Since the

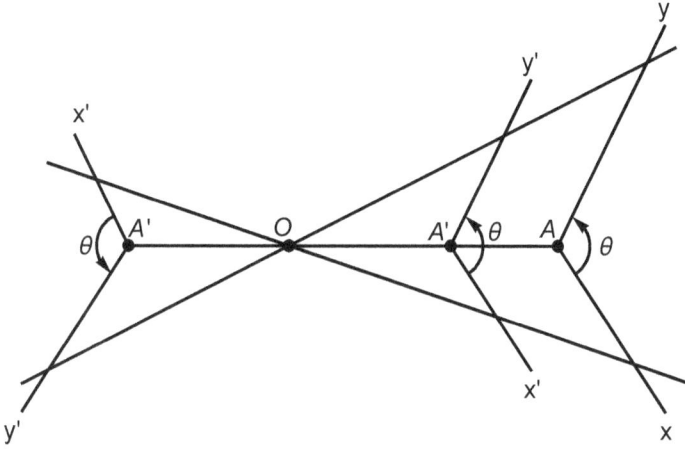

Figure 4.10: A homothety preserves oriented angles

half-turns and the positive homotheties preserve oriented angles, so does any negative homothety.

(h) Given any homothety $H_{[O,\rho]}$, $(\rho \neq 0)$, we can also consider the homothety $H_{[O,\frac{1}{\rho}]}$. The composition of these two homotheties in either order is the identity mapping. (Prove this directly from the definition!) Therefore, these two homotheties are inverse of each other. That is

$$H_{[O,\rho]}^{-1} = H_{[O,\frac{1}{\rho}]} \quad \Longleftrightarrow \quad H_{[O,\frac{1}{\rho}]}^{-1} = H_{[O,\rho]}.$$

(This also implies that any homothety is a one to one and onto function on the Euclidean plane $\mathcal{P}$.)

(i) A homothety $H_{[O,\rho]}$ maps:

(1) A straight line k to a parallel straight line $l = H_{[O,\rho]}(k)$, $(l \parallel k)$, and $k = l$ set-wise iff $O \in k$.

(2) A vector $\overrightarrow{u}$ to a parallel vector $\overrightarrow{v}$ of the same or opposite direction if $\rho > 0$ or $\rho < 0$, respectively, and the $\|\overrightarrow{v}\|$ is equal to $|\rho| \cdot \|\overrightarrow{u}\|$.

(3) A half line (open or closed) $\overrightarrow{Ax}$ to a parallel half line (open or closed respectively) $\overrightarrow{A'x'}$ of the same or opposite direction if $\rho > 0$ or $\rho < 0$, respectively.

(4) A straight segment to a straight segment and the interior of a straight segment to the interior of a straight segment.

(5) Since a homothety preserves oriented angles, $H_{[O,\rho]}$ is direct similarity and maps a triangle to a directly similar triangle with ratio of similarity $|\rho|$.

The proofs of these claims are either indicated above, (e)-(h), or are straightforward by using the so far presented material on homotheties. But, as we have proved analogous results for isometries in the previous chapter, we also prove them for similarities in **Section 4.3**, **Property (9)**, and therefore for homotheties, since, by **(e)** above, homotheties are similarities.

(j) **Circles and homotheties:** See **Figure 4.11** and fill in the details of the proof of the **Result** stated next, as a straightforward exercise.

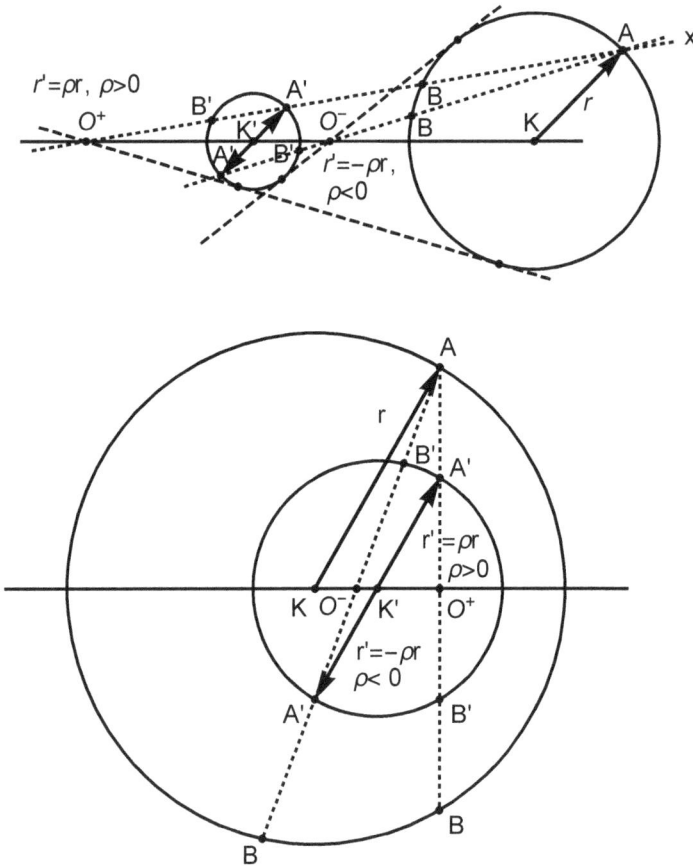

Figure 4.11: Homothetic circles. The homothetic ratio satisfies
$$\rho = \pm\frac{r'}{r} = \pm\frac{OK'}{OK} \quad \text{and} \quad \rho = \pm\frac{r'}{r} = \pm\frac{O^{\pm}K'}{OK}$$

Result: *A homothety $H_{[O,\rho]}$ maps a circle $C[K,r]$ to a circle $C[K',r']$ of radius $r' = |\rho|r$ and its center K' is on the ray OK, if $\rho > 0$, and on the opposite ray, if $\rho < 0$ and satisfies the relation $OK' = |\rho|OK$. So the center K' is the homothetic image of the center K.*

This result has several **consequences** some of which are stated next. Fill in the missing details in their proofs as straightforward exercises, using basic facts of the classical plane geometry and properties of similar triangles.

We consider any two corresponding vector-radii $\overrightarrow{KA}$ and $\overrightarrow{K'A'}$, such that $H_{[O,\rho]}(A) = A'$ and $H_{[O,\rho]}(K) = K'$. Necessarily, $\overrightarrow{KA} \parallel \overrightarrow{K'A'}$, these vectors are parallel and have: the same direction, if $\rho = \dfrac{r'}{r} > 0$; opposite direction, if $\rho = -\dfrac{r'}{r} < 0$. Then:

(1) The center of the homothety is the point of intersection of the center straight line KK' with a straight line AA' defined by two corresponding points A and A'. If $\rho > 0$, the homothety is positive and the center O^+ is closer to the center of the smaller circle but not between K and K'. Otherwise, $\rho < 0$, the homothety is negative and the center O^- is between K and K' and closer to the center of the smaller circle. (2) In particular, O is the intersection of KK' and the common tangents of the two circles, when common tangent(s) exist(s).

So, working backward, if we know the radii r and r' and the position of the centers K and K' of the circles $C[K, r]$ and $C[K', r']$, we can find the two homotheties that map $C[K, r]$ onto $C[K', r']$. The centers O^+ or O^- of the homotheties are on the straight line KK', and the ratios are $\rho = \pm\dfrac{r'}{r}$. The centers K and K' of the circles are corresponding homothetic points and must satisfy $OK' = |\rho|OK$. For convenience, we consider the line KK' as an axis and then we can compute O^+ and O^- as follow:

(1) For the positive homothety that maps the circle $C[K, r]$ to the circle $C[K', r']$, the ratio $\rho = \dfrac{r'}{r} > 0$. So, using **Lemma 4.1.1** and its **proof**, we find:

$$\overline{O^+K} = \frac{r}{r'-r}\overline{KK'} = \frac{r}{r-r'}\overline{K'K} \quad \text{and} \quad \overline{O^+K'} = \frac{r'}{r'-r}\overline{KK'} = \frac{r'}{r'-r}\overline{K'K}.$$

Each of these two relations determines the center O^+ of the positive homothety. (We could also use the similarity of the triangles KAO^+ and $K'A'O^+$ to prove this relation.)

(2) For the negative homothety that maps the circle $C[K, r]$ to the circle $C[K', r']$, the ratio is $\rho = -\dfrac{r'}{r} < 0$. So, by **Lemma 4.1.1** and its **proof**, we find:

$$\overline{O^-K} = \frac{-r}{r+r'}\overline{KK'} = \frac{r}{r+r'}\overline{K'K} \quad \text{and} \quad \overline{O^-K'} = \frac{r'}{r+r'}\overline{KK'} = \frac{-r'}{r+r'}\overline{K'K}.$$

Each of these two relations determines the center O^- of the negative homothety. (We could also use the similarity of the triangles KAO^- and $K'A'O^-$, to prove this relations.)

In this context, we also encounter the following convenient definition:

Definition 4.1.3 *We consider a homothety of two circles $C[K, r]$ and $C[K', r']$ whose center is O. Any homothetic ray Ox intersecting them, except for the tangent rays, intersects the two circles at two pairs of corresponding points, (A, A') and (B, B'), let us say. The corresponding points A and A', or B and B', are called* **homologous points with respect to** *O, and the points A and B', or B and A', (on the same ray Ox or its opposite) are called* **anti-homologous with respect to** *O. (The fixed point or center O of a homothety $H_{[O,\rho]}$ is sometimes called* **self-homologous point**.*)*

In **Figure 4.11** the pair of points (A, A') and (B, B') are homologous and the pair of points (A, B') and (B, A') are anti-homologous. The points of contact of the tangent homothetic rays with the circles $C[K, r]$ and $C[K', r']$ are both homologous and anti-homologous. There are several properties that homologous and anti-homologous points exhibit. We study many of these properties in the sequel and in the exercises (check). Also, the relation of homologous and anti-homologous points plays an important role in the transformation of inversion, which we study in **Chapter 6**. [See **Subsection 6.4.2** and **Remark (2)** after **Theorem 6.4.4**.]

In the case of homothetic circles, we also notice that

$$-\frac{\overrightarrow{O^- K}}{\overrightarrow{O^- K'}} = \frac{r}{r'} = \frac{\overrightarrow{O^+ K}}{\overrightarrow{O^+ K'}} \qquad \text{or} \qquad -\frac{\overrightarrow{O^- K'}}{\overrightarrow{O^- K}} = \frac{r'}{r} = \frac{\overrightarrow{O^+ K'}}{\overrightarrow{O^+ K}}.$$

So, *the points K, K', O^-, and O^+ written in this order, and lying on the same line as in Figure 4.11, form what we call a harmonic quadruple with ratio $\dfrac{r}{r'}$ or $\dfrac{r'}{r}$, respectively.*

The general definition of a harmonic quadruple is the following:

Definition 4.1.4 *We say that four points A, B, C, and D on the same straight line (or better on an axis) form a **harmonic quadruple**, if they satisfy the relation*

$$-\frac{\overline{CA}}{\overline{CB}} = \frac{\overline{DA}}{\overline{DB}} \left(= -\frac{\overline{AC}}{\overline{BC}} = \frac{\overline{AD}}{\overline{BD}} = etc. := \rho \right) \qquad \Longleftrightarrow$$

$$\frac{\overline{CA}}{\overline{CB}} \div \frac{\overline{DA}}{\overline{DB}} = -1 \left(= \frac{-\rho}{\rho} = \frac{\overline{CA}}{\overline{CB}} \cdot \frac{\overline{DB}}{\overline{DA}} \right).$$

*In such a quadruple, we call the value ρ **ratio of the harmonic quadruple** and the points C **and** D **harmonic conjugates with respect to the points** A **and** B.*

Notice: (a) If $\rho = 1$, then C is the midpoint of AB and $D = \infty$. (b) The equation that defines the harmonic quadruple implies: If C is between A and B, then B is between C and D, etc. (c) The equation is equivalent to

$$-\frac{\overline{AC}}{\overline{AD}} = \frac{\overline{BC}}{\overline{BD}} (:= \rho) \qquad \Longleftrightarrow \qquad \frac{\overline{AC}}{\overline{AD}} \div \frac{\overline{BC}}{\overline{BD}} = -1 \left(= \frac{\overline{AC}}{\overline{AD}} \cdot \frac{\overline{BD}}{\overline{BC}} \right).$$

Therefore the points A **and** B **are also harmonic conjugates with respect to the points** C **and** D.

In **Figure 4:12** we see two orders of four points that make two equivalent harmonic quadruples. We denote the first harmonic quadruple by

$$\{A, B; C, D\} = -1 = \frac{\overline{CA}}{\overline{CB}} \div \frac{\overline{DA}}{\overline{DB}} = \frac{\overline{CA}}{\overline{CB}} \cdot \frac{\overline{DB}}{\overline{DA}},$$

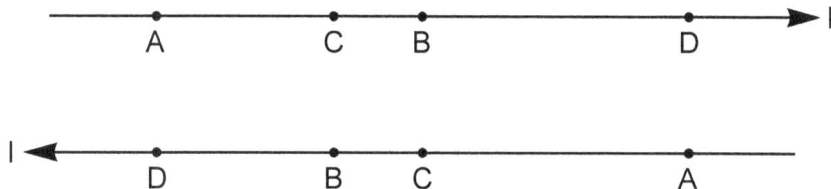

Figure 4.12: Harmonic quadruples $\{A, B; C, D\} = -1$ **on two axes** l's **of opposite directions**

(notice the orders of points) and the second equivalent harmonic quadruple by

$$\{C, D; A, B\} = -1 = \frac{\overline{AC}}{\overline{AD}} \div \frac{\overline{BC}}{\overline{BD}} = \frac{\overline{AC}}{\overline{AD}} \cdot \frac{\overline{BD}}{\overline{BC}}.$$

(Check the notation a book at hand uses, because this notation varies from book to book. Some properties, figures and construction of harmonic quadruples are presented in **Subsection 4.5.5.**)

Harmonic quadruples appear in many situations, e.g., the cross-ratio (or double ratio) of four points on an axis, as we shall see in **Definition 6.4.4**, and the applications that follow. Here, we remind the important theorem of the angle bisectors of a triangle, that we encounter in the basic plane geometry and we will invoke later in several situations. (Its proof is easy and read it in a relative textbook and draw a figure.)

Theorem 4.1.1 (Theorem of the angle bisectors of a triangle) *Let* P *and* Q *be the points of intersection of the bisectors of the interior and the exterior angle* $\widehat{A}$ *with the opposite side* $a = BC$ *of a triangle* ABC. *Then*

$$-\frac{\overline{PB}}{\overline{PC}} = \frac{\overline{QB}}{\overline{QC}} = \frac{AB}{AC} := \frac{c}{b} := \rho.$$

Hence, $\{B, P, C, Q\}$ *form a harmonic quadruple of ratio* $\rho = \dfrac{c}{b}$. *(AB* $= c$ *and* $AC = b$. *Notice:* $\rho = 1$ *iff* ABC *is isosceles, in which case* P *is the midpoint of* BC *and* $Q = \infty \Leftrightarrow AQ \parallel BC$. *Analogous relations hold with the other angles.)* *The converse is also true.*

We now consider two circles $C[K, r]$ and $C[K', r']$ and we examine the number of common tangents, since, as we saw above, the common tangents (when they exist) determine the centers of homotheties between them. In fact, when there are two external tangents, their point of intersection is on the center straight line KK' and is the center O^+. When there are two internal tangents, their point of intersection is on the center straight line $K'K$ and is the center O^-. If there is just one internal tangent, then the point of its intersection with the center straight line KK' is the point of tangency of the two circles. This point is the center O^-, if the two circles are externally tangent, and the center O^+ if the circles are internally tangent. So, there are the following cases:

No common tangents iff one circle is completely in the interior of the other circle. $\Longleftrightarrow KK' < |r - r'|$. See **Figure 4.13**.

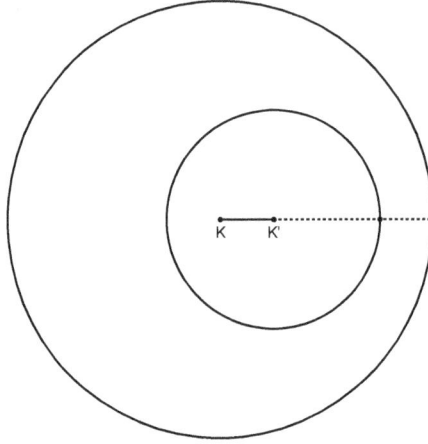

Figure 4.13: Circles with 0 common tangents

In such a case the centers O^+ and O^- of the two homotheties of the circles $C[K, r]$ and $C[K', r']$ cannot be found by means of tangents. We must use the method depicted in **Figure 4.11**.

One common tangent iff the circles are internally tangent. $\Longleftrightarrow KK' = |r - r'|$. See **Figure 4.14**.

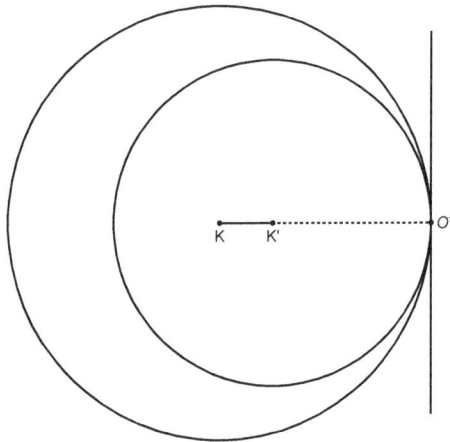

Figure 4.14: Circles with 1 common tangent

To find O^-, we must use the method depicted in **Figure 4.11**.

Two common tangents iff the circles intersect at two points.
$\Longleftrightarrow |r - r'| < KK' < r + r'$. See **Figure 4.15**.

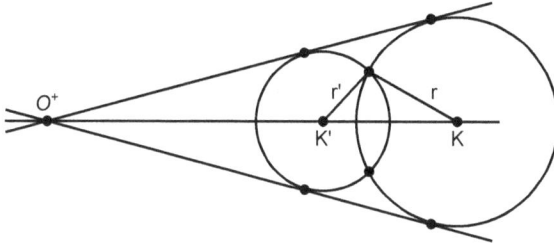

Figure 4.15: Circles with 2 common tangents

To find O^-, we must use the method depicted in **Figure 4.11**.

Three common tangents iff the circles are externally tangent.
$\Longleftrightarrow KK' = r + r'$. See **Figure 3.16**.

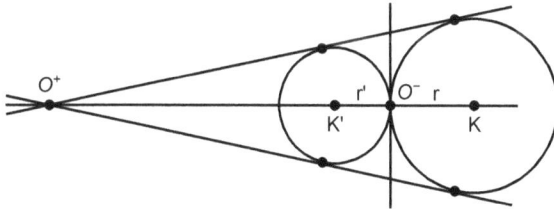

Figure 4.16: Circles with 3 common tangents

Four common tangents iff the circles do not intersect and they are outside each other. $\Longleftrightarrow r + r' < KK'$. See **Figure 4.17**.

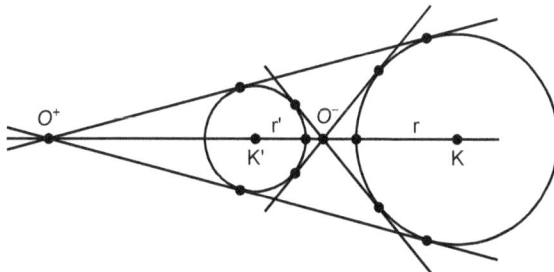

Figure 4.17: Circles with 4 common tangents

Infinitely many common tangents iff the two circles coincide. (A degenerate not interesting case.) $\iff K = K'$ and $r = r'$. See **Figure 4.18**.

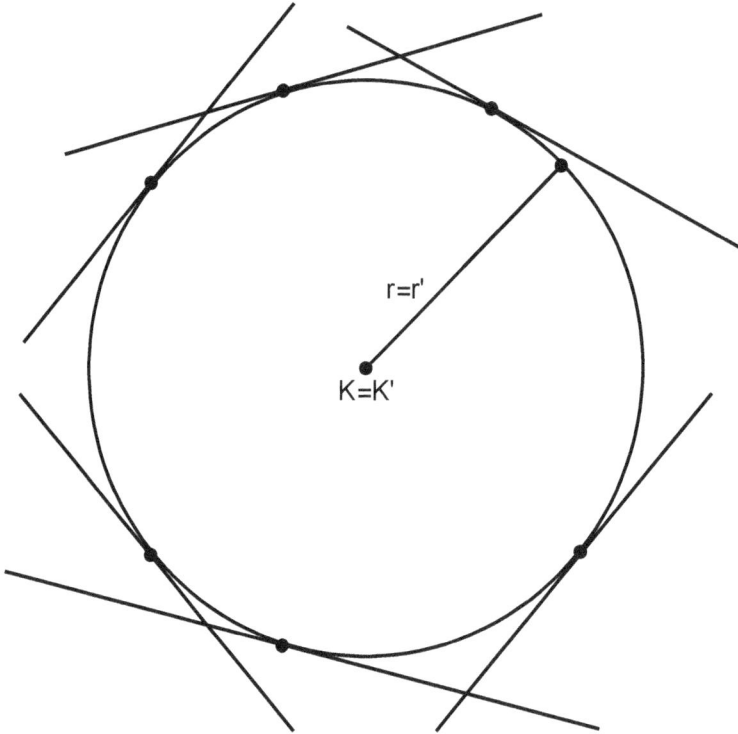

Figure 4.18: Circles with infinitely many common tangents

From the material exposed in **(j)** up to here, we can conclude the following important **Result:**

Theorem 4.1.2 *Any two circles of different radii are homothetic by two homotheties, one positive and one negative. The two centers of these homotheties and the two centers of the circles are on the same line and form a harmonic quadruple with ratio the ratio of the radii of the two circles. The centers of the positive and negative homotheties are harmonic conjugates with respect to the centers of the two circles.*

If the two circles have equal radii, then they are mapped to each other by the parallel translation $T_{\vec{u}}$ where $\vec{u}$ is the vector between their centers. Also, they are homothetic by a negative homothety (but no positive homothety, unless we consider the parallel translation as a positive homothety of center ∞ and ratio $\frac{\infty A'}{\infty A} = 1$, for any two corresponding points A and A').

Next, see **Figures 4.19, 4.20** and **4.21**, for additional illustration. We also summarize the ways of constructing the positive and negative homothety between two circles and examine two special situations for additional information.

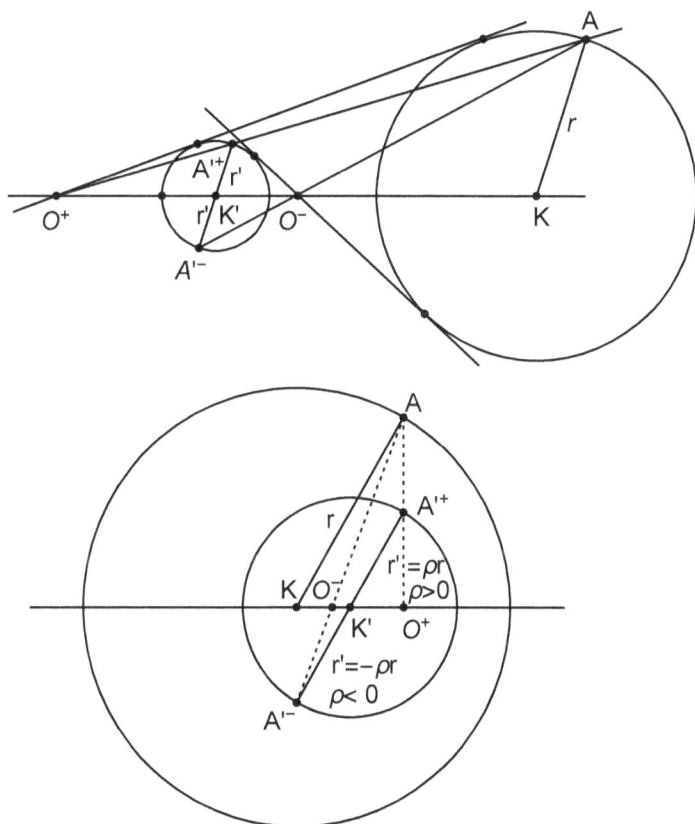

Figure 4.19: Centers of positive and negative homotheties of two unequal non-concentric circles are O^+ and O^-. $KA'^- \parallel KA \parallel K'A'^+$

To find the center O^+ of the positive homothety and the center O^- of the negative homothety, we pick a radius KA of the big circle and two opposite radii $K'A'^+$ and $K'A'^-$ of the small circle such that $\overrightarrow{KA}$ and $\overrightarrow{K'A'^+}$ have the same direction and $\overrightarrow{KA}$ and $\overrightarrow{K'A'^-}$ have opposite direction. Then, we draw AA'^+ and AA'^- and we have that $O^+ = KK' \cap AA'^+$ and $O^- = KK' \cap AA'^-$.

Also, O^+ is on the common external tangents whenever they exist, and O^- is on the common internal tangents whenever they exist. So, for two unequal and non-concentric circles the points O^+ and O^- are also located on the center-line of the two circles. When the unequal circles are externally or internally tangent, then the common point of tangency is O^- or O^+, respectively. (Draw two figures and provide the missing justifications as straightforward exercises.)

See **Figure 4.19**, and again **notice the harmonic quadruples of the points** K, K', O^-, **and** O^+, **so that**

$$-\frac{\overrightarrow{O^-K}}{\overrightarrow{O^-K'}} = \frac{r}{r'} = \frac{\overrightarrow{O^+K}}{\overrightarrow{O^+K'}}.$$

This is an important property to applications, and using it we can find one of the homotheties when we know the other one!

A **special case** of homothetic circles is the case of **circles with equal radii**, as in **Figure 4.20**. To find the center O^- of their negative homothety, we pick a radius KA of one of the circles and an opposite radius $K'A'$ of the other circle **parallel** to KA and we draw AA'. Then, $O^- = KK' \cap AA'$. It is easily seen that O^- is the midpoint of KK' and $\rho = -1$, in this situation. Therefore, in this situation, we have the negative homothety $H_{[O^-,-1]}$, which is equal to the half turn $R_{[O^-,\frac{\pi}{2}]}$.

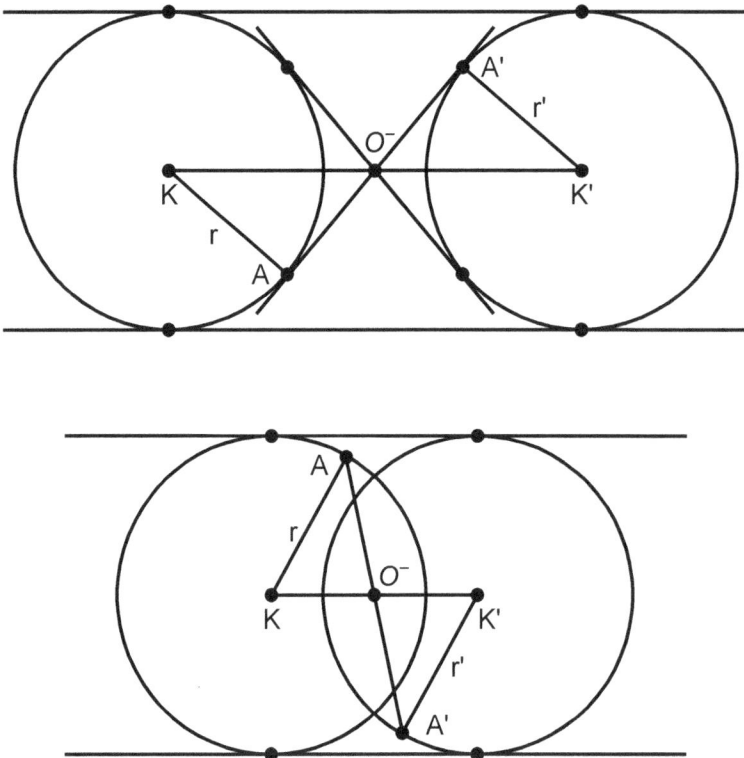

Figure 4.20: The center of the negative homothety of two different equal circles is O^-. $KA \parallel K'A'$

When the circles are externally tangent, then O^- becomes the common point of tangency. (Draw figure! Obviously, different equal circles cannot be internally tangent!) In this situation, the positive homothety (with ratio 1), has become the parallel translation $T_{\overrightarrow{KK'}}$.

Another **special case** of homothetic circles is the case of **two concentric circles**, as in **Figure 4.21**. There are no common tangents in this case. Parallel radial vectors of the same or opposite direction are on the same straight line through the common center O. Hence, all of these lines cannot determine any homothetic center as it was done before.

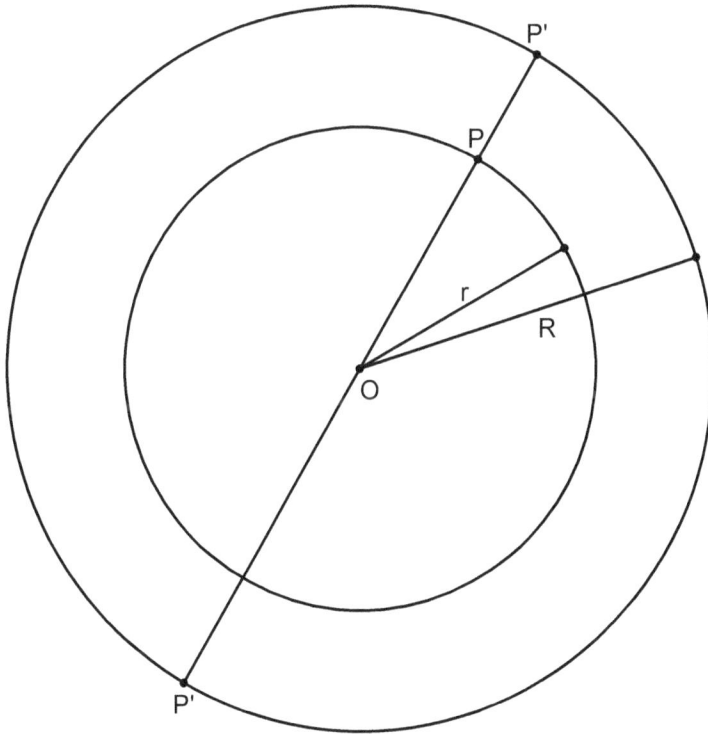

Figure 4.21: For the homotheties of two different concentric circles

However, it follows easily (justify it) that the two concentric circles $C[O, r]$ and $C[O, R]$ with $r < R$, as in **Figure 4.21**, are homothetic by the following positive and negative homotheties:

(a) $H_{[O, \pm \frac{R}{r}]}$ that map $C[O, r]$ to $C[O, R]$,   and

(b) their inverses $H_{[O, \pm \frac{R}{r}]}^{-1} = H_{[O, \pm \frac{r}{R}]}$ that map $C[O, R]$ to $C[O, r]$.

Note: As in (a) and (b) above, we always have that, if a figure $\mathfrak{F}_1$ is mapped to figure $\mathfrak{F}_2$ by a homothety $H_{[O, \rho]}$, then the figure $\mathfrak{F}_2$ is mapped to figure $\mathfrak{F}_1$ by the homothety $H_{[O, \rho]}^{-1} = H_{[O, \frac{1}{\rho}]}$.

4.2 Homotheties and their Compositions

We begin with stating and proving a theorem that characterizes the homotheties.

Theorem 4.2.1 (Characterization of Homotheties) *We consider a transformation of the plane* $\mathcal{H}$ $\mathcal{P} \longrightarrow \mathcal{P}$. *The following three statements are equivalent:*

   (a) $\mathcal{H}$ *is a homothety.*

   (b) There is a number $\rho \neq 0$, *such that, for any two pairs of corresponding points* (A, A'), $A' = \mathcal{H}(A)$, *and* (B, B'), $B' = \mathcal{H}(B)$, *we have*

$$\overrightarrow{A'B'} = \rho \cdot \overrightarrow{AB}.$$

(That is, any vector $\overrightarrow{AB}$ *is mapped to a vector* $\overrightarrow{A'B'}$, *such that* $\overrightarrow{A'B'} = \rho \cdot \overrightarrow{AB}$. *Hence, the vectors* $\overrightarrow{AB}$ *and* $\overrightarrow{A'B'}$ *are parallel and of the same or opposite direction if* $\rho > 0$ *or* $\rho < 0$, *respectively.)*

   (c) Given two corresponding points A *and* A', $A' = \mathcal{H}(A)$, *if the transformation* $\mathcal{H}$ *corresponds to any point* B *the point* B' *determined by the vector condition*

$$\overrightarrow{A'B'} = \rho \cdot \overrightarrow{AB},$$

where $\rho \neq 0$ *is some real constant, then* $\mathcal{H}$ *is a homothety of ratio* ρ *and center a point* O *that is completely determined by* A, A' *and* ρ.

In **(c)**, we will say that such a **homothety is defined by a pair of corresponding points** (A, A') **and the ratio** ρ.

Proof (a) $\implies$ (b) If $\mathcal{H}$ is a homothety, $\mathcal{H} = H_{[O, \rho]}$, ($\rho \neq 0$ is the homothetic ratio), then this claim follows by **property (f)** of homotheties proven in **Subsection 4.1.1**.

   (Notice that, given ρ, A, A' and B, the condition $\overrightarrow{A'B'} = \rho \cdot \overrightarrow{AB}$ determines the point B' on the straight line through the point A' and parallel to the straight line AB.)

   (b) $\implies$ (c) Since, by hypothesis,

$$\overrightarrow{A'B'} = \rho \cdot \overrightarrow{AB},$$

for any two pairs of corresponding points (A, A'), $A' = \mathcal{H}(A)$, and (B, B'), $B' = \mathcal{H}(B)$, we keep the pair (A, A'), $A' = \mathcal{H}(A)$, fixed and let B vary. Then, the claim follows immediately.

   (c) $\implies$ (a) **Case (1):** If $A = A'$ (A is fixed point) and $\rho = 1$ the vector condition of the Theorem implies that $B' = B$ and so, $\mathcal{H} = I$ is the identity, which is a special homothety.

Case (2): If $A = A'$ and $0 \neq \rho \neq 1$, then A is the unique fixed point and the vector condition in **(c)** is rewritten as

$$\overrightarrow{AB'} = \rho \cdot \overrightarrow{AB}.$$

By **Definition 4.1.2**, of homothety in general, $\mathcal{H} = H_{[A,\rho]}$, in this case. See **Figure 4.22**.

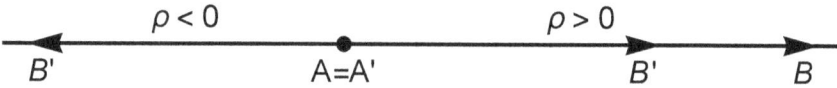

Figure 4.22: $A = A'$ **fixed point, so,** $H = H_{[A,\rho]}$

Case (3): In the general case, $A \neq A'$ and by hypothesis $\mathcal{H}(A) = A' \neq A$. In this case, we consider the homothety $H_{[O,\rho]}$ with homothetic ratio the given number ρ and corresponding to the point A the point A'. So, the center O is determined in **Lemma 4.1.1** and its **proof**, explicitly. So, by **Definition 4.1.2** of homothety, we have

$$H_{[O,\rho]}(A) = A' \quad \Longrightarrow \quad \overrightarrow{OA'} = \rho \cdot \overrightarrow{OA}.$$

See **Figure 4.23**.

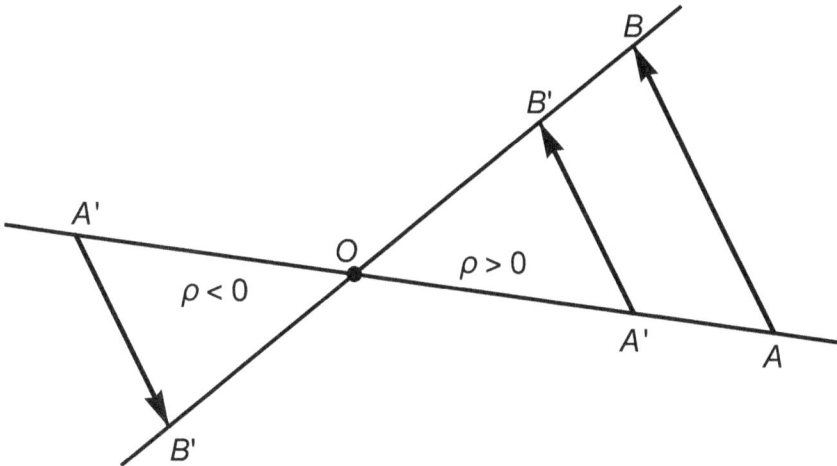

Figure 4.23: $A \neq A'$, **then** $H = H_{[O,\rho]}$, **where the** O **must be determined**

If now for any point B, we let $H_{[O,\rho]}(B) = B''$, then, as we have seen in properties **(e)** and **(f)** of **Subsection 4.1.1**, we have:

$$\overrightarrow{A'B''} = \rho \cdot \overrightarrow{AB}.$$

Also, by the vector condition of **(c)**, we have $\overrightarrow{A'B'} = \rho \cdot \overrightarrow{AB}$. Therefore, (by the last two vectors relations, $\overrightarrow{A'B''} = \overrightarrow{A'B'}$, and so, as A' is the same initial point of the two vectors, we get $B'' = B'$.

Finally, for any point $B \in \mathcal{P}$, we have proven $H_{[O,\rho]}(B) = \mathcal{H}(B)$. That is, $\mathcal{H} = H_{[O,\rho]}$, and so $\mathcal{H}$ is a homothety.

∎

Corollary 4.2.1 *If*

$$\overrightarrow{0} \neq \overrightarrow{AB} \parallel \overrightarrow{A'B'} \neq \overrightarrow{0} \quad \text{and} \quad \overrightarrow{AB} \neq \overrightarrow{A'B'},$$

or

$$\overrightarrow{AB} = \overrightarrow{A'B'} \quad \text{but with} \quad A = A' \quad \text{(and so } B = B' \quad \text{as well),}$$

then there exists one unique homothety $\mathcal{H}$, such that $\mathcal{H}(\overrightarrow{AB}) = \overrightarrow{A'B'}$.

Proof Let

$$\rho = \frac{\overrightarrow{A'B'}}{\overrightarrow{AB}}.$$

If $\overrightarrow{AB} \neq \overrightarrow{A'B'}$, as $\overrightarrow{0} \neq \overrightarrow{AB} \parallel \overrightarrow{A'B'} \neq \overrightarrow{0}$, $\rho \neq 1$. Then, the sought homothety $\mathcal{H}$ is determined by the pair of corresponding points (A, A') and the ratio ρ, as in **(c)** of the **Theorem**. (Also, see **Lemma 4.1.1** to determine the center O.) This homothety transforms $\overrightarrow{AB}$ to $\overrightarrow{A'B'}$.

If $\overrightarrow{0} \neq \overrightarrow{AB} = \overrightarrow{A'B'} \neq \overrightarrow{0}$ and $A = A'$ (and so $B = B'$), then $\rho = 1$ and $\mathcal{H}$ is the identity.

If $\overrightarrow{0} \neq \overrightarrow{AB} = \overrightarrow{A'B'} \neq \overrightarrow{0}$ and $A \neq A'$ (and so $B \neq B'$), then $\rho = 1$ and $\mathcal{H}$ is the parallel translation by the vector $\overrightarrow{AA'} = \overrightarrow{BB'} \neq \overrightarrow{0}$.

∎

Remarks: (1) If $\rho = 1$, then the relation $\overrightarrow{A'B'} = \overrightarrow{AB}$, in the conclusion of the **previous Theorem**, defines a parallel translation by the vector $\overrightarrow{AA'}$, as we have seen in **Subsection 2.5.1, (1)** and **(2)**.

(2) If (A, A') and (B, B') are two pairs of homothetic points, (and so, $\overrightarrow{AB}$ is parallel with $\overrightarrow{A'B'}$), the homothetic point of any point C that does not belong to the line AB, can be constructed as the intersection C' of the lines drawn from A' and B' and parallel to AC and BC, respectively. See **Figure 4.24**.

If the point D belongs to the line AB, then its homothetic point D' is constructed in the same way as long as we have constructed the homothetic of a point C that does not belong to the line AB. See **Figure 4.24** and prove the two claims by the similarity of the corresponding triangles.

(3) If $\rho \neq 1$ and the pairs of homothetic points (A, A') and (B, B') are not on the same straight line, then the center O of the homothety is the intersection of the straight lines AA' and BB'. See **Figures 4.6, 4.8, 4.10** and **4.23**.

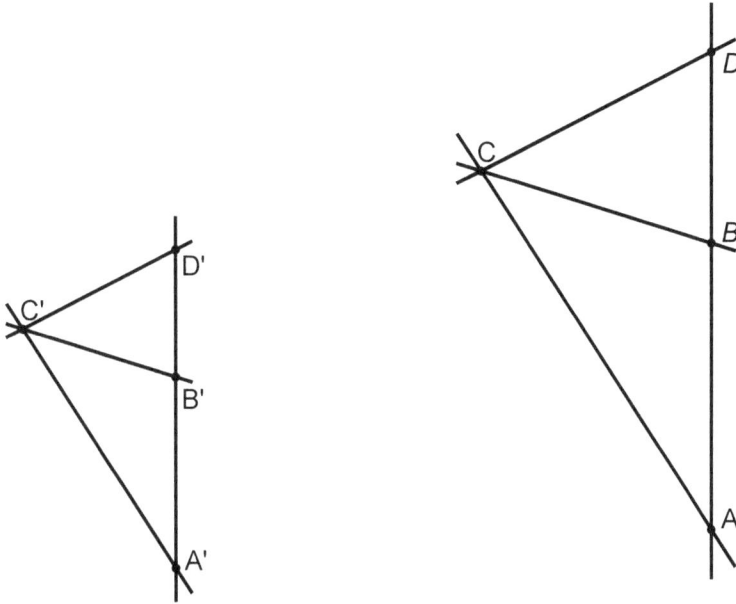

Figure 4.24: Geometrical construction of the homothetic points C' and D' of C and D, if we know two pairs of corresponding points (A, A') and (B, B') not on the same straight line

(4) If $\rho \neq 1$ and the pairs of homothetic points (A, A') and (B, B') are on the same straight line, then the center O of the homothety is on the straight line AA' (the same as BB') and determined by the algebraic relation

$$\frac{\overrightarrow{OA}}{\overrightarrow{OA'}} = \frac{\overrightarrow{OB}}{\overrightarrow{OB'}} = \frac{\overrightarrow{OB} - \overrightarrow{OA}}{\overrightarrow{OB'} - \overrightarrow{OA'}} = \frac{\overrightarrow{AB}}{\overrightarrow{A'B'}} := \rho.$$

The center O is on the same side of (A, A') and (B, B'), if $\rho > 0$, and between A and A' (and also between B and B'), if $\rho < 0$. See **Figure 4.25**. (Or, used **Lemma 4.1.1** to find O.)

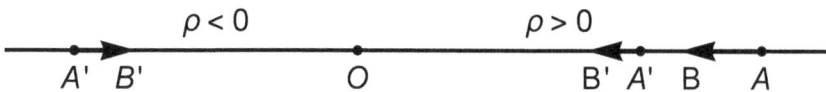

Figure 4.25: Determination of the center O of homothety if we know two pairs of corresponding points (A, A') and (B, B') on the same straight line

4.2.1 Composition of Homotheties

We consider two homotheties $F = H_{[O,\lambda]}$ and $G = H_{[Q,\rho]}$ and we would like to determine the composition $G \circ F$. To this end, any vector $\overrightarrow{AB}$ is mapped by F to $\overrightarrow{A'B'}$, and this new vector by G to $\overrightarrow{A''B''}$. So, the vector $\overrightarrow{AB}$ is mapped to $\overrightarrow{A''B''}$ by $G \circ F$. Hence we have

$$[\overrightarrow{A'B'} = \lambda\overrightarrow{AB}, \quad \text{and} \quad \overrightarrow{A''B''} = \rho\overrightarrow{A'B'}] \quad \Longrightarrow \quad \overrightarrow{A''B''} = \lambda\rho\overrightarrow{AB}.$$

Then, by **Theorem 4.2.1** and **Corollary 4.2.1**: (1) if $\lambda\rho \neq 1$, these compositions are homotheties with homothetic ratio $\lambda\rho \neq 1$, and we must determine their centers. (2) if $\lambda\rho = 1$, these compositions may be the identity or non-trivial parallel translations (and so the center is an infinity point). We distinguish two cases:

Case 1. $\lambda\rho \neq 1$.

For any points B and $B'' = G \circ F(B)$, as above, we have $\overrightarrow{A''B''} = \lambda\rho\overrightarrow{AB}$. Then, by **Theorem 4.2.1**, we have that $G \circ F$ **is a homothety of ratio** $\lambda\rho \neq 1$.

Next, **we must determine the center of this homothety**, $G \circ F$. There are two subcases to examine.

Subcase (I): The centers O and Q of F and G coincide, $(O \equiv Q)$. Then, $G \circ F(O) = G(O) = G(Q) = Q = O$. That is, the point $O \equiv Q$ is the fixed point of the homothety $G \circ F$. Therefore, in this case, $O \equiv Q$ **is the center of the homothety** $G \circ F$. We also notice that, in this case, $G \circ F = F \circ G = H_{[O,\lambda\rho]}$.

Hence: *The composition of two homotheties of the same center is commutative and is a new homothety of the same center and homothetic ratio the product of the two homothetic ratios.* See **Figure 4.26**.

Figure 4.26: The centers of the two homotheties coincide $O \equiv Q$.
$$F(A) = A_F, \; G(A) = A_G, \; G \circ F(A) = A''$$

Subcase (II): The centers O and Q are different, $(O \neq Q)$: Then, the straight line OQ is set-wise invariant by both F and G and so by $G \circ F$. Therefore it contains the center P of $G \circ F$ (and also the center of $F \circ G$).

To find P, we consider the line $l := OQ$ as axis with origin the point O and orientation $\overrightarrow{OQ}$ (as in **Figure 4.27**). Then, the vector $\overrightarrow{OP}$ is mapped to $\lambda\overrightarrow{OP}$ by the homothety $F = H_{[O,\lambda]}$. Subsequently, as O is chosen to be the origin, the vector $\lambda\overrightarrow{OP} - \overrightarrow{OQ}$ is mapped to $\rho(\lambda\overrightarrow{OP} - \overrightarrow{OQ}) + \overrightarrow{OQ}$, by the homothety $G = H_{[Q,\rho]}$. Since P is the fixed point of the new homothety $G \circ F$, we must have

$$\rho(\lambda\overrightarrow{OP} - \overrightarrow{OQ}) + \overrightarrow{OQ} = \overrightarrow{OP}.$$

Solving for $\overrightarrow{OP}$, since in this case $\lambda\rho \neq 1$, we find

$$\overrightarrow{OP} = \frac{\rho - 1}{\lambda\rho - 1}\overrightarrow{OQ}.$$

This vector equation determines P, the center of the new homothety $G \circ F$, on the line OQ, whenever $\lambda\rho \neq 1$. See **Figure 4.27**. Finally,

$$G \circ F = H_{[Q,\rho]} \circ H_{[O,\lambda]} = H_{[P,\lambda\rho]}, \quad \forall \ \lambda\rho \neq 1.$$

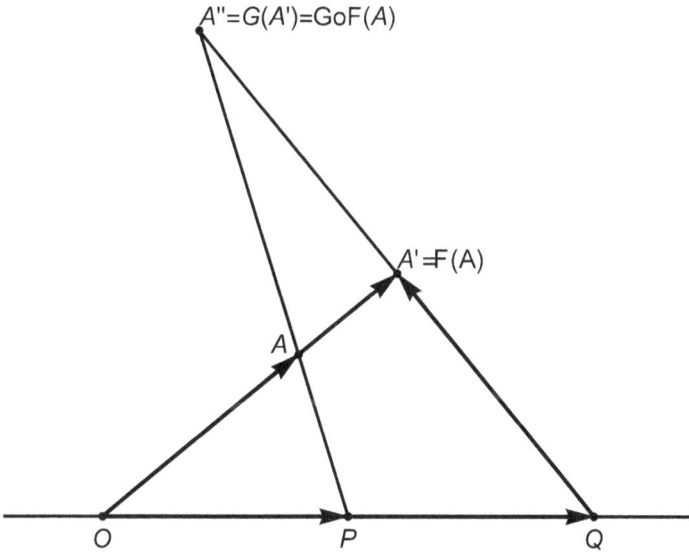

A"=G(A')=GoF(A)

A'=F(A)

A

O   P   Q

Figure 4.27: Geometrical construction of the center P of $G \circ F$,
when centers $O \neq Q$. $\overrightarrow{OP} = \dfrac{\rho - 1}{\lambda\rho - 1}\overrightarrow{OQ}$. **(Here $\lambda > 1$ and $\rho > 1$.)**

Remarks: (1) If $O \equiv Q$, then $\overrightarrow{OQ} = \overrightarrow{0}$ and so $\overrightarrow{OP} = \overrightarrow{0}$ and $O \equiv P \equiv Q$, yielding the result of **Subcase (I)**.

(2) $G \circ F \neq F \circ G$, because if P' is the center of $F \circ G$, then, as before with origin the point Q, we find $\overrightarrow{QP'} = \dfrac{\lambda - 1}{\lambda\rho - 1}\overrightarrow{QO}$. Thus, $P \neq P'$, in general.

Indeed: If we assume that for some λ and ρ, $P = P'$, since $\overrightarrow{OP} = \overrightarrow{OQ} + \overrightarrow{QP}$, we

have, $\quad \dfrac{\rho - 1}{\lambda\rho - 1}\overrightarrow{OQ} = \overrightarrow{OQ} + \dfrac{\lambda - 1}{\lambda\rho - 1}\overrightarrow{QO}, \quad$ and so $\quad \dfrac{\rho - 1}{\lambda\rho - 1} = 1 - \dfrac{\lambda - 1}{\lambda\rho - 1}.$

We simplify this equation to obtain $(\rho - 1)(\lambda - 1) = 0$. So, either $\lambda = 1$ or $\rho = 1$. This means that in order for the two compositions to have the same center, at least one of the homotheties must be the identity. In such a situation the compositions are rendered trivial.

So: If $P \neq Q$, $\lambda \neq 1$ and $\rho \neq 1$, the compositions $G \circ F$ and $F \circ G$ have different centers, and so they are not equal.

If $P \neq Q$, $\lambda \neq 1$ and $\rho \neq 1$, the composition $G \circ F$ is not commutative.

(3) *Both compositions $G \circ F$ and $F \circ G$ have the same ratio $\lambda \rho$. If $\lambda \rho = 1$ and $\lambda \neq 1$ (and so $\rho \neq 1$), then the point P is the infinity point of the straight line PQ ($P \neq Q$) and $G \circ F$ becomes a parallel translation by a vector parallel to PQ, as we will see in the next case below. Similarly for P'.*

(4) The center P can also be found by using **Lemma 4.1.1**, in the following way. The point O is mapped to itself by F and to a point O' by G (and so O is mapped to O' by $G \circ F$), such that $\overrightarrow{QO'} = \rho \overrightarrow{QO} = -\rho \overrightarrow{OQ}$. Therefore, as it is derived in the **proof of Lemma 4.1.1** and given that the homothetic ratio of $G \circ F$ is $\lambda \rho \neq 1$, we have

$$\overrightarrow{PO} = \frac{1}{\lambda \rho - 1} \overrightarrow{OO'} = \frac{1}{\lambda \rho - 1} \left(\overrightarrow{OQ} + \overrightarrow{QO'} \right)$$

$$= \frac{1}{\lambda \rho - 1} \left(\overrightarrow{OQ} - \rho \cdot \overrightarrow{OQ} \right) = \frac{1 - \rho}{\lambda \rho - 1} \overrightarrow{OQ}. \quad \text{So,} \quad \overrightarrow{OP} = \frac{\rho - 1}{\lambda \rho - 1} \overrightarrow{OQ}.$$

(5) **The center P can be constructed geometrically** as the intersection of the line OQ and the line AA'', defined by a pair of corresponding points A and $A'' = G \circ F(A)$. See **Figure 4.27**. Thus, in this case, **the fixed point or center P of the homothety $G \circ F$, is found as**

$$P = OQ \cap AA'' = OQ \cap A[(G \circ F)(A)], \quad \text{for any point } A \notin OQ.$$

Case 2. $\lambda \rho = 1 \Longleftrightarrow \rho = \dfrac{1}{\lambda}$.

For any points B and $B'' = G \circ F(B)$, as before, we have $\overrightarrow{A''B''} = \overrightarrow{AB}$. Hence, by **Subsection 2.5.1, (1)** and **(2)**, $G \circ F$ is the parallel translation $T_{\vec{u}}$, where $\vec{u} = \overrightarrow{AA''}$, where A is any point in the plane. To find the vector $\vec{u}$, we examine two subcases.

Subcase (I). The centers O and Q of F and G coincide, $(O = Q)$: We have that $F(A) = A'$ and $G(A') = A''$ and so $(G \circ F)(A) = A''$. Therefore, $\lambda \overrightarrow{OA} = \overrightarrow{OA'}$ and so $\overrightarrow{OA} = \dfrac{1}{\lambda} \overrightarrow{OA'}$. Also, $\rho \overrightarrow{QA'} = \dfrac{1}{\lambda} \overrightarrow{OA'} = \overrightarrow{OA''}$. Hence, $\overrightarrow{OA} = \overrightarrow{OA''}$, and so $A = A''$.

Therefore, in this subcase, $\overrightarrow{AA''} = \vec{u} = \vec{0}$ and the parallel translation

$$G \circ F = F \circ G = T_{\vec{0}} = I$$

is the identity, which can be considered as a homothety of center any point and homothetic ratio 1. [See also, **Subsection 4.1.1, (g)**.]

Subcase (II). The centers O and Q of F and G are different, $(O \neq Q)$: We pick any point $A \in \mathcal{P}$ and we will compute $G \circ F(A) = A''$ and the vector $\overrightarrow{AA''}$. See **Figure 4.28**.

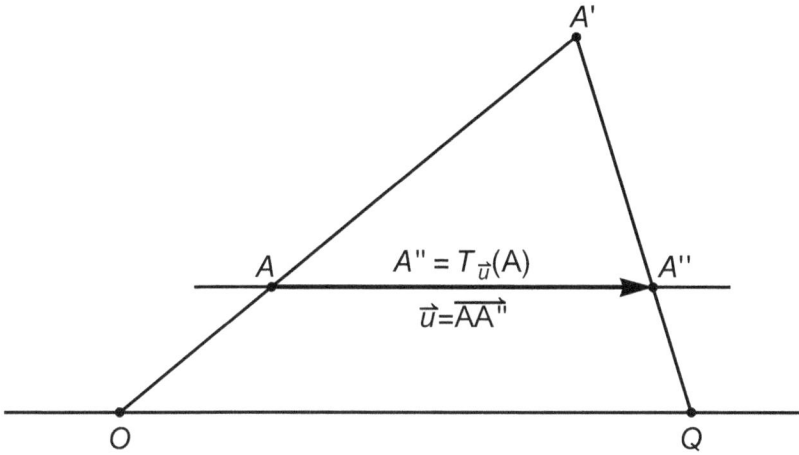

Figure 4.28: The centers O and Q of F and G are different,
$$G \circ F = T_{\vec{u}}, \text{ where } \vec{u} = \overrightarrow{AA''} = (\rho - 1)\overrightarrow{QO} = \left(\frac{\lambda - 1}{\lambda}\right)\overrightarrow{OQ}.$$
Here, $0 < \rho < 1$, $\lambda > 1$, $\lambda\rho = 1$.

We may choose a point A not on the straight line OQ. We let $F(A) = A'$ and $G(A') = G \circ F(A) = A''$. Then we have:
$$\overrightarrow{OA'} = \lambda\overrightarrow{OA}, \qquad \text{and} \qquad \overrightarrow{QA''} = \frac{1}{\lambda}\overrightarrow{QA'} \quad \left(\rho = \frac{1}{\lambda}\right).$$

Hence,
$$\overrightarrow{AA'} = \overrightarrow{OA'} - \overrightarrow{OA} = (\lambda - 1)\overrightarrow{OA}$$

and
$$\overrightarrow{A''A'} = \overrightarrow{QA'} - \overrightarrow{QA''} = \left(1 - \frac{1}{\lambda}\right)\overrightarrow{QA'} = \frac{\lambda - 1}{\lambda}\overrightarrow{QA'}.$$

Then, for the triangles $OA'Q$ and $AA'A''$ we have
$$\frac{\overrightarrow{AA'}}{\overrightarrow{OA'}} = \frac{\lambda - 1}{\lambda}, \qquad \text{and} \qquad \frac{\overrightarrow{A''A'}}{\overrightarrow{QA'}} = \frac{\lambda - 1}{\lambda}$$

and the angles $\widehat{OA'Q} = \widehat{AA'A''}$. Therefore, the triangles $OA'Q$ and $AA'A''$ are similar with ratio of similarity $s = \dfrac{\lambda - 1}{\lambda}$.

Therefore, $\overrightarrow{AA''} \parallel \overrightarrow{OQ}$ and $\overrightarrow{AA''} = \dfrac{\lambda - 1}{\lambda}\overrightarrow{OQ}$, which is a fixed vector independent of the point A. So, the composition $G \circ F$ is the parallel translation by the vector $\vec{u} = \dfrac{\lambda - 1}{\lambda}\overrightarrow{OQ}$. That is,
$$G \circ F = T_{\frac{\lambda - 1}{\lambda} \cdot \overrightarrow{OQ}}.$$

Remarks: (1) We can also find the vector $\vec{u}$ by choosing $A = O$ (on the straight line OQ). Then, we must find $A'' = (G \circ F)(O) = G[F(O)] = G(O)$ and the vector $\overrightarrow{AA''}$. By the definition of $G = H_{[Q,\rho]}$, we get $\overrightarrow{QA''} = \rho\overrightarrow{QO}$. Hence (as $A = O$), we find

$$\vec{u} = \overrightarrow{AA''} = \overrightarrow{QA''} - \overrightarrow{QA} = \overrightarrow{QA''} - \overrightarrow{QO} = \rho\overrightarrow{QO} - \overrightarrow{QO}$$

$$= (\rho - 1)\overrightarrow{QO} = \left(\frac{1}{\lambda} - 1\right)\overrightarrow{QO} = \left(\frac{1-\lambda}{\lambda}\right)\overrightarrow{QO} = \left(\frac{\lambda-1}{\lambda}\right)\overrightarrow{OQ}.$$

(2) The same result is obtained if, instead of O, we pick another point $A \neq O$ on the straight line OQ.

(3) The above outcome also gives the correct answer, $\vec{u} = \vec{0}$ of the previous case, when $O = Q$, and / or when $\lambda = \rho = 1$, that is, both F and G are identities.

After this exposition, we could say that a parallel translation behaves like a homothety with ratio $\rho = 1$ and center a point at infinity. We now conclude with the following two results.

Result 1. *The composition of two homotheties F and G with reciprocal ratios is a parallel translation, $G \circ F = T_{\vec{u}}$, whose vector $\vec{u}$ has been determined above. (If their centers coincide then this composition is the identity.)*

Result 2. *The set of homotheties of the same center is an algebraic commutative group under composition. (The identity is obtained when the homothetic ratio is 1.)*

Also: **Subcase (II) of Case 2,** $\left(\rho = \dfrac{1}{\lambda}\right)$ above, yields also the following:

Theorem 4.2.2 *Given two homotheties of the **same ratios and different centers**, each one of them is equal to the composition of the other one with a parallel translation.*

Proof Let $H_{[O,\lambda]}$ and $H_{[Q,\lambda]}$ be the two given homotheties ($\lambda \neq 0$). Then, we have

$$H_{[O,\lambda]} = H_{[O,\lambda]} \circ I = H_{[O,\lambda]} \circ \left\{H_{[Q,\frac{1}{\lambda}]} \circ H_{[Q,\lambda]}\right\}$$

$$= \left\{H_{[O,\lambda]} \circ H_{[Q,\frac{1}{\lambda}]}\right\} \circ H_{[Q,\lambda]} = T_{\vec{u}} \circ H_{[Q,\lambda]},$$

where $\vec{u} = (\lambda - 1)\overrightarrow{OQ}$, as we have proven above in **Subcase (II)** of **Case 2**. (Hence, $H_{[Q,\lambda]} = T_{-\vec{u}} \circ H_{[O,\lambda]}$.)

By analogous work, we also get

$$H_{[O,\lambda]} = H_{[Q,\lambda]} \circ T_{\vec{v}},$$

where $\vec{v} = \dfrac{\lambda - 1}{\lambda}\overrightarrow{OQ}$. (Hence, $H_{[Q,\lambda]} = H_{[O,\lambda]} \circ T_{-\vec{v}}$.) ∎

Remarks: (1) If the centers are the same, $O = Q$, then the two homotheties are also the same (since they have the same ratio) and the parallel translation is the identity, ($\vec{u} = \vec{0}$, since $\overrightarrow{OQ} = \vec{0}$).

(2) The **converse** of **this Theorem** is studied in the **next Subsection**.

Corollary 4.2.2 *The homothetic images $\mathfrak{F}_O$ and $\mathfrak{F}_Q$ of a figure $\mathfrak{F}$ by two homotheties $H_{[O,\lambda]}$ and $H_{[Q,\lambda]}$ of the same ratio $\lambda \neq 0$ can be mapped onto each other by the parallel translation $T_{\vec{u}}$ or $T_{-\vec{u}}$, with*

$$\vec{u} = (\lambda - 1)\overrightarrow{OQ}.$$

See **Figure 4.29**.

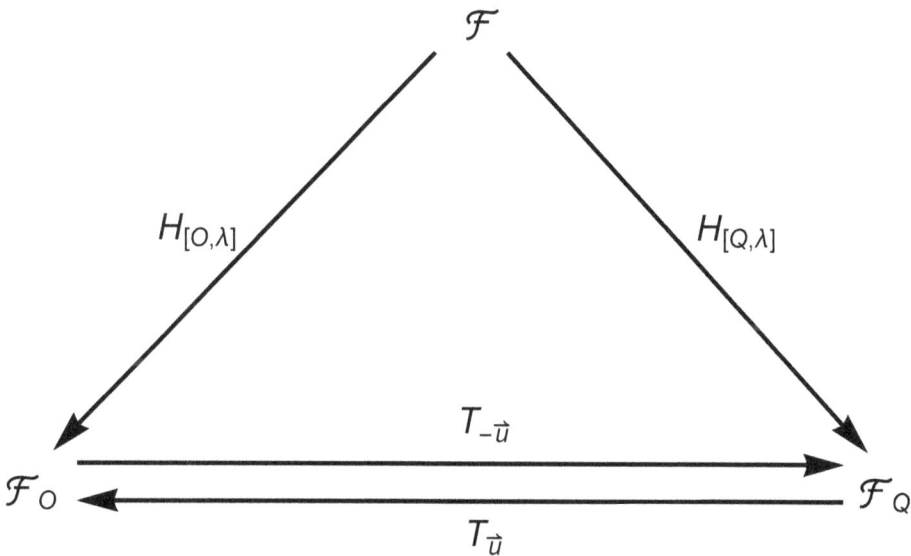

Figure 4.29: Homothetic figures and parallel translations
$$T_{\vec{u}} = T_{(\lambda-1)\overrightarrow{OQ}} \quad \text{and} \quad T_{-\vec{u}} = T_{(1-\lambda)\overrightarrow{OQ}}$$

4.2.2 Compositions of Homotheties and Parallel Translations and their Algebraic Group

We consider a parallel translation $T_{\vec{u}}$ and a homothety $H_{[O,\lambda]}$. By **Subsection 2.5.1** with the characterization of the parallel translations and **Theorem 4.2.1**, we obtain that the compositions

$$T_{\vec{u}} \circ H_{[O,\lambda]} \quad \text{and} \quad H_{[O,\lambda]} \circ T_{\vec{u}}$$

are both homotheties with **ratio** λ. So, **this Subsection** essentially deals with the converse of **Theorem 4.2.2**. Indeed:

(a) For $T_{\vec{u}} \circ H_{[O,\lambda]}$, we choose $A := O$ the fixed point of $H_{[O,\lambda]}$, in **Theorem 4.2.1**, and then

$$A' := T_{\vec{u}} \circ H_{[O,\lambda]}(A) = T_{\vec{u}} \circ H_{[O,\lambda]}(O) = T_{\vec{u}}(O) := O',$$

such that $\overrightarrow{AA'} = \overrightarrow{OO'} = \vec{u}$.

Consequently, for any point B in the plane, we let $C := H_{[O,\lambda]}(B)$ and then $T_{\vec{u}} \circ H_{[O,\lambda]}(B) = T_{\vec{u}}(C) := B'$. Then, we have: $\overrightarrow{OC} = \lambda \overrightarrow{OB}$ and $\overrightarrow{CB'} = \vec{u}$. Therefore,

$$\overrightarrow{A'B'} = \overrightarrow{O'B'} = \overrightarrow{O'O} + \overrightarrow{OC} + \overrightarrow{CB'} = -\vec{u} + \lambda \overrightarrow{OB} + \vec{u} = \lambda \overrightarrow{OB} = \lambda \overrightarrow{AB}.$$

[Or, we can say that any vector $\overrightarrow{AB}$ is mapped by $H_{[O,\lambda]}$ to a vector $\overrightarrow{A_1B_1}$, such that $\overrightarrow{A_1B_1} = \lambda \overrightarrow{AB}$. Then, this vector is mapped by $T_{\vec{u}}$ to an equal vector $\overrightarrow{A'B'} = \overrightarrow{A_1B_1} = \lambda \overrightarrow{AB}$.]

Hence, by **Theorem 4.2.1**, $\overrightarrow{A'B'} = \lambda \overrightarrow{AB}$ proves that

$$T_{\vec{u}} \circ H_{[O,\lambda]} \quad \text{is a homothety of ratio } \lambda.$$

(b) For $H_{[O,\lambda]} \circ T_{\vec{u}}$, we choose $A := T_{\vec{u}}(O)$, in **Theorem 4.2.1**, such that $\overrightarrow{OA} = \vec{u}$. Then

$$A' := H_{[O,\lambda]} \circ T_{\vec{u}}(A) = H_{[O,\lambda]} \circ T_{\vec{u}} \circ T_{-\vec{u}}(O) = H_{[O,\lambda]}(O) = O.$$

Consequently, for any point B in the plane, we let $C := T_{\vec{u}}(B)$, and so $\overrightarrow{BC} = \vec{u}$. Then $H_{[O,\lambda]} \circ T_{\vec{u}}(B) = H_{[O,\lambda]}(C) := B'$, and so $\overrightarrow{OB'} = \lambda \overrightarrow{OC}$, or $\overrightarrow{A'B'} = \lambda \overrightarrow{OC}$. Therefore,

$$\overrightarrow{A'B'} = \lambda \overrightarrow{OC} = \lambda \left(\overrightarrow{OA} + \overrightarrow{AB} + \overrightarrow{BC} \right) = \lambda \left(-\vec{u} + \overrightarrow{AB} + \vec{u} \right) = \lambda \overrightarrow{AB}.$$

[Or, we can say that any vector $\overrightarrow{AB}$ is mapped by $T_{\vec{u}}$ to an equal vector $\overrightarrow{A_1B_1} = \overrightarrow{AB}$. Then, this vector is mapped by $H_{[O,\lambda]}$ to a vector $\overrightarrow{A'B'}$, such that $\overrightarrow{A'B'} = \lambda \overrightarrow{A_1B_1} = \lambda \overrightarrow{AB}$.]

Hence, by **Theorem 4.2.1**, $\overrightarrow{A'B'} = \lambda \overrightarrow{AB}$ proves that

$$H_{[O,\lambda]} \circ T_{\vec{u}} \quad \text{is a homothety of ratio } \lambda.$$

(c) If $\lambda = 1$, then $H_{[O,\lambda]} = I$ is the identity, and so these compositions yield just the parallel translation $T_{\vec{u}}$.

Similarly, if $\vec{u} = \vec{0}$, $T_{\vec{u}} = I$ is the identity, and so, these compositions yield just the homothety $H_{[O,\lambda]}$.

(d) So, we consider $0 \neq \lambda \neq 1$ and $\vec{u} \neq \vec{0}$ and then, we must determine the **centers** (unique fixed points) of these two new homotheties.

Let P be the fixed point of the first composition, i.e., $T_{\vec{u}} \circ H_{[O,\lambda]}(P) = P$. Let also $H_{[O,\lambda]}(P) = P'$. Then we have: $\overrightarrow{OP'} = \lambda\overrightarrow{OP}$ and $T_{\vec{u}}(P') = P$ and so, $\overrightarrow{P'P} = \vec{u}$. Therefore,

$$\vec{u} = \overrightarrow{PP'} = \overrightarrow{OP} - \overrightarrow{OP'} = \overrightarrow{OP} - \lambda\overrightarrow{OP} = (1 - \lambda)\overrightarrow{OP}.$$

Hence, P is determined by the vector

$$\overrightarrow{OP} = \frac{1}{1 - \lambda}\vec{u}, \qquad (0 \neq \lambda \neq 1).$$

So, P is located on a straight through O and line parallel to the vector $\vec{u} \neq \vec{0}$, etc. See **Figure 4.30**. Finally,

$$T_{\vec{u}} \circ H_{[O,\lambda]} = H_{[P,\lambda]}.$$

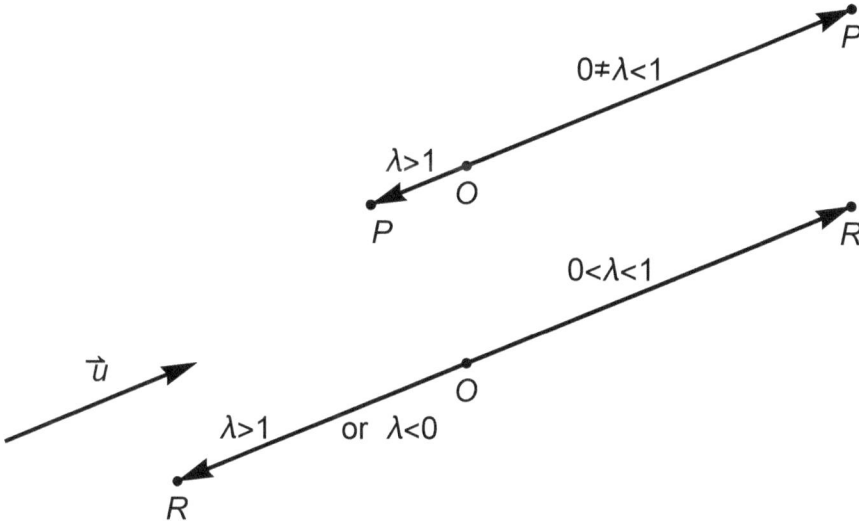

Figure 4.30: Centers P and R of the compositions of a parallel translation $T_{\vec{u}}$ and a homothety $H_{[O,\lambda]}$, in either order.

$$\overrightarrow{OP} = \frac{1}{1 - \lambda}\vec{u}, \text{ and } \overrightarrow{OR} = \frac{\lambda}{1 - \lambda}\vec{u}$$

(e) Now, let R be the fixed point of the second composition $H_{[O,\lambda]} \circ T_{\vec{u}}$, i.e., $H_{[O,\lambda]} \circ T_{\vec{u}}(R) = R$. Let also $T_{\vec{u}}(R) = R'$ and so, $\overrightarrow{RR'} = \vec{u}$. Then we have: $H_{[O,\lambda]}(R') = R$, and so $\overrightarrow{OR} = \lambda\overrightarrow{OR'}$. Hence,

$$\vec{u} = \overrightarrow{RR'} = \overrightarrow{OR'} - \overrightarrow{OR} = \frac{1}{\lambda}\overrightarrow{OR} - \overrightarrow{OR} = \frac{1 - \lambda}{\lambda}\overrightarrow{OR}.$$

Therefore, R is determined by the vector

$$\overrightarrow{OR} = \frac{\lambda}{1-\lambda}\vec{u}.$$

So, R is located on a straight through O and line parallel to the vector $\vec{u} \neq \vec{0}$, etc. See **Figure 4.30**. Finally,

$$H_{[O,\lambda]} \circ T_{\vec{u}} = H_{[R,\lambda]}.$$

Remarks: (1) We notice that these results are also valid for $\lambda = 0$ or 1. If $\lambda = 1$, the compositions are the parallel translation by $\vec{u}$ and the centers P and R become infinity points.

(2) We see that this R is not the same as P above. Therefore,

$$T_{\vec{u}} \circ H_{[O,\lambda]} \neq H_{[O,\lambda]} \circ T_{\vec{u}},$$

and so these compositions are not commutative.

(3) The centers P and R can also be found by using **Lemma 4.1.1 and its proof**.

For finding P, we observe that O is mapped to itself by $H_{[O,\lambda]}$ and subsequently to O' by $T_{\vec{u}}$, such that $\overrightarrow{OO'} = \vec{u}$. Then, since $T_{\vec{u}} \circ H_{[O,\lambda]}$ is a homothety of ratio λ by **Lemma 4.1.1 and its proof**, we find

$$\overrightarrow{PO} = \frac{1}{\lambda - 1}\overrightarrow{OO'},$$

and so

$$\overrightarrow{OP} = \frac{1}{1-\lambda}\overrightarrow{OO'} = \frac{1}{1-\lambda}\vec{u}.$$

For finding R, we observe that O is mapped to O_1 by $T_{\vec{u}}$, such that $\overrightarrow{OO_1} = \vec{u}$. Subsequently $\overrightarrow{OO_1} = \vec{u}$ is mapped by $H_{[O,\lambda]}$ to $\overrightarrow{OO'} = \lambda\vec{u}$. Then, since $H_{[O,\lambda]} \circ T_{\vec{u}}$ is a homothety of ratio λ by **Lemma 4.1.1 and its proof**, we find

$$\overrightarrow{RO} = \frac{1}{\lambda - 1}\overrightarrow{OO'},$$

and so

$$\overrightarrow{OR} = \frac{1}{1-\lambda}\overrightarrow{OO'} = \frac{\lambda}{1-\lambda}\vec{u}.$$

(f) The results proven above, clinch the following important **Result**:
The union of the homotheties and the parallel translations of the Euclidean plane is an algebraic non-commutative group under composition. (This is a subgroup of the group of all similarities.)

[In **Sections 4.4** and **4.5**, we will study the compositions of homotheties with rotations and reflections (and glides), respectively. See also some of the exercises. We shall see that these compositions yield other general similarities.]

4.2.3   Applications of Homothety

Example 4.2.1 Application 1.

In a given triangle ABC, inscribe a square $DEFG$ with side DE on BC.

Solution. See **Figure 4.31**.

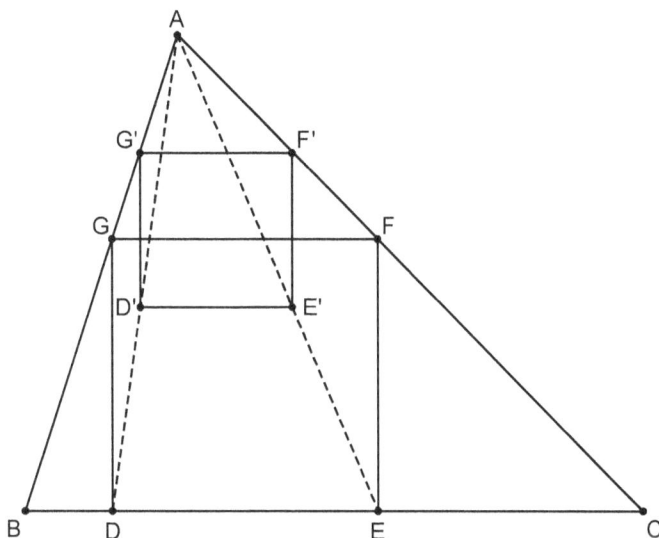

Figure 4.31: Example, A.HO. 1

We first construct any square $D'E'F'G'$ with side $F'G'$ parallel to BC. Then we draw the lines AD' and AE' and let the points of their intersection with BC be D and E, respectively. Then, we construct a square on DE, so that the side FG is closer to A. This square $DEFG$ is the solution.

Indeed, the points E and F are on the sides AC and AB, respectively, since the square $DEFG$ is homothetic to square $D'E'F'G'$, under the homothety with center A and ratio $\lambda = \dfrac{DE}{D'E'}$. (Supply the details!)

Remark: If angle $\widehat{B}$ is right then D is on B. If it is obtuse, then either D or D and E are outside BC and on the side of B that C does not lie.

▲

Example 4.2.2 Application 2.

Construct a circle tangent to a given circle of center O and a given straight line l, and given one of the points of tangency, either A on the circle, or B on the line.

Solution. See **Figure 4.32**.

As we know and see in the figure, the given circle must be homothetic to the sought circle, under a homothety with center A (their common point of tangency). This homothety must map the point B to one of the end-points of the diameter $DD' \perp l$ of the given circle. Thus, we have:

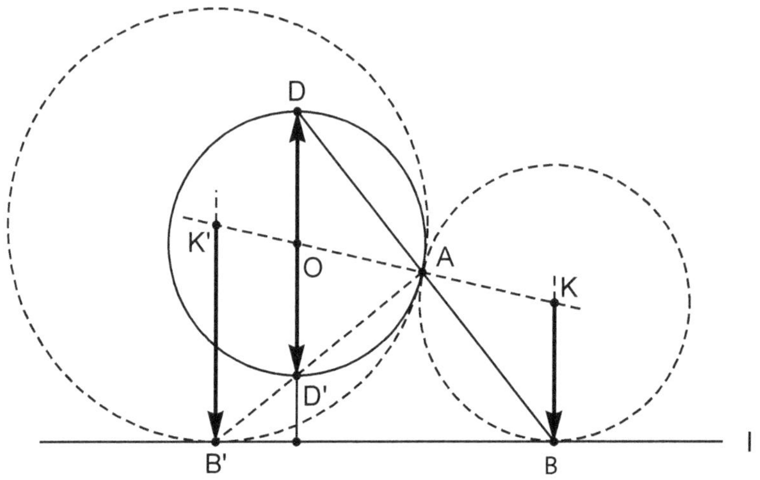

Figure 4.32: Example, A.HO. 2

a) If A is given, we determine B as a point of intersection of DA or $D'A$ with the line l.

b) If B is given, we define A as the second point of intersection of BD or BD' and the given circle.

With the second point of tangency determined, the center K of the sought after circle is the intersection of the lines OA and the line through B and perpendicular to l.

We have two solutions, in general. If $A = B$ and l is tangent to the circle, then there are infinitely many solutions. If $A = B$ and l is not tangent to the circle, then there are no solutions. Investigate more and draw figures for these cases!

Note: By choosing one point of tangency in all possible positions, we construct all the circles which are tangent to the given circle and to the given line. There are two families of such circles, each containing infinitely many circles. Draw some of these circles!

▲

Example 4.2.3 Application 3.

Theorem 4.2.3 (Euler's line and the nine-point circle of a triangle.)
*(1) In any triangle, the midpoints of the sides, the feet of the altitudes, and the midpoints of the segments joining the orthocenter and the vertices are nine points on the same circle, called **the nine-point circle of the triangle**.*

*(2) The circumcenter, the center of the nine-point circle, the center of gravity (centroid), and the orthocenter of the triangle are on the same line, called the **Euler's line of the triangle**. In this order, these four points are a harmonic quadruple with ratio 2.*

(3) The center of the nine-point circle is the midpoint of the segment joining the circumcenter and the orthocenter of the triangle and its radius is one-half of the circumradius of the triangle.

The **nine-point circle** is also called **Euler's circle**, and **Feuerbach's**[1] circle.

Proof. See **Figure 4.33**.

We consider a triangle ABC with circumcircle (ABC), circumcenter O and circumradius R. First: We denote:

(a) With O_1, O_2 and O_3 the midpoints of the sides BC, CA, and AB, respectively. Then, O_1O, O_2O, and OO_3 are the perpendicular bisectors of the corresponding sides of ABC and they are concurrent at the **circumcenter** O, thus determining it.

(b) With H_1, H_2 and H_3 the feet of the altitudes (heights) on these sides BC, CA, and AB, respectively. The three altitudes meet at the **orthocenter** H of the triangle.

(c) With P_1, P_2 and P_3 the midpoints of the segments joining H to the vertices, i.e., the segments HA, HB, and HC, respectively.

Next: We know that the medians AO_1, BO_2 and CO_3 of the triangle ABC pass through the same point G, called the **barycenter** or **center of gravity** or **centroid** of the triangle. G satisfies: $AG = \frac{2}{3}AO_1$ and so $GO_1 = \frac{1}{3}AO_1$, $BG = \frac{2}{3}BO_2$ and so $GO_2 = \frac{1}{3}AO_2$, and $CG = \frac{2}{3}AO_3$ and so $GO_3 = \frac{1}{3}CO_3$.

We consider the triangle $O_1O_2O_3$ (**medial triangle of** ABC) and its circumscribed circle $(O_1O_2O_3)$ with center E and radius R'. We know that $O_2O_3 \parallel BC$ and $O_2O_3 = \frac{1}{2}BC$, $O_3O_1 \parallel CA$ and $O_3O_1 = \frac{1}{2}CA$, and $O_1O_2 \parallel AB$ and $O_1O_2 = \frac{1}{2}AB$.

(I) Now, we have:

$$\frac{\overrightarrow{O_1O_2}}{\overrightarrow{AB}} = \frac{\overrightarrow{O_2O_3}}{\overrightarrow{BC}} = \frac{\overrightarrow{O_3O_1}}{\overrightarrow{CA}} = -\frac{1}{2}.$$

Therefore, the triangle $O_1O_2O_3$ is homothetic to ABC by a negative homothety of ratio $-\frac{1}{2}$ and center G, the common point of the medians of ABC.

Hence, the circle $(O_1O_2O_3)$ is homothetic to (ABC), by the same negative homothety, and so

$$\frac{R'}{R} = \left|-\frac{1}{2}\right| = \frac{1}{2} \qquad \text{and} \qquad \overrightarrow{GE} = -\frac{1}{2}\overrightarrow{GO}.$$

(II) Under the above homothety the image of the orthocenter H of ABC is the orthocenter of $O_1O_2O_3$, which is easily observed to be the point O. Hence,

$$\overrightarrow{GO} = -\frac{1}{2}\overrightarrow{GH}.$$

---

[1] Karl Wilhelm Feuerbach, German mathematician (high school teacher), 1800-1834.

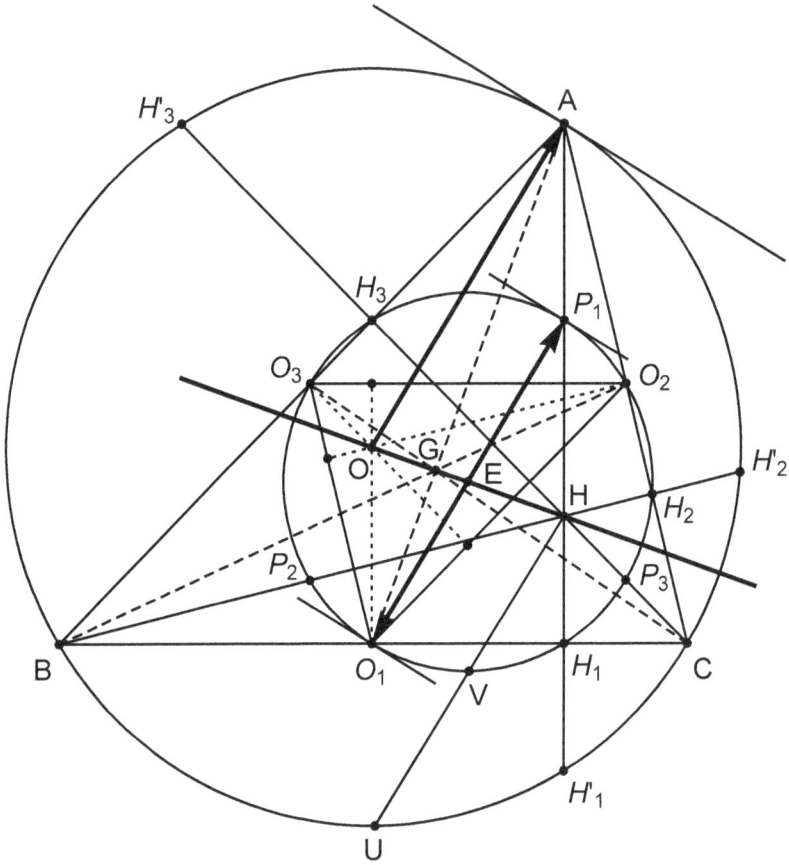

Figure 4.33: Example, A.HO. 3
The nine-point circle and Euler's line of the triangle ABC

From the last two equations we conclude that: **The points O, G, E, H are on the same straight line, called Euler's line, and also**

$$\overrightarrow{OE} = -\frac{1}{2}\overrightarrow{HO} = \frac{1}{2}\overrightarrow{OH},$$

that is, **the point E is the middle point of the segment OH.** Also,

$$-\frac{\overrightarrow{GO}}{\overrightarrow{GE}} = 2 = \frac{\overrightarrow{HO}}{\overrightarrow{HE}}.$$

Therefore the points O, E, G, and H form a harmonic quadruple of ratio 2, on Euler's line. (See **Definition 4.1.4.**)

Since the point E is the middle point of OH, E is on the perpendicular bisector of the segment O_1H_1. Therefore, the circle $(O_1O_2O_3)$ passes through the points H_1, H_2 and H_3 the feet of the altitudes.

(III) The two circles (ABC) and $(O_1O_2O_3)$ have also a positive homothety with ratio $\frac{1}{2}$. As we have seen (**Theorem 4.1.2** and **Definition 4.1.4**), its center is the conjugate of G (the center of the negative homothety), which, by the above equation, is H. Then, under this homothety the images of the vertices A, B, and C are the midpoints P_1, P_2 and P_3 of the segments sides HA, HB, and HC, respectively. Therefore, P_1, P_2 and P_3 belong to the circle $(O_1O_2O_3)$.

Finally, the nine points: O_1, O_2 and O_3, and H_1, H_2 and H_3, and P_1, P_2 and P_3 belong to the circle $(O_1O_2O_3)$, **the nine-point circle.**

∎

There are many properties of the nine-point circle. Use known facts of plane geometry and facts developed here to prove the following properties listed below. (In **Figure 4.33**, we have depicted some of them. Provide new figures when necessary. You may also observe properties not listed here or in the exercises.)

(a) If U is any point of the circle (ABC), then the midpoint V of HU is on the circle $(O_1O_2O_3)$. If $V = O_1$, then U, O and A are on the same diameter of the circle (ABC). Also, H_1 is the midpoint of HH_1', etc.

(b) The points E, O_1 and P_1 are on the same diameter of the circle $(O_1O_2O_3)$. (Similar result for the other two analogous triples of points.)

(c) The lines OA and O_1P_1 are parallel, and $OB \parallel O_2P_2$ and $OC \parallel O_3P_3$.

(d) The tangent of (ABC) at A and the tangents of $(O_1O_2O_3)$ at O_1 and P_1 are parallel (and similarly for the other two vertices.)

(e) $OO_1 = \frac{1}{2}HA = HP_1 = P_1A$. Analogous result with OO_2 and OO_3.

(f) The angle of the tangent of the circle $(O_1O_2O_3)$ at O_1 with the side BC has absolute value equal to $|\widehat{B} - \widehat{C}|$ (and similarly for O_2 and O_3).

(g) The projections B' and C' of B and C, respectively, on the bisector of the angle $\widehat{A}$ of the triangle ABC, and the points O_1 and H_1 are on the same circle. The center of this circle is *the midpoint of the minor arc $\overset{\frown}{O_1H_1}$* of the circle $(O_1O_2O_3)$. Similarly, if B'' and C'' are the projections of B and C on the external bisector of $\widehat{A}$, respectively, then B'', C'', O_1 and H_1 are on the same circle. The center of this circle is *the midpoint of the major arc $\overset{\frown}{O_1H_1}$* of the circle $(O_1O_2O_3)$. (Make figures.)

(h) Let I_a, I_b and I_c be the excenters of the triangle ABC. The nine-point circle of the triangle $I_aI_bI_c$ is the circle (ABC).

(i) From the vertices of the triangle ABC we draw straight lines parallel to the opposite sides. They form the triangle $A_1B_1C_1$, the **anti-medial triangle** of ABC. The nine-point circle of the triangle $A_1B_1C_1$ is the circle (ABC).

(j) The projections of the orthocenter H of ABC onto the internal and external bisectors of the angle $\widehat{A}$, the center of the nine-point circle of ABC and the midpoint of the side BC are collinear points.

(k) In any straight segment through the point G, with one end-point on the circle $(O_1O_2O_3)$ and the other end-point on the circle (ABC), the point G is one of the two trisecting points of this segment.

▲

Example 4.2.4 Application 4

Problem: Consider an isosceles triangle ABC, $AB = AC$, with the side AB fixed and the side AC rotating around A. Find the geometrical locus of the centers of gravity of the isosceles triangles ABC.
   Solution. See **Figure 4.34**.

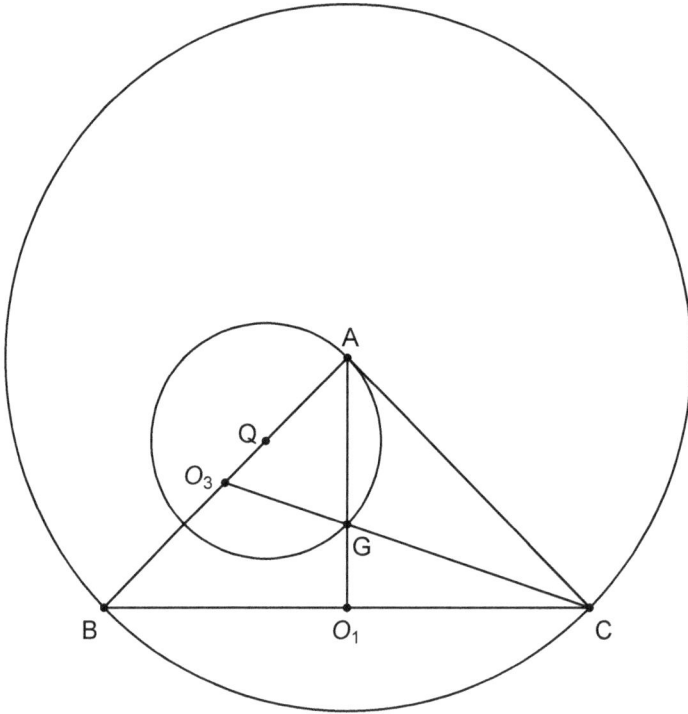

Figure 4.34: Example, A.HO. 4

   The vertex C moves on the circle with center A and radius AB. If O_3 is the midpoint of AB, fixed point, the center of gravity (centroid) G of ABC is the image of C under the homothety

$$H_{\left[O_3,\frac{1}{3}\right]} = H_{\left[A,\frac{2}{3}\right]} \circ H_{\left[B,\frac{1}{2}\right]}.$$

(See **Subsection 4.2.1** for the composition of homotheties.)
   Therefore, the geometrical locus of G is the circle homothetic to the circle with center A and radius AB, under the above homothety. This is the circle with radius $\dfrac{1}{3}AB$, and center Q, the homothetic of A, which is the point Q of the segment AO_3, determined by $\overrightarrow{QO_3} = \dfrac{1}{3}\overrightarrow{AO_3}$.

▲

Example 4.2.5 Application 5

Theorem 4.2.4 (Menelaus's Theorem) *Consider any triangle ABC and three points A', B' and C' on the straight lines of its sides BC, CA and AB, respectively. Then: A', B' and C' are on the same straight line if and only if they satisfy the condition*

$$\frac{\overrightarrow{A'B}}{\overrightarrow{A'C}} \cdot \frac{\overrightarrow{B'C}}{\overrightarrow{B'A}} \cdot \frac{\overrightarrow{C'A}}{\overrightarrow{C'B}} = 1.$$

Proof. See **Figure 4.35**.

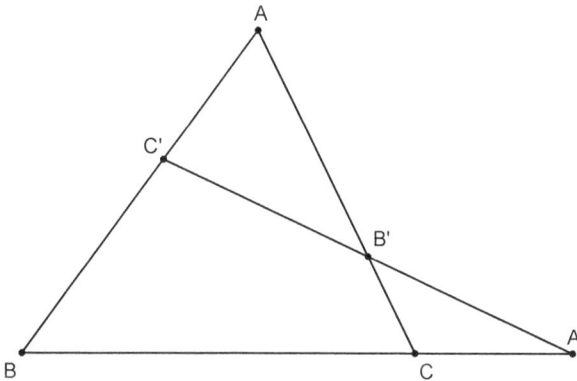

Figure 4.35: Example, A.HO. 5

We consider the homothety H_1 with center A' and mapping C to B, and the homothety H_2 with center C' and mapping B to A. Then $H_2 \circ H_1$ maps C to A. Also, as we know in any homothety, the center of the homothety and two homothetic points are collinear.

By **Subsection 4.2.1, (II)**, $H_2 \circ H_1$ is a new homothety with ratio the product of the ratios of H_1 and H_2, $\lambda = \dfrac{\overrightarrow{A'B}}{\overrightarrow{A'C}} \cdot \dfrac{\overrightarrow{C'A}}{\overrightarrow{C'B}}$, and center the common point of AC and $A'C'$.

So, B' is the center of $H_2 \circ H_1$ if and only if

$$\frac{\overrightarrow{B'A}}{\overrightarrow{B'C}} = \frac{\overrightarrow{A'B}}{\overrightarrow{A'C}} \cdot \frac{\overrightarrow{C'A}}{\overrightarrow{C'B}}, \quad \text{which is equivalent to} \quad \frac{\overrightarrow{A'B}}{\overrightarrow{A'C}} \cdot \frac{\overrightarrow{B'C}}{\overrightarrow{B'A}} \cdot \frac{\overrightarrow{C'A}}{\overrightarrow{C'B}} = 1.$$

∎

(See and prove the **sine form of Menelaus's Theorem** in the exercises. This form is convenient in many situations.)

The dual of **Menelaus's**[2] **Theorem** is **Ceva's**[3] **Theorem**, in the next example.   ▲

---

[2]Menelaus of Alexandria, Greek mathematician, c.70-c.130.
[3]Giovanni Benedetto Ceva, Italian mathematician, 1647-1734.

Example 4.2.6 Application 6

Theorem 4.2.5 (Ceva's Theorem) *Consider any triangle ABC and three points A', B', and C' on its sides BC, CA, AB, respectively. Then: the straight lines AA', BB', and CC' are concurrent (i.e., they pass through the same point) if and only if they satisfy the condition*

$$\frac{\overrightarrow{A'B}}{\overrightarrow{A'C}} \cdot \frac{\overrightarrow{B'C}}{\overrightarrow{B'A}} \cdot \frac{\overrightarrow{C'A}}{\overrightarrow{C'B}} = -1.$$

Proof. See **Figure 4.36**.

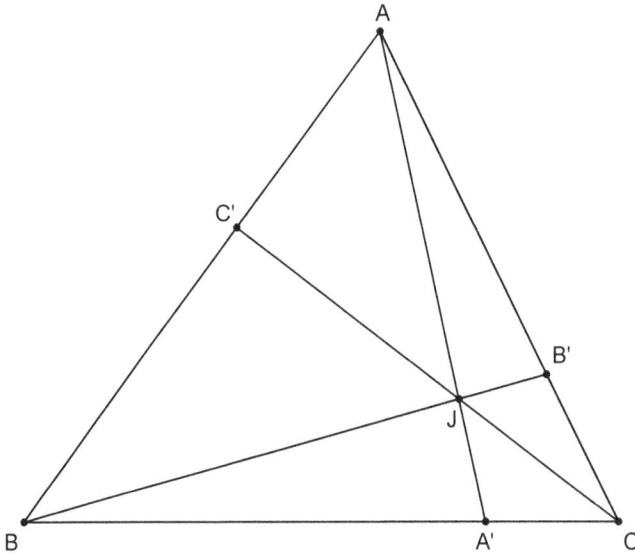

Figure 4.36: Example, A.HO. 6

The proof follows by applying **Menelaus's Theorem, 4.2.5**, to some to the triangles formed. Try to give a proof, or find and study it in another book.
∎

Any straight lines AA', BB', and CC', as above, concurrent or non-concurrent, are called **Cevians** for the triangle ABC.

Remark: Using **this Theorem** we can prove that many of the important triads of lines in a triangle are concurrent cevians. [For example: medians, symmedians, altitudes, angle bisectors (interior / exterior), and many more.]

(See and prove the **sine form of Ceva's Theorem** in the exercises. This form is convenient in many situations.)   ▲

Example 4.2.7 Application 7

Problem: Consider two lines l and l' in the plane and a point A. Construct a circle through A and tangent to both lines.
   Solution. See **Figure 4.37**.

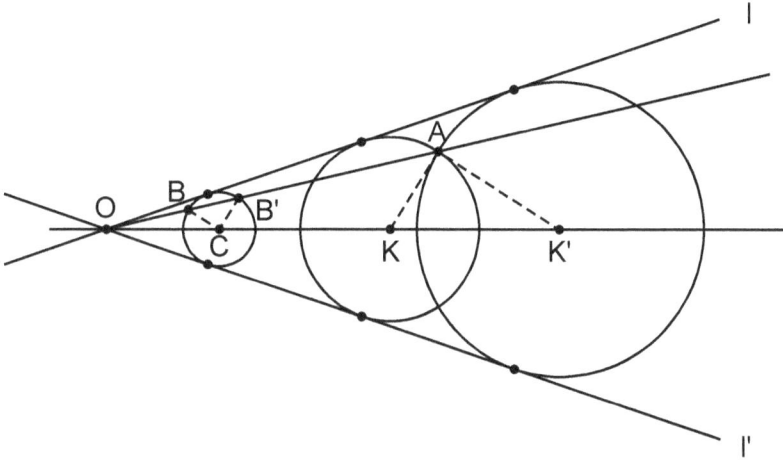

Figure 4.37: Example, A.HO. 7

The point A is in an angle formed by the lines l and l'. Any two circles inscribed in this angle are homothetic with center of homothety O the common point of l and l'. So, each of the inscribed circles can be considered as homothetic to a fixed circle with center C on the bisector of the angle of l and l' that contains A and tangent to the lines.
   So if a circle with center K is a solution, K is on OC. Let B and B' be the common points of OA with the fixed circle with center C and tangent to L and l'. Then, KA is parallel to CB or CB'. Now, the construction proceeds in the following steps:

   (a) We make a circle with center C on the bisector of the angle of l and l' that contains A and tangent to the lines. (Easy construction.)
   (b) We draw the line OA and let B and B' the common points with the above circle.
   (c) We draw from A lines parallel to CB and CB'. These lines intersect the bisector of this angle at the points K and K' respectively.
   (d) With centers K and K' and radii KA and $K'A$ we draw circles. These are two solutions of the problem.

   Note: Examine the cases in which A is on one of the lines l or l', or when the two lines are parallel ($l \parallel l'$).

Example 4.2.8 Application 8

In **Figures 4.38-4.41**, that follow, we show how to find circles tangent to two given circles using their positive and negative homotheties that always exist.

Suppose that the given circles C and C' have centers K and K' and a third circle $\mathcal{T}$, with center C, is tangent to both of them at the points T and T', respectively. Then, T is a center of homothety of C, and $\mathcal{T}$ and T' is a center of homothety of C' and $\mathcal{T}$.

If the contacts of $\mathcal{T}$ with C and C' are both external or both internal, then the points T and T' are centers of either positive or negative homotheties and the line TT' passes through the center of the positive homothety of the circles C and C'. [By the similar isosceles triangles formed in **Figures 4.38, 4.39** and **Subsection 4.1.1, (j)** (explained at the final two paragraphs of this example), or by **Application 9** that follows next.]

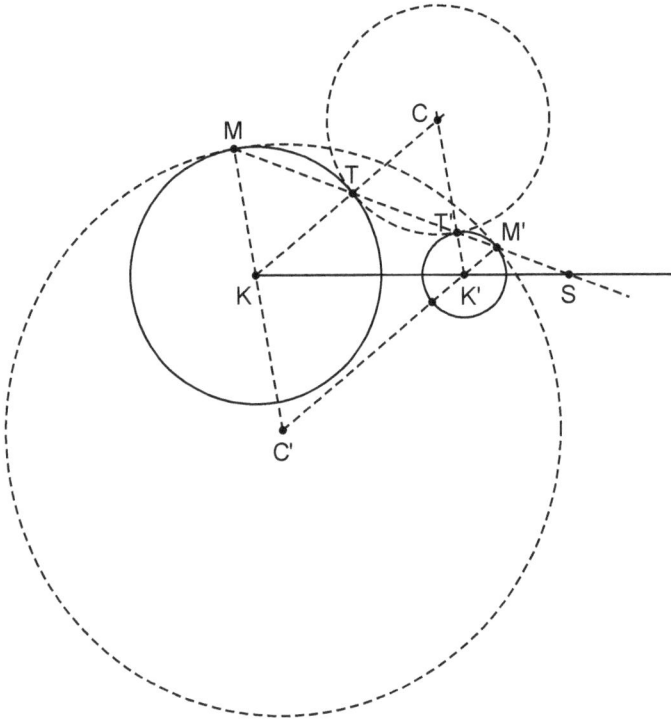

Figure 4.38: Example, A.HO. 8

If the contacts of $\mathcal{T}$ with C and C' are one external and the other internal, then the points T and T' are centers of one positive and one negative homotheties and the line TT' passes through the center of the negative homothety of the circles C and C'. [By the similar isosceles triangles formed in **Figures 4.40, 4.41** and **Subsection 4.1.1, (j)**, or by **Application 9** that follows next.]

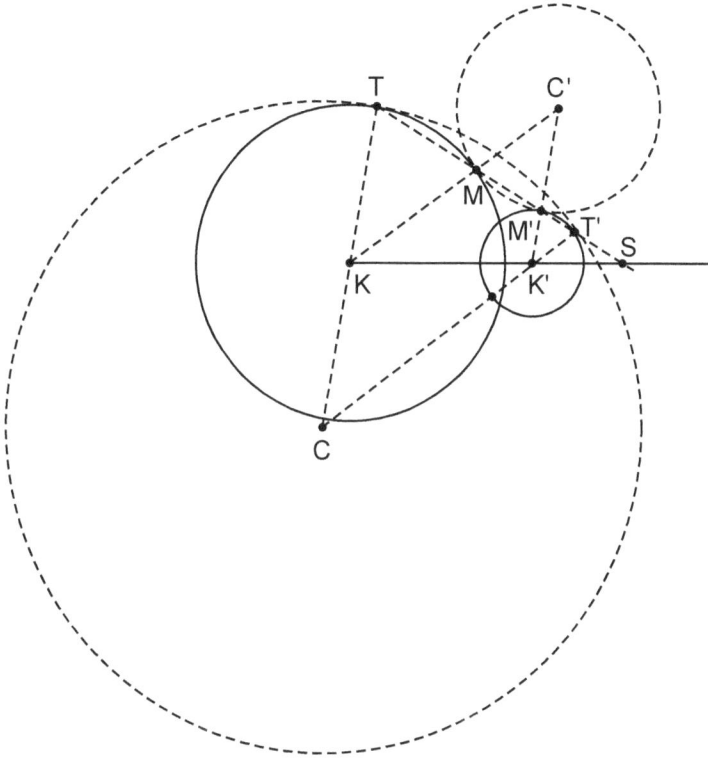

Figure 4.39: Example, A.HO. 8

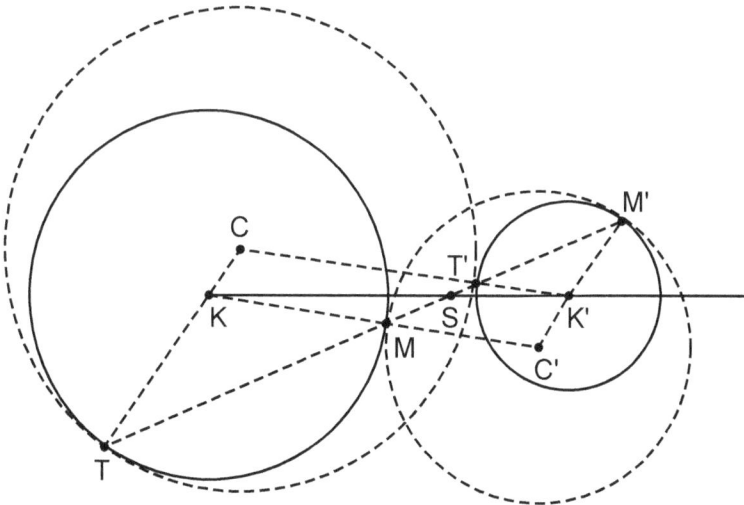

Figure 4.40: Example, A.HO. 8

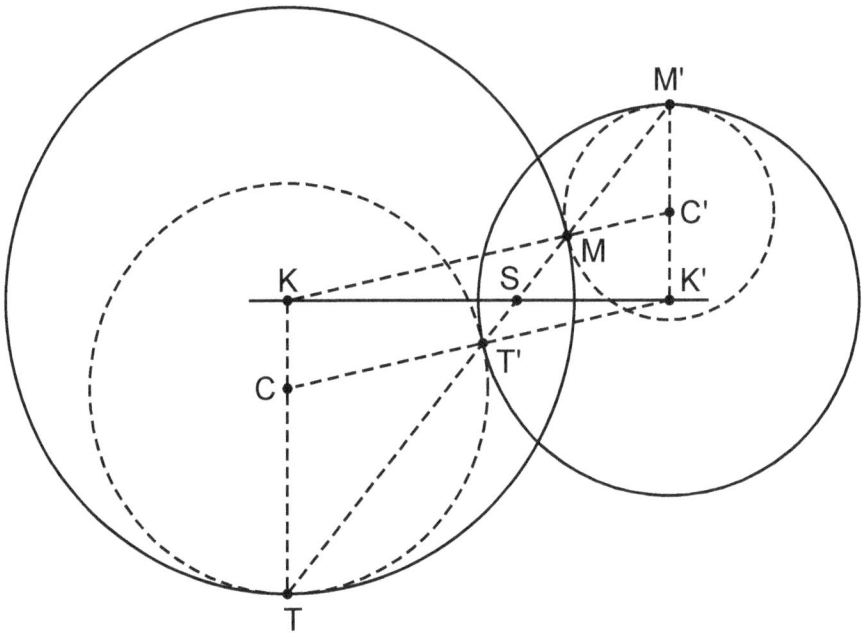

Figure 4.41: Example, A.HO. 8

So, we can construct a circle $\mathcal{T}$ tangent to the circles $\mathcal{C}$ and $\mathcal{C}'$ by drawing a line through a center S of one of the two homotheties and intersecting both circles. Let T and M be the points of intersection with the circle $\mathcal{C}$ and T' and M' the points of intersection with the circle $\mathcal{C}'$. If T' is not homothetic to T but to M, then the line KT' is not parallel to KT (because $K'T' \parallel KM$), and so the lines $K'T'$ and KT intersect at a point C.

The circle with center C and radius CT, is tangent to $\mathcal{C}$ at T and passes through T'. This is so, because

$$KM \parallel CT'$$

and so the triangles TKM and TOT' are similar. Also, the triangle KTM is isosceles and so similar to CTT'. Therefore,

$$CT = CT'.$$

Then, the circle $\mathcal{T}$ is tangent to both $\mathcal{C}$ and $\mathcal{C}'$.

Remark: In **Figures 4.38, 4.39, 4.40**, and **4.41**, the quadrilateral $CKK'C'$ is a parallelogram. Why?

Note: Examine the cases in which the two given circles are either externally or internally tangent and draw figures.

▲

Example 4.2.9 Application 9 (D' Alembert.[4])

Consider the six centers of homotheties of three circles. Any three of them in which there are either 0 or 2 centers of negative homotheties are collinear. That is, the centers of positive homotheties are collinear and any two centers of negative homotheties are collinear with the center of the appropriate positive homothety.

Solution. See **Figure 4.42**.

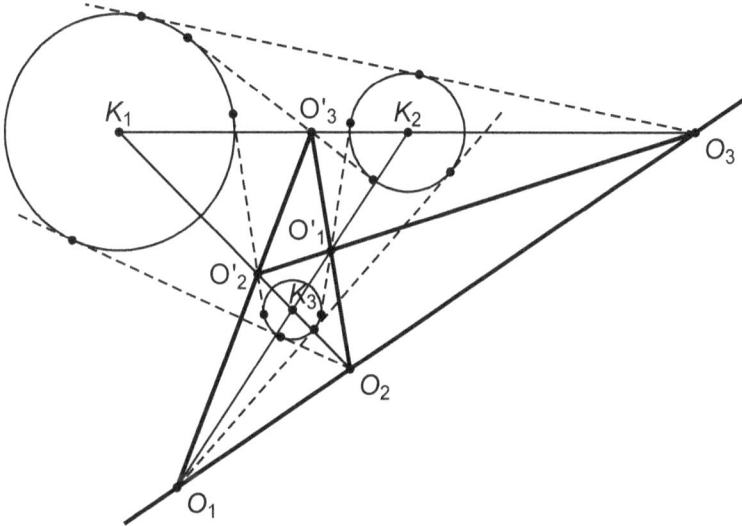

Figure 4.42: Example, A.HO. 9

Let K_1, K_2 and K_3 be the centers of the three circles, O_1, O_2 and O_3 the centers of the positive homotheties, and O'_1, O'_2 and O'_3 the centers of the negative homotheties, as in **Figure 4.42**. These, as we know, are determined by the center lines of the circles intersected by the external and internal tangents to the circles, respectively. By the results on compositions of homotheties in **Subsection 4.2.1**, the center of the composition of two homotheties is on the straight line defined by their centers or is an infinity point. Also, two inverse homotheties have the same center. Hence, each of the following triples of points

$$\{O_1, O_2, O_3\}, \qquad \{O'_3, O'_1, O_2\}, \qquad \{O'_2, O'_3, O_1\}, \qquad \{O'_1, O'_2, O_3\}.$$

is on the same straight line. (Complete the justification.)

Remarks: (1) Use **Menelaus's Theorem, 4.2.4**, to provide a different proof.

(2) Use **Ceva's Theorem, 4.2.5**, to prove that the triples of straight lines $\{K_1O'_1, \ K_2O'_2, \ K_3O'_3\}$, $\{K_1O'_1, \ K_2O_2, \ K_3O_3\}$, $\{K_1O_1, \ K_2O'_2, \ K_3O_3\}$, and $\{K_1O_1, \ K_2O_2, \ \text{and} \ K_3O'_3\}$, (not drawn in **Figure 4.42**) are concurrent.

---

[4]D' Alembert Jean-Batiste le Rond, French mathematician, 1717-1783.

(3) If two of the circles are equal, some of the above lines are parallel. See **Figure 4.43**. This is related to the results on compositions of homotheties in **Subsection 4.2.1**, when the two composed homotheties have reciprocal ratios.

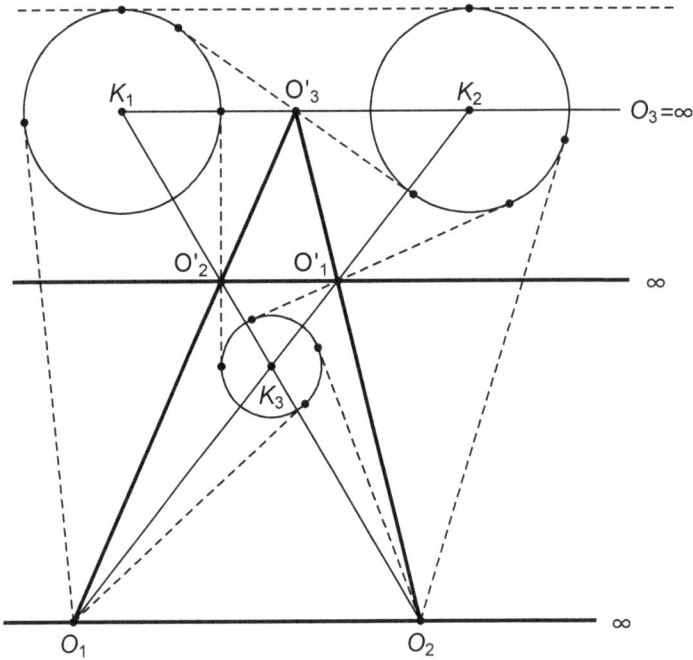

Figure 4.43: Example, A.HO. 9

(Make another figure with all three circles equal. Describe and explain what you observe!)

▲

Example 4.2.10 Application 10

In a given circular sector AOB of an angle $\widehat{AOB} < 90°$, inscribe a square $WXYZ$ such that the side WZ is on the radius OA, the vertex X is on the radius OB and the vertex Y is on the arc $\overarc{AB}$.

Solution. See **Figure 4.44**.

If we assume that we have constructed such a square $WXYZ$, then any homothetic square, e.g., $CDEF$, under a homothety $H_{[O,\lambda]}$, for any $0 < \lambda \neq 1$ has the side CF is on the radius OA, the vertex D is on the radius OB and the vertex E is on the line OY.

So, to find the demanded square, we construct any square $CDEF$ with vertex C chosen on the radius OA, and then D on the radius OB, such that $CD \perp OA$. The segment DC is the side of this preliminary square. Then, we find E inside the circular sector and finally F on the radius OA, as we know.

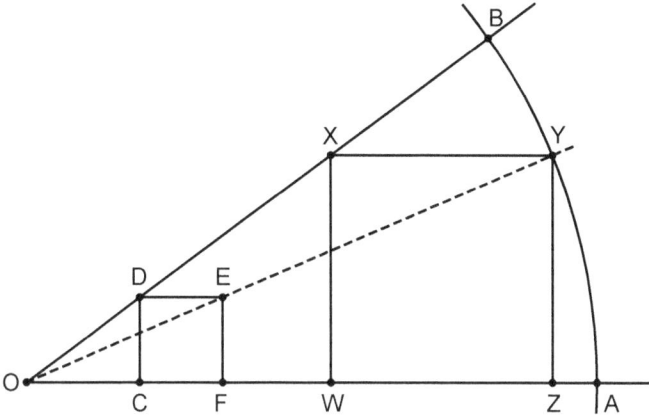

Figure 4.44: Example, A.HO. 10

Now, the construction and its justification are done easily. We draw the line OE and let Y be its intersection with the arc $\overset{\frown}{AB}$. Then, we draw $YZ \perp OA$. The segment YZ determines the side of the sought square. We complete the square by finding W and X, as we know, etc.

▲

Example 4.2.11 Application 11

A Problem of Archimedes[5] We consider a straight segment AB and an interior point C. With diameters AB, AC and CB, we draw three semicircles in the same half plane of the straight line AB. The plane region between these three semicircles is called **arbelo** (= knife of the shoemaker, in ancient Greek).

If O, F and E are the centers of these semicircles, then the corresponding radii are $R_O = \dfrac{AB}{2}$, $R_F = \dfrac{AC}{2}$ and $R_E = \dfrac{CB}{2}$. So, $R_O = R_F + R_E$. We also draw the half line starting at C and perpendicular to AB and let G the point of intersection with the semicircle on AB. See **Figure 4.45**.

We consider the two circles tangent to CG and two of the three semicircles, with center K_1 and K_2, as in **Figure 4.45**. We will prove that these circles have equal radii: $r_1 = r_2 = \dfrac{AC \cdot CB}{2 \cdot AB} = \dfrac{R_F \cdot R_E}{R_O} = \dfrac{R_F \cdot R_E}{R_F + R_E}$.

Solution. See **Figure 4.45**.

We prove this for the circle with center K_2 and similar work proves the same result for the circle with center K_1.

---

[5] Archimedes of Syracuse, Sicily (today in Italy). Greek mathematician, 287-212 BCE. He is considered by the great majority of mathematicians to be the greatest mathematician of all times. He is the first who conceived the process of integration by which he found the area of the circle of radius r to be $A = \pi r^2$, the length of its circumference $c = 2\pi r$ and also $\int_{-1}^{1} x^2 dx = \frac{2}{3}$. He also found the first approximation of π, that is, π was less than $3 + \frac{1}{7} = \frac{22}{7}$ but greater than $3 + \frac{10}{71} = \frac{223}{71}$. In the decimal notation we use today, this value is between 3.1429 and 3.1408. That is pretty close to the approximate value usually used 3.1416.

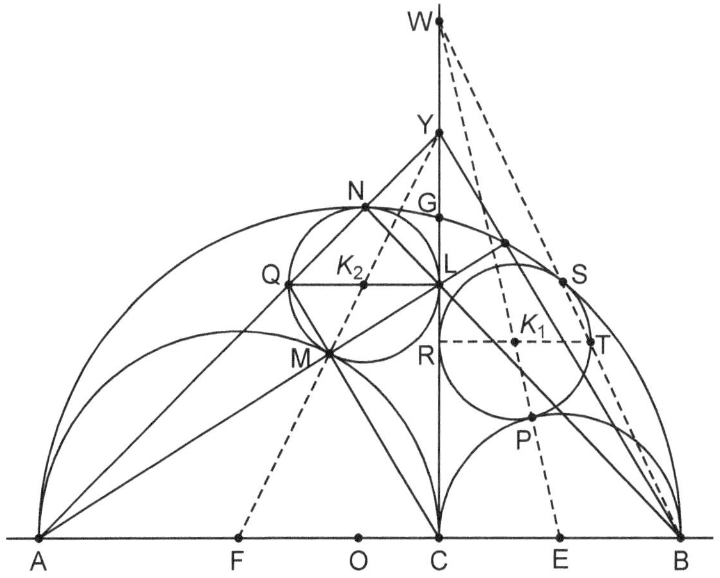

Figure 4.45: Example, A. HO. 11

Let this circle touch CG, and arcs AC and AB, at points L, M and N respectively, and let QL be the diameter $\perp CG$. The points of tangency M and N are centers of respective homotheties. Hence, the parallel diameters $QL \parallel AC$ correspond to each other, AL and CQ meet at M, and AQ and BL meet at N. Since QL is a diameter, the respected angles at M and N are right. We extend AN to meet CG at Y and observe that the altitudes CY and BN of the triangle ABY meet at L. Therefore, the (extended) AL is an altitude and so $BY \parallel QC$, both being perpendicular to AL.

Then, using **Thales'**[6] **Theorem** or similar triangles, we find

$$\frac{QL}{AC} = \frac{QY}{AY} = \frac{CB}{AB} \qquad \text{and so} \qquad 2\,r_2 = QL = \frac{AC \cdot CB}{AB}$$

and the result follows.

Remark: Observe that the points F, M, K_2 and Y are on the median of the right triangle AYC and form a harmonic quadruple with ratio

$$-\frac{\overline{MK_2}}{\overline{MF}} = \frac{\overline{YK_2}}{\overline{YF}} = \frac{r_2}{R_F} = \frac{R_E}{R_O} = \frac{R_E}{R_F + R_E}.$$

State the analogous result for the points E, P, K_1 and Y and the corresponding right triangle CBW of this case.

(See also **Example 6.6.6**, for two different solutions.)

▲

---

[6]Thales of Miletus, pre-Socratic Greek philosopher, mathematician and astronomer, c.624-c.546 BCE.

4.3 General Properties of Plane Similarities

The set of similarities of the plane has the following properties.

1. Every similarity is continuous at every point of the plane and it is uniformly continuous on the whole plane.

2. Every similarity is one to one.

3. Every similarity is onto.

4. The inverse of a similarity is a similarity with ratio the reciprocal of the ratio of the given similarity.

5. The composition of two similarities is a similarity with ratio the product of the two ratios. (Therefore, the composition of a similarity with an isometry is a new similarity with similarity ratio equal to the similarity ratio of the initial similarity.)

6. The identity mapping is a similarity (with ratio 1).

7. The set of the plane similarities is an algebraic group under the operation of composition, which is non-commutative.

8. Every similarity with ratio $\neq 1$ (i.e., not an isometry) has one unique fixed point. (Isometries have ratio 1 and, as we have seen, they have either 0, or 1, or ∞ fixed points.)

9. A similarity maps a straight line to a straight line (a straight segment to a straight segment and the interior of a straight segment or a vector to the interior of a straight segment or a vector), a circle to a circle (not necessarily equal), a triangle to a directly or oppositely similar triangle, an oriented angle to an equal or opposite oriented angle.

Proofs: We consider $S : \mathcal{P} \longrightarrow \mathcal{P}$ a similarity of the plane with ratio $\lambda > 0$.

Properties (1), (2), and **(3)** are proved in exactly the same way as the corresponding properties of the isometries **(1), (2),** and **(3)** as proven in **Section 2.2,** but adjusted to the factor $\lambda > 0$.

(4) Since S is one to one and onto the set-theoretic inverse S^{-1} is a well-defined function. Suppose $S^{-1}(Y) = X$ and $S^{-1}(Z) = W$. Then $S(X) = Y$ and $S(W) = Z$. Since S is a similarity, we get

$$d[S(X), S(W)] = \lambda \cdot d(X, W) \quad \text{and so} \quad d(Y, Z) = \lambda \cdot d[S^{-1}(Y), S^{-1}(Z)].$$

Thus,

$$d[S^{-1}(Y), S^{-1}(Z)] = \frac{1}{\lambda} \cdot d(Y, Z),$$

proving that S^{-1} is a similarity with ratio $\dfrac{1}{\lambda}$.

(5) We consider S, R : $\mathcal{P} \longrightarrow \mathcal{P}$ two similarities of the plane with ratios λ and ρ, respectively. For any two points A, $B \in \mathcal{P}$ we have:

$$d[(R \circ S)(A), (R \circ S)(B)] = d\{R[S(A)], R[S(B)]\}$$
$$= \lambda \cdot d[F(A), F(B)] = \lambda\rho \cdot d(A, B),$$

proving that $R \circ S$ is a similarity with ratio $\lambda\rho$.

(If R is an isometry, then $\rho = 1$ and $R \circ S$ is a new similarity with ratio $\lambda\rho = \lambda \cdot 1 = \lambda$.)

(6) Since for the identity mapping we have $I(A) = A$, for all $A \in \mathcal{P}$, the result follows immediately: I is an isometry and a similarity with ratio 1.

(7) Let $\mathcal{S}$ be the set of the plane-similarities of all ratios. **Properties (4), (5)**, and **(6)** prove that $(\mathcal{S}, \circ)$ is an algebraic group, since the associative property is always satisfied by the operation of composition. But, $(\mathcal{S}, \circ)$ is not a commutative group. We have already seen many compositions, and we will see more, that do not commute. (Composition is not commutative, in general.)

(8) Let S be a similarity with ratio $\lambda \neq 1$. It is enough to prove the claim for any $0 < \lambda < 1$. For, if $\lambda > 1$, then S^{-1} is a similarity with ratio $0 < \dfrac{1}{\lambda} < 1$ and would have one unique fixed point O, i.e., $S^{-1}(O) = O$. But then, $S(O) = O$ and so O would also be the unique fixed point of S. For the proof, we need to know the necessary results from analysis and topology.

(By the way, in mathematical analysis, a similarity S with ratio $0 < \lambda < 1$ is called a **contraction mapping** and if $\lambda > 1$ is called an **expansion mapping**.)

Uniqueness: First of all, for any similarity with ratio $0 < \lambda < 1$, the fixed point (if it exists) must be unique. If we assume that O and Q are fixed points, we must show that $O = Q$. Indeed:

$$d(O, Q) = d[S(O), S(Q)] = \lambda \cdot d(O, Q).$$

Then,

$$(1 - \lambda) \cdot d(O, Q) = 0.$$

Since, $1 - \lambda > 0$, we conclude that $d(O, Q) = 0$. Therefore, $O = Q$.

Existence: We pick any point $P \in \mathcal{P}$ and we define the sequence of points $P_0 = P$ and $P_n = S(P_{n-1})$, for $n = 1$, 2, 3, ...

If for some $n \in \mathbb{N}$ we have $P_{n-1} = P_n$, that is, if $P_{n-1} = S(P_{n-1})$, then the point $O := P_{n-1}(= P_n = P_{n+1} = \ldots)$ is the fixed point.

In general though, for the sequence of points (P_n), $n \in \mathbb{N}_0$, in the plane, we notice that: For any $n \in \mathbb{N}$, we have

$$0 \leq d(P_{n+1}, P_n) = d[S(P_n), S(P_{n-1})] = \lambda \cdot d(P_n, P_{n-1})$$
$$= \lambda \cdot d[S(P_{n-1}), S(P_{n-2})] = \lambda^2 d(P_{n-1}, P_{n-2}) = \ldots = \lambda^n d(P_1, P_0).$$

So, for any $n \geq m$ in $\mathbb{N}$, we get

$$0 \leq d(P_m, P_n)$$
$$\leq d(P_m, P_{m+1}) + d(P_{m+1}, P_{m+2}) + \ldots + d(P_{n-1}, P_n)$$
$$= \left(\lambda^m + \lambda^{m+1} + \ldots + \lambda^{n-1}\right) \cdot d(P_1, P_0)$$
$$= \frac{\lambda^m - \lambda^n}{1 - \lambda} \cdot d(P_1, P_0) \longrightarrow \frac{0 - 0}{1 - \lambda} \cdot d(P_1, P_0) = 0, \quad \text{as} \quad n \geq m \longrightarrow \infty,$$

because $0 < \lambda < 1$ and $d(P_1, P_0)$ is constant.

This proves that the sequence (P_n), $n \in \mathbb{N}_0$, is a Cauchy sequence and therefore it converges to a point $O \in \mathcal{P}$, because the plane, $\mathcal{P}$, is a complete metric space.[7] That is, $P_n \longrightarrow O$, as $n \longrightarrow \infty$.

But then,

$$S(P_n) = P_{n+1} \longrightarrow O, \quad \text{as} \quad n \longrightarrow \infty.$$

By, the continuity of S, we have:

$$O = \lim_{n \to \infty} P_{n+1} = \lim_{n \to \infty} S(P_n) = S\left(\lim_{n \to \infty} P_n\right) = S(O).$$

This proves that O is the fixed point of S, i.e., $S(O) = O$.

Remark: In the sections that follow, we will show how to construct (find) the fixed point by straight edge and compass. This, of course, will prove its existence synthetically.

(9) Consider any straight line k and two different points A and B on k. We must prove that $F(k)$ is the straight line l through the points $S(A)$ and $S(B)$. We pick any point $C \in k$. Then we have,

$$\overrightarrow{AC} + \overrightarrow{CB} = \overrightarrow{AB}$$

with A, B, and C on the straight like k. Also,

$$\overrightarrow{S(A)S(C)} + \overrightarrow{S(C)S(B)} = \overrightarrow{S(A)S(B)}.$$

Since S is a similarity, we have

$$\lambda \cdot d(A, B) = d[S(A), S(B)],$$
$$\lambda \cdot d(A, C) = d[S(A), S(C)],$$
$$\lambda \cdot d(B, C) = d[S(B), S(C)].$$

---

[7]In the Axiomatic Euclidean Geometry the completeness of the plane is an axiom interpreted in various equivalent ways. It is called Axiom of Dedekind, or continuity, or completeness. (Julius Wilhelm Richard Dedekind, German mathematician, 6 October 1831 - 12 February 1916.) A complete metric space is also called a Banach space (after Stefan Banach, Polish mathematician, 30 March 1892 - 31 August 1945, who distinguished the importance of complete spaces in all mathematics).

The axiom of Dedekind is a rephrase of the Axiom of Archimedes. Archimedes credited it to Eudoxus and so is also called the Axiom of Eudoxus or of Archimedes-Eudoxus. (Eudoxus of Cnidus, Greek mathematician, 390-337 BCE.)

So, the signed lengths of the pertinent segments satisfy

$$\overline{S(A)S(C)} + \overline{S(C)S(B)} = \overline{S(A)S(B)}.$$

See **Figures 4.46** and **4.47**.

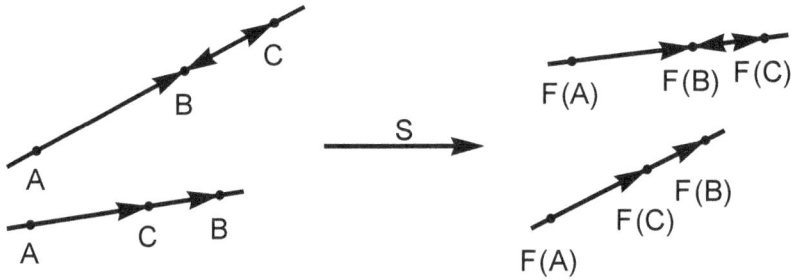

Figure 4.46: Similarities map straight lines to straight lines

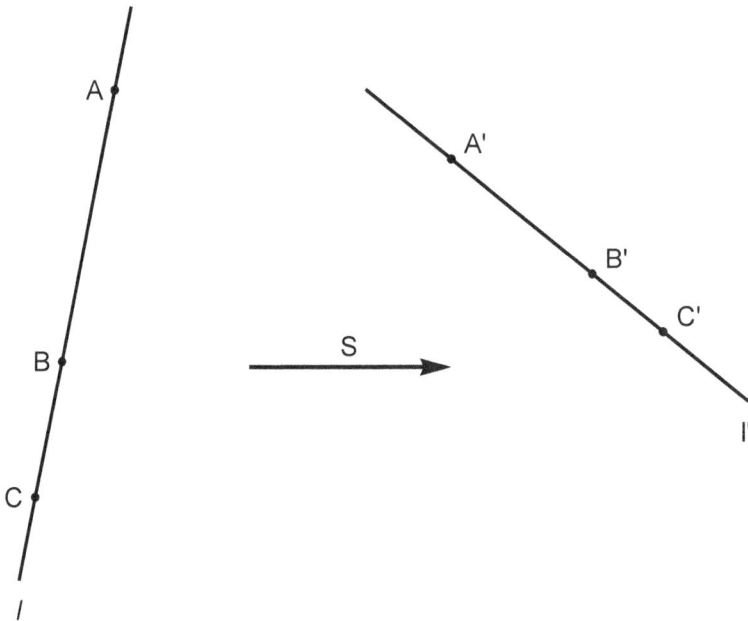

Figure 4.47: Similarities map straight lines to straight lines

Therefore, any one of the points $S(A)$, $S(B)$ and $S(C)$ lies on the straight line defined by the other two. Thus, $S(k) \subseteq l$. Since S is one to one and onto, and preserves a fixed proportion of distances, it must be $F(k) = l$.

Also, a circle $C[K, r]$ with center K and radius r has image

$$S\{C[K, r]\} = C[S(K), \lambda r],$$

the circle with center $S(K)$ and radius λr. See **Figure 4.48**.

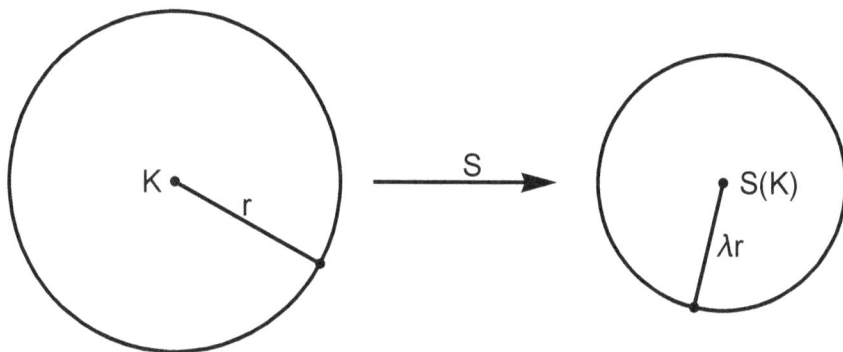

Figure 4.48: Similarities map circles to circles

Since a similarity S maps a triangle ABC to a triangle $A'B'C'$ with corresponding sides proportional, the two triangles are similar. (The ratio of proportion is the ratio $\lambda > 0$ of the similarity S.) These triangles are either directly similar, i.e., their corresponding oriented angles are equal, or are oppositely similar, i.e., their corresponding oriented angles are opposite. See **Figure 4.49**.

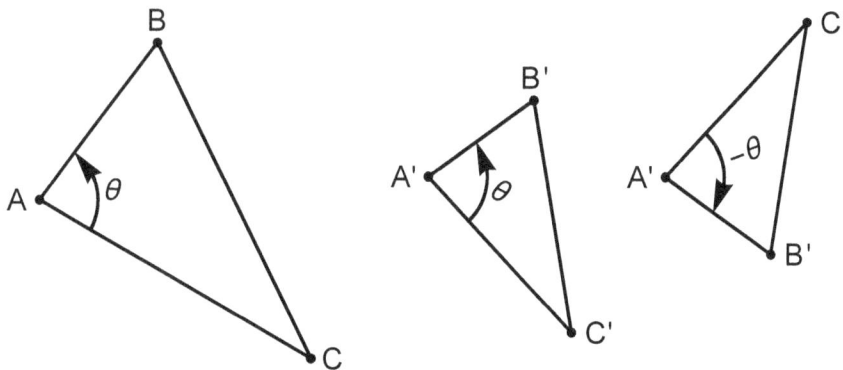

Figure 4.49: Similarities map triangles to directly or oppositely similar triangles

Note: The uniqueness part of **Property (8)** above, can also be proven by the following Lemma:

Lemma 4.3.1 *If a similarity S has more than one fixed points, then $\lambda = 1$ and so S is an isometry, which must be either the identity or a reflection. Therefore, a similarity with $0 < \lambda \neq 1$ has at most one fixed point.*

Proof If $P \neq Q$ are two fixed points of S then $S(P) = P$ and $S(Q) = Q$. Then

$$\lambda = \frac{d[S(P), S(Q)]}{d(P,Q)} = \frac{d(P,Q)}{d(P,Q)} = 1.$$

Therefore, S is an isometry.

If an isometry S is not the identity and has two fixed points, then, by **Theorem 2.2.2**, S is a reflection and the Lemma follows.

∎

The existence of the unique fixed point of a similarity S with similarity ratio $0 < \lambda \neq 1$ was also proven above, in **Property (8)**, using analysis and point-set topology. On the basis of the theory that follows until the end of **Subsection 4.5.2**, this can also be verified synthetically. This takes a combination of many results. (The converse with respect to the ratio λ is not true, since any non-trivial rotation has $\lambda = 1$ and exactly one fixed point.) We also conclude that, if S has no fixed points, then S must be a non-trivial parallel translation or a glide and so S is an isometry and $\lambda = 1$. So, a similarity with no or at least two fixed points must be an isometry. But, with exactly one fixed point, S may be either a rotation or a similarity with $0 < \lambda \neq 1$.

Theorem 4.3.1 (Determination of a similarity by 3 pairs of points) *A similarity $S : \mathcal{P} \longrightarrow \mathcal{P}$ is completely determined by the images A', B', and C' of three given points A, B, and C of the plane that do not lie on a straight line.*

Proof The proof of this Theorem is analogous to the **proof of Theorem 2.2.1**, adjusted by the factor of the ratio λ of the similarity S. See **Figure 4.50**.

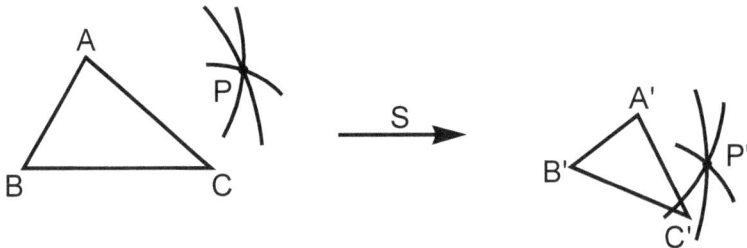

Figure 4.50: **A similarity is determined by a triangle and its image**

The ratio of the given similarity S, necessarily is

$$\lambda = \frac{A'B'}{AB} = \frac{B'C'}{BC} = \frac{C'A'}{CA}, \quad \text{and so it is known.}$$

(So, the triangles ABC and $A'B'C'$ are directly or oppositely similar.) Then, the image of any point $P \in \mathcal{P}$, $P' = S(P)$, is the intersection of the three circles

$$\mathcal{C}[A', \lambda AP], \qquad \mathcal{C}[B', \lambda BP], \quad \text{and} \quad \mathcal{C}[C', \lambda CP].$$

So, we state and prove a Lemma analogous to **Lemma 2.2.1** and then we prove this Theorem in a way analogous to the **proof of Theorem 2.2.1**.

■

By this **Theorem** and **property (9)** above, we obtain the Corollary:

Corollary 4.3.1 *If a similarity S maps **one** triangle to a directly or an oppositely similar triangle, then S maps every triangle to a directly or oppositely equal triangles, respectively. Therefore, the similarity S preserves or reverses every oriented angles, respectively.*

The **proof** is analogous to the **proofs of Corollaries 2.2.1** and **2.2.2**, in view of **Property (9)** and the **previous Theorem**. (Reproduce it as an exercise.)

■

So, as we did with the isometries, according to the **last Property, (9)**, and the **above Corollary**, we give the following definition:

Definition 4.3.1 *A **similarity** is called **direct**, if it preserves the oriented angles. It is called **opposite**, if it maps the oriented angles to their opposites.*

That is, given a similarity S and **any two non-zero** vectors $\overrightarrow{AB}$ and $\overrightarrow{CD}$, we let $\overrightarrow{A'B'} = S(\overrightarrow{AB})$ and $\overrightarrow{C'D'} = S(\overrightarrow{CD})$. Then, we have:
   (a) S is **direct** $\iff$ $\angle(\overrightarrow{AB}, \overrightarrow{CD}) = \angle(\overrightarrow{A'B'}, \overrightarrow{C'D'})$.
   (b) S is **opposite** $\iff$ $\angle(\overrightarrow{AB}, \overrightarrow{CD}) = -\angle(\overrightarrow{A'B'}, \overrightarrow{C'D'})$.
This definition can also be drawn from the **following Theorem** and its **Corollary**.

Theorem 4.3.2 (Decompositions of Similarities) *Every plane similarity can be decomposed into a composition of a positive homothety with center any point in the plane and homothetic ratio the ratio of the similarity and an isometry, in either order. Such a composition is not commutative, in general.*

Proof Consider any similarity S of ratio $\lambda > 0$, any point O of the plane $\mathcal{P}$ and the homothety $H_{[O,\lambda]}$. Since $H_{[O,\lambda]}^{-1} = H_{[O,\frac{1}{\lambda}]}$ is a similarity, [**Properties (h)** and **(g), Subsection 4.1.1**], the composition

$$S \circ H_{[O,\lambda]}^{-1} = S \circ H_{[O,\frac{1}{\lambda}]}$$

is a new similarity with ratio $\lambda \cdot \dfrac{1}{\lambda} = 1$ [**Properties (4) and (5), Section 4.3**].
   Therefore, $S \circ H_{[O,\lambda]}^{-1} := F$ is an isometry. Thus,

$$S = F \circ H_{[O,\lambda]}, \quad \text{with } F \text{ an isometry.}$$

If we consider $H_{[O,\lambda]}^{-1} \circ S$, we similarly prove that this is also an isometry G and therefore

$$S = H_{[O,\lambda]} \circ G, \quad \text{with } G \text{ an isometry,}$$

and $G \neq F$, in general.

∎

Remarks: (1) If $\lambda = 1$, then S is an isometry and $H_{[O,1]} = I$ identity and so the decompositions are trivial and they commute.

(2) Since O is any point, there are infinitely many such decompositions, which are not commutative, in general.

(3) Remarks analogous to the two **Remarks** that follow the **proof of Theorem 2.2.1** hold. That is:

(a) If the points A, B and C are on a straight line, the similarity may not be completely determined.

(b) If we assign the images of only two points, then there are exactly two similarities, one direct and one opposite, that map A to A' and B to B'. This conclusion is established through this Chapter. (See, **General Conclusion Based on Chapter 4**, at the end of this Chapter, before **Section 4.6**.)

Since homotheties preserve the oriented angles, from the **above Theorem** and the results on isometries, we obtain the following corollary.

Corollary 4.3.2 *In the* **previous Theorem**:

(a) S is a direct or an opposite similarity iff F (or G) is a direct or an opposite isometry, respectively.

(b) If S is a direct similarity, then F is a direct isometry and so, either a parallel translation or a rotation.

If F is a parallel translation, then S is a homothety.

(c) If S is an opposite similarity, then F is an opposite isometry and so, either a reflection or a glide.

If F is a glide, then S can be reduced to a composition of another homothety and a reflection.

Proof (a) This follows from the fact that any homothety is a direct similarity [**Property (e) of Subsection 4.1.1**], and so $S = F \circ H_{[Q,\lambda]}$ preserves or reverses the oriented angles in accordance to F preserving or reversing the oriented angles, respectively.

(b) If F is a direct similarity, then F is equal to a parallel translation or a rotation.

If $F = T_{\vec{u}}$ is a parallel translation, then $S = T_{\vec{u}} \circ H_{[Q,\lambda]} = H_{[P,\lambda]}$. This is a new homothety found in **Subsection 4.2.2**.

(c) If F is an opposite similarity, then F is equal to a reflection or a glide.

If F is a glide, that is, $F = T_{\vec{u}} \circ R_l = R_l \circ T_{\vec{u}}$, (with $\vec{u} \parallel l$), then

$$S = F \circ H_{[Q,\lambda]} = R_l \circ T_{\vec{u}} \circ H_{[Q,\lambda]}.$$

But, $T_{\vec{u}} \circ H_{[Q,\lambda]} = H_{[P,\lambda]}$ is a new homothety found in **Subsection 4.2.2**. Hence, F is reduced to a composition of another homothety and a reflection

$$F = R_l \circ H_{[P,\lambda]}.$$

(Similar work is done, if $S = H_{[Q,\lambda]} \circ G$. $G \neq F$, in general.) ■

Remark: This Corollary gives another proof of **Corollary 4.3.1**.

In the next Theorem, using **Property (8)** in **Section 4.3**, we prove that when the similarity is not an isometry (so, $0 < \lambda \neq 1$), then there is always a special unique commutative decomposition. That is, we have:

Theorem 4.3.3 *Every plane similarity which is not an isometry (so, the similarity ratio $0 < \lambda \neq 1$) can be decomposed into a unique composition of a homothety of the same ratio (so, the homothety is not the identity) and a rotation, or a reflection, in which the center of the rotation coincides with the center of the homothety, or the axis of the reflection passes through the center of the homothety. Such a composition is commutative.*

Proof Let $0 < \lambda \neq 1$ be the ratio of a similarity S and O its fixed point $[O' = S(O) = O$, according to **Property (8)** in **Section 4.3**]. We then consider the positive homothety $H_{[O,\lambda]}$ and the isometry F (or G), as claimed in **Theorem 4.3.2**. That is,

$$S = F \circ H_{[O,\lambda]}, \iff F = S \circ H_{[O,\frac{1}{\lambda}]}.$$

Then, O is a fixed point of F since it is fixed by both S and $H_{[O,\frac{1}{\lambda}]}$.

So we have: If O is the only fixed point of F, then F is a *rotation* with center O, by **Theorem 2.4.1** and S is a direct similarity. Otherwise, i.e., if F has more than one fixed points, then F is either a *reflection* or the *identity*, by **Theorem 2.2.2**.

If $F = I$, then $S = H_{[O,\lambda]}$, is a homothety with center O and S is a direct similarity.

If $F = R_l$ is a reflection, then S is an opposite similarity. The axis l of R_l contains O and we easily see that l is invariant by $S = F \circ H_{[O,\lambda]}$, since it is invariant by both F and $H_{[O,\lambda]}$. Also, the straight line k perpendicular to l at O is another invariant straight line of $S = F \circ H_{[O,\lambda]}$, since it is invariant by both F and $H_{[O,\lambda]}$. The straight line l is called the **principal axis** of S and k the **second axis** of S. See **Figure 4.52**.

Also, we easily observe in **Figures 4.51** and **4.52** that either $F = R_{[O,\alpha]}$ is a rotation or $F = R_l$ is a reflection, this composition is commutative, i.e.,

$$S = F \circ H_{[O,\lambda]} = H_{[O,\lambda]} \circ F.$$

By the uniqueness of the fixed point O, when the similarity ratio $0 < \lambda \neq 1$ and so the similarity is not an isometry [**Property (8)** in **Section 4.3**], the determined homothety $H_{[O,\lambda]}$ is unique. Then, the rotation

$$R_{[O,\alpha]} = H_{[O,\lambda]}^{-1} \circ S = H_{[O,\frac{1}{\lambda}]} \circ S$$

or the reflection

$$R_l = H_{[O,\lambda]}^{-1} \circ S = H_{[O,\frac{1}{\lambda}]} \circ S$$

is also unique. Therefore, the above decomposition is unique.

■

The importance of this Theorem forces us to illustrate **Figures 4.51** and
4.52 in more detail. Consider two points A and B and let

$$S(A) = A', \quad S(B) = B', \quad H_{[O,\lambda]}(A) = A'', \quad H_{[O,\lambda]}(B) = B''.$$

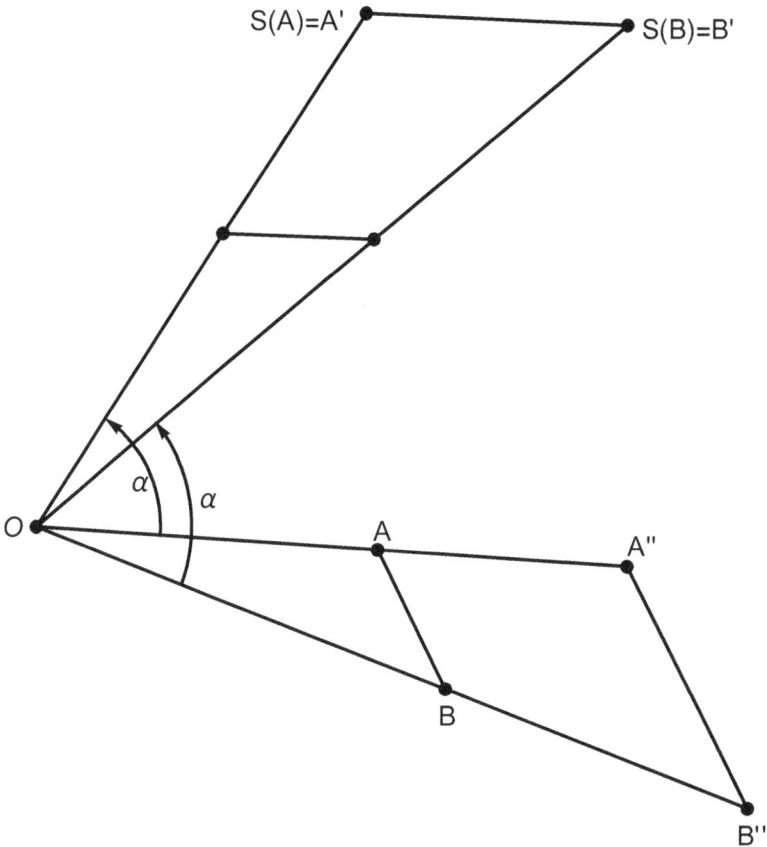

**Figure 4.51: This similarity is the unique commutative composition
of a homothety and a rotation about a common center O.**
$$S = R_{[O,\alpha]} \circ H_{[O,\lambda]} = H_{[O,\lambda]} \circ R_{[O,\alpha]}$$

Then, the following lengths satisfy

$$A'B' = S(AB) = \lambda AB = H_{[O,\lambda]}(AB) = A''B'',$$
$$OA' = S(OA) = \lambda OA = H_{[O,\lambda]}(OA) = OA'',$$
$$OB' = S(OB) = \lambda OB = H_{[O,\lambda]}(OB) = OB''.$$

That is, the triangles $OA'B'$ and $OA''B''$ have equal sides. So, if they are directly equal, by **Corollary 4.3.1**, S is a direct similarity. Hence $A'B'$ is obtained by rotating $A''B''$ by angle $\angle A''OA' = \angle B''OB'$. See **Figure 4.51**.

On the other hand, if the triangles $OA'B'$ and $OA''B''$ are oppositely equal, by **Corollary 4.3.1**, S is an opposite similarity. Then $A'B'$ is obtained by reflecting $A''B''$ on the line l, the common bisector of the angles $\angle A''OA'$ and $\angle B''OB'$. See **Figure 4.52**.

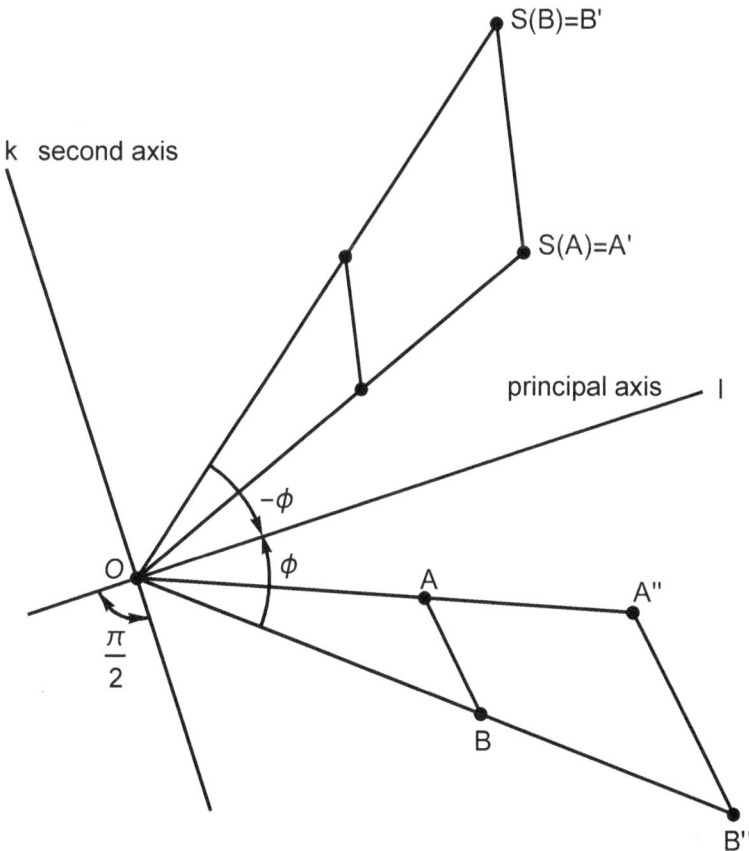

Figure 4.52: **This similarity is the unique commutative composition of a homothety and a reflection in l through the common center O. $S = R_l \circ H_{[O,\lambda]} = H_{[O,\lambda]} \circ R_l$**

We keep the point A fixed and therefore A' and A'' are also fixed and the angle $\angle A''OA' = \angle \alpha$ is kept constant. We then let B vary, and so for any B to obtain B', the B'' must be rotated by the fixed angle $\angle \alpha$ or reflected on the fixed line l. That is,

$$S = R_{[O,\alpha]} \circ H_{[O,\lambda]} \qquad \text{or} \qquad S = R_l \circ H_{[O,\lambda]}.$$

In the latter case the axis of the reflection l passes through the center O of the homothety. In this situation we observe that in both cases the homothety is positive; $(\lambda > 0)$.

Since O is a common fixed point of both the homothety and the rotation or the reflection, we also have that

$$S = H_{[O,\lambda]} \circ R_{[O,\alpha]}, \quad \text{or} \quad S = H_{[O,\lambda]} \circ R_l.$$

That is, *when the center of the homothety is the center of the rotation or a fixed point of the reflection, such a composition is commutative,* as can be easily checked and seen in **Figures 4.51** and **4.52**.

As we have stated in **Definition 4.3.1**, we see that: *A composition of a homothety and a rotation is a **direct similarity** (i.e., it preserves oriented angles since both the homothety and the rotation preserve oriented angles) and a composition of a homothety and a reflection is an **opposite similarity** (i.e., it reverses oriented angles, since the homothety preserves oriented angles but the reflection reverses the oriented angles).*

When S is direct and decomposed as

$$S = R_{[O,\alpha]} \circ H_{[O,\lambda]} = H_{[O,\lambda]} \circ R_{[O,\alpha]},$$

the homothety maps any vector $\overrightarrow{AB}$ to a vector

$$\overrightarrow{A'B'} = \lambda \overrightarrow{AB},$$

and so the vectors $\overrightarrow{AB}$ and $\overrightarrow{A'B'}$ are parallel $(\overrightarrow{AB} \parallel \overrightarrow{A'B'})$ and have the same direction $(\lambda > 0)$. So, $\angle(\overrightarrow{AB}, \overrightarrow{A'B'}) = 0$. Consequently, the rotation maps the vector $\overrightarrow{A'B'}$ to a vector $\overrightarrow{A''B''}$ of equal length and $\angle(\overrightarrow{A'B'}, \overrightarrow{A''B''}) = \alpha$. Therefore,

$$\angle(\overrightarrow{AB}, \overrightarrow{A''B''}) = 0 + \alpha = \alpha.$$

(See also **Lemma 4.4.1**.)

Thus, in case of a **direct similarity** S, we call the angle $\angle \alpha$ the **angle of the direct similarity** S or simply **similarity angle**. The **similarity ratio** is $\lambda > 0$. In general, we define:

Definition 4.3.2 *The **similarity angle** of a direct similarity S is the fixed angle $\angle[\overrightarrow{AB}, S(\overrightarrow{AB})]$, where $\overrightarrow{AB} \neq \overrightarrow{0}$ is any non-zero vector.*

We also observe the following corollary, which is a consequence of the commutativity of the composition of the previous Theorem.

Corollary 4.3.3 *The **commutative decomposition of a similarity** S, as in the Theorem, with $0 \neq \lambda \neq 1$, satisfies the following **inverse** and **composition** rules:*

$$(1) \quad S^{-1} = R_{[O,\alpha]}^{-1} \circ H_{[O,\lambda]}^{-1} = R_{[O,-\alpha]} \circ H_{[O,\frac{1}{\lambda}]}$$
$$= H_{[O,\lambda]}^{-1} \circ R_{[O,\alpha]}^{-1} = H_{[O,\frac{1}{\lambda}]} \circ R_{[O,-\alpha]}.$$

$$(2) \qquad S^{-1} = R_l^{-1} \circ H_{[O,\lambda]}^{-1} = R_l \circ H_{[O,\frac{1}{\lambda}]} = H_{[O,\lambda]}^{-1} \circ R_l^{-1} = H_{[O,\frac{1}{\lambda}]} \circ R_l.$$

(3) When S is direct, $\forall\ n \in \mathbb{Z}$,

$$S^{(\circ,n)} = R_{[O,n\alpha]} \circ H_{[O,\lambda^n]} = H_{[O,\lambda^n]} \circ R_{[O,n\alpha]}.$$

(4) When S is opposite, $\forall\ n \in \mathbb{Z}$,

$$S^{(\circ,n)} = \begin{cases} H_{[O,\lambda^n]}, & \textit{if}\ \ n\ \textit{is even,} \\[2mm] R_l \circ H_{[O,\lambda^n]} = H_{[O,\lambda^n]} \circ R_l, & \textit{if}\ \ n\ \textit{is odd.} \end{cases}$$

The **proof** is immediate from the Theorem and the properties of rotations, reflections and homotheties.

∎

Remarks: In particular we have the following important remarks:

A parallel translation is a direct similarity (i.e., preserves oriented angles), with ratio 1 and similarity angle 0, (mod 2π).

A rotation is a direct similarity (i.e., preserves oriented angles), with ratio 1 and similarity angle the rotation angle.

A reflection is an opposite similarity (i.e., reverses oriented angles), with similarity ratio 1. We do not define similarity angles for reflections and opposite similarities S, since with such similarity, $\angle[\overrightarrow{AB}, S(\overrightarrow{AB})]$ **is not constant**.

Any positive homothety is a direct similarity (i.e., preserves oriented angles), with similarity angle 0, (mod 2π), and similarity ratio the ratio of the homothety, as we have already seen in **Subsection 4.1.1, (e)** and **(g)**.

Any negative homothety is, also, a direct similarity (i.e., preserves oriented angles), with similarity angle π, (mod 2π), and similarity ratio the absolute value of the ratio of the negative homothety, as we have already seen in **Subsection 4.1.1, (c), (e)** and **(g)**.

Also, if

$$S = R_{[O,\alpha]} \circ H_{[O,\lambda]} = H_{[O,\lambda]} \circ R_{[O,\alpha]}$$

and $\angle\alpha = 0$, (mod 2π), in **Theorem, 4.3.3**, then $R_{[O,0]} = I$ is the identity and so

$$S = H_{[O,\lambda]} \quad \text{(the homothety)}.$$

(See also **Theorem 4.4.2**, for a characterization of a **direct similarity**.)

Given the **opposite similarity**

$$S = R_l \circ H_{[O,\lambda]} = H_{[O,\lambda]} \circ R_l$$

as claimed in **Theorem 4.3.3**, the bisector l of the angle $\angle A''OA'$ (and $\angle B''OB'$) is called the **principal axis** of S. The line $k \perp l$ at the point O is called the **second axis** of S. See **Figure 4.52**. These straight lines are set-wise fixed sets of S (invariant sets). This follows immediately from the fact that both lines pass through O and so they are fixed sets of the homothety $H_{[O,\lambda]}$. They are also fixed sets for the reflection R_l, since l is the axis of the reflection and k is a line perpendicular to l. [See **Subsections 2.2.1, (3), 2.2.2, (3)**, and **4.1.1, (d)**]. Apart from this fact, that is, both lines are set-wise fixed sets of S through the center O, these two lines are distinguished because they play an important role in understanding the structure of an opposite similarity.

From **Theorems 4.3.2, 4.3.3, Corollary 4.3.2, Subsection 4.1.1, (e)**, **Section 4.3, (5)**, and the **properties of glides** in **Section 2.3, (6)**, we conclude that in order to study the group $(\mathcal{S}, \circ)$ of similarities of the plane, we must study the homotheties and their compositions with other homotheties and isometries (parallel translations, rotations and reflections). The compositions of two homotheties was taken care of in **Subsection 4.2.1** and the compositions of homotheties and parallel translations in **Subsection 4.2.2**. Both yield new homotheties. In the sequel, we must examine the compositions of homotheties and rotations and the compositions of homotheties and reflections. In all cases, we must find the centers of these compositions and the similarity ratios.

We can also compose a homothety with a rotation whose center is different from the center of the homothety, or a homothety with a reflection whose axis does not pass through the center of the homothety. By the definitions of homotheties, similarities and isometries, these compositions are directly proven to be similarities whose ratios are the absolute values of the ratios of the homotheties (prove this!), but their centers (their unique fixed points, when the ratio $\neq 1$) are not necessarily the centers of the homotheties or the rotations and therefore they must be determined. Such compositions are not commutative, in general. Now, we continue with the direct similarities and their algebraic group.

4.4   Direct Similarities and their Algebraic Group

4.4.1   Compositions of Homotheties and Rotations

In **Subsection 4.2.2**, we have found that the compositions of homotheties with parallel translations, in any order, are new homotheties. In **this Subsection**, we study the compositions of homotheties and rotations. We will see that these compositions are neither homotheties nor rotations, in general. So, we consider a homothety $H_{[O,\lambda]}$ and a rotation $R_{[Q,\phi]}$, and we want to study the compositions:

$$R_{[Q,\phi]} \circ H_{[O,\lambda]} \quad \text{and} \quad H_{[O,\lambda]} \circ R_{[Q,\phi]}.$$

By **Consequences (a)** and **(c)** of **Subsection 4.1.1**, and compositions **(3)**, **(4)** and **(5)** of **Section 2.3**, without loss of generality, we consider $\lambda > 0$. These compositions are not commutative, in general. They are commutative iff the two centers coincide, $O \equiv Q$, as we have stated in the proof of **Theorem 4.3.3** or can be directly checked.

If $0 < \lambda \neq 1$, these compositions are neither homotheties nor rotations. We have seen that some compositions in **Subsections 4.2.1** and **4.2.2** were homotheties. Here, however, the vectors $\overrightarrow{AB}$ and $\overrightarrow{A'B'}$ that appear in **Theorem 4.2.1** are not parallel anymore, as this Theorem requires. Therefore, they cannot be homotheties. On the other hand, they cannot be rotations since the lengths of vectors (or straight segments) are multiplied by λ and therefore not preserved. These compositions yield more general similarities unless one of the two components is the identity. As we have seen, a homothety is the identity iff $\lambda = 1$ and a rotation is the identity iff $\angle\phi = 0$, (mod 2π). In such a case, homotheties and / or rotations become trivial cases of these general similarities, and their centers are the known points O and Q, respectively.

When $\lambda \neq 1$ and $\angle\phi \neq 0$, by **Section 4.3, (5)**, any such composition is a new similarity with similarity ratio $0 < \lambda \neq 1$. So, by **Theorem 4.3.3**, the above compositions have unique fixed points P and R, respectively. These points become the centers of the resulting homotheties and the resulting rotations in the following commutative compositions, which are equal to the original above compositions. That is,

$$R_{[Q,\phi]} \circ H_{[O,\lambda]} = H_{[P,\lambda]} \circ R_{[P,\phi]} = R_{[P,\phi]} \circ H_{[P,\lambda]},$$
$$H_{[O,\lambda]} \circ R_{[Q,\phi]} = H_{[R,\lambda]} \circ R_{[R,\phi]} = R_{[R,\phi]} \circ H_{[R,\lambda]}.$$

We begin with the following Lemma:

Lemma 4.4.1 *Consider the compositions*

$$R_{[Q,\phi]} \circ H_{[O,\lambda]} \quad \text{and} \quad H_{[O,\lambda]} \circ R_{[Q,\phi]}.$$

(a) If $\lambda > 0$, these compositions are direct similarities (i.e., preserve oriented angles), with similarity ratio λ and similarity angle ϕ.

(b) If $\lambda < 0$, these compositions are also direct similarities with similarity ratio $|\lambda| = -\lambda > 0$ and similarity angle $\phi + \pi$, (mod 2π).

Proof Case $\lambda > 0$.

We consider any vector $\overrightarrow{AB}$ and let $\overrightarrow{A'B'} = R_{[Q,\phi]} \circ H_{[O,\lambda]}(\overrightarrow{AB})$. Let also $H_{[O,\lambda]}(\overrightarrow{AB}) = \overrightarrow{A_1 B_1}$. Then, since $\lambda > 0$, by **Theorem 4.2.1**, we have

$$\overrightarrow{A_1 B_1} = \lambda \overrightarrow{AB} \qquad \text{or equivalently,}$$

$A_1 B_1 = \lambda AB$ (as segments) and the angle $\angle(\overrightarrow{AB}, \overrightarrow{A_1 B_1}) = 0$, (mod 2π).

Next, we have that $\overrightarrow{A'B'} = R_{[Q,\phi]}(\overrightarrow{A_1 B_1})$ and so, by **Theorem 2.5.3**, we get

$A_1 B_1 = A'B'$ (as segments) and the angle $\angle(\overrightarrow{A_1 B_1}, \overrightarrow{A'B'}) = \phi$, (mod 2π).

Hence, by the above equalities, we find

$A'B' = \lambda AB$ (as segments) and the angle $\angle(\overrightarrow{AB}, \overrightarrow{A'B'}) = \phi$, (mod 2π).

So, the similarity ratio is λ and the similarity angle is ϕ.

If now we consider another non-zero vector $\overrightarrow{CD}$ and we let $\overrightarrow{C'D'} = R_{[Q,\phi]} \circ H_{[O,\lambda]}(\overrightarrow{CD})$, then again

$$\angle(\overrightarrow{CD}, \overrightarrow{C'D'}) = \phi, \text{ (mod } 2\pi).$$

$$\text{Therefore,} \quad \angle(\overrightarrow{AB}, \overrightarrow{CD}) = \angle(\overrightarrow{AB}, \overrightarrow{A'B'}) + \angle(\overrightarrow{A'B'}, \overrightarrow{C'D'}) + \angle(\overrightarrow{C'D'}, \overrightarrow{CD})$$

$$= \phi + \angle(\overrightarrow{A'B'}, \overrightarrow{C'D'}) - \phi = \angle(\overrightarrow{A'B'}, \overrightarrow{C'D'}),$$

that is, $R_{[Q,\phi]} \circ H_{[O,\lambda]}$ preserves oriented angles.

Case $\lambda < 0$. We have the subcases:

(a) If $\phi = 0$, (mod 2π), then $R_{[Q,\phi]} = I$ the identity, and so $S = H_{[O,\lambda]} = R_{[O,\pi]} \circ H_{[O,-\lambda]} = H_{[O,-\lambda]} \circ R_{[O,\pi]}$, with similarity ratio $-\lambda > 0$ and similarity angle π.

(b) If $\phi = \pi$, (mod 2π), then $R_{[Q,\phi]} \circ R_{[O,\pi]}$ is either the identity, if $O = Q$, or a parallel translation [by **Section 2.3, (2)** and **(4)**], if $O \neq Q$. So, S is: either the homothety $H_{[O,\lambda]} = R_{[O,\pi]} \circ H_{[O,-\lambda]} = H_{[O,-\lambda]} \circ R_{[O,\pi]}$ viewed as a similarity with similarity ratio $-\lambda > 0$ and similarity angle π, or a new homothety (according to **Subsection 4.2.2**), viewed as a similarity with similarity ratio $-\lambda > 0$ and similarity angle 0.

(c) Otherwise, the composition of the two rotations $R_{[O,\pi]}$ and $R_{[Q,\phi]}$, in either order, is a new rotation about a new center by angle $\phi + \pi$, as proven in **Section 2.3, (2)** and **(4)**. Thus, this subcase reduces to the first case above, with similarity ratio positive.

Analogous work is done with $H_{[O,\lambda]} \circ R_{[Q,\phi]}$, finishing the proof of the Lemma. ∎

The centers of the similarities claimed in the above **Lemma** will be determined in the next **Subsection 4.4.3**.

Now, we prove the following theorem, in which we prove a relation that exists between the two compositions stated at the beginning of **this Subsection**.

Theorem 4.4.1 *For any* $Q_1 \in \mathcal{P}$, *there is* an $O_1 \in \mathcal{P}$, *such that, the similarity given by*

$$S = R_{[Q,\theta]} \circ H_{[O,\lambda]}, \quad with \quad \lambda > 0,$$

is equal to

$$S = H_{[O_1,\lambda]} \circ R_{[Q_1,\theta]},$$

and so the composition $R_{[Q,\theta]} \circ H_{[O,\lambda]}$ *can be written in infinitely many ways in the form of the second composition and vice-versa.*

(If in this Theorem we replace the rotation by a parallel translation, then see **Subsection 4.2.2** for the analogous result.)

Proof By the **above Lemma**, the similarity $S = R_{[Q,\theta]} \circ H_{[O,\lambda]}$ is direct. The similarity angle is θ and the similarity ratio is $\lambda > 0$, without loss of generality.

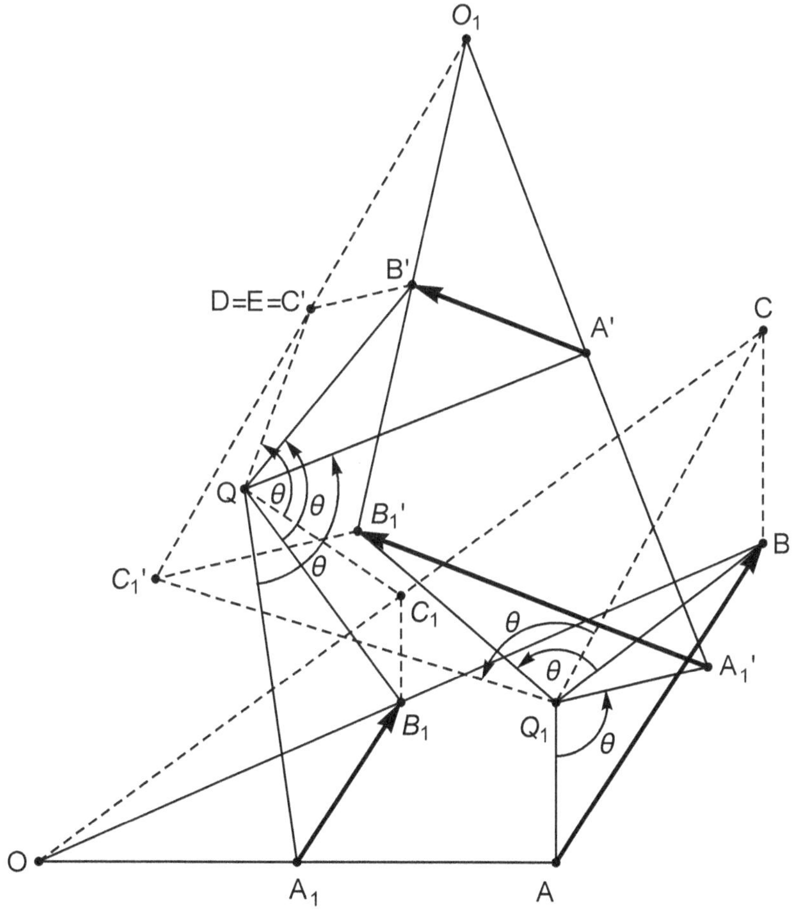

Figure 4.53: Decomposition of a similarity

We choose any $Q_1 \in \mathcal{P}$ and we consider the rotation $R_{[Q_1,\theta]}$. Then, the composition

$$R_{[Q,\theta]} \circ H_{[O,\lambda]} \circ R_{[Q_1,-\theta]} := H_{[O_1,\lambda]}$$

is a homothety $H_{[O_1,\lambda]}$, by **Theorem 4.2.1**; its ratio is readily λ and its center O_1 must be determined. We construct the center O_1 geometrically below. See **Figure 4.53**. Then, we would have

$$R_{[Q,\theta]} \circ H_{[O,\lambda]} = H_{[O_1,\lambda]} \circ R_{[Q_1,\theta]},$$

thus proving the Theorem.

We have: If $\overrightarrow{A'B'} = S(\overrightarrow{AB})$, then, by the **previous Lemma** (and its proof, or the results of **Theorem 4.3.3**), we have

$A'B' = \lambda AB$ (as segments) and the angle is $\angle(\overrightarrow{AB}, \overrightarrow{A'B'}) = \theta$, (mod 2π).

We let

$$\overrightarrow{A_1'B_1'} = R_{[Q_1,\theta]}(\overrightarrow{AB}).$$

Then, by **Theorem 2.5.3**, we have

$A_1'B_1' = AB$ (as segments) and the angle is $\angle(\overrightarrow{AB}, \overrightarrow{A_1'B_1'}) = \theta$, (mod 2π).

Hence, we have

$A'B' = \lambda A_1'B_1'$ (as segments) and the angle is $\angle(\overrightarrow{A'B'}, \overrightarrow{A_1'B_1'}) = 0$, (mod 2π).

Therefore,

$$\overrightarrow{A'B'} = \lambda \cdot \overrightarrow{A_1'B_1'}.$$

Then, by **Theorem 4.2.1** and its **Corollary**, there is a homothety $H_{[O_1,\lambda]}$ such that

$$H_{[O_1,\lambda]}(\overrightarrow{A'B'}) = (\overrightarrow{A_1'B_1'}).$$

Now, we keep AB fixed and for any point $C \in \mathcal{P}$, as shown in **Figure 4.53**, we first let

$$D := R_{[Q,\theta]} \circ H_{[O,\lambda]}(C) \quad \text{and} \quad E := H_{[O_1,\lambda]} \circ R_{[Q_1,\theta]}(C).$$

Then, using equalities and similarities of triangles, we can easily prove that $D = E$ and so,

$$S(C) = R_{[Q,\theta]} \circ H_{[O,\lambda]}(C) = H_{[O_1,\lambda]} \circ R_{[Q_1,\theta]}(C).$$

Therefore,

$$S = R_{[Q,\theta]} \circ H_{[O,\lambda]} = H_{[O_1,\lambda]} \circ R_{[Q_1,\theta]},$$

where $Q_1 \in \mathcal{P}$ is chosen arbitrarily and then O_1 is determined. (In general, $O \neq O_1$.)

■

[See also **Corollary 4.4.1**, (3).]

4.4.2 General Characterizations of Direct Similarities

Theorem 4.4.2 (Characterization of Direct Similarities by $\{A, A', \rho, \phi\}$)
A plane transformation $S : \mathcal{P} \longrightarrow \mathcal{P}$ with $S(A) = A'$ given, is a direct simi-larity of similarity ratio $\rho > 0$ and similarity angle ϕ if and only if
$\forall \, B \in \mathcal{P}$, $B' = S(B)$ *is determined by the following two conditions*

$$A'B' = \rho AB \quad \text{(as segments) and the angle} \quad \measuredangle(\overrightarrow{AB}, \overrightarrow{A'B'}) = \phi, \; (\text{mod } 2\pi).$$

Proof ($\Longrightarrow$) Suppose S is a direct similarity of similarity ratio $\rho > 0$ and similarity angle ϕ and $S(A) = A'$ given. For any $B \in \mathcal{P}$, we let $B' = S(B)$.
The first condition

$$A'B' = \rho AB,$$

follows from the **Definition of a similarity, 4.1.1**.
Next, by **Theorem 4.3.2** and **Corollary 4.3.2**, we get that $S = H_{[O,\rho]} \circ F$, where $H_{[O,\rho]}$ is a homothety of center some point $O \in \mathcal{P}$ and F is a direct isom-etry. Then, the similarity angle ϕ is equal to zero, if F is a parallel translation, or equal to the rotation angle, if F is a rotation.
Now, we consider any non-zero vector $\overrightarrow{AB} \neq \overrightarrow{0}$ and let $\overrightarrow{A_1 B_1} = F(\overrightarrow{AB})$ and $\overrightarrow{A'B'} = H_{[O,\rho]}(\overrightarrow{A_1 B_1}) = H_{[O,\rho]} \circ F(\overrightarrow{AB})$. Then, we have

$$\measuredangle(\overrightarrow{A_1 B_1}, \overrightarrow{A'B'}) = 0, \; (\text{mod } 2\pi), \quad \text{and} \quad \measuredangle(\overrightarrow{AB}, \overrightarrow{A_1 B_1}) = \phi, \; (\text{mod } 2\pi).$$

Consequently, we obtain

$$\begin{aligned}
\measuredangle(\overrightarrow{AB}, \overrightarrow{A'B'}) &= \measuredangle(\overrightarrow{AB}, \overrightarrow{A_1 B_1}) + \measuredangle(\overrightarrow{A_1 B_1}, \overrightarrow{A'B'}) \\
&= \measuredangle(\overrightarrow{AB}, \overrightarrow{A_1 B_1}) + 0 = \measuredangle(\overrightarrow{AB}, \overrightarrow{A_1 B_1}) = \phi, \; (\text{mod } 2\pi).
\end{aligned}$$

($\Longleftarrow$) Suppose S is a plane transformation such that $A' = S(A)$. Also, for every $B \in \mathcal{P}$, we let $B' = S(B)$, and we assume that the following two conditions

$$A'B' = \rho AB \quad \text{(as segments) and the angle} \quad \measuredangle(\overrightarrow{AB}, \overrightarrow{A'B'}) = \phi, \; (\text{mod } 2\pi)$$

are fulfilled. We will prove that S is a direct similarity of similarity ratio $\rho > 0$ and similarity angle ϕ.
If $\phi = 0$ (mod 2π), then the two conditions are equivalent to $\overrightarrow{A'B'} = \rho \overrightarrow{AB}$ and so, by **Theorem 4.2.1** and **Subsection 4.1.1, (g)**, or **Lemma 4.4.1**, S is a direct similarity of ratio $\rho > 0$ and angle $\phi = 0$ (mod 2π).
If $\rho = 1$, then the conditions

$$A'B' = AB \quad \text{(as segments) and the angle} \quad \measuredangle(\overrightarrow{AB}, \overrightarrow{A'B'}) = \phi, \; (\text{mod } 2\pi)$$

prove that S is a rotation by angle ϕ, by **Theorem 2.5.3**, and so S is a direct similarity of angle ϕ.

So, we assume that $\phi \neq 0$ and $\rho \neq 1$. Choose any point $Q \in \mathcal{P}$ and consider the homothety $H_{[Q,\rho]}$, and let $A_1 = H_{[Q,\rho]}(A)$. If for any $B \in \mathcal{P}$, we let $B_1 = H_{[Q,\rho]}(B)$, then, since $\rho > 0$ by **Theorem 4.2.1**, we have

$$\overrightarrow{A_1 B_1} = \rho \overrightarrow{AB}.$$

So, $A_1 B_1 = AB$ (as segments) and the angle $\angle(\overrightarrow{AB}, \overrightarrow{A_1 B_1}) = 0$, (mod 2π).
Therefore, by hypotheses, we get

$$A_1 B_1 = A'B', \quad \text{as segments}, \quad \text{and}$$

$$\angle(\overrightarrow{A_1 B_1}, \overrightarrow{A'B'}) = \angle(\overrightarrow{A_1 B_1}, \overrightarrow{AB}) + \angle(\overrightarrow{AB}, \overrightarrow{A'B'})$$

$$= 0 + \angle(\overrightarrow{AB}, \overrightarrow{A'B'}) = \angle(\overrightarrow{AB}, \overrightarrow{A'B'}), \quad (\text{mod } 2\pi).$$

These last two relations prove that there is a rotation $R_{[O,\phi]}$ determined by (A, A') and angle ϕ such that $R_{[O,\phi]}(A_1 B_1) = A'B'$. [See **Subsections 2.5.2** and **2.5.3, (I)**.]

Thus, $S = R_{[O,\phi]} \circ H_{[Q,\rho]}$, which is a direct similarity of ratio ρ and angle ϕ.
∎

We now easily obtain:

Corollary 4.4.1 *(1) Let $S : \mathcal{P} \longrightarrow \mathcal{P}$ be a direct similarity as in the Theorem. Then, S^{-1} is determined by the conditions*
$AB = \frac{1}{\rho} A'B'$ *(as segments) and the angle* $\angle(\overrightarrow{A'B'}, \overrightarrow{AB}) = -\phi$, *(mod 2π).*
So, S^{-1} is a direct similarity with ratio $\dfrac{1}{\rho}$ *and angle $-\phi$.*

(2) The composition of a rotation and a homothety, in either order, is a direct similarity of ratio the absolute value of the ratio of the homothety and angle the rotation angle.

(3) A direct similarity S can be decomposed into

$$S = R_{[O,\phi]} \circ H_{[Q,\rho]} \quad or \quad S = H_{[Q,\rho]} \circ R_{[O,\phi]},$$

in infinitely many ways.

[Provide the necessary arguments. **(1)** was also proved in **Corollary 4.3.3, (1)**. For **(3)**, you may need **Lemma 4.4.1** and / or **Theorem 4.4.1**.]
∎

Next, we prove the following:

Theorem 4.4.3 (Characterization of Direct Similarities by Vectors)
If $\overrightarrow{AB}$ and $\overrightarrow{A'B'}$ are any two non-zero vectors, then there exists a unique direct similarity S, such that $\overrightarrow{A'B'} = S(\overrightarrow{AB})$.

Proof We let

$$\rho = \frac{A'B'}{AB} > 0 \quad \text{and} \quad \phi = \angle(\overrightarrow{A'B'}, \overrightarrow{A'B'}).$$

Then, S is the unique direct similarity defined by (A, A'), ρ and ϕ, claimed by the **previous Theorem**. This similarity maps the vector $\overrightarrow{AB}$ to the vector $\overrightarrow{A'B'}$, as done in the **previous Theorem**.

■

Geometrical Construction of the similarity claimed by the **Theorem**: For any chosen point P, we let $P' = S(P)$. Then P' is found by constructing the triangle $A'B'P'$ directly similar to the triangle ABP, since we know one of its sides, namely $A'B' = \rho AB$, and:

either its two angles

$$\angle(\overrightarrow{A'B'}, \overrightarrow{A'P'}) = \angle(\overrightarrow{AB}, \overrightarrow{AP}), \quad \text{and} \quad \angle(\overrightarrow{B'A'}, \overrightarrow{B'P'}) = \angle(\overrightarrow{BA}, \overrightarrow{BP}),$$

or its other two sides $\quad A'P' = \rho AP \quad$ and $\quad B'P' = \rho BP$.

(Illustrate the above constructions of the triangle $A'B'P'$ by drawing figures!)

By **the Theorem** and the **Geometric Construction**, we immediately get:

Corollary 4.4.2 *Let $S : \mathcal{P} \longrightarrow \mathcal{P}$ be a plane transformation.*

(a) If S preserves the oriented angles, then S maps straight segments to straight segments and is a direct similarity.

(b) The composition of two direct similarities is a direct similarity. (The ratio of the composition is the product of the two ratios, as this is so for any two similarities.)

(c) If S is a direct similarity and has two fixed points, then it is the identity. So, a direct similarity has at most one fixed point. If it does not have fixed points, then it must be a non-trivial parallel translation.

(d) The set of all direct similarities $\mathcal{D}$ is an algebraic group under composition.

[Provide the proving arguments.]

By all results proven so far (Theorems, Corollaries, Lemmata, and Remarks) we conclude the **Result**:

The set $\mathcal{D}$ of all direct similarities of the plane $\mathcal{P}$ is equal to the union of the set of the direct isometries (parallel translations and rotations) with the set of homotheties and the set of all compositions of homotheties with direct isometries.

All results discussed and proven in the **last two Subsections** along with the compositions of parallel translations, rotations, and homotheties (that we have studied before) prove that the composition of direct similarities is a direct similarity and the inverse of a direct similarity is also a direct similarity. Therefore we have the **Result**:

The set $\mathcal{D}$ of all direct similarities of the plane $\mathcal{P}$ is a non-commutative algebraic group under composition and therefore, is a subgroup of the group of similarities $(\mathcal{S}, \circ)$.

4.4.3   Determination of Direct Similarities

In **Definition 4.3.1**, we have defined a **similarity** as **direct**, if it preserves the oriented angles. (It is **opposite**, if it maps the oriented angles to their opposites, i.e., it reverses the oriented angles.)

Let S be a direct similarity with similarity ratio $\lambda > 0$ and $\overrightarrow{AB}$ and $\overrightarrow{CD}$ be any two vectors. We let $\overrightarrow{A'B'} = S(\overrightarrow{AB})$ and $\overrightarrow{C'D'} = S(\overrightarrow{CD})$. Then, $A'B' = \lambda AB$ and $C'D' = \lambda CD$. If $\overrightarrow{AB} \neq \overrightarrow{0}$ and $\overrightarrow{CD} \neq \overrightarrow{0}$, then $\overrightarrow{A'B'} \neq \overrightarrow{0}$ and $\overrightarrow{C'D'} \neq \overrightarrow{0}$, and since S preserves oriented angles, we get

$$\sphericalangle(\overrightarrow{AB}, \overrightarrow{CD}) = \sphericalangle(\overrightarrow{A'B'}, \overrightarrow{C'D'}).$$

Now: **Without using Theorem 4.3.3**, we prove:

Theorem 4.4.4 *Any direct similarity with similarity ratio $\lambda > 0$ and similarity angle θ, is either a parallel translation, or a rotation, or a homothety of ratio $\lambda > 0$, or it can be decomposed into a composition, in either order, of a homothety of ratio $\lambda > 0$ and a rotation both of which are not identities, in infinitely many ways in general.*

If the similarity is a parallel translation or a homothety, then the similarity angle is $\theta = 0$.

If the similarity is a rotation or the composition of a rotation with a homothety, then the similarity angle is θ, the angle of the rotation.

Proof We consider a direct similarity S of ratio $\lambda > 0$. By **Theorem 4.3.2**, $S = F \circ H$, the composition of a homothety H of ratio $\lambda > 0$ and a direct isometry F, in infinitely many ways in general. (Or $S = H \circ F$.) We have the following three cases.

(1) If $\lambda = 1$, then H is the identity and so S is a direct isometry. Therefore, it is either a parallel translation, or a rotation. (The identity is included in these isometries.)

(2) If S is a composition of a homothety and a parallel translation (in either order), then S is either the same homothety (if the parallel translation is the identity) or a new homothety as found in **Subsection 4.2.2**.

(3) Otherwise, S is a composition of a homothety of ratio $0 < \lambda \neq 1$ and an appropriate rotation (in either order) in which neither the homothety nor the rotation is identity.

The similarity angles, as referred above, have been discussed after **Definition 4.3.2**, in **Lemma 4.4.1**, etc.   ■

By the **above Theorem**, we could replace **Definition 4.3.1** (as some books, especially of lower level, do) by the following **Definition**:

*A **direct similarity** is a composition (in either order) of a **positive homothety** with a **direct isometry**. (The direct isometry is either the identity, or a parallel translation, or a rotation.)*

The image $\mathfrak{F}'$ of a figure $\mathfrak{F}$ by a direct similarity is called **directly similar figure** to the original figure $\mathfrak{F}$.

4.4.4   Determination of the Center of a Direct Similarity

We first remind the reader of the **Apollonius[8] circle**. In the sequel, we are going to refer to it often and so the reader must have it fresh in memory.

Theorem 4.4.5 *The set of all points P of the plane, whose distance from two given points A and B have ratio $0 < \lambda \neq 1$, i.e., $\dfrac{PA}{PB} = \lambda$, is a circle whose diameter is determined by the points C and D that satisfy the condition*

$$-\frac{\overrightarrow{AC}}{\overrightarrow{BC}} = \frac{\overrightarrow{AD}}{\overrightarrow{BD}} = \lambda > 0,$$

so, C is between A and B, and D is outside the interval AB.

If $\lambda = 1$, then the set of points that satisfy $\dfrac{PA}{PB} = 1$, is the perpendicular bisector of the segment AB (which can be viewed as a circle of infinite radius and center the infinity point of the direction determined by the straight line AB).

This circle is called the **Apollonius circle** determined by the points A and B and the ratio $\lambda > 0$. Notice that the points A, B, C, and D, in this Theorem, form a harmonic quadruple with ratio λ. (In the limiting cases $\lambda = 0$ and $\lambda = \infty$ the circle degenerates to the points A and B, respectively, and therefore these cases are not interesting.)

Proof See **Figure 4.54**.
    If $\lambda = 1$, then

$$\frac{PA}{PB} = 1 \quad \Longleftrightarrow \quad PA = PB,$$

and therefore, the set of points sought, is the perpendicular bisector of the segment AB, at the midpoint M.

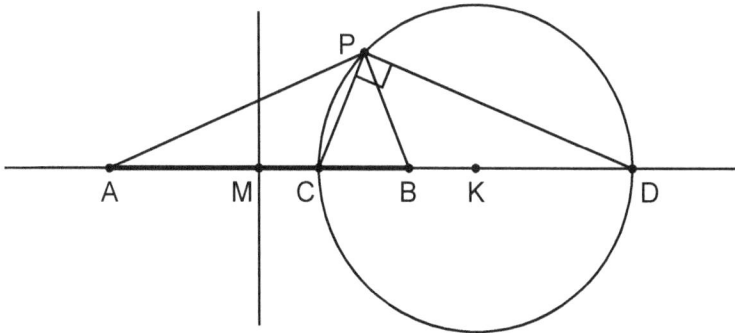

Figure 4.54: Apollonius circle determined by A, B, and λ

---

[8]Apollonius of Perga, Greek mathematician, c.262-c.190 BCE.

If $0 < \lambda \neq 1$, then there exists a point C between A and B and a point D exterior to AB, such that,

$$\frac{\overrightarrow{CA}}{\overrightarrow{CB}} = -\lambda, \qquad \text{and} \qquad \frac{\overrightarrow{DA}}{\overrightarrow{DB}} = \lambda.$$

These points, B and D, satisfy the required condition and so, they belong to the sought geometrical locus.

Let now P a point in the sought set not on the line AB. Then,

$$\lambda = \frac{PA}{PB} = \frac{CA}{CB} = \frac{DA}{DB},$$

and so, by the known theorem of the angle bisectors of a triangle (**Theorem 4.1.1**), the lines PC and PD are the bisectors of the angles of the triangle CPD at P. Therefore, $PC \perp PD$.

So, P belongs to the circle with diameter CD, which is determined by A, B, and λ.

Conversely: Let P be a point of the above circle. By construction, the points A, C, B, and D form a harmonic quadruple with ratio λ, and the lines PC and PD are perpendicular to each other, as the angle $\widehat{CPD}$ is inscribed on a semicircle. Then, the lines PC and PD are the bisectors of the angles of the triangle AMB at M. Then,

$$\frac{PA}{PB} = \frac{CA}{CB} = \frac{DA}{DB} = \lambda.$$

So, the point P satisfies the required condition.

∎

(Many properties of the Apollonius circles are stated in the exercises.)

Next: We continue with the **determination of the center of a direct similarity**.

First Case: We would like to prove the existence of the center (fixed point) of a non-trivial direct similarity S characterized in **Theorem 4.4.2**. That is, the direct similarity $S \neq I$ is not the identity and is defined by a pair of corresponding points (A, A'), i.e., $A' = S(A)$, the similarity ratio $\lambda > 0$ and the similarity angle θ. So, the data are:

(1) The points A and A', such that $A' = S(A)$.
(2) The similarity ratio $\lambda > 0$.
(3) The similarity angle θ.

Without loss of generality, we also assume:

(a) $A \neq A'$. {If $A = A' = S(A)$, by **Lemma 4.3.1** or **Corollary 4.4.2, (c)**, implies that A is the center and then $S = R_{[A,\theta]}$.}
(b) $0 \leq \theta < 2\pi$.
(c) $0 < \lambda \neq 1$. {If $\lambda = 1$, S is a direct isometry, and so either the parallel translation $T_{\overrightarrow{AA'}}$ or the rotation $R_{[O,\theta]}$ found in **Subsection 2.5.3**.}

Under the above data and conditions, there are the following four subcases:

Subcase 1.1: $\theta = 0$.

Then $S = H_{[O,\lambda]}$ is a positive homothety. The center O is on the straight line AA' determined, as in **Lemma 4.1.1**, by the relation

$$\frac{\overrightarrow{OA'}}{\overrightarrow{OA}} = \lambda > 0, \quad \text{etc.}$$

So, $\overrightarrow{OA'}$ and $\overrightarrow{OA}$ have the same direction and located outside the straight segment AA' either to the left or to the right of it. See **Figure 4.2**. This point is unique and its geometrical construction is basic.

Subcase 1.2: $0 < \theta < \pi$.

If we preliminarily assume that under the above conditions S has a fixed point O, then $S(O) = O$ and O is unique by **Lemma 4.3.1** or **Corollary 4.4.2, (c),** (proven synthetically). Moreover,

$$S(\overrightarrow{OA}) = \overrightarrow{OA'} \quad \text{or equivalently} \quad OA' = \lambda OA \quad \text{and} \quad \angle(\overrightarrow{OA}, \overrightarrow{OA'}) = \theta.$$

The first relation shows that O lies in the Apollonius circle, whose points P satisfy

$$\frac{PA'}{PA} = \lambda.$$

The second relation shows that O belongs to the circular arc, whose points subtend angle θ with the segment AA', i.e.,

$$\angle(\overrightarrow{OA}, \overrightarrow{OA'}) = \theta.$$

See **Figure 4.55**.

But, these two geometrical loci are completely determined by the data á-priori. Their intersection defines a point O.

This point is now proven to be fixed and unique under the similarity S, as follows:

By construction, O satisfies the relations, $OA' = \lambda OA$ and $\angle(\overrightarrow{OA}, \overrightarrow{OA'}) = \theta$. Let us suppose that $O' = S(O)$. We must show $O = O'$.

We have $A' = S(A)$, and so

$$A'O' = S(AO).$$

Then, since the similarity ratio is $\lambda > 0$ and the similarity angle is θ, we have

$$A'O' = \lambda AO \quad \text{and} \quad \angle(\overrightarrow{AO}, \overrightarrow{A'O'}) = \theta.$$

From the second relation along with

$$\angle(\overrightarrow{AO}, \overrightarrow{A'O}) = \angle(\overrightarrow{OA}, \overrightarrow{OA'}) = \theta,$$

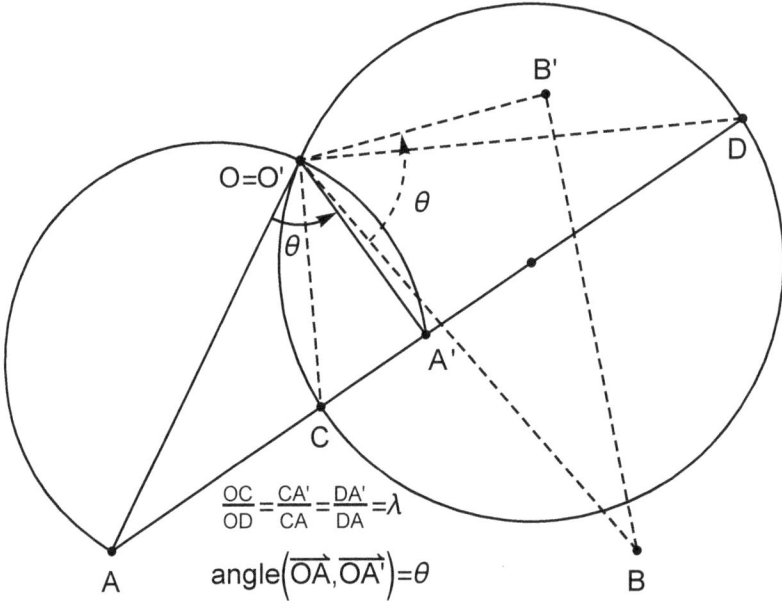

Figure 4.55: The Apollonius circle and arc determine the fixed point
O **of a similarity determined by** (A, A'), λ, **and** $0 < \theta < \pi$

we get that O' is a point of the half-line $A'O$. Then, by the first relation along
with $A'O = OA' = \lambda OA$, we get that $O' = O$. This proves synthetically the
existence of the fixed point and its construction by straight edge and compass!

So, by **Theorem 4.4.4**, we get the decomposition of the similarity S as

$$S = R_{[O,\theta]} \circ H_{[O,\lambda]},$$

and because the center of the homothety and center of the rotation coincide, as
we have also seen before, this composition is commutative, i.e.,

$$S = H_{[O,\lambda]} \circ R_{[O,\theta]}.$$

Subcase 1.3: $\theta = \pi$.

Then $S = R_{[O,\pi]} \circ H_{[O,\lambda]} = H_{[O,-\lambda]}$ is a negative homothety. The center O
is on the straight line AA' determined, as in **Lemma 4.1.1**, by the relation

$$\frac{\overrightarrow{OA'}}{\overrightarrow{OA}} = -\lambda < 0.$$

So, $\overrightarrow{OA'}$ and $\overrightarrow{OA}$ have opposite direction. The center O is located between
A and A'. See **Figures 4.3** and **4.5**. This point is unique and its geometrical
construction is basic.

Subcase 1.4: $\pi < \theta < 2\pi$.

We let $0 < \phi = \theta - \pi < \pi$. As in **Subcase 1.2, Figure 4.55**, we complete the circle that contained the arc of points that subtended angle ϕ with the segment AA'. See **Figure 4.56**.

The other point of intersection of the two circles (other from the one of **Subcase 1.2**), is the sought center now.

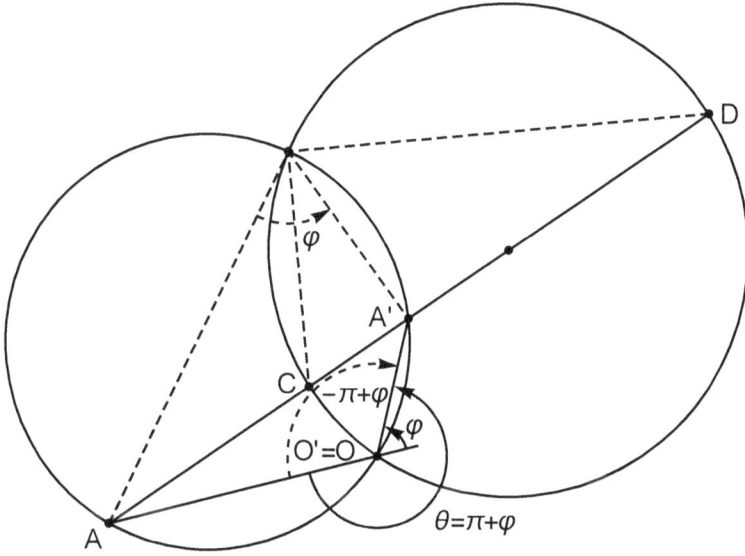

Figure 4.56: The Apollonius circle and arc determine the fixed point O of a similarity determined by (A, A'), λ, and $\pi < \theta < 2\pi$

The uniqueness of O is proven in the same way as in **Subcase 1.2**.

Remarks: (1) The above results are not valid when S is a parallel translation ($\lambda = 1$ and $\theta = 0$, but its center is an infinity point). But, they are valid when S is a non-trivial rotation [$\lambda = 1$ and $\theta \neq 0$, (mod 2π)] or a non-trivial homothety ($\lambda \neq 1$ and $\theta = 0$ or π). Then, the center of S is the center of the rotation or the homothety and the similarity angle is θ.

(2) For any $B \in \mathcal{P}$, we let $B' = S(B)$ and so

$$OB' = \lambda OB \quad \text{and} \quad \measuredangle(\overrightarrow{OB}, \overrightarrow{OB'}) = \theta.$$

Then, the triangles OAA' and OBB' are directly similar. That is, any two triangles obtained in this way are directly similar. See **Figure 4.55**.

(3) Observe that in **Figures 4.55** and **4.56**, the points D and C are the centers of **Subcases 1.1** and **1.3**, respectively. So, **Figure 4.56** determines the centers (fixed points) in all the above four subcases.

Second Case: We would like to know how to find the center of a direct similarity S that is not the identity as characterized in **Theorem 4.4.3**. That is, $S \neq I$ (not the identity) is defined by two different pairs of corresponding points (A, A') and (B, B'), i.e., $A' = S(A)$ and $B' = S(B)$, or equivalently by **Section 4.3, (9)**, defined by two vectors

$$\overrightarrow{AB} \neq \vec{0} \quad \text{and} \quad \overrightarrow{A'B'} \neq \vec{0} \quad \text{such that} \quad \overrightarrow{A'B'} = S(\overrightarrow{AB}),$$

[moving from $A' = S(A)$ to $B' = S(B)$]. Then, the similarity ratio and similarity angle are

$$\lambda = \frac{A'B'}{AB} > 0, \quad \text{and} \quad \theta = \angle(\overrightarrow{AB}, \overrightarrow{A'B'}).$$

The similarity angle without loss of generality can be considered $0 \leq \theta < 2\pi$.

Then, **the general case** can be treated a the **First Case** above, if we consider the data:

Pair of points $[A, A' = S(A)]$, $\lambda = \dfrac{A'B'}{AB} > 0$, and $0 \leq \theta = \angle(\overrightarrow{AB}, \overrightarrow{A'B'}) < 2\pi$.

(Redraw figures in this **Second Case** according to the results of the **First Case**.)

We also notice:
(1) If $A = A'$ and $B \neq B'$, then the center O is the fixed point A.
(2) If $B = B'$ and $A \neq A'$, then the center O is the fixed point B.
(3) If $A = A'$ and $B = B'$, then $S = I$ is the identity, by **Lemma 4.3.1** or **Corollary 4.4.2**.
(4) If $A \neq A'$ and $B \neq B'$, then there are the following **five special subcases** in which we can also give other equivalent geometrical constructions of the center (fixed point) O, as we can straightly check.

After disposing of these trivial cases, **we assume $A \neq A'$ and $B \neq B'$.** Then we have:

Subcase 2.1: (a) The vectors $\overrightarrow{AB}$ and $\overrightarrow{A'B'}$ have the same direction but are not equal and the lines AB and $A'B'$ do not coincide. See **Figure 4.57**. (The common point of the parallel lines $AB \parallel A'B'$ is a point at infinity.)

Then, the sought direct similarity is

$$S = H_{[O,\lambda]},$$

the homotety with center O the intersection of the lines AA' and BB'. The similarity ratio and the similarity angle are:

$$0 < \lambda = \frac{A'B'}{AB} \neq 1 \quad \text{and} \quad \theta = 0.$$

[See also **Figures 4.6, 4.7, 4.8, 4.9, Subsection 4.1.1, (f), Theorem 4.2.1, Corollary 4.2.1** and **Figure 4.41**.]

(b) If the lines AB and $A'B'$ coincide, then see **Section 4.2, (4), Figure 4.25**, with similarity ratio

$$0 < \rho = \lambda = \frac{A'B'}{AB} \neq 1.$$

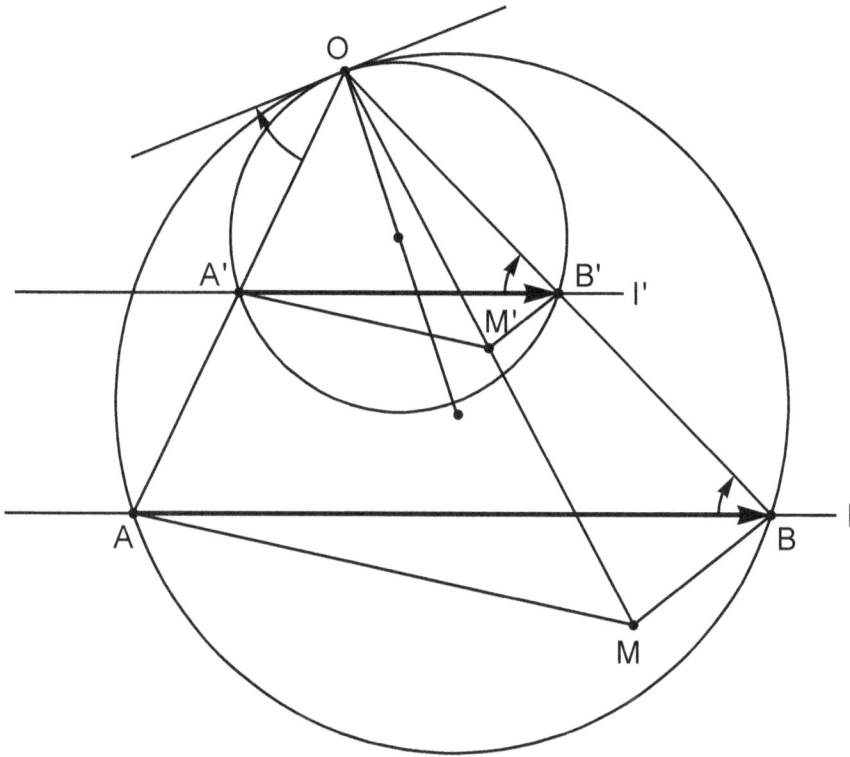

Figure 4.57: Determination of the center of similarity in the special case that $\overrightarrow{AB}$ and $\overrightarrow{A'B'}$ are parallel, have the same direction and $AB \neq A'B'$

Subcase 2.2: (a) The vectors $\overrightarrow{AB}$ and $\overrightarrow{A'B'}$ have opposite direction, their lengths are not equal and the lines AB and $A'B'$ do not coincide. See **Figure 4.58**. (The common point of the parallel lines $AB \parallel A'B'$ is a point at infinity.) Then, the sought direct similarity is

$$S = R_{[O,\pi]} \circ H_{[O,\lambda]} = H_{[O,\lambda]} \circ R_{[O,\pi]},$$

where the center O is again the intersection of the lines AA' and BB', (since the lines AB and $A'B'$ do not coincide). The similarity ratio is

$$\lambda = \frac{A'B'}{AB} > 0 \quad \text{and the similarity angle is} \quad \pi.$$

[See also **Figures 4.6, 4.7, 4.8, 4.9, Subsection 4.1.1, (f), Theorem 4.2.1, Corollary 4.2.1**.]

(b) If the lines AB and $A'B'$ coincide, then see **Section 4.2, (4), Figure 4.25**, with $\rho = -\lambda = -\dfrac{A'B'}{AB} < 0$.

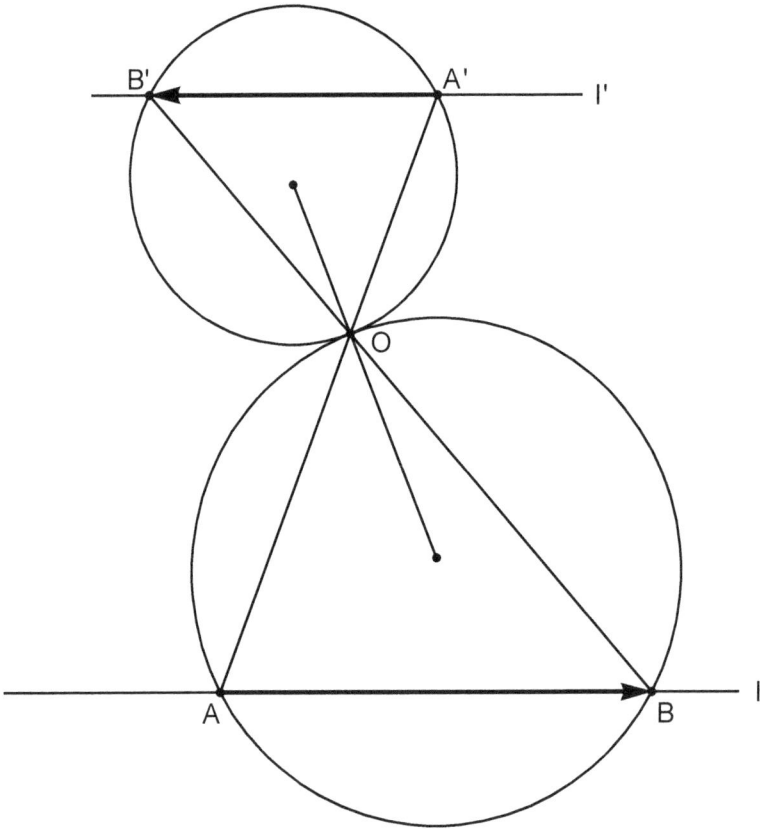

Figure 4.58: Determination of the center of similarity in the special case that $\overrightarrow{AB}$ and $\overrightarrow{A'B'}$ are parallel, have opposite direction and $AB \neq A'B'$

Subcase 2.3: (a) The vectors are opposite $\overrightarrow{AB} = -\overrightarrow{A'B'}$ and the lines AB and $A'B'$ do not coincide. See **Figure 4.59**. (The common point of the parallel lines $AB \parallel A'B'$ is a point at infinity.)

Then, the sought direct similarity is the rotation

$$S = R_{[O,\pi]},$$

with center O the intersection of the lines AA' and BB' and rotation angle π. The similarity ratio is

$$\lambda = \frac{A'B'}{AB} = 1 \quad \text{and the similarity angle is} \quad \pi.$$

[See also **Figure 2.57, Subsection 2.5.3 (III)** and compare with **Figure 4.59**.]

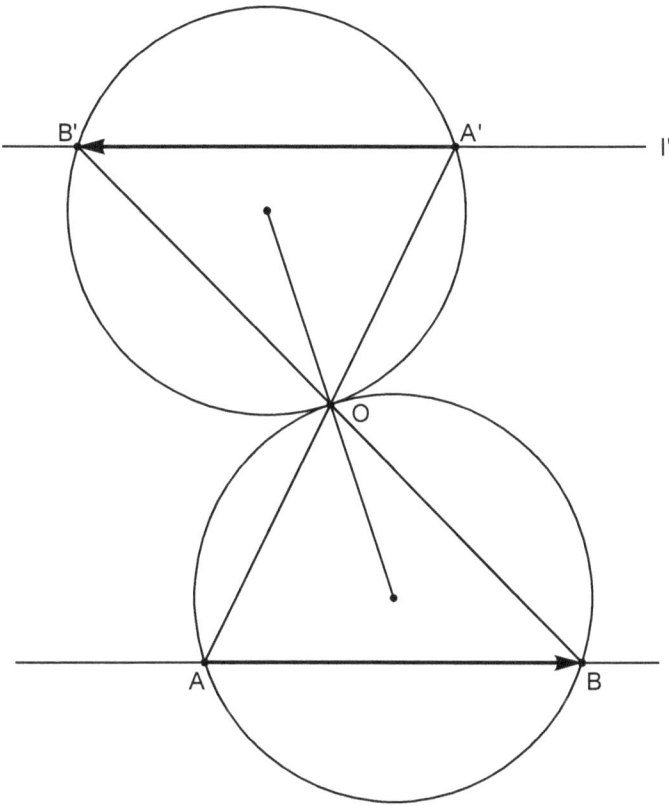

Figure 4.59: Determination of the center of similarity
in the special case $\overrightarrow{AB} = -\overrightarrow{A'B'}$

(b) If the lines AB and $A'B'$ coincide, then $S = R_{[O,\pi]}$ is the half-turn whose center is the common midpoint on AA' and BB'. See **Subsection 2.1.1, (2)** and draw a figure.

Subcase 2.4: The vectors are equal $\overrightarrow{AB} = \overrightarrow{A'B'}$ and the lines AB and $A'B'$ may or may not coincide. Then, the sought direct similarity is the parallel translation

$$S = T_{\vec{u}}, \quad \text{with} \quad \vec{u} = \overrightarrow{AA'} = \overrightarrow{BB'}.$$

The similarity ratio and the similarity angle are

$$\lambda = \frac{A'B'}{AB} = 1 \quad \text{and} \quad \theta = 0.$$

[See **Figure 2.52, Subsection 2.5.1** and **Theorem 2.5.1**.]

Special Subcase 2.5: The vectors are not parallel $(\overrightarrow{AB} \nparallel \overrightarrow{A'B'})$ and have the same length $AB = A'B'$.

We let K be the intersection of the lines AB and $A'B'$. The sought direct similarity is the rotation

$$S = R_{[O,\theta]},$$

with center O defined by the intersection of the circles (KAA') and (KBB'), or the common point of the midpoint perpendiculars of the segments AA' and BB' and $\theta = \measuredangle(\overrightarrow{AB}, \overrightarrow{A'B'})$.

The similarity ratio is

$$\lambda = \frac{A'B'}{AB} = 1,$$

and the similarity angle is equal to the rotation angle θ for which

$$0 < \theta = \measuredangle(\overrightarrow{AB}, \overrightarrow{A'B'}) < 2\pi, \quad \text{without loss of generality.}$$

[See **Figures 2.53, 2.54** and **Subsection 2.5.2** and compare with **Figure 4.61** in which the point K is labeled by I.]

Next, we develop two general geometrical constructions of the center O of a direct similarity S.

Three General Geometrical Constructions of the Center O

We assume $A \neq A'$ and $B \neq B'$ (and nothing else). We want to find the similarity S such that

$$A' = S(A) \quad \text{and} \quad B' = S(B).$$

As we know, the similarity ratio and the similarity angle are

$$\lambda = \frac{A'B'}{AB} > 0, \quad \text{and} \quad 0 \leq \measuredangle(\overrightarrow{AB}, \overrightarrow{A'B'}) < 2\pi.$$

First General Construction of the Center O.

We consider the following important general case under the four hypotheses:

(1) $AB \nparallel A'B'$,

(2) $AB \neq A'B'$,

(3) so $0 < \lambda = \dfrac{A'B'}{AB} \neq 1$,

(4) **and the points** A, B, A', B', **and** I, **the point of intersection of the two lines** $AB \nparallel A'B'$, **are five distinct points.**

Then, the center O of the similarity S, lies on the following four geometrical loci: See **Figures 4.60** and **4.61**.

(1) The circle defined by IAA'. (Shown next!)

(2) The circle defined by IBB'. (Shown next!)

(3) The circular arc, whose points subtend angle θ with the segment AA'. (Already seen, **Figure 4.55**.)

(4) The circular arc, whose points subtend angle θ with the segment BB'. (Already seen, **Figure 4.55**.)

If we assume that the fixed point O exists $(O' = O)$, then the triangles OAB and $OA'B'$ have proportional sides (of ratio λ) and so, they are similar. [See **Figures 4.60** and **4.61**. See also **Remark (2)** of the **First Case** before.] So,

$$\measuredangle(BO, BA) = \measuredangle(B'O, B'A') \quad \text{or} \quad \measuredangle(BO, BI) = \measuredangle(B'O, B'I).$$

Hence, by the properties of the inscribable quadrilaterals, $BIB'O$ is an in-

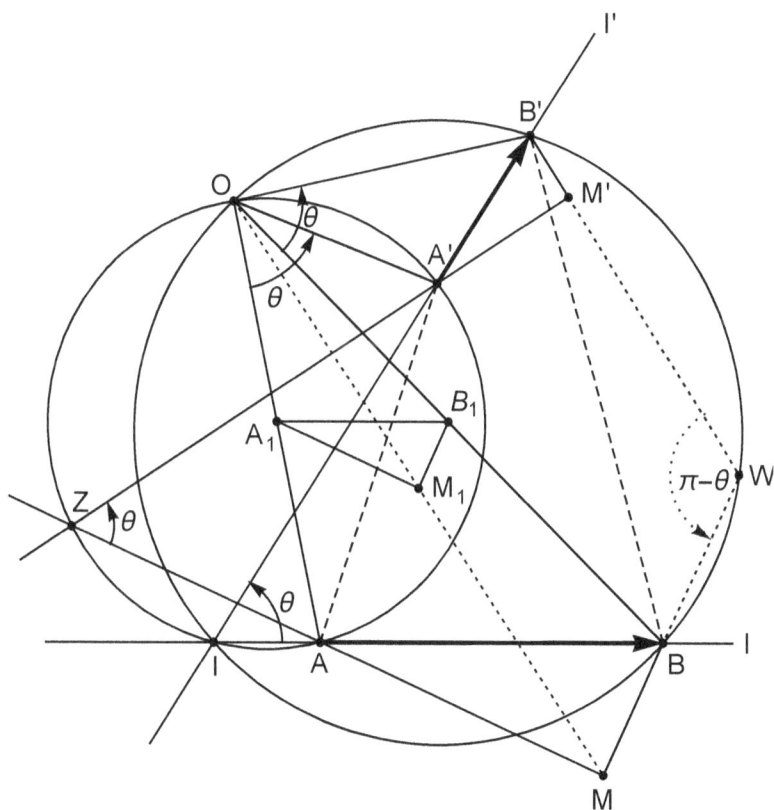

Figure 4.60: **Finding the center of a similarity determined by two pairs of corresponding points (A, A') and (B, B'), in the general case in which A, A', B, B', and I are different from one another and $\overrightarrow{AB} \nparallel \overrightarrow{A'B'}$ and $AB \neq A'B'$**

scribable quadrilateral and so O is on the circle circumscribed to the triangle IBB'.

Similarly,

$$\measuredangle(AB, AO) = \measuredangle(A'B', A'O) \quad \text{or} \quad \measuredangle(AO, AI) = \measuredangle(A'O, A'I).$$

Hence, $AIA'O$ is an inscribable and O is on the circle of the triangle IAA'.

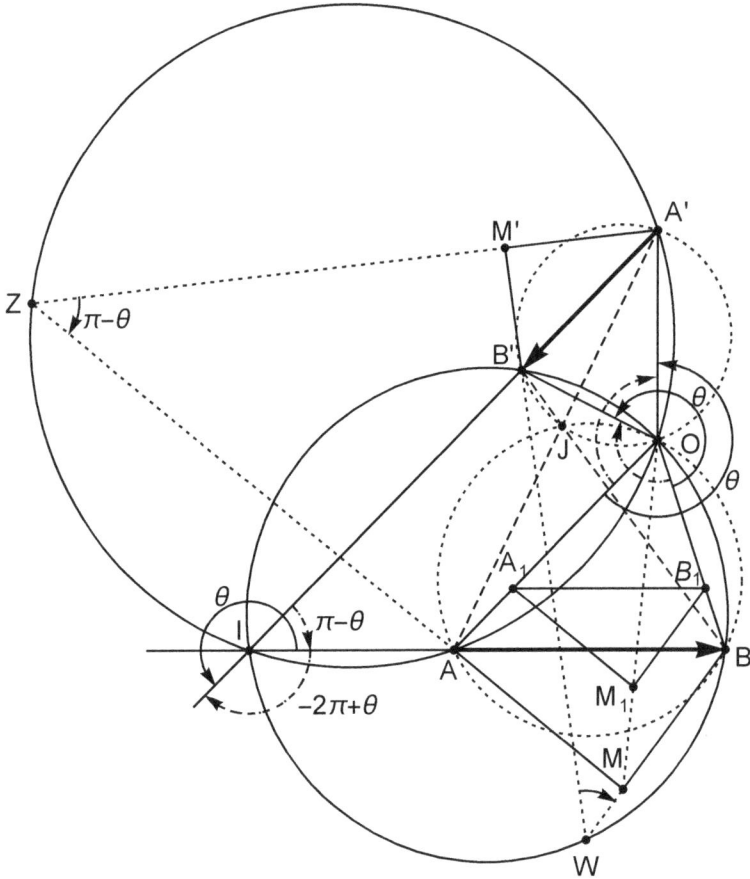

Figure 4.61: **Finding the center of a similarity determined by two pairs of corresponding points (A, A') and (B, B'), in the general case in which A, A', B, B', and I are different from one another and $\overrightarrow{AB} \nparallel \overrightarrow{A'B'}$ and $AB \neq A'B'$**

Now, thinking conversely, O is determined as the second point of intersection of the two circles (IAA') and (IBB'), which in this case are defined by these triples of points. (The first common point is I, of course. Sometimes $O \equiv I$, as explained in **Subsection 4.5.6.**)

Then, we see that

$$\measuredangle\theta = \measuredangle(\overline{IAB}, \overline{IA'B'}) = \measuredangle(OA, OA') = \measuredangle(OB, OB').$$

If now to any chosen point M, we associate the point M' such that the triangles ABM and $A'B'M'$ are directly similar, we easily prove that

$$\measuredangle\theta = \measuredangle(OM, OM') \qquad \text{and} \qquad \frac{OM'}{OM} = \frac{A'B'}{AB}.$$

Therefore,
$$M' = S(M).$$

Since the triangles OAB and $OA'B'$ are directly similar, by construction, we have that $O = S(O)$, and so O is the fixed point of the similarity S. Thus, the similarity S is completely determined as

$$S = R_{[O,\theta]} \circ H_{[O,\lambda]} = H_{[O,\lambda]} \circ R_{[O,\theta]}.$$

The inverse if S is

$$S^{-1} = R_{[O,-\theta]} \circ H_{[O,\frac{1}{\lambda}]} = H_{[O,\frac{1}{\lambda}]} \circ R_{[O,-\theta]}.$$

[Prove also that, if AM intersects the circle $(IAA'O)$ at a point Z, then Z is on the line $A'M'$. Analogous result for the lines MB, $M'B'$ and the circle $(IBB'O)$. See **Figures 4.60** and **4.61**. Also, the centers of the circles AIA' and BIB' are on the mid-perpendicular line of the segment IO.]

Remark: If $AB \parallel A'B'$ then I is a point at infinity and so the circles IAA' and IBB' become the straight lines AA' and BB' and O is their point of intersection. See **Figures 4.57, 4.58** and **4.59**.

Other Facts: (a) **The center O found above, in Figures 4.60 and 4.61, is also the center of the similarity S' that maps the segment AA' onto BB'** [or defined by these segments in this order. So now, the pairs of corresponding points are: (A, B) and (A', B')]. See **Figure 4.62**. The similarity ratio is

$$\lambda' = \frac{B'B}{A'A}.$$

This is proven in the same way as in the case of the segments AB and $A'B'$. Let J be the intersection of the lines AB and $A'B'$. The triangles OAA' and OBB' are directly similar [the corresponding sides at O are proportional, the contained angles are equal, and they have the same orientation (why?)] and so

$$\measuredangle(OA, AJ) = \measuredangle(OB, BJ) := \phi.$$

Therefore, the circle (JAB) passes through O. Also,

$$\measuredangle OA'J = \phi + \theta = \measuredangle OB'J.$$

Therefore, the circle $(JA'B')$ also passes through O.

So, besides the four loci found previously, we see that **the center O also belongs to the following two additional loci**:

(5) the circle (ABJ) and   **(6)** the circle $(A'B'J)$.

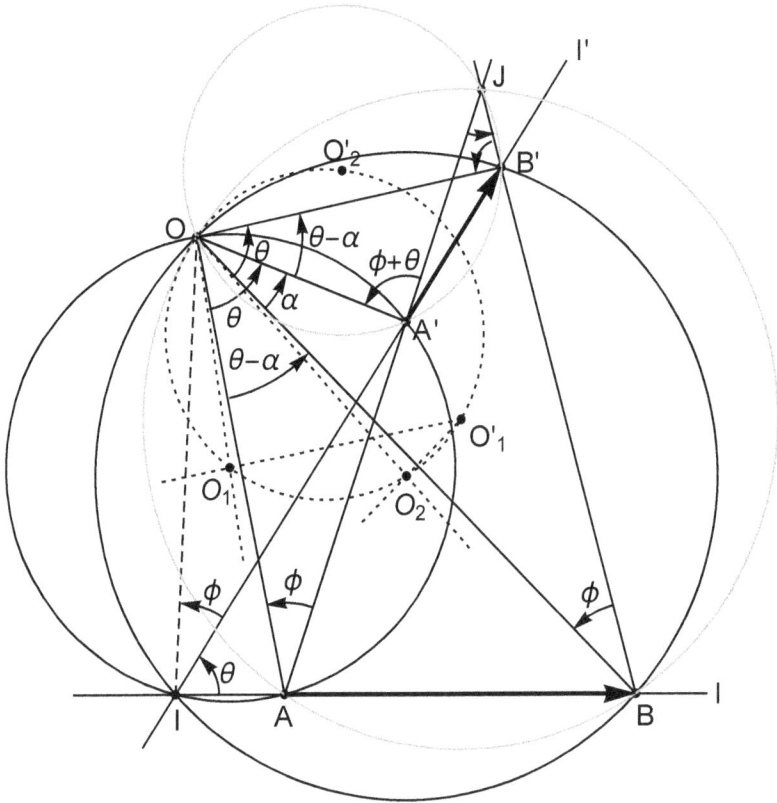

Figure 4.62: "O" is also the center of a similarity determined by the pairs of corresponding points (A, B) and (A', B'), in the general case examined before

We also let $\alpha := \angle BOA'$ and observe that the similarity angle is

$$\angle AJB = \theta - \alpha.$$

Thus, the similarity S' is

$$S' = R_{[O, \theta - \alpha]} \circ H_{[O, \lambda']} = H_{[O, \lambda']} \circ R_{[O, \theta - \alpha]}.$$

Then, the inverse of S' is

$$S'^{-1} = R_{[O, -\theta + \alpha]} \circ H_{[O, \frac{1}{\lambda'}]} = H_{[O, \frac{1}{\lambda'}]} \circ R_{[O, -\theta + \alpha]}.$$

Now, using the properties of the inscribable quadrilaterals and the fact that the center line of two circles is perpendicular to their common chord, we can fairly easily prove the following two facts:

(b) The centers O_1, O_2, O_1', O_2' of the above four circles and O are on the same circle, called the **Miquel**[9] **circle of the complete quadrilateral** $IABB'A'J$. (See also **Example 4.6.3**. Check: $O_1O_1' \perp OA$, $O_2O_1' \perp OB$ and $\angle AOB = \angle O_1OO_2 = \angle O_1O_1'O_2$, etc.) The point O is called the **Miquel point of the complete quadrilateral** $IABB'A'J$.

Note: The Miquel point of a complete quadrilateral should not be confused with the infinitely many Miquel points of a triangle. But, this point can be also demonstrated by using the Miquel points of triangles. Find this topic in bibliography.

(c) If the quadrilateral $ABB'A'$ is inscribable, then the point O is on the line IJ and vice-versa.

Now, we examine how to solve the problem in the limiting subcases of the **previous general case**.

Subcase I.1: $A = I \neq A'$.

If $A = I \neq A'$, then the center C of the limiting circle (IAA') is the common point of the line perpendicular to AB at A, and the midpoint perpendicular of the segment $AA' = IA'$. The radius is $CA = CA'$. See **Figure 4.63**.

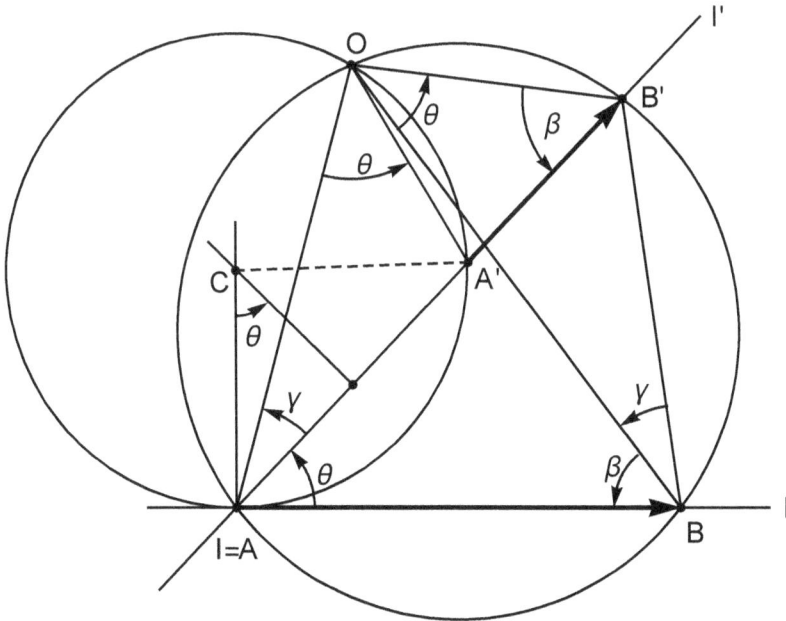

Figure 4.63: Special subcase $A = I \neq A'$

---

[9] Auguste Miquel, French mathematician, 1816-1851.

Subcase I.2: $A \neq I = A'$.

If $A \neq I = A'$, then the center C of the limiting circle $(IA'B')$ is the common point of the line perpendicular to $A'B'$ at A', and the midpoint perpendicular of the segment $A'A = IA$. The radius is $CA' = CA$. See **Figure 4.64**.

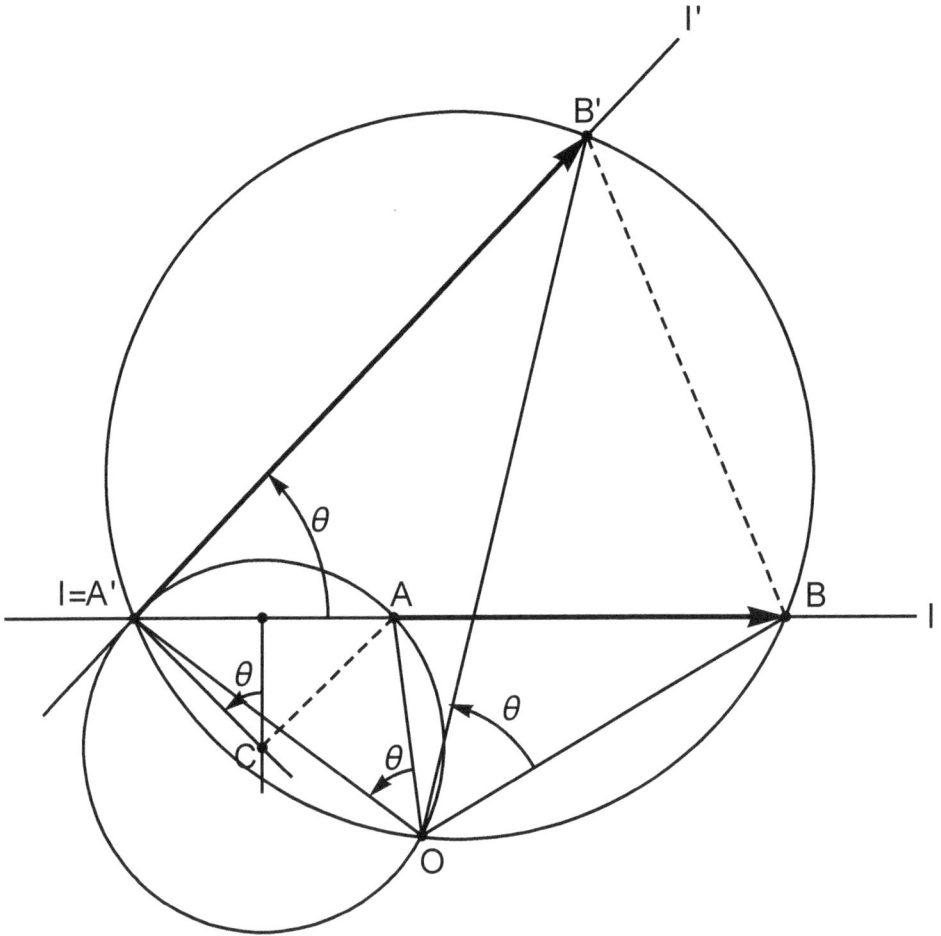

Figure 4.64: Special subcase $A \neq I = A'$

The two subcases that follow are similar to the previous two. They are all limiting cases of the **first general construction** of O.

Subcase I.3: $B = I \neq B'$.

If $B = I \neq B'$, then the center C of the limiting circle (IBB') is the common point of the line perpendicular to AB at B, and the midpoint perpendicular of the segment $BB' = IB'$. The radius is $CB = CB'$. Make a **Figure**.

Subcase I.4: $B \neq I = B'$.

If $B \neq I = B'$, then, the center C of the limiting circle is the common point of the line perpendicular to $A'B'$ at B', and the midpoint perpendicular of the segment $B'B = IB$. The radius is $CB' = CB$. Make a **Figure**.

Second General Construction of the Center O:

The center O, besides the six loci found so far, lies on two additional loci, which are the following two Apollonius circles:

(7) The Apollonius circle, whose points P satisfy $\dfrac{PA'}{PA} = \lambda$.

(8) The Apollonius circle, whose points P satisfy $\dfrac{PB'}{PB} = \lambda$.

The point O as found in the previous general construction **(I)** satisfies

$$\frac{OA'}{OA} = \frac{OB'}{OB} = \frac{A'B'}{AB} = \lambda,$$

by the direct similarity of the triangles OAB and $OA'B'$. Therefore, O belongs to both of the above Apollonius circles and is the center (fixed point) of the direct similarity S, as proven before. See **Figure 4.65**.

[As in the **first general construction**, to prove that O is the fixed point of S, we notice again that for any point $X \in \mathcal{P}$ we have: $X' := S(X)$ if and only if the triangles ABX and $A'B'X'$ are directly similar. (Provide proof.) Since, by construction, the triangles ABO and $A'B'O$ are directly similar because they have proportional sides of ratio λ and are directly oriented, we have that $O = O'$ or $O = S(O)$.

It also follows that the above Apollonius circles are either tangent at O (see **Subsection 4.5.6**) or they intersect at a second point, Q in **Figure 4.65**. When we study the opposite similarities in the next section, we shall see that the point Q is the center of the opposite similarity that maps A to A' and B to B'.]

We must also prove that,

$$\widehat{AOB} = \widehat{A'OB'} = \sphericalangle(AB, A'B') := \theta.$$

We let I be the common point of the straight line AB and $A'B'$. By the similarity of the triangles OAB and $OA'B'$, we get

$$\widehat{OAB} = \widehat{OA'B'}.$$

Therefore, the quadrilateral $AIA'O$ is inscribable. Hence,

$$\widehat{AOA'} = \widehat{BIA'} = \theta.$$

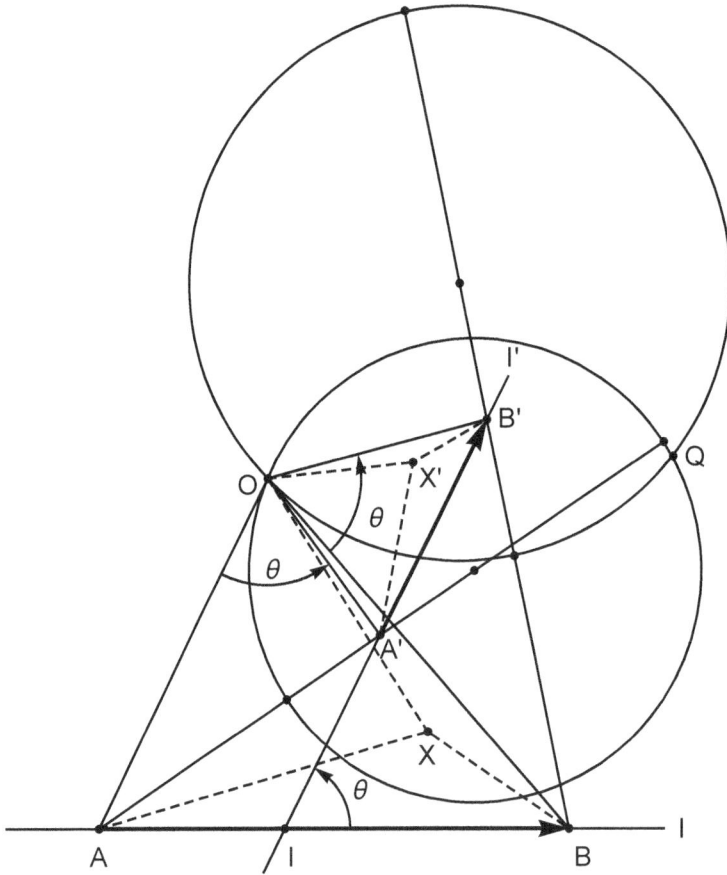

Figure 4.65: **The Apollonius circles on AA' and BB' with ratio the similarity ratio $\lambda = \dfrac{A'B'}{AB}$ intersect at the center O of the direct similarity. The other point of intersection Q is the center of the opposite similarity that maps AB to $A'B'$.**

(Similarly the $B'BIO$ is inscribable. We have used this fact in the previous construction. Also, the second point of intersection of the two Apollonius circles, Q, will play the role of O, as the center of the opposite similarity that maps $\overrightarrow{AB}$ to $\overrightarrow{A'B'}$, in **Subsection 4.5.2**. See **Figures 4.77** and **4.78** and compare. The letters O and Q have been interchanged. The case $O = Q$ is studied in **Subsection 4.5.6**.)

Third General Construction of the Center O.

Next, we prove that O is also on two additional loci, [(**9**) and (**10**) below], which are á-priori determined straight lines.

(**9**) The first straight line is **one of the two lines of proportional distances** from the two lines $\overline{IAB}$ and $\overline{IA'B'}$ with proportion λ. For this we must give some preliminaries.

We firstly remind the reader of the **definition** and the **geometrical construction of the two lines of proportional distances from two given lines**, either the two given lines intersect or are parallel. So, we have:

Construction of the two straight lines of proportional distances from two given lines.

Given two different lines l_1 and l_2, we would like to find all the points P whose distances PS and PT from the given lines l_1 and l_2, respectively, have constant ratio λ, given. That is, $\dfrac{PS}{PT} = \lambda$.

Case 1: The lines l_1 and l_2 intersect at a point I. See Figure 4.66.

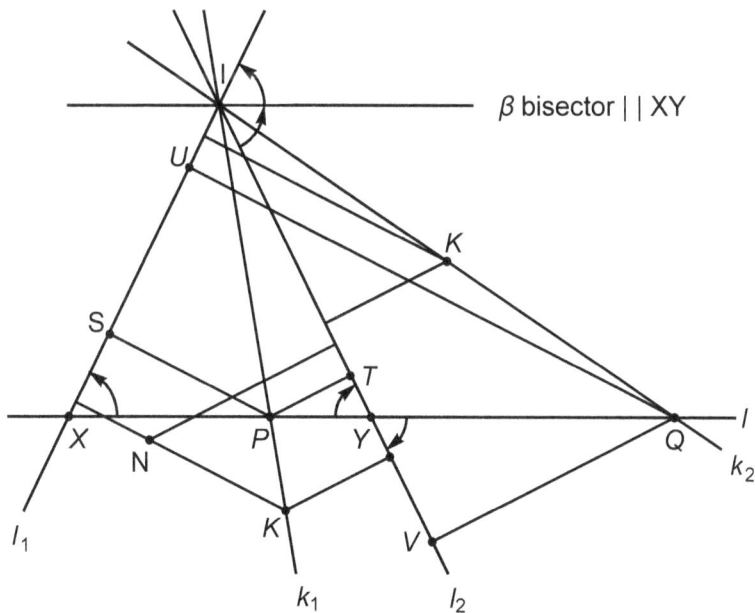

Figure 4.66: Line k of proportional distances for two given intersecting lines l_1 and l_2

We draw the bisector β of one of the angles of l_1 and l_2 at I. Then, we draw a line $l \parallel \beta$ that intersects l_1 and l_2 at the points X and Y, respectively.

We divide the segment XY internally and externally by two points P and Q, respectively, such that $\dfrac{PX}{PY} = \dfrac{QX}{QY} = \lambda$ (harmonic division with ratio λ). Let the distances of P from l_1 and l_2 be PS and PT and similarly the distances of Q from l_1 and l_2 be QU and QV, respectively.

Since the triangle IXY is isosceles (why?), the triangles PXS, PYT, QXU, and QYV are all similar to one another (why?). Then,

$$\frac{PS}{PT} = \frac{PX}{PY} = \lambda \qquad \text{and} \qquad \frac{QU}{QV} = \frac{QX}{QY} = \lambda.$$

Now, any point K of the two lines $k_1 := IP$ and $k_2 := IQ$ has distances from the given lines l_1 and l_2 whose ratio is constant λ. (Prove!) No other point N of the plane exists satisfying this property. (Prove!)

The lines k_1 and k_2 are called **the lines of proportional distances from the lines** l_1 and l_2, respectively, with proportion ratio λ.   ■

Case 2: The lines l_1 and l_2 are parallel ($l_1 \parallel l_2$). See Figure 4.67.

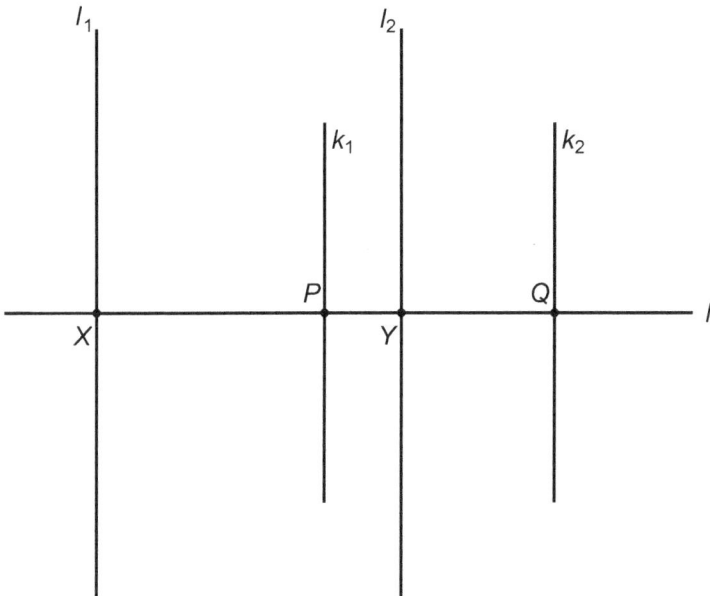

Figure 4.67: Lines k_1 and k_2 of proportional distances
from two parallel lines $l_1 \parallel l_2$

We draw a line $l \perp l_1 \parallel l_2 \perp l$ that intersects l_1 and l_2 at the points X and Y, respectively. We divide the segment XY internally and externally by two points P and Q, respectively, such that

$$\frac{PX}{PY} = \frac{QX}{QY} = \lambda$$

(harmonic division with ratio λ). The segments PX, PY, QX, and QY coincide with the distances of P and Q from l_1 and l_2, respectively.

In this case, the lines of proportional distances are easily proved to be: The line $k_1 \perp l$ at P and the line $k_2 \perp l$ at Q (so, $k_1 \parallel l_1 \parallel l_2 \parallel k_2$). Again, no other point N of the plane exists satisfying this property. (Prove!)

■

Now, for finding the center O (fixed point) of a direct similarity of ratio λ, the above geometrical locus is applied as follows. If we draw the projections M and M' of O on the two lines $\overline{IAB}$ and $\overline{IA'B'}$, respectively, then the segments OM and OM' are heights of the two (directly) similar triangles OAB and $OA'B'$. By the similarity of the two triangles, we get

$$\frac{OM'}{OM} = \frac{A'B'}{AB} = \lambda.$$

Therefore, O is on one of the two lines of the proportional distances from the two lines $\overline{IAB}$ and $\overline{IA'B'}$ with proportion λ. This line, as we have seen, passes through the common point I of the two straight lines AB and $A'B'$, and so it is the line IO. See, **Figure 4.68**.

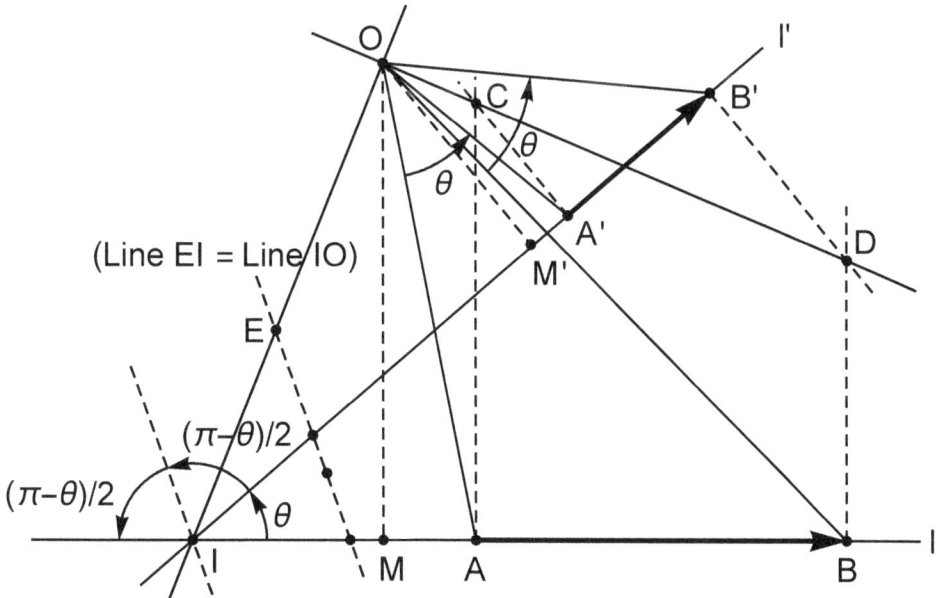

Figure 4.68: **The center of a direct similarity is the point of intersection of the line of proportional divisions** CD **with of one of the lines of the proportional distances** OI, **determined by the segments** AB **and** $A'B'$ **and their lines.**

(10) The second straight line that contains the fixed point O is

constructed in the following way: At A and A', we draw lines perpendicular to lines AB and $A'B'$, respectively. They intersect at a point C. At B and B', we draw lines perpendicular to lines AB and $A'B'$, respectively. They intersect at a point D. Then, the point O is on the straight line $k := CD$. See **Figures 4.69** and **4.70**.

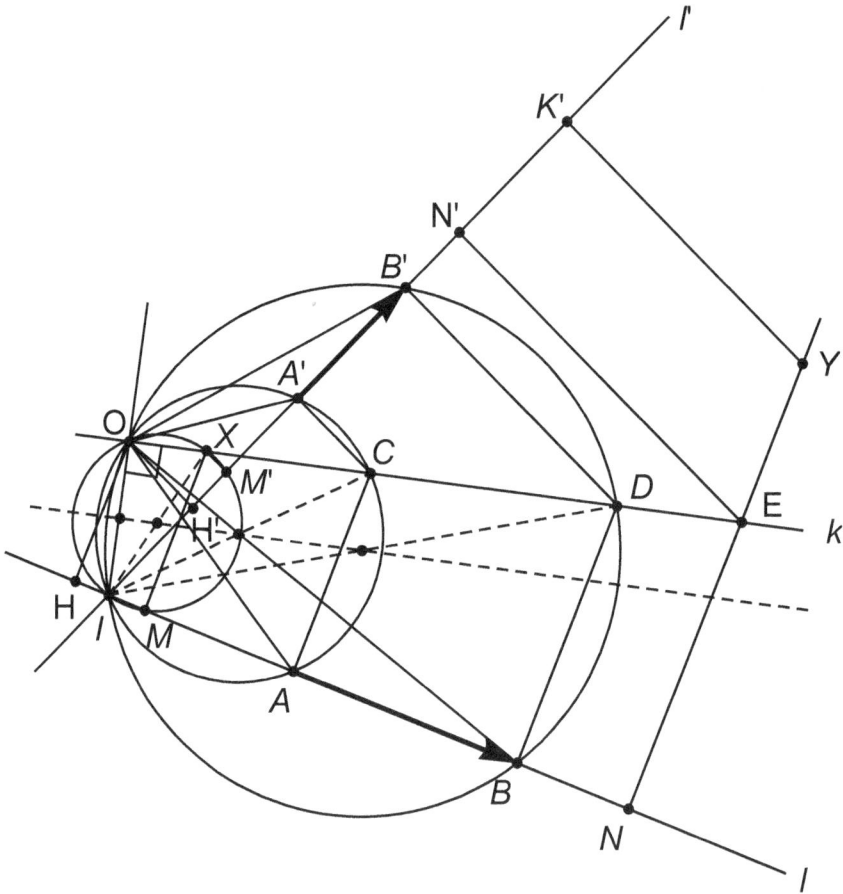

Figure 4.69: Line k of proportional divisions for two given intersecting lines l and l' and with respect to the vectors $\overrightarrow{AB}$ and $\overrightarrow{A'B'}$ on them, respectively

The proof of this is based on the fact that, for any point X of the line $k := CD$, its projections M and M' on the lines AB and $A'B'$, respectively, satisfy the proportion (obtained as we indicate in the **Result** that follows next)

$$\frac{M'A'}{MA} = \frac{M'B'}{MB} = \frac{A'B'}{AB} := \lambda,$$

and no other point Y of the plane satisfies this proportion!

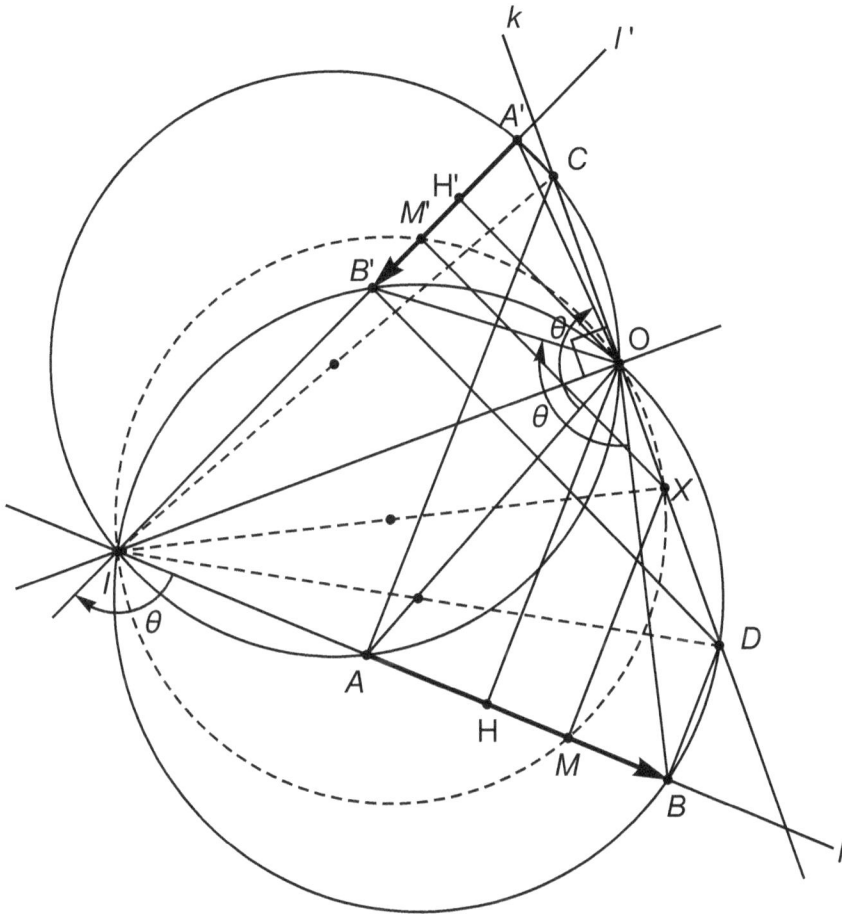

Figure 4.70: **Line k of proportional divisions for two given intersecting lines l and l' and with respect to the vectors $\overrightarrow{AB}$ and $\overrightarrow{A'B'}$ on them, respectively**

The point O satisfies this proportion, because in the three pairs of similar triangles $[OAB$ and $OA'B']$, $[OHA$ and $OH'A']$, and $[OHB$ and $OH'B']$ the heights corresponding to the vertex O are OH and OH', per pair and respectively. Therefore, by the similarity of the triangles, we get

$$\frac{H'A'}{HA} = \frac{H'B'}{HB} = \frac{OH'}{OH} = \frac{A'B'}{AB} = \lambda.$$

Hence, O is on **the line $k := CD$ of the proportional divisions,** which, as we have just seen, is á-priori determined.

The above line k is called **the line of proportional internal / external divisions of two given segments** AB and $A'B'$ **that do not lie on parallel lines.** Its general **geometrical construction** is based on the following:

Result: *On two given different lines l and l' that intersect at a point I, we fix two non-zero vectors $\overrightarrow{AB}$ and $\overrightarrow{A'B'}$, respectively. At A and A', we draw lines perpendicular to lines l and l', respectively; they intersect at a point C. At B and B', we draw lines perpendicular to lines l and l', respectively; they intersect at a point D. See, **Figures 4.69** and **4.70**.*

Any point X of the line $k := CD$ has the following property: Let M and M' be the projections of X on the lines l and l'. Then,

$$\frac{M'A'}{M'B'} = \frac{XC}{XD} = \frac{MA}{MB} \implies \frac{M'A'}{MA} = \frac{M'B'}{MB} = \frac{M'B' - M'A'}{MB - MA} = \frac{A'B'}{AB} := \lambda, \text{ fixed.}$$

*No other point Y of the plane (as shown in **Figures 4.69** and **4.70**) exists satisfying this property. (Prove!)*

(This **result** follows straightforwardly by manipulating the proportions derived *when two or more lines are intersected by at least three parallel lines*, as claimed in the pertinent well-known **Thales' Theorem**. Review it once more.)

Notice that, such a line or points do not exist when $l \parallel l'$, unless the sizes and positions of AB and $A'B'$ on the lines are specially chosen. We observe that in such a case, in general, the lines that determined the points C and D, as in **Figures 4.69** and **4.70**, are parallel and so, either do not intersect or coincide and therefore, the line k cannot be determined. We also notice the **results**:

*For any point X on the straight line k of the proportional divisions and its projections M and M' on the lines l and l', respectively, (see **Figures 4.69** and **4.70**), the quadrilateral $IMXM'$ is inscribable. (Point I is the intersection of lines l and l'.) The diameter of the circle $(IMXM')$ is the segment IX. All the circles $(IMXM')$ pass through the center of the similarity O and so their centers are on the midpoint perpendicular to IO.*

Corollary 4.4.3 *(a) The straight lines of the proportional distances IO and of the proportional divisions DCO, with respect to the lines IA and IA' and the straight segments AB and $A'B'$ on them (see **Figure 4.66-4.70**), are perpendicular to each other.*

(b) If $\lambda = 1$, then the direct similarity S reduces to a rotation with $AB = A'B'$, center O and angle $\measuredangle(IA, IA')$. Then:

(1) The line IO is the bisector of one of the angles of l and l' at I. Then, DCO is parallel to the bisector of its supplementary angle at I. So, the center of a rotation is also the common point of one of the lines of equal distances and the line of equal divisions with respect to l and l' and the equal segments AB and $A'B'$ on them.

(2) The projections M and M' of a point X of the line of equal divisions on the lines IA and IB are corresponding points. Therefore their midpoint perpendiculars pass through the point O, the center of the rotation.

(3) All the circles MIM' pass through the center of the rotation O, just as the circles AIA' and BIB' do so.

(4) The centers of these circles are on the midpoint perpendicular of IO, which is parallel to DCO and to the bisector of the supplementary angle of $\angle(IA, IA')$ at I.

Proof (a) This follows from the fact that the circle (AIA') contains the points C and O and IC is a diameter. [Similarly the circle (BIB') contains D and O and ID is diameter.] Since IC is a diameter, $\angle IOC = \dfrac{\pi}{2}$.

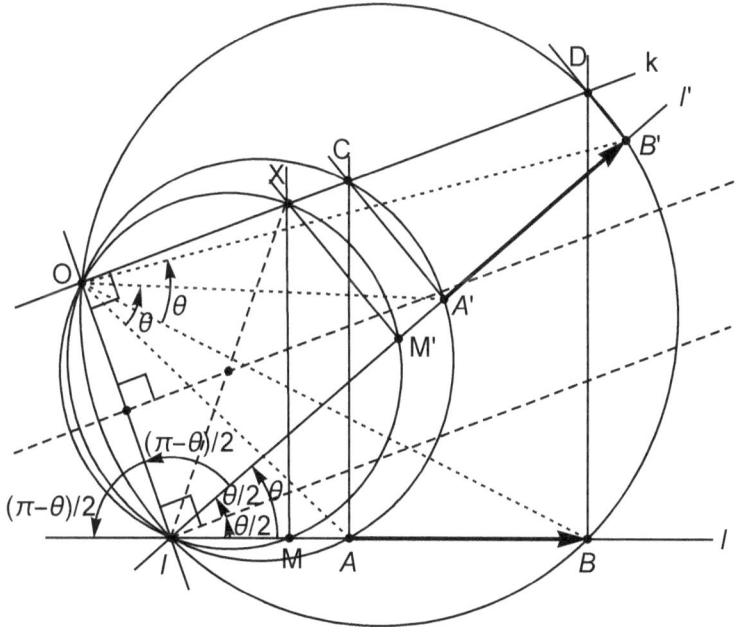

Figure 4.71: When $\lambda = 1$, the similarity is a rotation. This figure also shows the extra facts about its center.

(b) When $\lambda = 1$, the homothety is identity and so the direct similarity S reduces to a rotation with center O and angle $\angle(IA, IA')$. See **Figure 4.71**. Then, the line of proportional distances IO becomes line of equal distances, that is, the bisector of one of the angles $\angle AIA'$. [See also **Corollary 2.5.1**, (4).] Also, the line DCO becomes line of equal divisions and since $DCO \perp IO$, DCO is parallel to the bisector of the supplementary angle of $\angle(IA, IA')$ at I.

The remaining claims follow from all facts about the center O and the rotation angle $\angle(IA, IA')$ that we have studied so far.

As an exercise, **prove directly the fact**: *The straight line of equal divisions of two rays Il and Il' (as in* **Figure 4.71**) *is parallel to the bisector of the angle $\widehat{lIl'}$ and so perpendicular to the bisector of its supplementary angle.*

We continue with two important results concerning straight lines and circles:

Result 1: From all the constructions we have performed so far, we conclude that: **Given a ratio $\lambda > 0$ fixed, the geometrical locus of the centers of all direct similarities of ratio $\lambda > 0$ that map l to l' are the two straight lines of proportional distances, IE and IF.** See **Figure 4.72**.

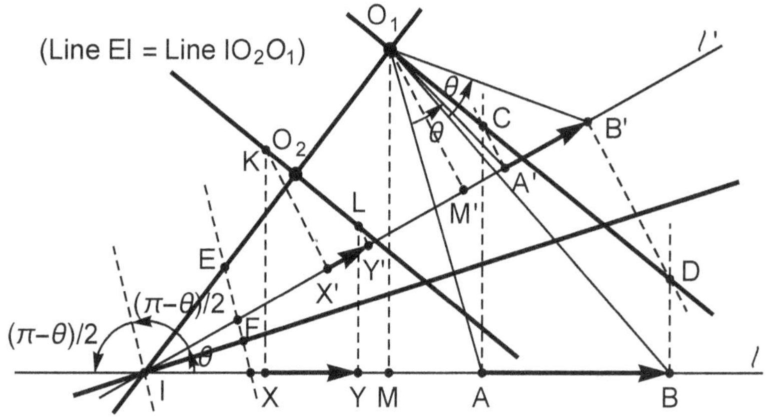

Figure 4.72: The geometrical locus of the centers of all similarities that map l to l' is the two lines of proportional distances IE and IF.

As in **Figures 4.69, 4.70** and **4.71**, we place any vectors $\overrightarrow{AB}$, $\overrightarrow{A'B'}$ or $\overrightarrow{XY}$, $\overrightarrow{X'Y'}$ on l and l', respectively, and such that the ratio of their lengths is

$$\frac{A'B'}{AB} = \frac{X'Y'}{XY} = \ldots = \lambda.$$

Then, for any such pair of vectors, we find the direct similarities, as we have constructed them in the previous paragraphs, that map $\overrightarrow{AB}$ to $\overrightarrow{A'B'}$, or $\overrightarrow{XY}$ to $\overrightarrow{X'Y'}$, and so on. All the centers O_1, O_2, ..., of these direct similarities are on the two lines of the proportional distances IE and IF.

The lines of proportional divisions, CD, KL, etc., with respect to these pairs of vectors, etc., as in **Figures 4.69, 4.70, 4.71** and **4.72**, depend not only on the ratio λ but also on the relative position of the two vectors in each pair. So, they are shifted parallel to themselves and perpendicular to one of the lines of the proportional distances IE, or IF, etc.

Therefore we conclude that: **With no restriction on the similarity ratio, two straight line l and l' are mapped to each other by infinitely many direct similarities whose centers cover the whole plane $\mathcal{P}$.**

[See also: "About opposite similarities between two straight lines" at the **end of this Chapter, (3)**, before **Section 4.6** of the Applications.]

Result 2: *Given two circles, there are infinitely many direct similarities that map one onto the other. Moreover: there is a unique such direct similarity that maps any given point on the first circle to any given point on the second circle.*

Consider any two circles $C[O, r]$ and $C[O', r']$. Then, all similarities that map $C[O, r]$ onto $C[O', r']$ must map O to O' and have the same similarity ratio $\lambda = \dfrac{r'}{r}$; only their centers are different. We have the following three cases:

Case (a) $O \neq O'$ and $r \neq r'$. See **Figure 4.73**. Since the center P of any similarity, as above, must satisfy $\dfrac{PO'}{PO} = \lambda = \dfrac{r'}{r}$, the point P must belong to the Apollonius circle $C[C, CP]$ on the segment OO' and of ratio $\lambda = \dfrac{r'}{r}$.

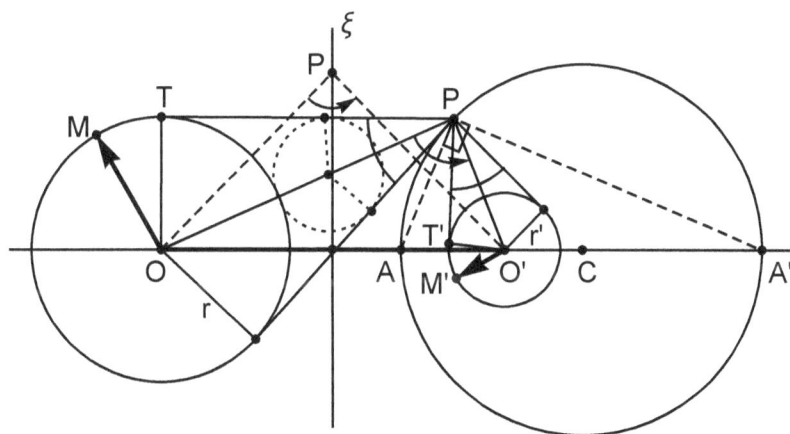

Figure 4.73: The geometrical locus of the centers of all similarities that map $C[O, r]$ to $C[O', r']$ belongs to the Apollonius circle on the segment OO' and of ratio $\lambda = \dfrac{r'}{r}$.

Finally, the geometrical locus of the centers P of all direct similarities that map $C[O, r]$ onto $C[O', r']$ is the Apollonius circle of ratio $\lambda = \dfrac{r'}{r}$ on OO'.

Let AA' be the diameter of the Apollonius circle on the straight line OO'. Then the points O, A, O', A' form a harmonic quadruple of ratio $\lambda = \dfrac{r'}{r}$. **The points A and A' are the centers of the negative and the positive homothety, respectively, that map the two circles onto each other.**

Definition 4.4.1 *The circle with diameter AA' is called **the circle of similarity (or similitude) of the two given circles** $C[O, r]$ and $C[O', r']$.*

So, the circle of similarity of $C[O, r]$ and $C[O', r']$ is the Apollonius circle on AA' with ratio $\lambda_1 = \dfrac{r'}{r}$, or $\lambda_2 = \dfrac{r}{r'}$, where A and A' are the centers of the two homotheties between the two circles. Therefore, there are two such circles, one containing O' and another containing O. (Make a figure for the second circle.)

Case (b) $O \neq O'$ and $r = r'$. Then, $\lambda = 1$, and the direct similarities are isometries. The geometrical locus of the centers P of all the isometries that map $C[O, r]$ onto $C[O', r']$ is the perpendicular bisector ξ of the center-segment OO', which may be considered as an infinite circle. See **Figure 4.73**. These isometries are the rotations about the points $P \in \xi$ by angle $\angle(PO, PO')$ and also the parallel translation $T_{\overrightarrow{OO'}}$, considered as a rotation with center the ∞ point of ξ.

In both cases **(a)** and **(b)**, the following similarities

$$S_P = H_{[P, \pm \frac{r'}{r}]} \circ R_{[P, \theta]} = R_{[P, \theta]} \circ H_{[P, \pm \frac{r'}{r}]}$$

map the circle $C[O, r]$ onto the circle $C[O', r']$, where $\theta = \angle(PO, PO')$ (with the $+\frac{r'}{r}$) or $\theta = \angle(PO, PO') + \pi$ (with the $-\frac{r'}{r}$), and P any point of the Apollonius circle with diameter AA' (found above), when $r \neq r'$, or any point of the perpendicular bisector of the segment OO', when $r = r'$.

Now, any such similarity can also be composed from the right side with the rotations $R_{[O, \phi]}$, for any angle ϕ, since the circle $C[O, r]$ is invariant under them and from the left side with $R_{[O', \phi']}$, for the same reason. In the way we find all of the direct similarities between $C[O, r]$ and $C[O', r']$. (Elaborate!)

Extra Property: *The two pairs of tangents drawn from any point of the geometrical loci, found in either case **(a)** or **(b)**, on the two circles, form equal angles.* (The triangles PTO and $PT'O'$ in **Figure 4.73** are similar. Justify!)

Case (c) $O = O'$. Then, the circles are concentric and the Apollonius circle degenerates to the point $O = O'$. All direct similarities that map $C[O, r]$ to $C[O', r']$ have center the point $O = O'$ and ratio $\lambda = \frac{r'}{r}$. They are compositions of homotheties $H_{[O, \pm \frac{r'}{r}]}$ or $H_{[O, \pm \frac{r}{r'}]}$ and rotations $R_{[O, \phi]}$ and $R_{[O', \phi']}$, for any angles ϕ and ϕ'. (Elaborate!)

In all three cases above, one way to map a point $M \in C[O, r]$ to a point $M' \in C[O'r']$ (or vice-versa) by a direct similarity, is to find, in any way, the direct similarity that maps O to O' and M to M'. For instance, we can use appropriate rotations $R_{[O, \phi]}$ and $R_{[O', \phi']}$, or find the unique direct similarity that maps the vector $\overrightarrow{OM}$ to the vector $\overrightarrow{O'M'}$, as we have studied before. See **Figure 4.73** and elaborate!

{For **opposite similarities between two circles**, see the paragraph titled: "*About opposite similarities between two straight lines or two circles*", before **Subsection 4.5.6**, and combine it with the material developed here. For example: In **Figure 4.73**, the line AP is the bisector of the angle $\widehat{OPO'}$ (why?) and the dotted circle is the image of $C[O, r]$ under the homothety $H_{[P, \frac{r'}{r}]}$. Then, $C[O', r']$ is the reflection of the dotted circle in the line PA. This results to an opposite similarity with center P and principal axis PA, between the two circles. (Develop this part in more details!)}

Some Additional Results

(I) With the help of **Figure 4.74**, we show that two non-parallel lines l and l', which intersect at the point I, a pair of corresponding points (A, A'), $[A \in l$ and $S(A) = A' \in l']$, and the similarity ratio $\lambda > 0$, determine two similarities S_1 and S_2. So, on the basis of what we have studied thus far we can justify that:

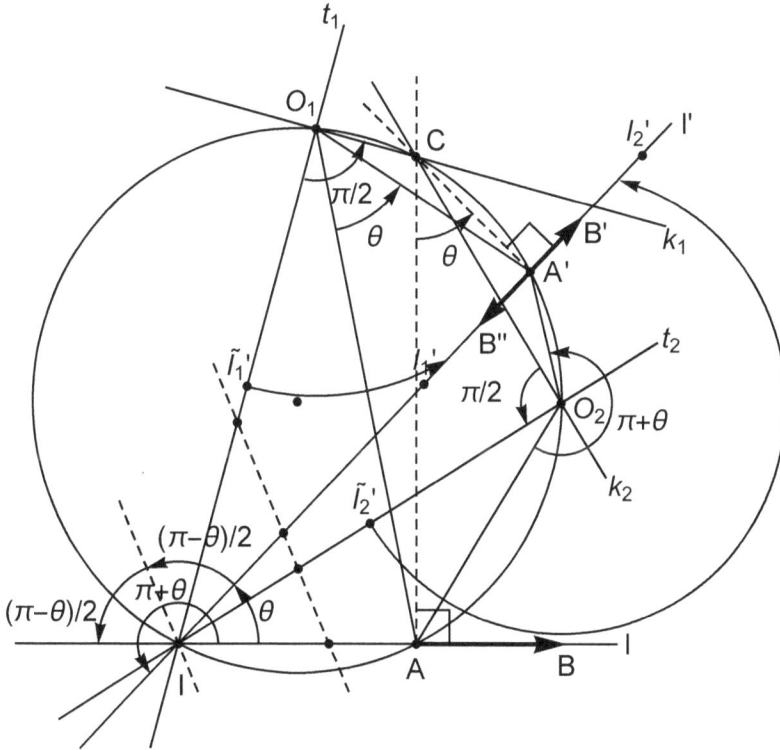

Figure 4.74: **Two direct similarities (rotations, if $\lambda = 1$) can be determined by the ratio λ and a pair of corresponding points (A, A') on two non-parallel lines l and l'.**

(1) The similarity S_1 has center O_1 and similarity angle θ. The line of proportional distances is $t_1 := IO_1$ and the line of proportional divisions $k_1 := CO_1$. These lines are perpendicular to each other at O_1. The point I is mapped to I'_1.

(2) The similarity S_2 has center O_2 and similarity angle $\pi + \theta$. The line of proportional distances is $t_2 := IO_2$ and the line of proportional divisions $k_2 := CO_2$. These lines are perpendicular to each other at O_2. The point I is mapped to I'_2.

(3) Both centers O_1 and O_2 also lie on the circle (IAA').

(4) Both similarities can also be determined by fixing a non-zero vector $\overrightarrow{AB}$ on l, of length 1, and two opposite vectors $\overrightarrow{A'B'} = -\overrightarrow{A'B''}$ on l', of length $\lambda > 0$. Then, we determine S_1 by the vectors $\overrightarrow{AB}$ and $\overrightarrow{A'B'}$ and S_2 by the vectors $\overrightarrow{AB}$ and $\overrightarrow{A'B''}$, in the way we did earlier.

(II) In **Figures 4.75** and **4.76**, we studied again the rotation we have examined in **Figure 2.63**. Here again, we have an isosceles triangle ABC with vertex A and base BC. Now, checking the two figures, we can justify that:

(1) The rotation ($\lambda = 1$) determined by the vectors $\overrightarrow{AB}$ and $\overrightarrow{CA}$ has center the circumcenter O of the triangle ABC, and rotation angle $\pi - \widehat{A}$.

(2) Line of proportional distances, here equal distances, is the bisector on the angle $\widehat{A}$ and the line of proportional divisions, here equal divisions, the line perpendicular to it at O. Therefore, the latter is parallel to the bisector of the external angle of the triangle at A.

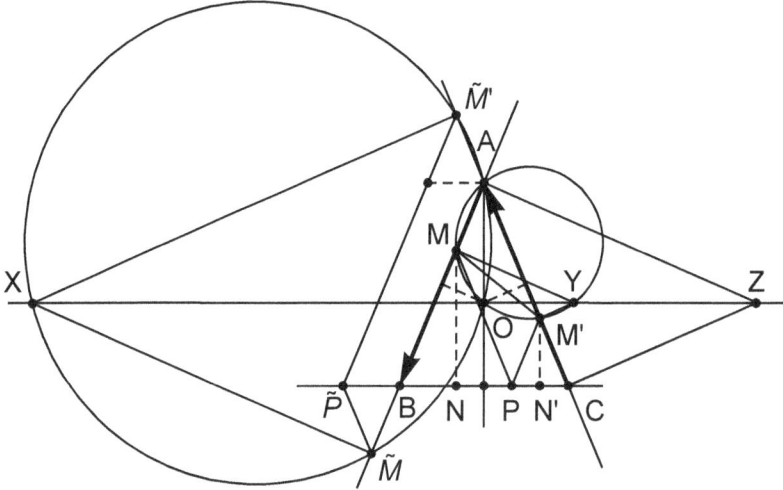

Figure 4.75: Isosceles triangle ABC with vertex A and its rotation determined by the vectors $\overrightarrow{AB}$ and $\overrightarrow{CA}$.

(3) The pair of corresponding, by the rotation, points (M, M') and $(\widetilde{M}, \widetilde{M'})$ form equal divisions on the lines AB and AC with respect to the vectors $\overrightarrow{AB}$ and $\overrightarrow{CA}$. These divisions can be achieved by picking any point P or $\widetilde{P}$ anywhere on the base line and drawing lines parallel to the lines AC and AB, respectively. Then, M and M' are the points of intersections, respectively.

(4) For any $P \in AB$, the corresponding quadrilateral $AMOM'$ (or $\widetilde{A}\widetilde{M}O\widetilde{M'}$, etc.) is inscribable.

(5) For any $P \in AB$, the projection of the corresponding segment MM' (or $\widetilde{M}\widetilde{M'}$, etc.) on the line AB (or on the line of equal divisions), has fixed length equal to $\dfrac{AB}{2}$.

(6) The midpoint perpendiculars of the segments MM' all pass through the fixed point O.

(7) The geometrical locus of all midpoints S of the segments MM' is the common **Simson line** of all triangles AMM', corresponding to the fixed point O. This is the mid-parallel of the triangle ABC and so parallel to the bisector of the external angle of the triangle at A and to the base AB.

(8) The geometrical locus of all centers K of the circles $AMOM'$ is a fixed line parallel to the bisector of the external angle of the triangle at A and to the base AB.

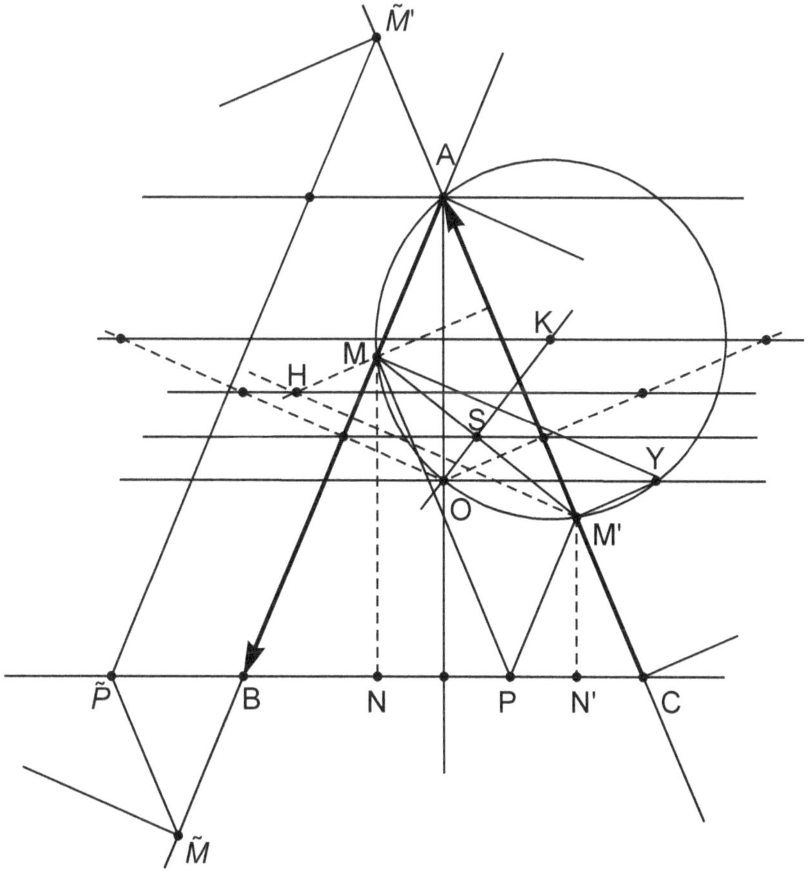

Figure 4.76: Isosceles triangle ABC with vertex A and its rotation determined by the vectors $\overrightarrow{AB}$ and $\overrightarrow{CA}$.

(9) The geometrical locus of the orthocenters H of the triangles AMM' is the common **Steiner line** of all triangles AMM', corresponding to the fixed point the point O. This is also parallel to the bisector of the external angle of the triangle at A and to the base AB.

4.5   Opposite Similarities

4.5.1   Compositions of Homotheties and Reflections

Given a homothety $H_{[Q,\lambda]}$, and a reflection R_l, we want to study the compositions

$$H_{[Q,\lambda]} \circ R_l \quad \text{and} \quad R_l \circ H_{[Q,\lambda]}.$$

By **Consequences (a)** and **(c)** of **Subsection 4.1.1**, without loss of generality, we consider $0 < \lambda \neq 1$. Otherwise, if $\lambda = 1$, the homothety is the identity and so the above compositions are equal to reflections. If $\lambda < 0$, the homothety $H_{[Q,\lambda<0]}$ is the commutative composition of the half-turn $R_{[Q,\pi]}$ and the positive homothety $H_{[Q,|\lambda|>0]}$. But, the composition of a half-turn and a reflection is a glide [see **Section 2.3, (7)**]. Consequently, by **Remark (3)** after **Theorem 4.3.2**, etc., the composition of a glide and a homothety (in either order) reduces to a composition of a new homothety and a reflection.

By **Section 4.3, (5)**, any composition as the above is a new similarity S with similarity ratio $0 < \lambda \neq 1$. The vectors $\overrightarrow{AB}$ and $\overrightarrow{A'B'}$ that appear in **Theorem 4.2.1** are not parallel anymore, as this Theorem requires. Therefore, these compositions cannot be homotheties. On the other hand, they cannot be reflections since the lengths of vectors (or straight segments) are multiplied by λ and therefore not preserved.

By **Theorem 4.3.2** and **Corollary 4.3.2**, any such similarity can be written as the composition of a homothety and an isometry (here reflection or glide) in infinitely many ways, and these compositions are not commutative, in general. But, as **Theorem 4.3.3** proves, such a similarity S has a center O (unique fixed point) and two unique perpendicular axes ξ and ξ', the principal axis and the second axis, through O. Also and very importantly, S is equal to the following commutative composition

$$\text{either} \quad S = H_{[Q,\lambda]} \circ R_l = H_{[O,\lambda]} \circ R_\xi = R_\xi \circ H_{[O,\lambda]},$$
$$\text{or} \quad S = R_l \circ H_{[Q,\lambda]} = H_{[O,\lambda]} \circ R_\xi = R_\xi \circ H_{[O,\lambda]},$$

with $H_{[O,\lambda]}$ and R_ξ unique. We are mostly interested in this unique and convenient composition for S, in which the unique elements O, ξ and ξ' must be determined. This is done in **Subsections 4.5.2** and **4.5.4**.

So, these compositions are neither homotheties nor reflections nor glides, in general. They yield other more general similarities, unless the homothety is the identity, when $\lambda = 1$. In such a case, the reflections become trivial cases of these general similarities.

Since homotheties preserve oriented angles and reflections and glides reverse them, the compositions of homotheties and reflections or glides are opposite similarities. (As it has been already defined, a similarity is opposite if it reverses the oriented angles.)

If, in the above compositions, we replace the homothety by a parallel translation $T_{\vec{u}}$, then the outcome of each composition is a glide which is an opposite similarity with $\lambda = 1$. See **Section 2.3, (6)**.

4.5.2 Determination of an Opposite Similarity

In **this Subsection**, without loss of generality, we consider compositions of positive homotheties of ratio $0 < \lambda \neq 1$ and reflections. [See **Subsections 2.3.1, 2.4.1, Section 2.3, (6), (7), (8), Consequence (c) of Subsection 4.1.1.**] We do not need to consider the glides separately, since they are special commutative compositions of parallel translations with reflections and then, by **Subsection 4.2.2**, the composition of a parallel translation and a homothety is a new homothety.

The image $\mathfrak{F}'$ of a figure $\mathfrak{F}$ by an opposite similarity is called **oppositely similar** to the figure $\mathfrak{F}$.

We would like **to determine an opposite similarity** S (claimed by **Theorem 4.4.3**), which is defined by two pairs of corresponding points (A, A') and (B, B'), i.e., $A' = S(A)$ and $B' = S(B)$, or equivalently by **Section 4.3, (9)**, defined by the two vectors $\overrightarrow{AB}$ and $\overrightarrow{A'B'}$ such that $\overrightarrow{A'B'} = S(\overrightarrow{AB})$ and moving from $A' = S(A)$ to $B' = S(B)$. This **is done by determining the similarity ratio $0 < \lambda \neq 1$, the center O and the principal axis ξ of S.**

General Case: We assume that the vectors are not parallel and do not have the same length. That is, $\overrightarrow{AB} \nparallel \overrightarrow{A'B'}$ and $AB \neq A'B'$. The ratio of such similarity ought to be $\lambda = \dfrac{A'B'}{AB}$. So, we know (A, A'), (B, B'), and $0 < \lambda \neq 1$. We call I the common point of the lines AB and $A'B'$.

In this situation, the unique center O of an opposite similarity lies on two Apollonius circles, the straight line of proportional distances and the straight line of proportional divisions. The intersection of any two of these loci may consist of one or two points. The center O is either the unique point of intersection or the correct one from the two points of intersection. More concretely we have:

(1) See **Figures 4.77** and **4.78**. The triangle OAB and $OA'B'$ are suppose to be oppositely similar. So,

$$\frac{OA'}{OA} = \frac{OB'}{OB} = \frac{A'B'}{AB} = \lambda.$$

Therefore, O is one point of the intersection of the two Apollonius circles whose points X satisfy

$$\frac{XA'}{XA} = \lambda = \frac{XB'}{XB}.$$

Next, notice that the other point of intersection of these Apollonius circles, Q, is the center of the direct similarity that maps $\overrightarrow{AB}$ to $\overrightarrow{A'B'}$, studied in **Section 4.4.4**, before. (See **Figure 4.65** and compare. The letters O and Q have been interchanged. The case $O = Q$ is studied in **Subsection 4.5.6.**)

Now, we consider the two points A_1 and B_1 such that

$$\overrightarrow{OA_1} = \lambda \overrightarrow{OA} \quad (\text{so, } OA_1 = OA') \quad \text{and} \quad \overrightarrow{OB_1} = \lambda \overrightarrow{OB} \quad (\text{so, } OB_1 = OB').$$

Since the triangles OAB and $OA'B'$ are oppositely similar, the same thing is true of the triangles OA_1B_1 and $OA'B'$. (In fact, the later are oppositely equal). Then, we easily prove that the bisectors of the angles $\angle(OA_1, OA')$ and $\angle(OB_1, OB')$ are opposite rays that form a straight line ξ and the pairs of points A and A_1 and B and B_1 are symmetrical with respect to ξ. Hence, we have obtained the commutative decomposition of S

$$S = R_\xi \circ H_{[O,\lambda]} = H_{[O,\lambda]} \circ R_\xi,$$

with ξ the *principal axis* of S and *second axis* $\xi' \perp \xi$ at the center O. The

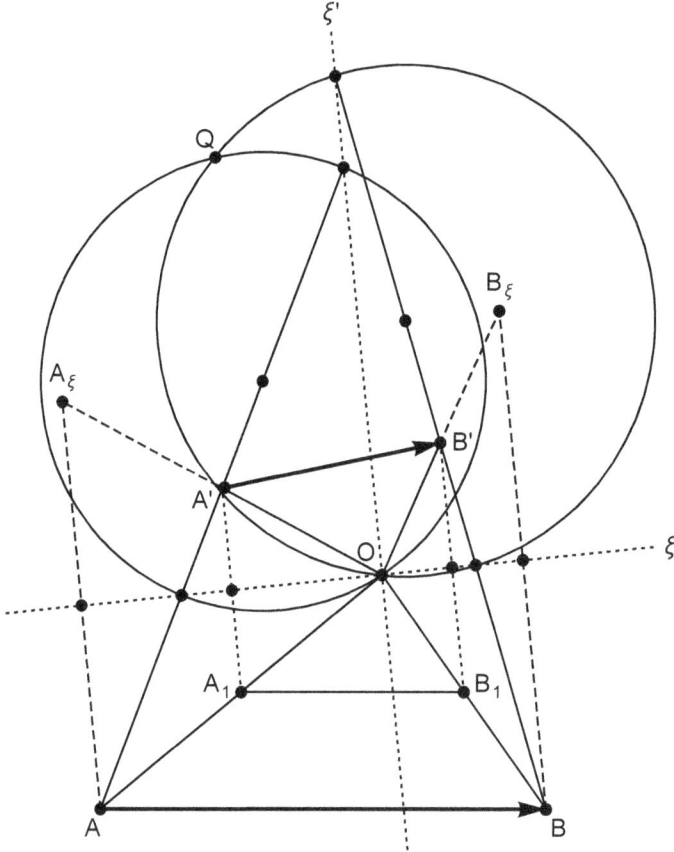

Figure 4.77: Determination of the center O of an opposite similarity by the Apollonius circles of the segments AA' and BB', as arranged in the figure, and ratio the similarity ratio $\lambda = \dfrac{A'B'}{AB}$. The other point of intersection Q is the center of the direct similarity. Notice A_ξ and B_ξ the reflections of A and B in ξ, the principal axis of S. Notice $H_{[O,\lambda]}(AB) = A_1B_1 \parallel AB \parallel H_{[Q,\lambda]}(AB)$.

commutativity of this decomposition can be easily proven directly (as indicated by the points A_ξ and B_ξ in **Figure 4.77**) or follows from **Theorem 4.3.3**.

In conclusion, we find O and then, *the principal axis of S is the common bisector of the known angles $\angle(OA, OA')$ and $\angle(OB, OB')$.*

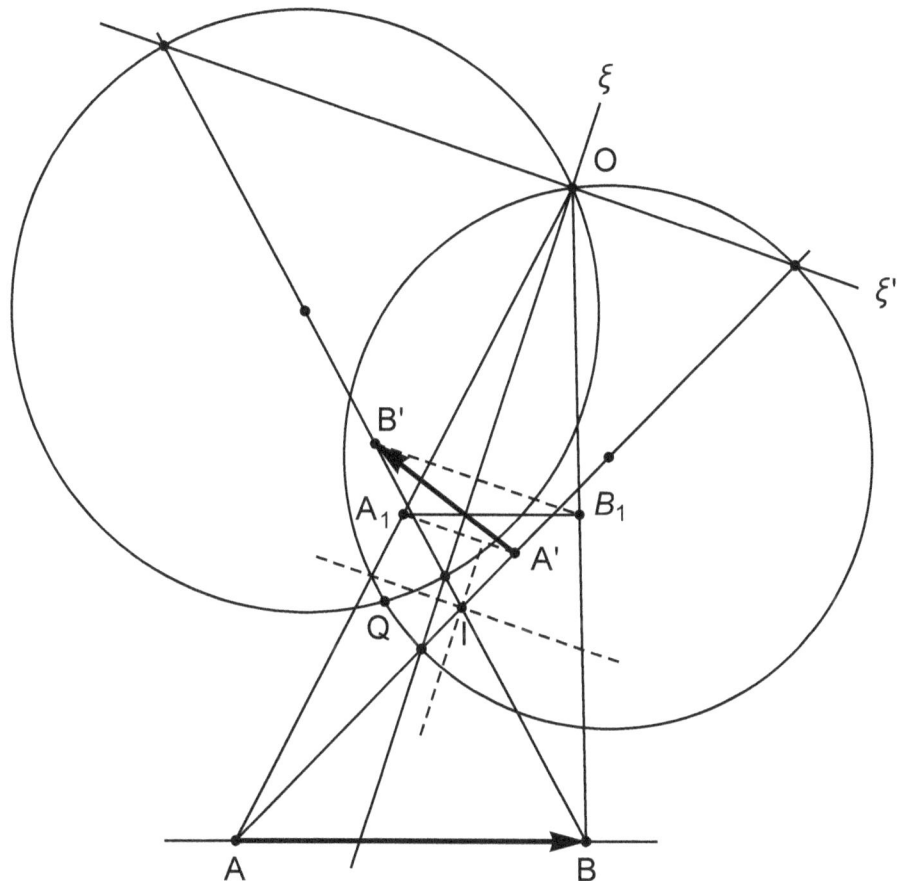

Figure 4.78: Determination of the center O of an opposite similarity by the Apollonius circles of the segments AA' and BB', as arranged in the figure, and ratio the similarity ratio $\lambda = \dfrac{A'B'}{AB}$. The other point of intersection Q is the center of the direct similarity. Notice $H_{[O,\lambda]}(AB) = A_1B_1 \parallel AB \parallel H_{[Q,\lambda]}(AB)$.

Redraw **Figures 4.77** and **4.78** in the following cases: (a) $A = A'$. (b) $B = B'$. (c) $A = B'$. (d) $B = A'$. (e) The vectors $\overrightarrow{AB}$ and $\overrightarrow{A'B'}$ intersect at interior point. Describe the resulting opposite similarities. [Cases **(a)** and **(b)** are simple. In **(c)** and **(d)** can be done directly, or you may use a parallel translation first and then a homothety and a reflection. But, as proven in **Subsection 4.2.2**, homothety composed with a parallel translation is a new homothety. See also the exercises.]

(2) As in the case of the direct similarity, O lies on the line k of the proportional divisions of the lines $l := AB$ and $l' := A'B'$ and with respect to the segments AB and $A'B'$, as shown before. See **Figure 4.79**.

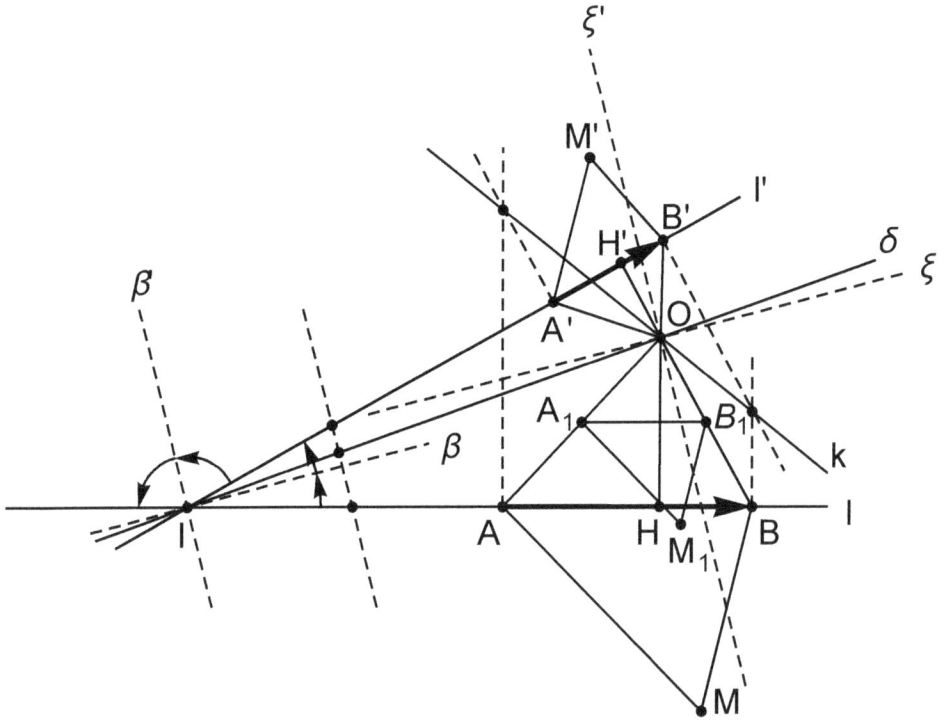

Figure 4.79: **The center of an opposite similarity as the intersection of the lines of proportional distances and proportional divisions**

(3) As in the case of the direct similarity, O lies on the line δ of the proportional distances of the lines $l := AB$ and $l' := A'B'$ with proportions $\dfrac{A'B'}{AB} = \lambda$, as exposed before. See **Figure 4.79**.

Finally, the proof that the point O, as constructed above, is fixed, is analogous to the proof given in the **General Constructions (I) and (II)** in **Subsection 4.4.4**, for the direct similarity. For any point $X \in \mathcal{P}$, we have that: $X' := S(X)$ if and only if the triangles ABX and $A'B'X'$ are oppositely similar. (Provide proof. Easy! By construction the triangles ABO and $A'B'O$ are oppositely similar because they have proportional sides of ratio λ and are oppositely oriented.) So, we must have $O = S(O)$. **This proves that O is the fixed point.** Another proof is analogous to the proof in **Subcase 1.2** of the **First Case of Subsection 4.4.4**.

Special cases: (1) We assume $\overrightarrow{AB} \parallel \overrightarrow{A'B'}$ and $AB \neq A'B'$ (the segments do not have the same length). That is, the given vectors are parallel and do not have the same length. The ratio of such a similarity is $\lambda = \dfrac{A'B'}{AB}$. So, we know: (A, A'), (B, B'), $0 < \lambda \neq 1$ and the lines AB and $A'B'$ are parallel.

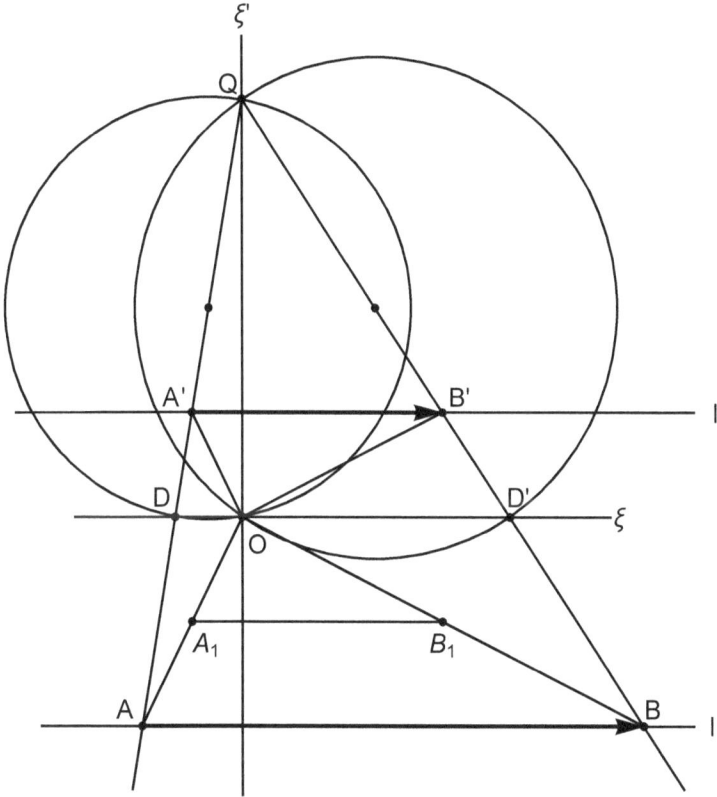

Figure 4.80: Determination of the center O of an opposite similarity by the Apollonius circles of the segments AA' and
BB', as arranged in the figure. The similarity ratio is
$\lambda = \dfrac{A'B'}{AB}$. The other point of intersection Q is the center of the direct similarity. Here the case is special sin-
nce $\overrightarrow{AB} \parallel \overrightarrow{A'B'}$ and the vectors have the same direction.

Then, as before, the center O is one point of the intersection of the two Apollonius circles whose points X satisfy $\dfrac{XA'}{XA} = \lambda = \dfrac{XB'}{XB}$, respectively. See Figures 4.80 and 4.81.

The center O also lies on one of the straight lines of proportional distances from the parallel lines $l := AB$ and $l' := A'B'$.

When $\overrightarrow{AB}$ and $\overrightarrow{A'B'}$ have the same direction, this line coincides with the principal axis of the similarity ξ and is located between the lines l and l', such that the ratio of its distances from l' and l is λ. Also, $\xi \parallel l \parallel l'$ and so we can draw through O the line $\xi \parallel l \parallel l'$ and the second axis is $\xi' \perp \xi$ through O. See **Figure 4.80**.

When $\overrightarrow{AB}$ and $\overrightarrow{A'B'}$ have opposite directions, the line of proportional distances is perpendicular to the principal axis of the similarity ξ and coincides with the second axis of the similarity. It is located outside the tape located between the lines $l \parallel l'$ and on the side of the smaller segment, such that the ratio of its distances from l' and l is λ. Also, $\xi \perp l$ and $\xi \perp l'$ and so we can draw through O the line $\xi \perp l$ and the second axis is $\xi' \perp \xi$ through O. See **Figure 4.81**.
(The line of proportional divisions of l and l' is not defined when $l \parallel l'$.)

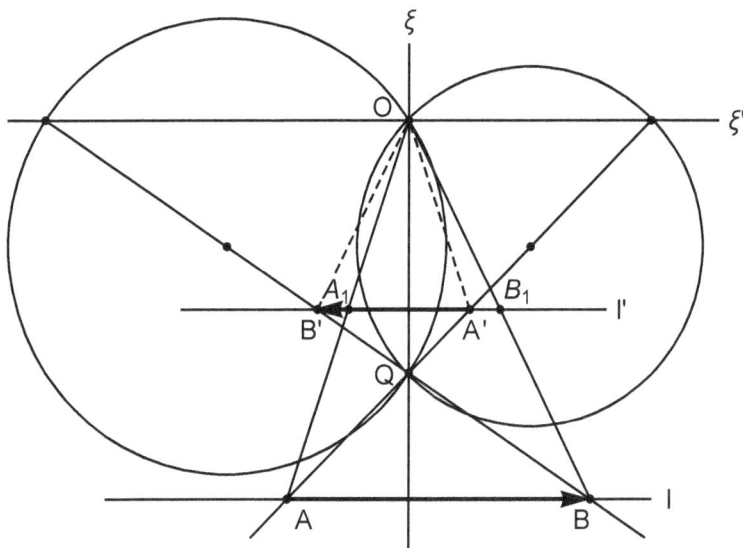

Figure 4.81: **Determination of the center O of an opposite similarity by the Apollonius circles of the segments AA' and BB', as arranged in the figure. The similarity ratio is $\lambda = \dfrac{A'B'}{AB}$. The other point of intersection Q is the center of the direct similarity. Here the case special since $\overrightarrow{AB} \parallel \overrightarrow{A'B'}$ and the vectors have opposite directions.**

Therefore, in this special case, O is found in the three geometrical loci just explained, not in all four as was the situation in the previous case.

Remark: We can easily prove that when $AB \parallel A'B'$ and $AB \neq A'B'$, then one of the points of intersection of the two Apollonius circles is the common point of the straight lines AA' and BB'. (See **Figures 4.80** and **4.81**.)

In **Figure 4.82** below, (as in **Figures 2.46, 2.54**, and others), we have equal vectors $\overrightarrow{AB} = \overrightarrow{A'B'}$ on different parallel straight lines l and l'. Then, $\overrightarrow{AA'} = \overrightarrow{BB'}$, and $\overrightarrow{AB}$ can be mapped onto $\overrightarrow{A'B'}$ by either the parallel translation $T_{\overrightarrow{AA'}} = T_{\overrightarrow{BB'}}$ (direct isometry) or the glide $G = R_\xi \circ T_{\overrightarrow{AA''}} = T_{\overrightarrow{AA''}} \circ R_\xi$ or a reflection R_ξ when $\overrightarrow{AA''} = \overrightarrow{0}$ (opposite isometry), where A'' and B'' are the projections of A and B on the line l', and the straight line ξ is the mid-parallel line of l and l'. The second axis ξ' is the line at infinity.

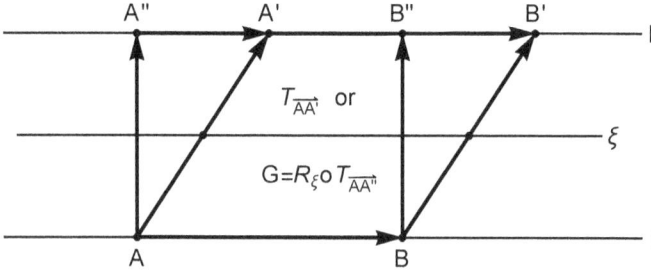

Figure 4.82: A special case of a parallel translation or a glide

This conclusion is valid even if $\overrightarrow{AB} = \overrightarrow{A'B'}$ are on the same line. In such a case, $\xi = l = l'$ and $A'' = A'$ and $B'' = B'$, i.e., $\overrightarrow{AA''} = \overrightarrow{AA'}$ and $\overrightarrow{BB''} = \overrightarrow{BB'}$. (Draw figure! We can find more opposite similarities that map $\overrightarrow{AB}$ onto $\overrightarrow{A'B'}$, in these situations. Find some of them as an exercise.)

(2) In **Figures 4.83-4.88**, we see the special cases in which A, B, A', B' are on the same straight line l. Then, the similarity ratio $\lambda = \dfrac{A'B'}{AB} > 0$. In this situation, the two Apollonius circles, with respect to AA' and BB' and $\lambda = \dfrac{A'B'}{AB} > 0$, are internally or externally tangent or become the mid-perpendicular lines to the pertinent segments. This can be proven by a direct computation (left as an exercise). The common point O is the sought center of the opposite similarity S. One of the two axes is l and the other is the perpendicular to l at the common point O.

But, we can also calculate the position of O (without using the Apollonius circles) in the following way: For instance, in **Figure 4.83**, we have:

$$\lambda = \frac{A'B'}{AB} > 0 \quad \text{and} \quad OA' = \lambda OA. \quad \text{Then,}$$

$$\overrightarrow{A'A} = \overrightarrow{OA} - \overrightarrow{OA'} = \overrightarrow{OA} - \lambda\overrightarrow{OA} = (\lambda - 1)\overrightarrow{OA} = (1 - \lambda)\overrightarrow{AO}.$$

So, O $(= A_2 = B_2$, in **Figure 4.83**,) is determined by the vector

$$\overrightarrow{AO} = \frac{\overrightarrow{A'A}}{1 - \lambda} = \frac{\overrightarrow{AA'}}{\lambda - 1}.$$

The principal axis is $\xi = l$ and the second axis is $\xi' \perp l$ at O. Finally, S is:

$$S = R_{\xi=l} \circ H_{[O,\lambda]} = H_{[O,\lambda]} \circ R_{\xi=l}.$$

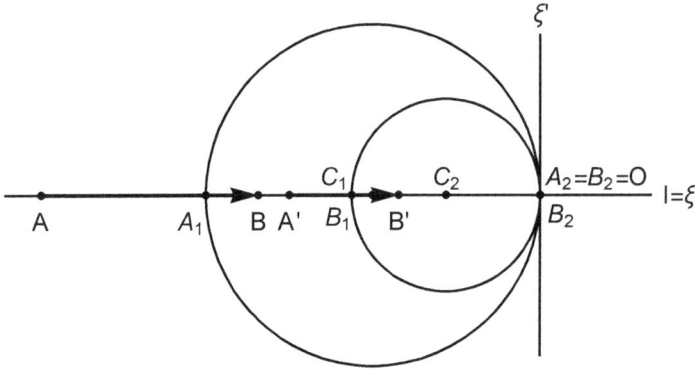

Figure 4.83: Special case of an opposite similarity in which the points A, B, A', B' are on the same straight line

In this case, if $\overrightarrow{AB} = \overrightarrow{A'B'}$, then $\lambda = 1$ and O is a point at infinity. Hence, the homothety becomes the parallel translation $T_{\vec{u}}$ by the vector $\vec{u} = \overrightarrow{AA'} = \overrightarrow{BB'}$ and the opposite similarity S is the glide

$$S = R_{\xi=l} \circ T_{\vec{u}} = T_{\vec{u}} \circ R_{\xi=l}.$$

The second axis ξ' is the line at infinity.

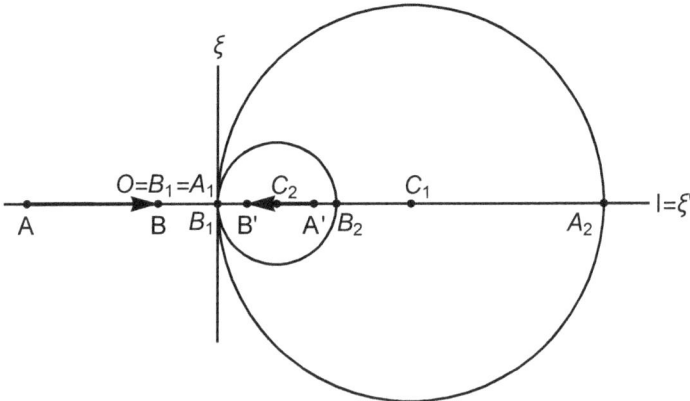

Figure 4.84: Special case of an opposite similarity in which the points A, B, B', A' are on the same straight line

Similarly, in **Figure 4.84**, we have:

$$\lambda = \frac{A'B'}{AB} > 0 \quad\text{and}\quad \overrightarrow{OB_1} = \lambda\overrightarrow{OB} = -\lambda\overrightarrow{OB'}. \quad\text{Then,}$$

$$\overrightarrow{BB'} = \overrightarrow{BB_1} + 2\overrightarrow{B_1O} = \overrightarrow{BB_1} - 2\lambda\overrightarrow{OB} = \overrightarrow{OB_1} - \overrightarrow{OB} - 2\lambda\overrightarrow{OB}$$
$$= \lambda\overrightarrow{OB} - \overrightarrow{OB} - 2\lambda\overrightarrow{OB} = -(1+\lambda)\overrightarrow{OB} = (1+\lambda)\overrightarrow{BO}.$$

So, O ($= A_1 = B_1$, in **Figure 4.84**,) is determined by the vector $\overrightarrow{BO} = \dfrac{\overrightarrow{BB'}}{1+\lambda}$.

The principal axis is $\xi \perp l$ at O and the second axis is $\xi' = l$. When $\lambda = 1$, ξ is the midpoint perpendicular of the segment BB' (and AA'). Finally,

$$S = R_\xi \circ H_{[O,\lambda]} = H_{[O,\lambda]} \circ R_\xi.$$

(3) Analogously, we have **Figures 4.85** and **4.86**. Explain what happens there!

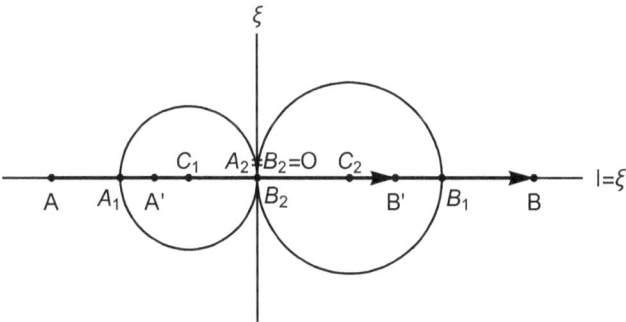

Figure 4.85: The special case of an opposite similarity

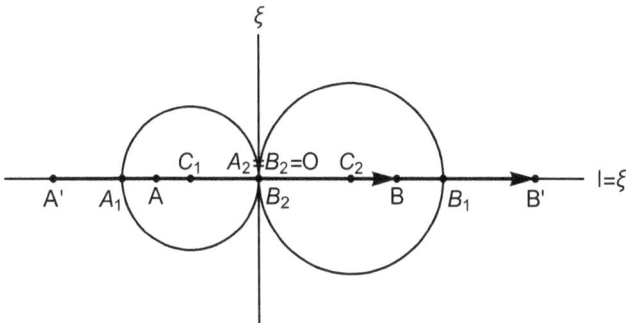

Figure 4.86: The special case of an opposite similarity

(4) If the points are situated in the order A, B, B', A' or B, A, A' B', and
$AB = A'B'$, then the ratio is $\lambda = 1$ and the two Apollonius circles become
the midpoint perpendiculars to BB' or AA'. The similarity S reduces to the
reflection R_ξ where the principal axis ξ is the midpoint perpendiculars to BB'
or AA'. See **Figures 4.87** and **4.88**. {The direct similarity in this case is the
half-turn $R_{[O,\pi]}$.}

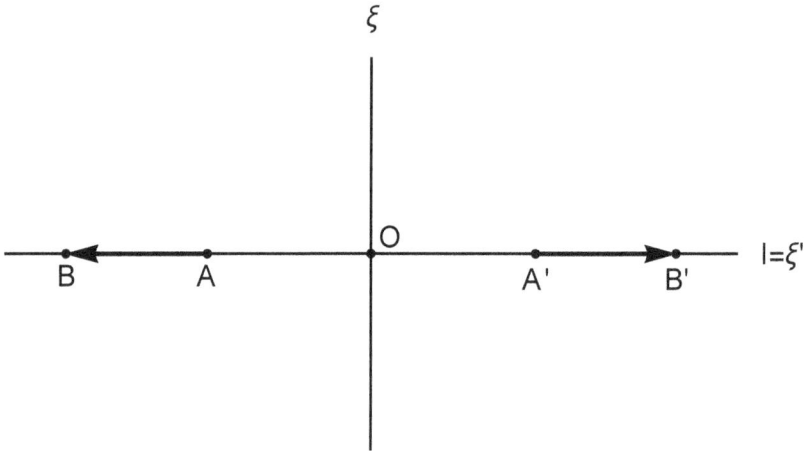

Figure 4.87: The special case of an opposite similarity

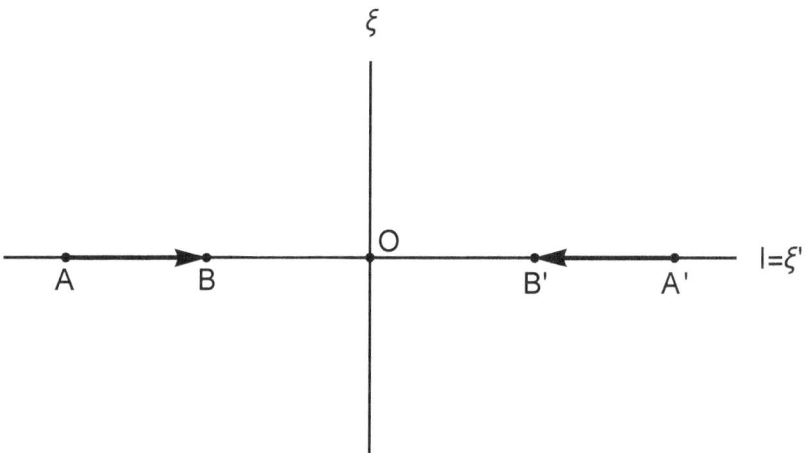

Figure 4.88: The special case of an opposite similarity

4.5.3 Other Facts on the Center of an Opposite Similarity.

(a) The ratio of the distances OH' and OH of O from $l' := A'B'$ and $l = AB$, respectively, is $\dfrac{OH'}{OH} = \dfrac{A'B'}{AB} = \lambda$. This follows from the fact that the triangles OAB and $OA'B'$ are oppositely similar. See **Figures 4.77, 4.78**, especially **4.79** and **4.89**.

(b) For any two corresponding points M and $M' := S(M)$, it holds $\dfrac{OM'}{OM} = \lambda$, since the triangles OAM and $OA'M'$ are oppositely similar.

(c) The point O subtends opposite angles with any two corresponding segments MN and $M'N'$, since the triangles OMN and $OM'N'$ are oppositely similar.

Remarks: (1) If $\overrightarrow{AB} = \overrightarrow{A'B'}$ (so, $l := AB \parallel l' := A'B'$ and $\lambda = 1$), we consider m the mid-parallel line of the lines $l := AB \parallel l' := A'B'$, the projection A'' of A onto l' and the vector $\overrightarrow{u} = \overrightarrow{A''A'}$. [See **Subsection 2.5.4, (b)** and draw figure.] Then, the glide $T_{\overrightarrow{u}} \circ R_m = R_m \circ T_{\overrightarrow{u}}$ maps $\overrightarrow{AB}$ onto $\overrightarrow{A'B'}$. This is an opposite isometry. (We have also seen that if $\overrightarrow{v} = \overrightarrow{AA'}$, then the parallel translation $T_{\overrightarrow{v}}$ maps $\overrightarrow{AB}$ onto $\overrightarrow{A'B'}$. See **Subsection 2.5.1, Figure 2.53**. This is a direct isometry.)

(2) If $\overrightarrow{AB} = -\overrightarrow{A'B'}$ (so, $l := AB \parallel l' := A'B'$ and $\lambda = 1$), we have seen that the half-turn $R_{[O,\pi]}$, where O is the point of intersection of AA' and BB'. [See **Subsection 2.5.2, (III), Figure 2.58**.] Here, the triangles ABO and $A'B'O$ are equal and so their circumscribed circles have equal radii and are easily proven to be externally tangent at O. This a direct isometry. But, the glide found in **Subsection 2.5.4, (c), Figure 2.70**, and **Figure 4.84**, maps $\overrightarrow{AB}$ to $\overrightarrow{A'B'}$. This is an opposite isometry.

4.5.4 Other Facts on the Axes of an Opposite Similarity.

Characteristic Property of the Axes: In the general case, in order to determined the opposite similarity S with similarity ratio $0 < \lambda \neq 1$ from two given vectors, we assume

$$\overrightarrow{AB} \nparallel \overrightarrow{A'B'} \qquad \text{and} \qquad AB \neq A'B'.$$

So, we know two pairs of corresponding points, (A, A') and (B, B'), the similarity ratio is

$$0 < \lambda = \frac{A'B'}{AB} \neq 1$$

and the straight lines AB and $A'B'$ are not parallel, $AB \nparallel A'B'$. We have called I, the common point of the lines AB and $A'B'$.

In **Subsection 4.5.2**, we have seen how we find the center O and the axes of S using the appropriate Apollonius circles and/or other geometric loci. Here, we study again **characteristic property of the axes of** S, given that we have found its center O. On the rays OA and OB, we pick the points A_1 and B_1 such that $OA' = OA_1$ and $OB' = OB_1$. Then, the triangles $OA'B'$ and OA_1B_1 are oppositely equal. See **Figures 4.89, 4.77, 4.78** and **4.79**. We have:

$$\frac{OA_1}{OA} = \frac{OA'}{OA} = \frac{OB'}{OB} = \frac{OB_1}{OB} = \lambda,$$

and since $\angle(OA_1, B_1) = -\angle(OA', OB')$, we have $A_1B_1 \parallel AB$ and so $\dfrac{A_1B_1}{AB} = \lambda.$

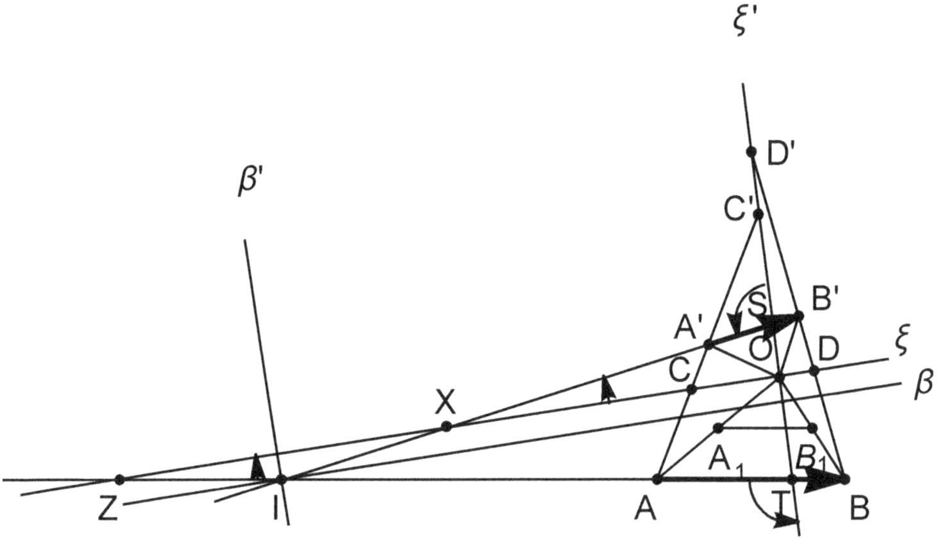

Figure 4.89: Determination of the axes of an opposite similarity

Since A' and B' are reflections of A_1 and B_1, respectively, and the triangles $OA'A_1$ and $OB'B_1$ are isosceles, the bisectors of the angles

$$\angle(OA, OA') = \angle(OA, OA_1) \quad \text{and} \quad \angle(OB, OB') = \angle(OB, OB_1)$$

are opposite rays (prove!) and their union is a straight line ξ, which is **the axis of the reflection** R_ξ, the reflection in the decomposition of S. This ξ is **the principal axis of the opposite similarity** S, since it is fixed by both the homothety $H_{[O,\lambda]}$ and the reflection R_ξ and so by S.

In conclusion, in this case we find O and then, *the principal axis of S is the common bisector of the known angles* $\angle(OA, OA')$ *and* $\angle(OB, OB')$.

Since ξ passes through O, which is also the center of $H_{[O,\lambda]}$, S is equal to the commutative composition

$$S = H_{[O,\lambda]} \circ R_\xi = R_\xi \circ H_{[O,\lambda]}.$$

Now, the line $\xi' \perp \xi$ at O is also fixed by both $H_{[O,\lambda]}$ and the reflection R_ξ, and so by S. This is the **second axis of the opposite similarity** S. Then, $\xi' \parallel \beta'$ the bisector of $\widehat{ZIX}$ the supplementary angle of $\widehat{AIA'} = \widehat{BIB'}$, since $\xi \perp \beta$, $\beta \perp \beta'$ and $\xi' \perp \xi$.

In the special case, we assume

$$\overrightarrow{AB} \parallel \overrightarrow{A'B'} \qquad \text{and} \qquad AB \neq A'B'.$$

Then, we know (A, A'), (B, B'), the ratio of similarity $0 < \lambda = \dfrac{A'B'}{AB} \neq 1$ and the lines AB and $A'B'$ are parallel, $l := AB \parallel A'B' := l'$.

In this case, *the principal axis ξ coincides with the straight line of the proportional distances* (described earlier, e.g., **Figure 4.81, 4.80, 4.72, 4.69, 4.68**, etc.), *and therefore the second axis is the line $\xi' \perp \xi$ at O*. Hence, $\xi \parallel l \parallel l'$, and $\xi' \perp l$ and $\xi' \perp l'$.

Now we continue with some **useful and additional facts of the axes of** S.

(1) No straight line besides ξ and ξ' are fixed by S. (Justify!) So, the two axes of an opposite similarity are unique (and they pass through its unique center).

(2) Let β be the bisector of the angle $\angle(\overrightarrow{AB}, \overrightarrow{A'B'})$ (angle of two vectors) at the point I and β' the bisector of its supplementary angle at I. Since $A_1B_1 \parallel AB$, we get $\angle(\overrightarrow{AB}, \overrightarrow{A'B'}) = \angle(\overrightarrow{A_1B_1}, \overrightarrow{A'B'})$ and so $\xi \parallel \beta$ and $\xi' \parallel \beta'$. See **Figures 4.79** and **4.89**.

Using what we already know and **Figures 4.89**, we prove that the straight lines $A'B'$, the axis ξ and A_1B_1 are concurrent at the point X, ξ bisects the angle $\widehat{A_1XA'}$, and $\widehat{A_1XA'} = \widehat{AIA'}$. (Check!) The angles $\widehat{A_1XA'}$ and $\widehat{AIA'}$ have one common side and the other sides parallel. Therefore they have parallel bisectors. Hence, $\xi \parallel \beta$.

Therefore, the two axes intersect the lines AA' and BB' at points C, C' and D, D', respectively. These points divide the vectors $\overrightarrow{AA'}$ and $\overrightarrow{BB'}$ internally and externally in ratio λ (harmonic divisions, by the theorem of the bisectors of the interior and exterior angles of a triangle). That is,

$$\frac{CA'}{CA} = \frac{C'A'}{C'A} = \lambda = \frac{DB'}{DB} = \frac{D'B'}{DB}.$$

Looking at it backward, we can perform these harmonic divisions to find the points C, C' and D, D', á-priori and afterward the axes are the lines CD and $C'D'$. (These points are points of the two Apollonius circles that we studied earlier.)

(3) Each of any two corresponding lines n and n' $[n' = S(n)]$, makes equal angles with either axis of S. Therefore, each of the bisectors of the two angles

of n and n', is parallel to one of the axes. See **Figures 4.79** and **4.89** and read off the proof. (If necessary draw some auxiliary lines.)

(4) For any point A and its corresponding $A' = S(A)$, the segment AA' is divided by one axis internally and by the other axis externally in the similarity ratio λ. This is so because the axis are the bisectors of the angles of OA and OA'. See **Figures 4.79** and **4.89** and read off the proof. (If necessary draw some auxiliary lines.)

(5) The ratio of the distances of a point A and its corresponding point $A' = S(A)$ from either axis is equal to the similarity ratio λ. See **Figures 4.79** and **4.89**, use (4) and read off the proof. (If necessary draw some auxiliary lines.)

(6) An opposite similarity that is not a glide has at least one fixed point. If it has two fixed points, then it is a reflection and has a whole axis point-wise fixed.

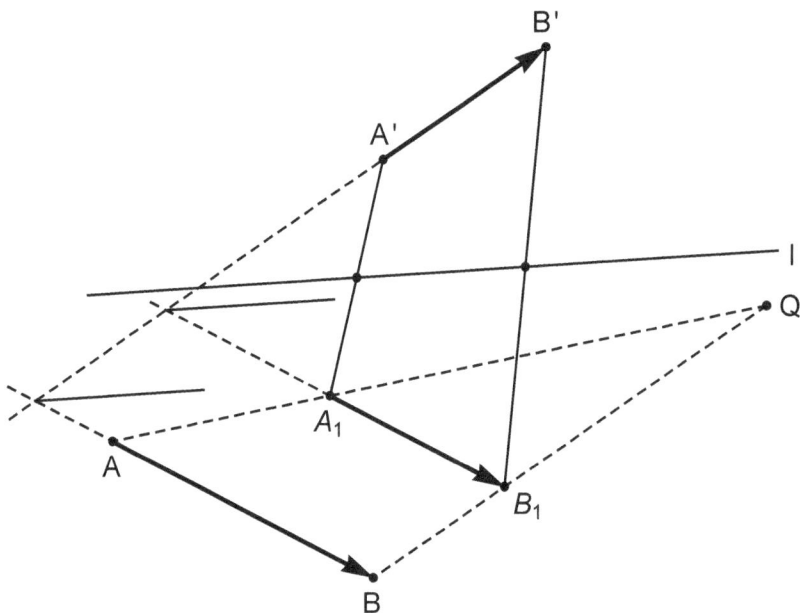

Figure 4.90: The axes of reflections in a decomposition of an opposite similarity have the same direction

(7) By **Theorem 4.3.2** and **Corollary 4.3.2**, any opposite similarity S **can be written as the composition of a homothety and an isometry, here reflection or glide, in infinitely many ways, and these compositions are not commutative, in general.** Each of these compositions is eventually reduced to a composition of a homothety and a reflection and, in

general, does not have principal and second axes. But, S itself and each reflection or glide has its own axis (principal) and we have the **Result**:

> *The axes of the reflections or glides in these decompositions of*
> *S are parallel to each other, i.e., they have the same direction.*

Proof Let $0 < \lambda \neq 1$ be the similarity ratio of S and consider any point $Q \in \mathcal{P}$ and the homothety $H_{[Q,\lambda]}$. See **Figure 4.90**. We consider a vector $\overrightarrow{AB} \neq \vec{0}$ and its image $\overrightarrow{A'B'} = S(\overrightarrow{AB})$. Let also $\overrightarrow{A_1 B_1} = H_{[Q,\lambda]}(\overrightarrow{AB})$ and consider the reflection R_l or glide G_l determined by the vector $\overrightarrow{A_1 B_1}$ and $\overrightarrow{A'B'}$. Then, as we know from **Theorem 4.3.2**, there is a reflection R_l or a glide G_l, such that

$$S = R_l \circ H_{[Q,\lambda]} \qquad \text{or} \qquad S = G_l \circ H_{[Q,\lambda]}.$$

As we have proved, the axis l of R_l or glide G_l is parallel to the bisector of one angle $\angle(\overrightarrow{A_1 B_1}, \overrightarrow{A'B'})$. (For example, see **Lemma 2.4.2**, etc.) But, $\overrightarrow{A_1 B_1} \parallel \overrightarrow{AB}$ and so l is parallel to the bisector of one angle $\angle(\overrightarrow{AB}, \overrightarrow{A'B'})$. Therefore, l is always parallel to the principal axis of S which is unique. [See **additional facts** (1) and (2), above.]

∎

4.5.5 Additional Properties of the Center and the Axes of an Opposite Similarity and Harmonic Quadruples

Let S be an opposite similarity of the plane $\mathcal{P}$, and let A, A' and A'' be three points in $\mathcal{P}$. If we assume that $A' = S(A)$ and $A'' = S(A') = S \circ S(A)$ (the two successive images of A by S), we would like to determine the opposite similarity S from the point A and its two successive images A' and A'', whenever possible.

Here, we essentially want to determine the opposite similarity S (that is, to find the ratio, center, and the two axes of S) that maps the vector $\overrightarrow{AA'}$ onto the vector $\overrightarrow{A'A''}$, such that the point A is mapped to point A'. We are going to solve this problem with a method different form the methods used so far. To this end, we examine the following cases and subcases:

Case (I) *Suppose the points A, A', and A'' are not on the same straight line. Therefore, they are distinct one another. Then, the similarity ratio is*

$$\frac{A'A''}{AA'} = \frac{S(A)S(A')}{AA'} = \frac{S(AA')}{AA'} =: \lambda.$$

We are going to prove the following:

Subcase (1): *If $\lambda = 1$, then S is a glide completed determined below.*

Subcase (2): *If $\lambda \neq 1$, then S is completely determined by the following three results:*

(a) The center O of S is the point of intersection of the line AA'' and the tangent line of the circle $(AA'A'')$ at A'.

(b) O is the midpoint of the segment DE, where D and E are the points of intersection of the straight line AA'' and the internal and external bisector of the angle $\widehat{A'}$ of the triangle $AA'A''$, respectively.

*(c) The axes of S are the perpendicular bisectors of the segments $A'D$ and $A'E$, respectively, where D and E are the points defined in **(b)**.*

Proofs: Subcase (1): If $\lambda = 1$ ($\iff A'A'' = AA' \neq 0$), then the similarity S is an opposite isometry. Since, in this case, $A \neq A' = S(A) \neq S \circ S(A) = A'' \neq A$, the isometry S is the glide $G_l(\overrightarrow{MN})$, with axis l the straight line defined by the midpoints M and N of the segments AA' and $A'A''$, respectively, and translation vector $\overrightarrow{MN}$.

[Draw a correct figure and justify that $S =: G_l(\overrightarrow{MN}) = R_l \circ T_{\overrightarrow{MN}}$.]

Subcase (2): Now suppose $0 < l \neq 1$. See **Figure 4.91**.

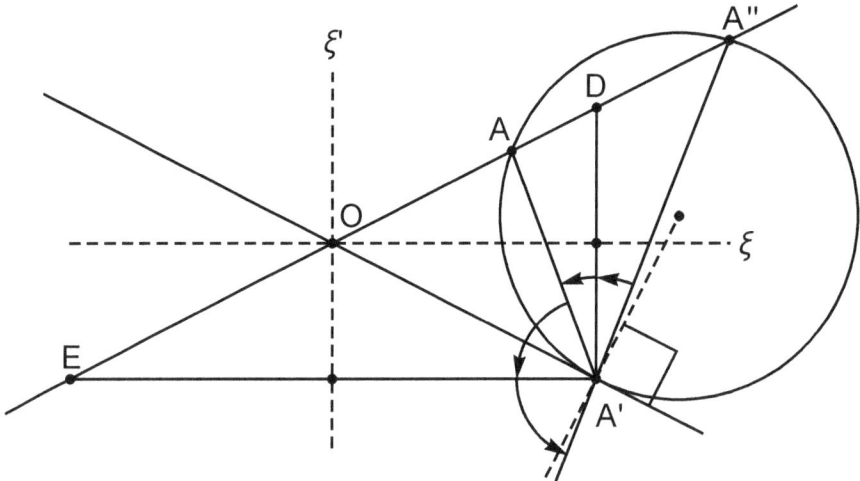

Figure 4.91: Other determinations of the center O
of an opposite similarity

(a) Let ξ be the principal axis of S and $S = R_\xi \circ H_{[O,\lambda]} = H_{[O,\lambda]} \circ R_\xi$. By **Corollary 4.3.3, (4)**, we have

$$S \circ S = H_{[O,\lambda]} \circ R_\xi \circ R_\xi \circ H_{[O,\lambda]} = H_{[O,\lambda]} \circ I \circ H_{[O,\lambda]} = H_{[O,\lambda]} \circ H_{[O,\lambda]} = H_{[O,\lambda^2]}.$$

Since $A'' = S \circ S(A) = H_{[O,\lambda^2]}(A)$, the points O, A, and A'' are collinear. Since O is fixed, $S(O) = O$, we also have $\dfrac{OA'}{OA} = \dfrac{OA''}{OA'} = \lambda$, and so $OA'^{\,2} = OA \cdot OA''$. This relation implies that the line OA' is tangent to the circle $(AA'A'')$ at the point A'.

(b) Let D and E be the points at which the internal and external bisectors of the angle $\widehat{A'}$ of the triangle $AA'A''$ intersect the line AA''. Then, by the known Theorem of the bisectors of an angle in a triangle, we have

$$\frac{DA''}{AD} = \frac{EA''}{EA} = \frac{A'A''}{A'A} = \lambda.$$

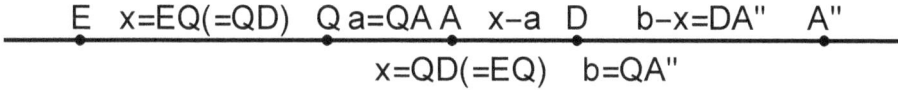

E x=EQ(=QD) Q a=QA A x–a D   b–x=DA"   A"
—————•————————•————————•————————————•———————•———————
              x=QD(=EQ)   b=QA"

Figure 4.92: Q **is the midpoint of** DE **in** $\{A'', A; D, E\} = -1,$
a harmonic quadruple

Suppose Q is the midpoint of DE. See **Figure 4.92**. We let $QE = x = QD$, $QA = a$, $QA'' = b$, and so, $AD = x - a$, and $DA'' = b - x$. Then, by the first part of the above equality we get

$$\frac{x+b}{x+a} = \frac{b-x}{x-a} = \lambda \iff [x + b = \lambda(x + a) \text{ and } b - x = \lambda(x - a).]$$

By adding and subtracting the last two equations (or the numerators and the denominators of this proportion) and simplifying, we find

$$\frac{b}{x} = \lambda \qquad \text{and} \qquad \frac{x}{a} = \lambda.$$

Multiplying these two equations sidewise, we find the relation

$$\frac{b}{a} = \lambda^2, \qquad \text{or} \qquad \frac{\overline{QA''}}{\overline{QA}} = \lambda^2 \left[= \left(\frac{DA''}{AD}\right)^2 = \left(\frac{EA''}{EA}\right)^2 \right].$$

By the way, we have obtained the following **General Result**: *This relation constitutes a characteristic property of the midpoint* Q *of* DE *in any harmonic quadruple* $\{A'', A; D, E\} = -1$ (as depicted in **Figures 4.91** and **4.92**).

Since $\dfrac{OA'}{OA} = \dfrac{OA''}{OA'} = \lambda$, we also find by multiplying, $\dfrac{OA''}{OA} = \lambda^2$. Therefore, $Q \equiv O$, proving the claim, since O is the unique point with this property, given that $\{A'', A; D, E\} = -1$ is a harmonic quadruple (defined by the theorem of the internal and external bisectors of an angle of a triangle).

Remark: From $\dfrac{b}{x} = \dfrac{x}{a}$, we also get the useful equivalent characteristic relation $x^2 = ab$, for a harmonic quadruple $\{A'', A; D, E\} = -1$, namely:

$$QE^2 = QD^2 = \overline{QA} \cdot \overline{QA''}.$$

(c) The triangle $A'DE$ is right at $\widehat{A'}$ and $A'O$ is the median to its hypotenuse. Therefore, $A'O = DO = EO$. We have seen [in the **General Case** of **Subsections 4.5.2** and **4.5.4**] that the axes are the bisectors of the angles $\angle AOA'$, since $A' = S(A)$. Therefore, the axes are the midpoint perpendiculars of $A'D$ and $A'E$, since $A'D$ and $A'E$ are the bases of the isosceles triangles $OA'D$ and $OA'E$.

[Now, determine S in **Subcase (2)** of **Case (I)**, by making the appropriate Apollonius circles on the segments AA' and $A'A''$, as done in **Subsection 4.5.2**.]

Case (II) *Suppose the points A, A', and A'' are on the same straight line.* Then, there are several situations to be considered.

(a) If $A = A' = S(A)$ (A is fixed), then $A' = S(A) = S(A') = S \circ S(A) = A''$. That is, $A = A' = A''$ and therefore this point is the center of S. Nothing else can be concluded about S.

(b) If $A' = A''$ (A' is fixed), then $S(A) = S \circ S(A)$, and so $A = S(A) = A'$, because S is one to one. Hence, this case is the same as the previous one.

(c) Suppose $A = A'' \neq A'$. Then, $AA' = S(AA') = S(A)S(A') = A'A''$ (equal lengths) and so $\lambda = \dfrac{A'A''}{AA'} = 1$. Therefore, the similarity S is an opposite isometry. Since $S(A) = A''$, $S = R_l$ is the reflection with axis l the midpoint perpendicular straight line of the segment AA'.

(d) We assume that all three points are distinct one another and lie on the same straight line l. Then, the center and the axes of S are determined, depending on the order in which the three points are situated on l and the distances between them.

Since AA' is mapped to $A'A''$, the line l is set-wise invariant by S and so is either the principal or the second axis of S. Hence [or, for the same reason as in **Case (I), (2), (a)**], the center O is on the line AA'' and so all four points are on the straight line l. Now, if A is not between A' and A'', then l is the principal axis of S. Otherwise, if A is between A' and A'', l is the second axis. (Justify this claim.)

Again:

$$\frac{OA'}{OA} = \frac{OA''}{OA'} = \frac{A'A''}{AA'} = \lambda, \quad \text{and so} \quad OA'^{\,2} = OA \cdot OA''.$$

Hence, the center O is determined on l by this relation.

The positions and distances of the points on l play a role on how to find O, by using the latter relation. We examine the configuration as in **Figure 4.93** and the reader can go ahead to examine the other possibilities, imitating this example and the construction that we state at the end.

Suppose, A, A', and A'' are given in this order on l (A' is located between A, and A''). As we have said above, l must be the principal axis of S. (There are 5 more possible permutated configurations left to the reader. Which ones?)

From **Figure 4.93**, we see that the relation

$$OA'^{\,2} = OA \cdot OA'' \quad \text{is equivalent to} \quad (OA + AA')^2 = OA \cdot (OA + AA'').$$

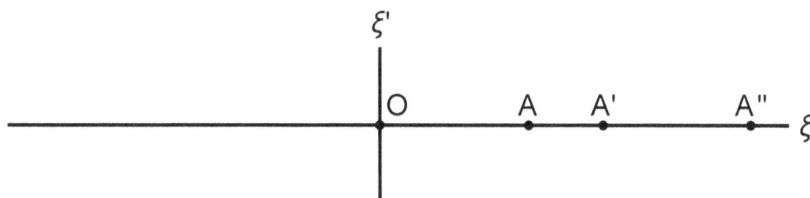

Figure 4.93: Determinations of the center O of a special opposite similarity

After solving for OA and adjusting to oriented lengths, we find

$$\overline{OA} = \frac{AA'^{\,2}}{\overline{AA''} - 2\overline{AA'}} \,.$$

($\overline{OA}$ is positive / negative, iff O is located to the left / right of A, respectively.)
    If $AA'' = 2AA'$, then $AA' = A'A''$, $\lambda = 1$, O is a point at infinity, and S is the glide

$$S = R_l \circ T_{\overrightarrow{AA'}} = R_l \circ T_{\overrightarrow{A'A''}} \,.$$

Conclusion: *An opposite similarity S can be determined by a point A and two successive images of it, $A' = S(A)$ and $A'' = S \circ S(A) = S(A')$, when $A \neq A'$.* (That is, the point A is not fixed.)

A Construction Concerning the Harmonic Quadruples

Three distinct points W, X, Y are given on a straight line l in this order (X is between W and Y). Find the point Q on l, such that, $\overline{QW}^{\,2} = \overline{QX} \cdot \overline{QY}$.
    See **Figure 4.94**. (The reader must be familiar with the harmonic divisions of straight segments and harmonic quadruples of points on a line.)

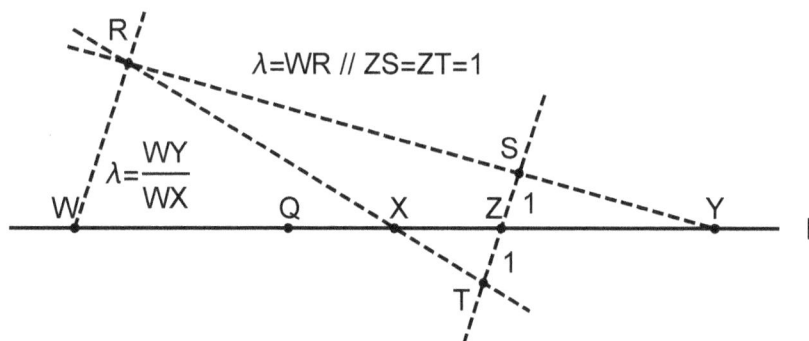

Figure 4.94: Construction of a harmonic quadruple $\{W, Z; X, Y\} = -1$ of points

Use the relation $\dfrac{WX}{XZ} = \dfrac{WY}{ZY} = \dfrac{WX + WY}{XZ + ZY} = \dfrac{WX + WY}{XY}$ (known ratio)
to find Z the harmonic conjugate point of W with respect to X and Y (lying between X and Y), (as in **Figures 4.94** or **4.92**). Then, as we have seen in the **proof of (1), (b)** and the **Remark** at the end of it, the relation

$$\overline{QW}^2 = \overline{QX} \cdot \overline{QY} = \overline{QZ}^2$$

characterizes the midpoint Q of the segment WZ. So, after finding Z we find the midpoint Q of WZ. Thus we have proved again the **Result**:

The relation $\overline{QW}^2 = \overline{QX} \cdot \overline{QY} = \overline{QZ}^2$ *is a characteristic property of the midpoint Q of WZ, in a harmonic quadruple $\{X, Y; W, Z\}$.*

Conversely: *This relation is equivalent to the quadruple $\{X, Y; W, Z\}$ (or equivalently the quadruple $\{W, Z; X, Y\}$) being harmonic.*

As we know from the theory of harmonic quadruples of points on a line or axis, given three different points W, X, Y on a line, **the harmonic conjugate point Z of W with respect to X and Y**, determined by definition by the relation $\dfrac{\overline{WY}}{\overline{WX}} = -\dfrac{\overline{ZY}}{\overline{ZX}}$, **is unique**. This implies that the point Q, as sought originally, is unique.

Also, as in the configuration of **Figure 4.94**, both X and Y are always on the same side (either left, or right) of Q on l. If Z were to be the midpoint of XY, then $\lambda = \left| \dfrac{\overline{ZX}}{\overline{ZY}} \right| = 1$ and W should have been a point at infinity. (Many properties of the harmonic quadruples are listed in the exercises.)

About opposite similarities between two straight lines or two circles.

As we have explained in **item (9) of Subsection 4.4.3** with the direct similarities (before the part of "Some Additional Results"), in the same way we prove that:
Given a ratio $\lambda > 0$ fixed, the geometrical locus of the centers of all opposite similarities of ratio $\lambda > 0$ that map l to l' are the two straight lines of proportional distances. With no restriction on the similarity ratio, two straight line l and l' are mapped to each other by infinitely many opposite similarities whose centers cover the whole plane $\mathcal{P}$. (Make new figure and elaborate analogously with the case of the direct similarities.)

Also: as we have explained in **item (10) of Subsection 4.4.3** with the direct similarities (before the part of "Some Additional Results"), in the same way we prove that:
Given two circles, there are infinitely many opposite similarities that map one onto the other.
(Again there are three cases and the geometrical loci of the centers of the opposite similarities are the same as in the case of the direct. Make figures and elaborate analogously with the cases of the direct similarities.)

Another Commutative Decomposition
of an Opposite Similarity

We consider an opposite similarity S with center O, similarity ratio $\lambda > 0$, principal axis ξ and second axis ξ'. In **Subsections 4.5.2** and **4.5.4**, we have proved the commutative decomposition of S with respect to the principal axis ξ and we have found that

$$S = R_\xi \circ H_{[O,\lambda]} = H_{[O,\lambda]} \circ R_\xi. \quad \text{(Depicted also in \textbf{Figure 4.95}.)}$$

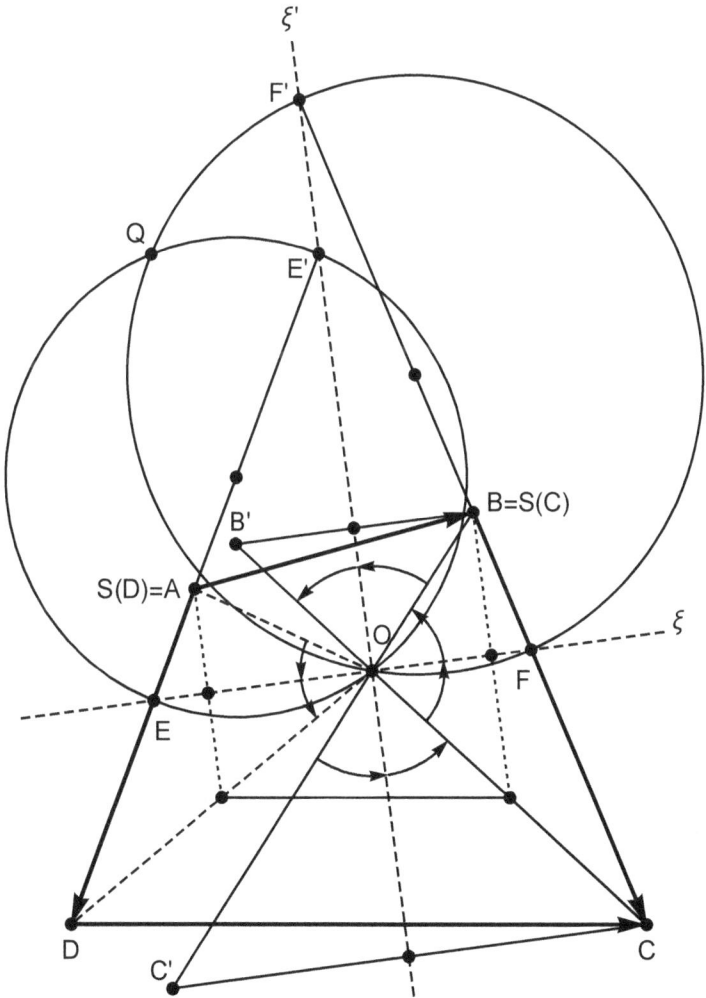

**Figure 4.95: A Commutative Decomposition of
an Opposite Similarity**

Here, we will prove the following commutative decomposition of S with respect to the second axis ξ':

$$S = H_{[O,-\lambda]} \circ R_{\xi'} = R_{\xi'} \circ H_{[O,-\lambda]}.$$

So, this commutative decomposition of S switches the roles of principal and second axes (but changes the homothetic ratio to apposite). See **Figure 4.95**.

We consider any pair of corresponding points (C, B), i.e., $B = S(C)$, as in **Figure 4.95**. We know that the axes ξ and ξ' are the bisectors of the angles (OC, OB). Then, C', the symmetrical of C in ξ', is a point of the line OB and is homothetic of B under the homothety $H_{[O,-\lambda]}$, because $OC' = OC$ and $OB = \lambda OC = \lambda OC'$. So, $\forall \; C \in \mathcal{P}$, we have $S(C) = B = H_{[O,-\lambda]} \circ R_{\xi'}(C)$. Therefore,

$$S = H_{[O,-\lambda]} \circ R_{\xi'}.$$

Similarly, B' the symmetrical of B in ξ' is a point of the line OC and is homothetic of C under the homothety $H_{[O,-\lambda]}$, because $OB' = OB = \lambda OC$. So,

$$\forall \; C \in \mathcal{P}, \quad \text{we have} \quad S(C) = B = R_{\xi'} \circ H_{[O,-\lambda]}(C).$$

Therefore,

$$S = R_{\xi'} \circ H_{[O,-\lambda]}.$$

Finally, we have obtained

$$S = H_{[O,-\lambda]} \circ R_{\xi'} = R_{\xi'} \circ H_{[O,-\lambda]}.$$

4.5.6   Coincidence of the Centers of the Direct and Opposite Similarities

As in the general cases of the direct and the opposite similarities (and isometries) studied before, we again consider two non-zero vectors $\overrightarrow{AB} \neq \overrightarrow{0}$ and $\overrightarrow{A'B'} \neq \overrightarrow{0}$ lying on different lines that intersect at a point I.

Usually the centers, O, of the direct and / or the opposite similarities that map $\overrightarrow{AB}$ onto $\overrightarrow{A'B'}$ are different one another and different from I. But, there is a case in which the three points coincide, $O = I$. See **Figures 4.96** and **4.97** describing two situations of this case.

This situations happens whenever $AA' \parallel BB'$. (It may happen $A = A' = I$, i.e., AA' is the zero segment.) In such a case, as we always have, the ratio of the direct or opposite similarity is

$$\lambda = \frac{A'B'}{AB} > 0.$$

Then using basic results and standard known theorems of the classical plane geometry, we can prove the following items. (Prove them as exercises! There are several other things you can observe and prove them too, in both **Figures 4.96** and **4.97**.):

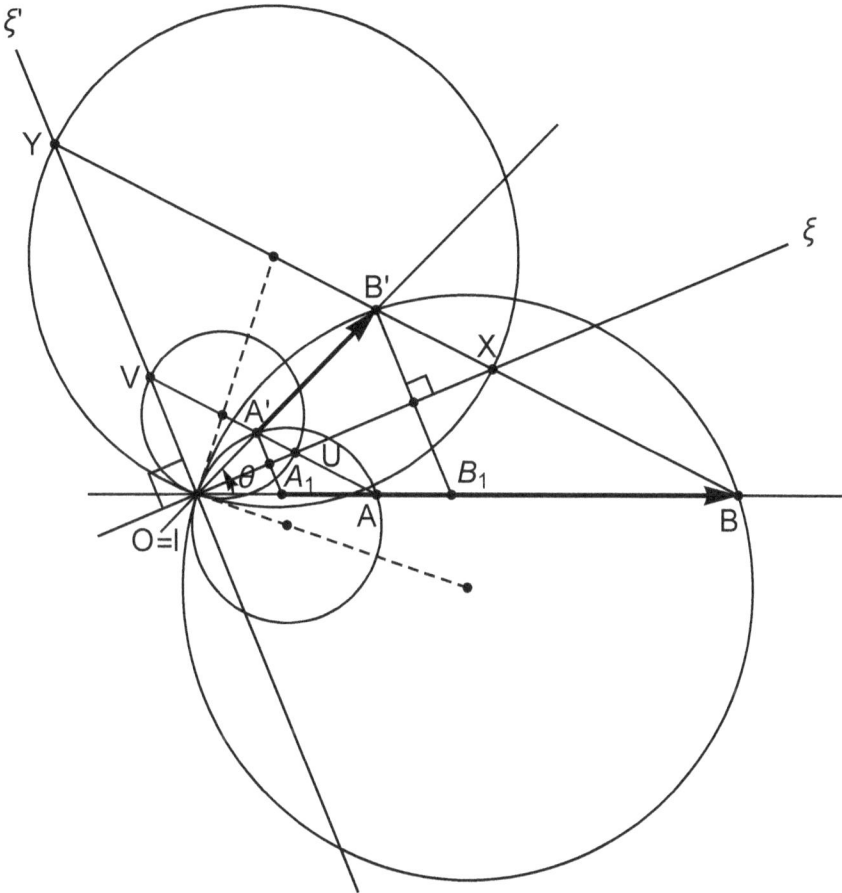

Figure 4.96: The center $O = I$, for either the direct or the opposite similarity

(1) $AA' \parallel BB'$ if and only if the triangles IAA' and IBB' are directly similar. Their similarity ratio is

$$\lambda = \frac{A'B'}{AB} > 0.$$

(2) $AA' \parallel BB'$ if and only if the circles (IAA') and (IBB') are tangent at I. So, their centers are collinear with I, and $O = I$.

(3) $AA' \parallel BB'$ if and only if the Apollonius circles (IUV) and (IXY), for ratio λ on the parallel straight segments AA' and BB', are tangent at I. So, their centers are collinear with I, and $O = I$. The triangles IUV and IXY are right, $\widehat{UIV} = \widehat{XIY} = 90°$.

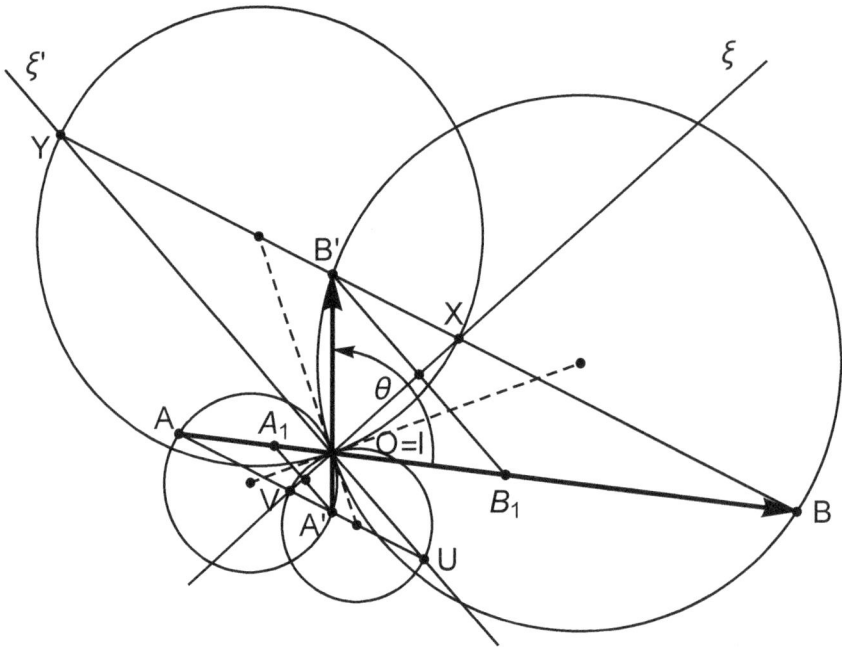

**Figure 4.97: The center $O = I$, for either the
direct or the opposite similarity**

(4) If we let

$$\overrightarrow{A_1 B_1} := \lambda \overrightarrow{AB} = H_{[I,\lambda]}(\overrightarrow{AB}),$$

then $\overrightarrow{A_1 B_1}$ is symmetrical to $\overrightarrow{A'B'}$ in the bisector of the angle

$$\measuredangle \theta = \measuredangle(\overrightarrow{AB}, \overrightarrow{A'B'}).$$

(5) The principal axis ξ of the opposite similarity is the bisector of the angle

$$\measuredangle \theta = \measuredangle(\overrightarrow{AB}, \overrightarrow{A'B'}).$$

The second axis ξ' is the straight line perpendicular to ξ at the point $I = O$.
(6) The direct similarity is

$$H_{[O,\lambda]} \circ R_{[O,\theta]} = R_{[O,\theta]} \circ H_{[O,\lambda]}.$$

(7) The opposite similarity is

$$H_{[O,\lambda]} \circ R_{\xi} = R_{\xi} \circ H_{[O,\lambda]}.$$

(Compare this general treatment of this topic with the partial results described in **Figures 2.64, 2.71, 4.57, 4.58, 4.59, 4.80, 4.81** and **4.83-4.88**.)

General Conclusion Based on Chapter 4

Given any two non-zero vectors $\overrightarrow{AB} \neq \vec{0}$ and $\overrightarrow{A'B'} \neq \vec{0}$ in the plane, there is a unique direct and a unique opposite similarity (which is an isometry if the two vectors have equal lengths) of the plane that maps $\overrightarrow{AB}$ onto $\overrightarrow{A'B'}$. The similarity ratio is $\lambda = \dfrac{A'B'}{AB} > 0$, and the direct-similarity-angle is $\theta = \angle(\overrightarrow{AB}, \overrightarrow{A'B'})$.

See also **Figure 4.98** below, where $A'B'BA$ is convex. (Make a figure with $A'B'BA$ non-convex.)

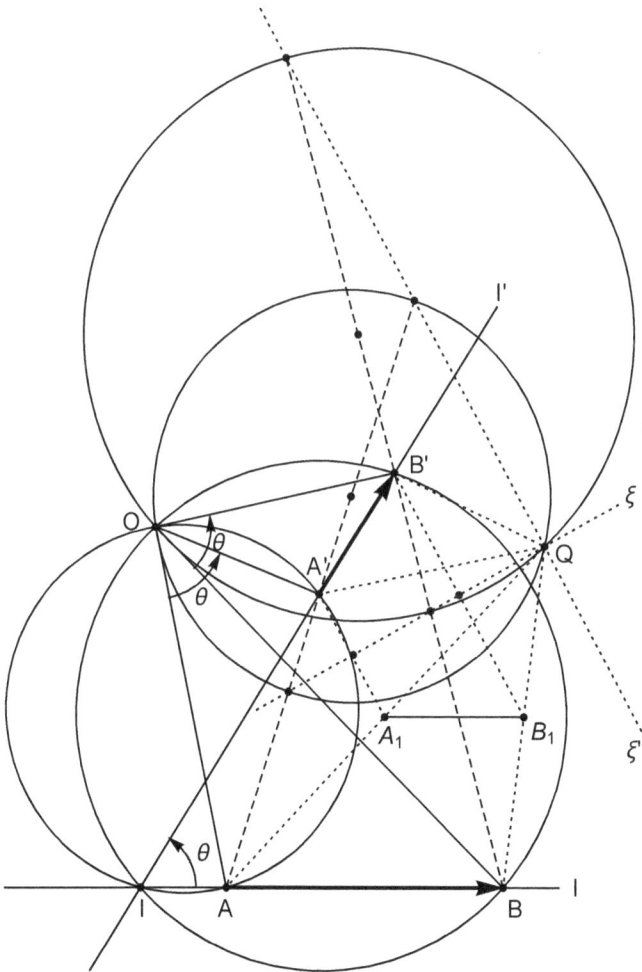

Figure 4.98: The centers O and Q of the direct and the opposite similarities that map $\overrightarrow{AB}$ to $\overrightarrow{A'B'}$

4.6 Applications of Similarities

Example 4.6.1 Application 1.

Consider two parallel lines l and l' and a fixed point O of their plane. A variable straight line through O intersects l and l' at points A and A', respectively. If point P is determined by the conditions

$$AP = AA' \quad \text{and} \quad \angle(\overrightarrow{AA'}, \overrightarrow{AP}) = \frac{\pi}{2},$$

then find the geometrical locus of P.

Solution. See **Figure 4.99**.

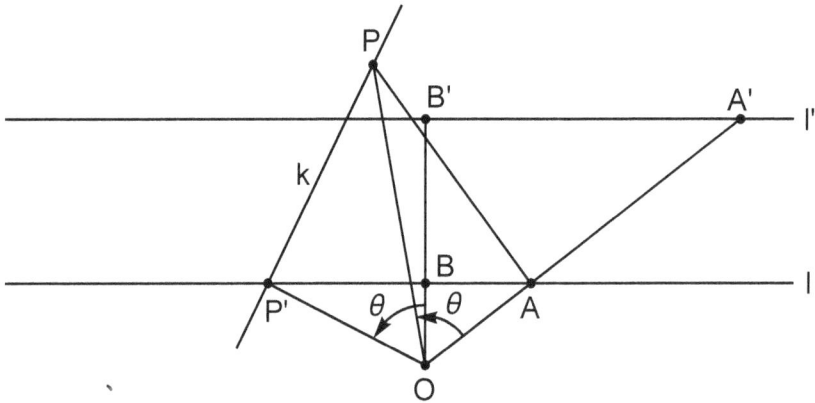

Figure 4.99: Example, A.SI. 1

The right triangles OAP are similar to one another because, as $AP = AA'$, we get $\dfrac{OA}{AP} = \dfrac{OA}{AA'}$ which is constant, because $l \parallel l'$.

If B and B' are the projections of O on the lines l and l' and P' is the point of the locus corresponding to OB ($BP' = BB'$), then, from the similar triangles OAP and OBP', we get

$$\frac{OP}{OA} = \frac{OP'}{OB} = \lambda \text{ constant}, \quad \text{and}$$

$$\angle(\overrightarrow{OA}, \overrightarrow{OP}) = \angle(\overrightarrow{OB}, \overrightarrow{OP'}) = \theta \text{ constant}.$$

We conclude that the point P is the image of A under the direct similarity

$$S = H_{[O,\lambda]} \circ R_{[O,\theta]} = R_{[O,\theta]} \circ H_{[O,\lambda]}.$$

So, as A moves on the line l, P moves on a line k. But, $P' = S(B)$. Therefore, by **Subsection 4.4.4, (7)**, the line k is the perpendicular to OP' at the point P'.

▲

Example 4.6.2 Application 2.

A point C traces a semicircle with center O and diameter AB. We construct a right isosceles triangle PBC with $\widehat{P} = 90°$ and P on the same side of AB as the semicircle. Find the geometrical locus of the vertex P.

Solution. See **Figure 4.100**.

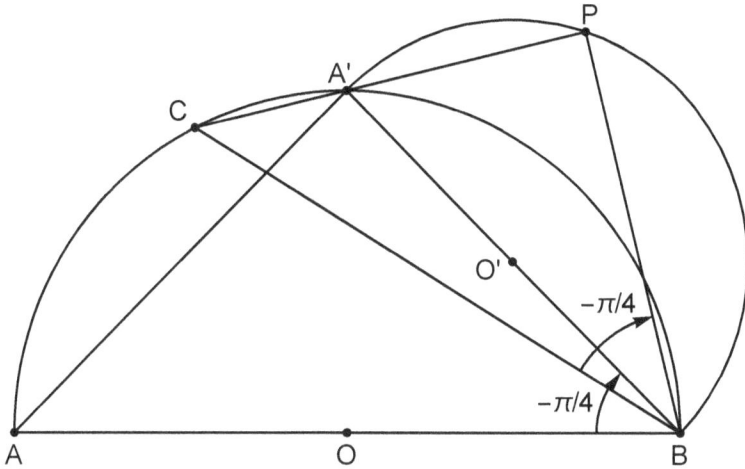

Figure 4.100: Example, A.SI. 2

It must be that $\measuredangle(\overrightarrow{BC}, \overrightarrow{BP}) = -\frac{\pi}{4}$ and $BP = BC \cdot \frac{\sqrt{2}}{2}$.

Therefore, P is the image of C under the direct similarity

$$S = H_{\left[B, \frac{\sqrt{2}}{2}\right]} \circ R_{\left[B, -\frac{\pi}{4}\right]} = R_{\left[B, -\frac{\pi}{4}\right]} \circ H_{\left[B, \frac{\sqrt{2}}{2}\right]}.$$

Therefore, the geometrical locus of P is the semicircle whose diameter is $A'B$, where $A' = S(A)$, or the vertex of the right isosceles triangle $A'BC$, and so A' is a known point.

▲

Example 4.6.3 Application 3.

Consider a quadrilateral $ABCD$ and the points $E = AB \cap CD$ and $F = BC \cap AD$.

Prove that:

(1) The circles (FAB), (FCD), (EBC), and (EAD) pass through the same point O, called the **Miquel point** of the quadrilateral $ABCD$ or the complete quadrilateral $ABCDEF$.

(2) The centers O_1, O'_1, O_2, and O'_2 of the circles in **(1)**, respectively, and the point O are on the same circle, called the **Miquel circle** of the quadrilateral $ABCD$ or the complete quadrilateral $ABCDEF$.

(3) If the quadrilateral $ABCD$ is inscribable, then O is on the line EF.

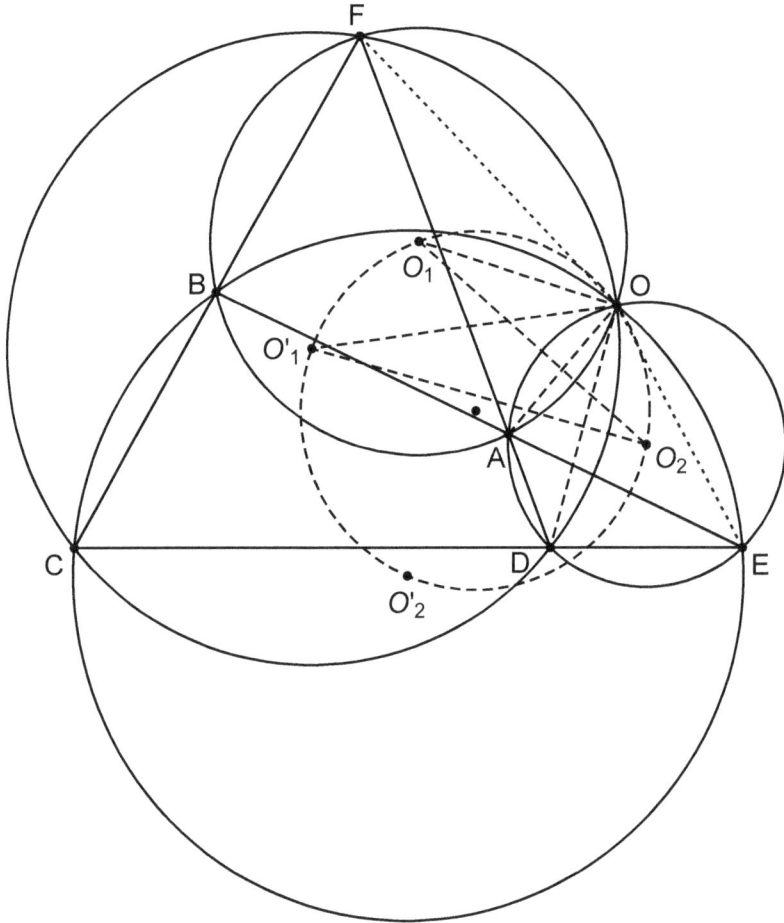

Figure 4.101: Example, A.SI. 3

Solution. See **Figure 4.101**. We must be familiar with **Subsection 4.4.4**.

(1) The vectors $\overrightarrow{AB}$ and $\overrightarrow{DC}$ determine a similarity S_1, whose center is O, the common point of the circles (EBC), and (EAD), other than E. But O is also the center of the similarity S_2 determined by the vectors $\overrightarrow{AD}$ and $\overrightarrow{BC}$. Therefore, O is also a common point of the circles (FAB) and (FCD).

(2) Under the similarity S_1 the circle (OAB) is mapped to (OCD). Therefore, the center O_1 and O_1' are corresponding points. Hence,

$$\angle(\overrightarrow{OO_1}, \overrightarrow{OO_1'}) = \angle(\overrightarrow{OA}, \overrightarrow{OD}).$$

Also, OA and OD are common chords of (EAD) with (ZAB) and (ZCD), respectively. Hence,

$$O_2O_1 \perp OA \qquad \text{and} \qquad O_2O_1' \perp OD.$$

Therefore,

$$\angle(O_2O_1, O_2O_1') = \angle(\overrightarrow{OA}, \overrightarrow{OD}).$$

By the above two angle equalities, we get that the center O_2 is ia point of the circle (OO_1O_1'). We prove the same for the center O_2'.

Note: This result can also be shown by using the **Miquel points of tri-angles**. Find this topic in bibliography.

(3) First, we have

$$\angle(OF, OA) = \angle(BF, BA)$$

and

$$\angle(OE, OA) = \angle(DE, DA).$$

If the quadrilateral $ABCD$ is inscribable, then

$$\angle(BF, BA) = \angle(DE, DA).$$

This implies that the straight lines OE and OF coincide.

▲

Example 4.6.4 Application 4.

Consider a quadrilateral $ABCD$. Let points E and E' be the points that divide $\overrightarrow{AD}$ harmonically in the ratio $\dfrac{AB}{CD}$ and F and F' the points that divide $\overrightarrow{BC}$ harmonically in the same ratio.

Prove that the straight lines EF and $E'F'$:

(1) Are parallel to the bisectors of the angles of the lines CD and AB.

(2) Intersect at the point O, a common point of the circles with diameters EE' and FF'.

Solution. See **Figure 4.102**. We must be familiar with **Subsections 4.5.2** and **4.5.4**.

The similarity S determined by the vectors $\overrightarrow{DC}$ and $\overrightarrow{AD}$ maps D to A and C to B. This similarity is opposite and we know that its axes ξ and ξ' are defined by the points that divide the vectors $\overrightarrow{AD}$ and $\overrightarrow{BC}$ harmonically in ratio $\dfrac{AB}{AC}$.

Therefore, the axes are the lines EF and $E'F'$. They are parallel to the bisectors of the angles of the corresponding lines CD and AB and they intersect at the center O of the similarity S. Also, O belongs to the Apollonius circles with diameters EE' and FF'.

(**This Example and Figure 4.102** can be viewed as a quick review of opposite similarities).

▲

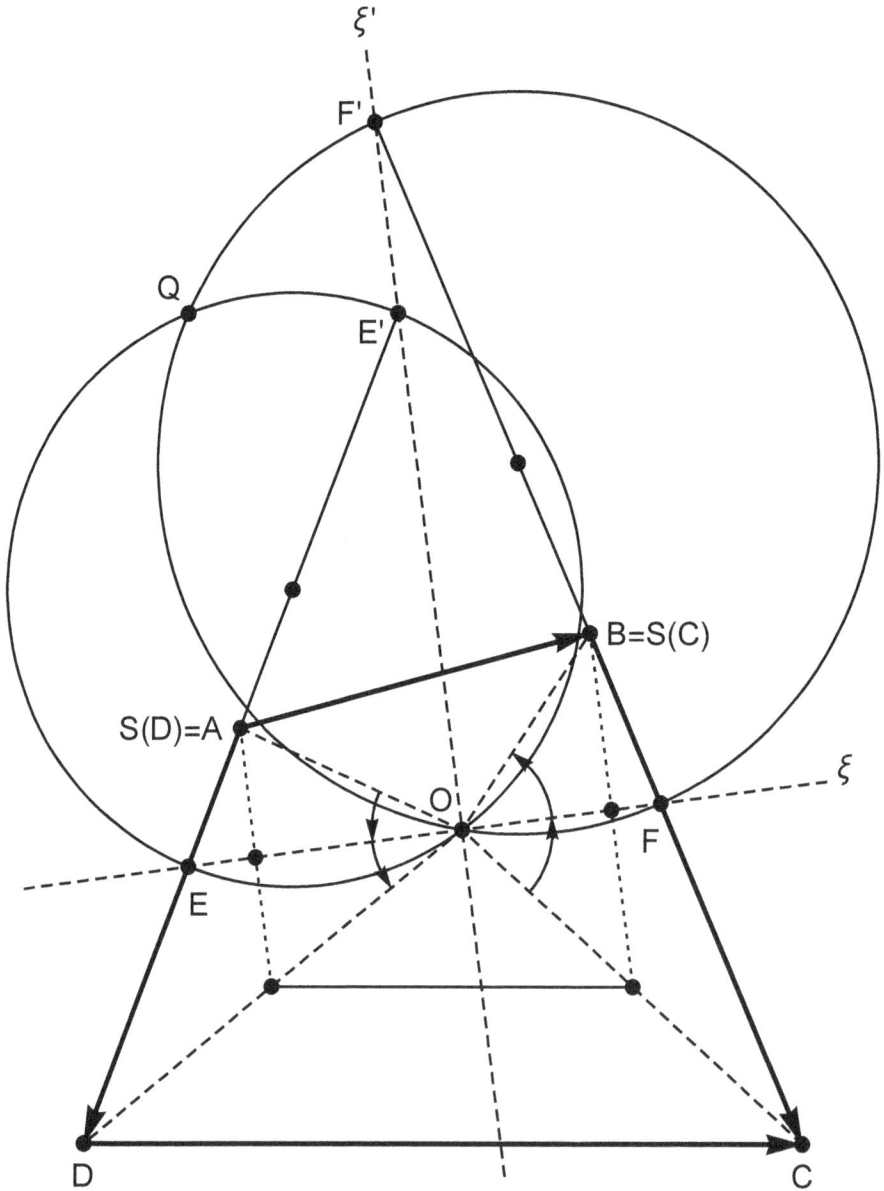

Figure 4.102: Example, A.SI. 4

Example 4.6.5 Application 5. The results proven in **Subsections 4.4.4, 4.5.2, 4.5.4** and **4.5.6**, imply the following interesting **Result**:

Consider any four points in the plane A, B, C and D and let $\lambda = \dfrac{AB}{CD}$.
Then, the two Apollonius circles on BC and AD such that their points P and Q satisfy $\dfrac{PB}{PC} = \lambda \neq 1$ *and* $\dfrac{QA}{QD} = \lambda \neq 1$, *either intersect at two points (the centers of the direct and opposite similarities that map* $\overrightarrow{DC}$ *to* $\overrightarrow{AB}$), *or are tangent to each other.*

These Apollonius circles are tangent to each other if and only if the centers of the direct and opposite similarities that map $\overrightarrow{DC}$ *to* $\overrightarrow{AB}$ *coincide, or if and only if BC* $\parallel$ *AD. Moreover, in such a case, we also have:*

(1) The point of tangency coincides with the common point of AB and CD.

(2) They are internally tangent if ABCD is convex.

(3) They are externally tangent if ABCD is not convex (if and only if the straight segments AB and CD intersect internally).

If $\lambda = 1$ ($\Longleftrightarrow$ *AB = CD*), *then the two Apollonius circles become the perpendicular bisectors of the straight segments BC and AD, which may coincide, or intersect at a point, or be parallel.*

So, the two special Apollonius circles constructed above are never disjoint, except when they degenerate to two parallel straight lines (or else, they intersect at the infinity point).

(See the corresponding figures in the referred subsections at the beginning above and **Figure 4.98** where the correspondence of points from that figure to the present situation is: $A' \longrightarrow A$, $B' \longrightarrow B$, $B \longrightarrow C$ and $A \longrightarrow D$ and $A'B'BA = ABCD$ is convex.)

Remark: In case the two Apollonius circles are tangent at the point $I = O$, then the same conclusions **(1)**, **(2)** and **(3)** above hold for the circles (ODA) and (OBC). (See **Figures 4.96** and **4.97**.)

It would be interesting to provide another synthetic proof of the above result, or a proof using analytic geometry with $4 \times 2 = 8$ real variables which are the coordinates of the four initial points. Also, the geometry of complex numbers could provide another way. (The pertinent computations with analytic geometry or complex numbers may be lengthy. In this way, this result would yield some geometrical inequalities.)

▲

Example 4.6.6 Application 6.

Suppose that two polygons $A_1 A_2 A_3 \ldots A_n$ and $B_1 B_2 B_3 \ldots B_n$, each having n sides, are oppositely similar. That is, the corresponding sides are proportional

$$\frac{B_1 B_2}{A_1 A_2} = \frac{B_2 B_3}{A_2 A_3} = \ldots = \frac{B_{n-1} B_n}{A_{n-1} A_n} = \frac{B_n B_1}{A_n A_1} = \lambda,$$

and the corresponding oriented angles are opposite

$$\angle(\overrightarrow{A_1 A_2}, \overrightarrow{A_1 A_n}) = -\angle(\overrightarrow{B_1 B_2}, \overrightarrow{B_1 B_n}), \quad \angle(\overrightarrow{A_2 A_3}, \overrightarrow{A_2 A_1}) = -\angle(\overrightarrow{B_2 B_3}, \overrightarrow{B_2 B_1}), \quad \ldots,$$

and so on.

To find the opposite similarity that maps $A_1 A_2 A_3 \ldots A_n$ onto $B_1 B_2 B_3 \ldots B_n$, it is enough to find the opposite similarity that maps one side onto its corresponding one, $A_1 A_2$ onto $B_1 B_2$, let us say. Then all the sides of $A_1 A_2 A_3 \ldots A_n$ will coincide with the corresponding sides of $B_1 B_2 B_3 \ldots B_n$, since the opposite similarities reverse the orientation of the angles and the angles of $A_1 A_2 A_3 \ldots A_n$ and $B_1 B_2 B_3 \ldots B_n$ are correspondingly opposite, and all the sides of $A_1 A_2 \ldots A_n$ will be multiplied by $\dfrac{B_1 B_2}{A_1 A_2} = \lambda$.

We can do this in the following way: We consider the vector $\overrightarrow{B_1' A_2}$ parallel to $\overrightarrow{B_1 B_2}$ and with length $B_1' A_2 = A_1 A_2$, the bisector β of the angle $\widehat{A_1 A_2 B_1'}$, and K the common point of the straight lines $A_2 B_2$ and $B_1' B_1$. **Figure 2.103**.

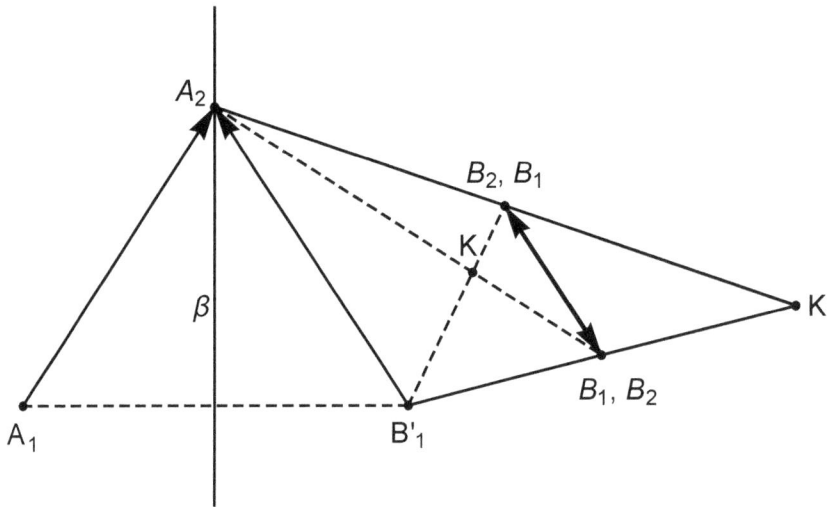

Figure 4.103: Example, A.SI. 6

Then, the opposite similarity sought is $S = H_{[K,\lambda]} \circ R_\beta$. Indeed, by the reflection in the straight line β, R_β, $\overrightarrow{A_1 A_2}$ will be mapped onto $\overrightarrow{B_1' A_2}$ and the homothety $H_{[K,\lambda]}$ will map $\overrightarrow{B_1' A_2}$ to $\overrightarrow{B_1 B_2}$. The homothety $H_{[K,\lambda]}$ is positive or negative if the vectors $\overrightarrow{B_1' A_2}$ and $\overrightarrow{B_1 B_2}$ have the same or opposite direction, respectively. S reverses the orientation of the angles, therefore it maps the oriented angles of $A_1 A_2 A_3 \ldots A_n$ onto the oriented angles of $B_1 B_2 B_3 \ldots B_n$.

This S is one of those claimed in **Theorem 4.3.2**. Otherwise, we can work directly as in **Figures 4.77** and **4.78**, etc., to find the axes and the center of the composition $S = H_{[K,\lambda]} \circ R_\beta$.

▲

For further results, applications, and examples study and solve the exercises that follow.

Chapter 5

Exercises on Similarities

We abbreviate "straight line", "-es" with "s.l.", "s.ls"; "geometrical locus", "-es" with "g.l.", "g.ls"; and "if and only if" with "iff".

5.1 Homotheties and Harmonic Quadruples

5.1.1 General

1. Consider two circles $C[K_1, r_1]$ and $C[K_2, r_2]$. Prove:

   (a) The circles can be mapped onto each other by one positive and one negative homothety. When $r_1 = r_2$ and $K_1 \neq K_2$, then the positive homothety degenerates to the parallel translation $T_{\overrightarrow{K_1 K_2}}$ or $T_{\overrightarrow{K_2 K_2}}$.

   (b) The common point of the external / internal tangents of the two circles (when they exist) is the center of the positive / negative homothety, respectively.

   (c) If a s.l. intersects the circles at points A and B, respectively, and the s.l $K_1 K_2$ at points C^+ / C^-, and also $\dfrac{\overline{C^{\pm} A}}{\overline{C^{\pm} B}} = \pm \dfrac{r_1}{r_2}$, then $C^{\pm}$ are the centers of the homotheties.

   (d) If a third circle is tangent to the circles at points A and B, respectively, then the s.l. AB passes trough the center of one of the two homotheties.

2. Prove that a homothety transforms a tangent of a circle to a tangent of the homothetic image of the circle.

3. Two circles are tangent at point P. A tangent line to one of them at a point F defines two points K and L on the other circle. Prove that PF is the bisector of one of the angles (PK, PL).

4. Prove that two parallel s.ls can be mapped onto each other by infinitely many homotheties. Find them all.

5. Find subsets of the plane which are set-wise fixed by a homothety (not listed in the text).

6. Prove that if a circle and a s.l. intersect at an angle ϕ, then their homothetic images intersect at the same angle ϕ.

7. Two circles are externally tangent a the point F. We consider any chord FA of one of them and the chord of the other FA' which is perpendicular to FA. Prove that the s.l. AA' passes through a fixed point.

8. (a) Prove that two triangles are homothetic iff their corresponding sides are parallel (and so the triangles are similar). This homothety is unique and is positive or negative iff the triangles are directly or oppositely similar, respectively.
(b) Prove that two polygons are homothetic iff their corresponding sides are parallel (either all in the same or all in opposite direction).
(c) Prove that two squares are homothetic by one positive and one negative homothety iff they have parallel corresponding sides.

9. **Properties of anti-homologous points**:
(a) Two pairs of anti-homologous points form oppositely similar triangles with the homothetic center.
(b) Prove that the product of the distance of the center of a homothety between two circles from two anti-homologous points is the same for all pairs of anti-homologous points.
(c) The four points of two pairs of anti-homologous points are concyclic.
(d) The tangents to the two circles at anti-homologous points make equal angles with the homothetic ray through them. Conversely:
If two tangent segments drawn to two circles from an outside point are equal, then the points of contact are anti-homologous.

10. Consider two intersecting circles $C[K_1, r_1]$ and $C[K_2, r_2]$ at A and B and let O and O' be their two homothetic centers. Prove that the lines AO and AO' are the angle bisectors of the lines $K_1 A$ and $K_2 A$.

11. Two given circles are tangent at a point A. From a fixed point F we draw a s.l. FPQ intersecting one of the circles at P and Q. The s.ls FP and FQ intersect the second circle at the points P' and Q'. Show that the s.ls $P'Q'$ go through a fixed point.

12. Two circles are tangent at the point A. A straight line intersects the first at M and N. The s.ls AM and AN intersect the second circle at M' and N'. Prove: if the s.l. MN passes through a fixed point, then the s.l. $M'N'$ passes through another fixed point.

13. Consider three pairwise tangent circles. Prove that the s.ls connecting one point of tangency with the other two, define in one of the circles a diameter parallel to the center line of the other two.

14. A triangle ABC has side BC and circumscribed circle fixed. (A moves on the circle.) Show that the nine-point circle of ABC is tangent to a constant circle.

15. Consider a triangle ABC, its altitudes AH_1, BH_2, CH_3, its circumscribed circle (ABC) and its nine-point circle $(H_1H_2H_3)$. Prove:
    (a) Circle $(H_1H_2H_3)$ is completely inside circle (ABC) iff ABC is scalene (that is, all its angles are acute).
    (b) Circle $(H_1H_2H_3)$ is internally tangent with circle (ABC) (at A) iff ABC is right (at A).
    (c) Circle $(H_1H_2H_3)$ intersects circle (ABC) (at two points) iff the triangle ABC has an obtuse angle.

16. Find all rectangles, parallelograms and isosceles trapeziums hidden in **Figure 4.33**. Also prove that the triangles $O_1O_2O_3$ and $P_1P_2P_3$ are equal and the half turn $R_{[E,\frac{\pi}{2}]}$ maps one onto the other.

17. We take four points A, B, C, D on a s.l. and in this order. We consider the three pairs of circles with diameters (AB,CD), (AC,BD) and (AD,BC), respectively. Prove the following three results:
    (a) The centers of the direct homotheties of the 1st and 2nd pairs of circles coincide on the radical axis of the 3rd pair.
    (b) The centers of the direct homotheties of the 1st pair of circles and the negative homotheties of the 3rd pair of circles coincide on the radical axis of the 2nd pair.
    (c) The centers of the opposite homotheties of the 2nd and the 3rd pairs of circles coincide on the radical axis of the 1st pair.

18. On an axis we consider the points $A = -3$, $B = 2$, and $C = -2$. Find the harmonic conjugate of each point with respect to the other two.

19. Suppose the four points A, B, C and D form a harmonic quadruple, i.e., $\{A, B; C, D\} = -1$, with C between A and B and B between C and D. Prove:
    (a) If $-\dfrac{\overline{CA}}{\overline{CB}} = k > 1$, then $\dfrac{\overline{CA}}{\overline{DA}} = \dfrac{k-1}{k+1}$.
    (b) If $0 < -\dfrac{\overline{CA}}{\overline{CB}} = k < 1$, then $\dfrac{\overline{CA}}{\overline{DA}} = \dfrac{1-k}{1+k}$.
    (c) If $\dfrac{\overline{CA}}{\overline{CB}} = k$ and O is the midpoint of AB, then $\dfrac{\overline{OC}}{\overline{OD}} = k^2$.

20. **Equivalent conditions for a harmonic quadruple.** On an axis we pick points A, B, C and D with abscissas a, b, c and d, respectively. Let also $M = m = \dfrac{a+b}{2}$ and $N = n = \dfrac{c+d}{2}$ be the midpoints of AB and CD, respectively.

The four points A, B, C and D form a harmonic quadruple, such that, $\{A, B; C, D\} = -1$, (C is between A and B, and B is between C and D) iff one of the following condition, **(a)-(j)**, holds:

(a) The points C and D divide the vector $\overrightarrow{AB}$, internally and externally respectively, in opposite ratios.

(b) $(a + b)(c + d) = 2(ab + cd)$.

(c) $(c - m)(d - m) = (a - m)^2 = (b - m)^2$.

(d) $(a - n)(b - n) = (c - n)^2 = (d - n)^2$.

(e) $(n - m)^2 = (a - m)^2 + (c - n)^2$.

(f) $(b - a)(c + d - 2a) = 2(c - a)(d - a)$ $\iff$ $\dfrac{2}{b - a} = \dfrac{1}{c - a} + \dfrac{1}{d - a}$.
Write three more analogous relations.

(g) $(b - a)(n - a) = (b - a) \cdot \left(\dfrac{c + d}{2} - a \right) = (c - a)(d - a)$.
Write three more analogous relations.

(h) $\dfrac{n - a}{b - n} = \dfrac{c - a}{b - c} \cdot \dfrac{d - a}{b - d}$.

(i) $\dfrac{n - a}{b - n} = -\left(\dfrac{c - a}{b - c} \right)^2 = -\left(\dfrac{d - a}{b - d} \right)^2$.

(j) $\overline{AB} \cdot \overline{CD} + 2\overline{AD} \cdot \overline{BC} = 0$.

(You may take as point of reference on the axis on of the given points, e.g., the point A, for convenience. Of course any point of reference would work!)

(k) Next, write all the relations (b)-(i) in terms of vectors (or oriented lengths), as done, for example, in (j).

21. If four points A, B, C and D form a harmonic quadruple on an axis, such that, $\{A, B; C, D\} = -1$, (C is between A and B, and B is between C and D). Let M and N ne the midpoint of AB and CD, respectively. Then

$$\overline{NA} \cdot \overline{BC} \cdot \overline{BD} = \overline{NB} \cdot \overline{AC} \cdot \overline{AD}.$$

Write and prove the analogous relation that involves M.

22. If the points A, B, C and D form a harmonic quadruple, $AB = s$, and $\dfrac{\overline{AC}}{\overline{CB}} = \lambda$, then compute the length of CD and of the other segments.

23. If the points A, B, C and D form a harmonic quadruple and $\dfrac{\overline{AC}}{\overline{CB}} = \lambda$, then compute the ratios $\dfrac{\overline{AD}}{\overline{DC}}$ and $\dfrac{\overline{BD}}{\overline{DC}}$.

24. Let $\{A, B; C, D\}$ be a harmonic quadruple and A' the harmonic conjugate of C with respect to A and D and B' the harmonic conjugate of C with respect to B and D. Prove that $\{A', C; B', D\}$ is also a harmonic quadruple.

25. Consider a triangle ABC and its heights AA', BB', CC' and its ortho-center H. Let Q be the point of intersection of the s.ls AA' and $B'C'$. Prove that the points A, H, Q and A' are a harmonic quadruple.

26. Consider a triangle ABC, its median AM, and the s.l. $AX \parallel BC$. Let a s.l. $l \nparallel AM$ intersect AB, AM, AC, and AX at the points P, Q, S, and T respectively. Prove that the quadruple $\{P, Q; S, T\}$ is harmonic.

27. Given a straight segment AB and $\lambda > 0$, we know that the corresponding Apollonius circle is the set of points X such that $\dfrac{XA}{XB} = \lambda$. Prove:

    (a) If $\lambda = 1$, then the Apollonius circle becomes the midpoint perpendicular of the straight segment AB. (So, its radius is ∞ and its center is the infinity point of the s.l. AB.)

    (b) Let the diameter of the Apollonius circle on the s.l. AB be CD with C between A and B and D outside AB. Then $\{A, C; B, D\}$ is a harmonic quadruple with ratio λ.

    (c) For different λ's the Apollonius circles are disjoint.

    (d) For each point P in the plane, there is an Apollonius circle through P.

28. Given a straight segment AB and $\lambda > 0$, consider the Apollonius circle of the points X such that $\dfrac{XA}{XB} = \lambda$. Let the diameter of the Apollonius circle on the s.l. AB be CD with C between A and B and D outside AB, and M the midpoint of AB and K the midpoint of CD, i.e., the center of the Apollonius circle. Prove the following relations (in which we must notice carefully the orientation of the straight segments):

    (a) $$\overline{CA} = \frac{-\lambda}{\lambda + 1}\overline{AB}, \qquad \overline{CB} = \frac{1}{\lambda + 1}\overline{AB},$$
    $$\overline{DA} = \frac{\lambda}{1 - \lambda}\overline{AB}, \qquad \overline{DB} = \frac{1}{1 - \lambda}\overline{AB}.$$

    (b) The radius of the Apollonius circle is $R_a = \dfrac{\lambda}{\lambda^2 - 1}\overline{AB} = \dfrac{\lambda}{|\lambda^2 - 1|}AB$.

    (c) $\overline{AK} = \dfrac{\lambda^2}{\lambda^2 - 1}\overline{AB}, \qquad \overline{BK} = \dfrac{1}{\lambda^2 - 1}\overline{AB}, \qquad \overline{MK} = \dfrac{\lambda^2 + 1}{2(\lambda^2 - 1)}\overline{AB}$.

    (d) $\overline{KA} \cdot \overline{KB} = R_a^2$. (Hence, A and B are inverse points in the Apollonius circle, as we define the inverse points in **Section 6.1**.)

    (e) The length of the tangent segment to the Apollonius circle from M is equal to $t = \sqrt{MK^2 - R_a^2} = \dfrac{\overline{AB}}{2}$, a constant independent of λ.

    (f) Transcribe the above formulae to the Apollonius circle on a triangle ABC with sides $a = \overline{BC}$, $b = \overline{CA}$ and $c = \overline{AB}$, that intersects the side $a = \overline{BC} > 0$ at its common points with the internal and external bisectors of the angle $\widehat{A}$. (So, $\lambda = \dfrac{c}{b} > 0$, or $\dfrac{b}{c} > 0$, and so $R_a = \dfrac{abc}{|b^2 - c^2|}$, etc.)

29. Consider any trapezium $ABCD$, where $AB \parallel CD$ and $AB > CD$. Let O be the common point of the diagonals AC and BD and P the common point of the non-parallel sides AD and BC. Let also M and N be the common points of the s.l. OP with the parallel sides AB and CD, respectively. Prove:
(a) The point O is the midpoint of the segment drawn through it and parallel to the two bases, and the points M and N are the midpoints of the two bases of the trapezium.
(b) The quadruple $\{M, O; N, P\}$ is harmonic.

30. Consider a circle of diameter AB and let CD be a chord perpendicular to AB and P any point of the circle. Let the s.ls PC and PD intersect the s.l AB at the points E, F. Prove:
(a) The quadruple $\{A, B, E, F\}$ is harmonic.
(b) $OE \cdot OF = r^2$, where r is the radius of the circle.

31. Consider a triangle ABC and the diameter $DE \perp BC$ of the circumscribed circle. Let the s.ls DA and EA intersect BC at the points P, Q. Prove that the quadruple $\{B, C, P, Q\}$ is harmonic.

32. Consider a triangle ABC and its median AM. Let I be the point defined by $\overrightarrow{AI} = \dfrac{1}{3}\overrightarrow{AB}$ and the point Q defined by $\overrightarrow{BQ} = 2\overrightarrow{BA}$. Prove:
(a) The quadruple $\{A, B, I, Q\}$ is harmonic.
(b) If the s.ls AM and CI intersect at the point S, prove that
$\overrightarrow{SA} + \overrightarrow{SM} = \overrightarrow{0}$ and $\overrightarrow{IS} = \dfrac{1}{4}\overrightarrow{IC}$.

33. In a triangle ABC, consider:
(1) M the midpoint of the side BC.
(2) The height $AH_1 = h_a$.
(3) The points E and F where the internal and external bisectors of the angle $\hat{A}$ intersect the s.l. BC.
(4) The inscribed and exscribed circles $C[I, \rho]$, $C[I_a, \rho_a]$, $C[I_b, \rho_b]$, and $C[I_c, \rho_c]$, and
(5) their points of tangency D, D_a, D_b, and D_c with the s.l. BC.
Prove:
(a) The quadruples $\{A, E, I, I_a\}$ and $\{A, F, I_b, I_c\}$ are harmonic.
(b) The projections of I_b and I_c on the s.l of the height $AH_1 = h_a$ divide the segment AH_1 harmonically. Then, given any two elements of $\{h_a, \rho_b, \rho_c\}$, we can find the third one.
(c) The distance of the point M from the s.l. AB is $\dfrac{h_c}{2}$, where h_c is the height CH_3.
(d) The midpoints of the segments II_a and I_bI_c are the points of intersection of the circle (ABC) and the perpendicular bisector of the side BC.

(e)

$$(1) \quad MD^2 = MI_a^2 = \overline{ME} \cdot \overline{MH_1},$$
$$(2) \quad MD_b^2 = MD_c^2 = \overline{MF} \cdot \overline{MH_1},$$
$$(3) \quad \frac{2}{h_a} = \frac{1}{\rho} - \frac{1}{\rho_a} = \frac{1}{\rho_b} + \frac{1}{\rho_c}.$$

(f) Write the above results with respect to the sides CA and AB.

(g) Construct a triangle ABC from the data $\{b + c = AC + AB, \rho_a, \rho_b\}$.

34. Consider a triangle ABC with all the angles acute. Prove that the tangent lines to the nine-point circle at the midpoints of the sides of ABC, form a triangle homothetic to the orthic triangle of ABC by a homothety whose center is on Euler's line of ABC.

35. Consider a triangle ABC, H its orthocenter, G the center of gravity (centroid) and O_1, O_2, O_3 the midpoints of the sides BC, CA and AB, respectively.

(a) Draw any two half lines Hx and Hy (rays emanating from H) intersecting the circumscribed circle of ABC at the points D and E, respectively and the nine-point circle of ABC, $(O_1O_2O_3)$, at the points D' and E', respectively. Prove that $DE \parallel D'E'$ and $D'E' = \frac{1}{2}DE$.

(b) Draw any s.l. through G intersecting the circumscribed circle of ABC on one side of G at the point D and the nine-point circle of ABC, $(O_1O_2O_3)$, at the point D' on the other side of G. Prove that G is a trisecting point of the straight segment $D'D$.

(c) Draw any two s.ls through G intersecting the circumscribed circle of ABC on one side of G at the points D and E, respectively, and the nine-point circle of ABC, $(O_1O_2O_3)$, at the points D' and E', respectively, on the other side of G. Prove that $DE \parallel D'E'$ and $D'E' = \frac{1}{2}DE$.

(d) Let K be any point of the nine-point circle $(O_1O_2O_3)$. Consider $AX \parallel KO_3$ with X on the circumcircle (ABC). Prove that $AX = 2\,KO_1$. Also prove that, $BX \parallel KO_2$, $CX \parallel KO_3$ and $BX = 2KO_2$, $CX = 2KO_3$.

36. Consider a convex quadrilateral $ABCD$ and the 4 triangles formed by its vertices when chosen by three. Prove that the centers of gravity of these triangles defined a quadrilateral homothetic to $ABCD$ under the homothety with center O the common midpoint of the medians of $ABCD$ and ratio $\lambda = -\frac{1}{3}$.

37. Consider an orthocentric quadruple $\{A, B, C, H\}$. Prove that the centers of gravity of the triangles HBC, HCA and HAB define a triangle homothetic to ABC under the homothety of center H and ratio $\lambda = \frac{3}{2}$. Also, the orthocenter of the new triangle coincides with the center of gravity

(centroid) of ABC. (**Definition**: Four points A, B, C, D form an **ortho-centric quadruple**, if each point is the orthocenter of the triangle with vertices the other three.)

38. Consider two circles $C[O_1, r_1]$ and $C[O_2, r_2]$ such that $r_1 < r_2$. We let $d := O_1 O_2$, H_d and H_o the centers of the direct and negative homotheties between the circles, respectively, and ξ an exterior tangent s.l. of $C[O_1, r_1]$ and $C[O_2, r_2]$ (both circles are in the same half-plane defined by ξ) and ε an interior tangent s.l. of $C[O_1, r_1]$ and $C[O_2, r_2]$ (the circles are in different half-planes defined by ε, when such an ε exists). Prove:

(a) $O_1 H_d = \dfrac{r_1 d}{r_2 - r_1}$ and $O_1 H_o = \dfrac{r_1 d}{r_2 + r_1}$.

(b) $O_2 H_d = \dfrac{r_2 d}{r_2 - r_1}$ and $O_2 H_o = \dfrac{r_2 d}{r_2 + r_1}$.

(c)   distance$(H_o, \xi) = \dfrac{2 r_1 r_2}{r_2 + r_1}$ and   distance$(H_d, \varepsilon) = \dfrac{2 r_1 r_2}{r_2 - r_1}$.
(Both independent of d.)

39. Consider a triangle ABC and $C[O, R]$ and $C[I, r]$ its circumcircle and incircle. Let I_a, I_b, I_c be the points of tangency of $C[I, r]$ with the sides $a = BC$, $b = CA$ and $c = AB$. Let H' be the orthocenter of the triangle $I_a I_b I_c$. Prove:

(a) O, I and H' are on a s.l.

(b) $\dfrac{\overline{IO}}{\overline{IH'}} = \dfrac{-R}{r}$.

40. Consider a triangle ABC and let P and P' be the projections of its ortho-center H on the internal and external bisectors of the angle $\hat{A}$. Let F be the center of the nine-point circle of ABC and M and N the midpoints of BC and HA. Prove the points P, N, P', F and M are on a s.l.

41. **The sine form of Menelaus's and Ceva's Theorems.** Consider a triangle ABC.

(a) Let points $A' \in BC$, $B' \in CA$, and $C' \in AB$. Make a figure and use the law of sines appropriately to prove: A', B', C' are collinear iff

$$\frac{\sin(\widehat{A'AB})}{\sin(\widehat{A'AC})} \cdot \frac{\sin(\widehat{B'BC})}{\sin(\widehat{B'BA})} \cdot \frac{\sin(\widehat{C'CA})}{\sin(\widehat{C'CB})} = 1,$$

where, without loss of generality, the angles are not oriented (positive).

(b) Make a figure and use the law of sines appropriately to prove: Cevians AA', BB', and CC' are concurrent iff

$$\frac{\sin(\widehat{A'AB})}{\sin(\widehat{A'AC})} \cdot \frac{\sin(\widehat{B'BC})}{\sin(\widehat{B'BA})} \cdot \frac{\sin(\widehat{C'CA})}{\sin(\widehat{C'CB})} = 1,$$

where, without loss of generality, the angles are not oriented (positive).

42. In **Figure 4.31 of Application 1** (of homotheties), let $a = BC$, $b = CA$, $c = AB$, $w =$ the side of the square $DEFG$ and $E =$ the area of the triangle ABC. Prove that

$$w = \frac{ac\sin(\widehat{B})}{a + c\sin(\widehat{B})} = \frac{ab\sin(\widehat{C})}{a + b\sin(\widehat{C})} = \frac{2aE}{a^2 + 2E} \leq \sqrt{\frac{E}{2}} \implies w^2 \leq \frac{E}{2}.$$

So, given a, the maximum area of such a square is $\dfrac{E}{2}$ and occurs if the height of the triangle ABC on a is $h_a = a$.

5.1.2 Geometrical Constructions

1. Pick in succession four points A, B, C, and D on a s.l. l. Construct:
   (a) A point P such that $\overline{PA} \cdot \overline{BP} = \overline{CP} \cdot \overline{PD}$.
   (b) A pair of points (E, Z) that divides both segments AB and CD harmonically. Prove that the pair (E, Z) is unique.

2. Given two points A and B construct two points C and D harmonic conjugate of A and B and such that the segment CD has given length $x > 0$.

3. Construct a circle through a given point F and intersecting two s.ls l_1 and l_2 at acute angles θ and ϕ.

4. Consider a circle, a point P and a s.l l. Construct a circle satisfying the following three conditions: (1) its center on the s.l. l, (2) passing through the point P, and (3) being tangent to the given circle.

5. Construct a square $CDEF$ inscribed in:
   (a) A semicircle of diameter AB such that the side CD is on AB and E and F are on the semicircle.
   (b) A circular section of chord AB with the side CD on AB.
   (c) A circular sector of arc $\overset{\frown}{AB}$ with the vertices C and D on the arc $\overset{\frown}{AB}$.
   (d) (i) A triangle with a side of the square is a segment of a side of the triangle.
   (ii) If the triangle is scalene or right, then prove that the largest square is the one corresponding to the smallest side of the triangle.
   (iii) If the triangle is equilateral of side a, then prove that such a square is unique and its area is $\dfrac{3a^2}{(2a + \sqrt{3})^2}$.
   (In each case investigate the possible number of solutions.)

6. Construct a rectangle similar to a given rectangle and inscribed in:
   (a) A triangle ABC with one side of the rectangle a segment of a side of the triangle. (There are two or six solutions, depending on the triangle having an obtuse angle or otherwise.)
   (b) Half a circle with one side of the rectangle a segment of the diameter. (In each case investigate the possibility solutions.)

7. Construct a s.l. intersecting the sides AB and AC of a triangle ABC at the points D and E, respectively, such that $BD = DE = EC$.

8. In a given triangle ABC inscribe a rhombus $DEFG$ such that:
   (a) One of its vertex is a vertex of the ABC. When is such a rhombus a square?
   (b) No vertex is a vertex of ABC. (Some solutions may be squares.)

9. From a given point, find a s.l. that cuts on two circles chords proportional to the radii of the two circles.

10. Two circles and a point F are given. Construct two parallel tangent lines of the circles whose distances from F have ratio $\frac{p}{q}$.

11. In a given circle construct a chord that is divided into three equal segments by two given radii.

12. Construct a triangle ABC from the data $\widehat{A} > 90°$, $a - b$ and $a - c$, where $a = BC$, $b = CA$ and $c = AB$.

13. We are given a s.l. l, two points A and B not on it, and an angle $0 \leq \theta \leq \pi$. On l find a point P, such that, $|\widehat{PAB} - \widehat{PBA}| = \theta$.

14. Consider two circles $C[O, r]$ and $C[O', r']$ outside of each other and a point F outside of them and not on their center line. Construct two parallel radial vectors $\overrightarrow{OA}$ and $\overrightarrow{O'A'}$ of the same direction that subtend equal oriented angles with F. That is, $\angle(\overrightarrow{FO}, \overrightarrow{FA}) = \angle(\overrightarrow{FK'}, \overrightarrow{FA'})$.

15. Consider a number r, a fixed point F and a circle not passing through F. Construct a s.l. OAB intersecting the circle at A and B, such that $\dfrac{\overline{FA}}{\overline{FB}} = r$.

16. Construct a circle tangent to a given circle at a point T and another given circle. Investigate the number of solutions.

17. Construct a circle tangent to two s.ls and a given circle. Investigate the number of solutions.

18. Construct a circle through two given points and tangent to a given s.l. Investigate the number of solutions.

19. Construct a circle through a given point and intersecting two s.ls at given angles θ and ϕ. Investigate the number of solutions.

20. Construct a triangle ABC, in the following six cases of given data, and investigate the number of solutions:
   (a) Angle $\widehat{A}$ and the medians through B and C.
   (b) The angles $\widehat{B}$ and $\widehat{C}$ and the Euler segment OH joining the circumcenter with the orthocenter.

(c) The vertex A, the center of gravity (centroid) G and the condition that the vertices B and C belong to two given s.ls.

(d) The vertex A, the center of gravity (centroid) G and the condition that the vertices B and C belong to a given circle.

(e) The ratio $\dfrac{AB}{AC}$, the angle $\widehat{A}$ and the perimeter of the triangle.

(f) The sides AB and AC and the segment AD of the bisector of the angle $\widehat{A}$, (D on the side BC).

21. Consider two intersecting circles and a s.l. l. Construct a s.l. parallel to l and intersecting the two circles at two pairs of points such that one pair divides the other pair harmonically.

22. Pick a point S on the extension of the diameter AB of a circle C (S is outside the circle). Construct a s.l. intersecting the circle at P and Q such that the projection of the chord PQ on AB is equal to a given segment a.

23. Consider two intersecting circles and a s.l. l. Construct a s.l. parallel to l and intersecting the two circles in two pairs of point which are harmonic conjugates of each other.

24. Consider two angles and a s.l. l. Construct a s.l. parallel to l and cutting equal segment on the two angles.

25. On the sides BC, CA, and AB of a triangle ABC find points D, E, F, respectively, such that $BD = DE = EF = FA$.

5.1.3 Geometrical Loci

1. Consider a circle $C[O,r]$ and a fixed point A not on it. Let $P \in C[O,r]$ be a moving point. Find the g.ls of the points of intersection of the s.ls AP with both bisectors of the angles $\widehat{AOP}$.

   Examine the same question if $A \in C[O,r]$.

2. Consider two fixed lines l and l' intersecting at O and variable point P projected on them at the points A and B, respectively. Find the g.l. of P if AB has a given direction.

3. Consider two fixed lines l and l' and a point A. A moving line of given constant direction intersects them at B and C, respectively. Find the g.l. of the centers of gravity of the triangles ABC.

4. Consider all the circles tangent to a straight line l at a point A of it. Find the g.l. of the points of tangency of all tangents to these circles that have a given direction.

5. Consider all the circles tangent to two s.ls Ox and Oy. Find the g.ls of the points of tangency of all tangents to these circles that have a given direction.

6. Consider two concentric circles and let F be a fixed point on the smaller one. Draw a chord FA of the smaller circle and then a chord BFC of the greater circle, perpendicular to FA.

   (a) Prove that the center of gravity (centroid) of the triangles ABC is constant as A moves.

   (b) Find the g.ls of the midpoints of the sides of the triangles ABC.

7. Take two circles that intersect at the points A and B. From a variable point P of one of them draw the s.ls FA and FB, which intersect the other circle at the points C and D. Find the g.l. of the centers of gravity of the triangles ACD.

8. Consider a triangle ABC with B and C fixed and A moving on a s.l. or a circle. In either case, find the g.ls of the centers of gravity of the triangles ABC.

9. Consider a circle and a fixed point A on it. Find the g.l. of the centers of gravity of the triangle ABC inscribed in the circle whose side BC is equal to a given segment.

10. For a triangle ABC, let D be the point of intersection of the bisector of the angle $\hat{A}$ and the side BC and E the point of intersection of this bisector with the circumscribed circle to ABC. If we consider A, D and E constant, find the g.l. of the centers of gravity of the triangles ABC.

11. We take two parallel s.ls $l \parallel l'$, another s.l. k intersecting them, and a circle $\mathcal{C}$. Find:

    (a) The g.l. of the centers of the homotheties that map l onto l' and $\mathcal{C}$ onto another circle $\mathcal{C}'$ which is tangent to k.

    (b) The g.l. of the centers of the circles $\mathcal{C}'$.

12. We take a circle $\mathcal{C}$ and two non-diametrical points A and A' of it. We consider pairs of circles $\mathcal{D}$ and $\mathcal{D}'$ tangent to each other at a point P and tangent to $\mathcal{C}$ at A and A', respectively.

    (a) Examine the kinds of tangency that may exist among the circles $\mathcal{C}$, $\mathcal{D}$ and $\mathcal{D}'$, depending on the position of P.

    (b) Find the g.l. of the points P.

    (c) Find the g.l. of the points of intersection of the common exterior tangents of $\mathcal{D}$ and $\mathcal{D}'$.

13. We pick two points B and C on the side Ox of an angle $\widehat{xOy}$ and a point A moving on Oy. What is the g.l. of the center of the square inscribed in the triangle ABC and having one side on BC.

14. Consider a circle and two points A and B on a s.l. passing through the center of the circle. For any diameter CD, the s.ls AC and BD intersect at a point P. Find the g.l. of the points P.

15. We take four points A, B, C, D on a s.l. and in this order. We consider the set of points P such that $(PA, PB) = \phi$, which is (a part of) a circle of center O, and the set of points Q such that $(QA, QB) = -\phi$, which is (a part of) a circle with center O'.

    (a) Prove that the s.l. OO' passes through a fixed point as ϕ varies.

    (b) Find the g.l. of the centers of the direct homotheties of the two circles.

16. We consider all the circles passing through two fixed points A and B. Let T be the point of tangency of any of these circles with a tangent s.l. perpendicular to the s.l. AB. Find the g.l. of the centers of the nine-point circles of the triangles ATB.

17. Consider a triangle ABC. Find the g.ls of:

    (a) The points P, for which the triangles PAB and PAC are equivalent (i.e., of equal area).

    (b) The points P, for which the triangles PAB, PAC and PBC are equivalent (i.e., of equal area).

18. Consider two parallel s.ls $l_1 \parallel l_2$, a point P interior to the tape of $l_1 \parallel l_2$ and point Q exterior to the tape. Variable s.l. through Q intersects l_1 at A and l_2 at B. The s.l. BP intersects l_1 at K. Find the g.l. of the intersection of the s.ls KQ and PA.

19. Consider a square $ABCD$ and the circle $C[D, DA]$. A variable s.l. through D intersects the s.l. AC at N, the s.l. AB at P and the circle $C[D, DA]$ at M and M' such that MM' is a diameter. Find:

    (a) The g.l. of N' the harmonic conjugate of N with respect to M and M'.

    (b) The g.l. of P' the harmonic conjugate of P with respect to M and M'.

20. Consider a circle, its tangent s.l. at a point B, and a point A on this tangent s.l. From A, we draw a s.l. intersecting the circle at C and D. Let C' and D' be the projections of C and D on the s.l. AB. Find the g.l. of the intersection of $C'D$ and CD'.

21. Circle of variable center O passes through two fixed points A and B. Let P be the point of OB such that $\dfrac{\overrightarrow{PB}}{\overrightarrow{PO}} = r$, where $r < 0$ is a given negative number.

    (a) Find the g.l. of the points P.

    (b) We draw the s.l. perpendicular to BO at P and let C and D be the points of its intersection with the circle. Find the g.l. of the center of gravity (centroid) G of the triangle ACD. Moreover, prove that the s.l. OG passes through a fixed point.

    (c) Find the g.l. of the center of the nine-point circle of the triangle ACD.

5.2 Direct Similarities

5.2.1 General

1. Prove that two intersecting circles are similar under a similarity with center one of the points of intersection. Then prove that any s.l. defined by a pair of corresponding points of this similarity, passes through the other point of intersection.

2. Prove that two circles can be mapped onto each other by infinitely many direct similarities. Describe them.

3. Consider two circles $C[O, r]$ and $C[O', r']$, on which the points M and M', correspondingly, move in such a way that, $\angle(\overrightarrow{OM}, \overrightarrow{O'M'}) = \theta$ (fixed given angle). The s.l. MM' intersects the circles at second points N and N', respectively. Prove that the point N' corresponds to N under a similarity and find this similarity.

4. Consider two circles $C[O, r]$ and $C[O', r']$, and two variable radii OA and OA' of them, correspondingly, such that $\angle(\overrightarrow{OA}, \overrightarrow{O'A'}) = \theta$ fixed given angle. If the s.ls OA and $O'A'$ intersect at a point P, prove that the circle PAA' passes through a fixed point.

5. We consider a triangle ABC and a point D. We construct triangles ADE and DBF similar to ABC. Prove that $CFDE$ is a parallelogram.

6. Consider two directly similar triangles ABC and $A'B'C'$ in the plane such that $\left(\dfrac{AB}{A'B'} = \dfrac{BC}{B'C'} = \dfrac{CA}{C'A'} \right)$. Prove the equality of the angles

$$\angle(AB, A'B') = \angle(BC, B'C') = \angle(CA, C'A').$$

7. Let OAB and $OA'B'$ be two similar triangles with heights OH and OH'. Let K, L, M be the midpoints of the segments AA', BA', $B'A$. Compare the triangles KLM and OHH' and prove that the s.ls LM and HH' are perpendicular.

8. Let K, L, M be the midpoints of the sides BC, CA and AB of a triangle ABC. Construct similarly oriented right isosceles triangles BRC, CQA, and ARB.

   (a) Find the image of the vector $\overrightarrow{KQ}$ by $S_2 \circ S_1$, where S_1 and S_2 are the similarities with centers C and B, similarity ratios $\sqrt{2}$ and $\dfrac{\sqrt{2}}{2}$, respectively, and rotation angles $\dfrac{\pi}{2}$ for both.

   (b) Compare the triangles BPQ and PCR and prove that the s.ls AP, BQ and CR pass through the same point.

9. Two circles $C[O, r]$ and $C[O', r']$ intersect at the points A and B. Point M traces the circle $C[O, r]$ and another point M' traces the circle $C[O', r']$, in such a way that $\angle(\overrightarrow{OA}, \overrightarrow{OM}) = \angle(\overrightarrow{O'A}, \overrightarrow{O'M'})$.
   (a) Prove that the triangles AMM' and AOO' are similar and the s.l. MM' passes through B.
   (b) Find the set of points that divide the vector $\overrightarrow{MM'}$ in a given ratio.

10. Consider the rotation $R_{[O, \phi]}$, $\phi \neq 0$, (mod 2π) and the homothety $H_{[Q, \lambda]}$, $\lambda \in \mathbb{R} - \{0\}$. Determine the centers, the ratios, and the similarity angles of the direct similarities $R_{[O, \phi]} \circ H_{[Q, \lambda]}$ and $H_{[Q, \lambda]} \circ R_{[O, \phi]}$. (Distinguish cases for some particular values of λ and the cases $O \neq Q$ and $O = Q$.)

11. Redraw **Figure 4.60** in the following cases:
   (a) $A = A'$.
   (b) $B = B'$.
   (c) $A = B'$.
   (d) $B = A'$.
   (e) The vectors $\overrightarrow{AB}$ and $\overrightarrow{A'B'}$ intersect at interior point. In each case describe the resulting direct similarities.

12. Consider two direct similarities $S_1 = R_{[O_1, \theta_1]} \circ H_{[O_1, \lambda_1]}$ and $S_2 = R_{[O_2, \theta_2]} \circ H_{[O_2, \lambda_2]}$, in the form claimed by **Theorem 4.3.3**.
   (a) Prove that $S_2 \circ S_1$ is a direct similarity and describe how it can be reduced to a composition of a rotation and a homothety, as in **Theorem 4.3.3**.
   (b) Find the center and the angle of the rotation and the center and the ratio of the homothety.
   (c) When this composition reduces to a rotation?
   (d) When this composition reduces to a homothety?

5.2.2   Geometrical Constructions

1. We consider a direct similarity for which the center O and a pair of corresponding points (A, A') are known. Construct a point M and its corresponding M' when:
   (a) MM' is on a given s.l. l.
   (b) $\overrightarrow{MM'} = \overrightarrow{V}$ is a given vector.
   (c) The midpoint of MM' is given.
   (d) The perpendicular bisector on MM' is given.

2. Two s.ls l and l' are given and a point A. Construct a point P on l and a point P' on l', such that the triangle APP' is isosceles and $\widehat{P} = 90°$.

5.2.3   Geometrical Loci

1. A triangle ABC has fixed side BC the vertex A is moving such that the angle $\widehat{A}$ remains constant. Find the g.l. of the projection of the midpoint

of the side AB on the side AC.

2. We consider two axes Ix and Iy and two points A and A' on them. We consider variable points P and P' on the axes, such that $\overline{AP} = \lambda \overline{A'P'}$, where $0 < \lambda \neq 1$.

   (a) Prove that the circles (IPP') pass through the same point O, which is different from I in general.

   (b) Find the g.l. of the projections of O on the s.ls PP'.

   (c) Given a straight segment PP', as above, we consider the Apollonius circle of the points X, defined by $\dfrac{XP}{XP'} = \lambda$. Prove:

   (i) The g.l. of the centers of these circles, as λ varies, is a s.l.
   (ii) All these circles pass through two fixed points.

3. Consider two equal (externally) tangent circles with centers O and O'. We consider all variable radii OM and $O'M'$ such that $\measuredangle(OM, O'M') = \dfrac{\pi}{2}$. Find the g.l. of the midpoint of the segment MM'.

4. (I) The vertex A of a triangle ABC is fixed and the vertex B traces a s.l. and ABC remains directly similar to a given fixed triangle $A'B'C'$. Find the g.ls of:
   (a) The vertex C.   (b) The center of gravity (centroid) of ABC.
   (c) The orthocenter of ABC.   (d) The circumcenter of ABC.

   (II) Repeat **(I)** if B traces a given circle.

5. We take a s.l. l and a fixed point B not on it. We consider variable isosceles right triangles ABC with vertex A moving on l, $\widehat{A} = 90°$ and $\measuredangle(\overrightarrow{BA}, \overrightarrow{BC}) = 45°$. Find the g.l. of:

   (a) The vertex C, as A moves along l.

   (b) The midpoints of the sides of the triangles ABC.

   (c) The center of gravity (centroid) of the triangles ABC.

6. Consider two parallel s.ls l and l' and a fixed point F. A variable s.l. through F intersects l and l' at points A and A'. Let O be the center of the circle with diameter AA'.

   (a) Find the g.ls of the points of tangency of the tangents drawn from F to the above circle.

   (b) Prove that the above circle with each g.l. found in **(a)** defines a chord of constant length.

7. Consider a s.l. l, a point O and its projection N on l. We consider all the similarities of center O that map a point N' of l to N.
   (a) Find the similarity angle, if the similarity ratio is a given $\lambda > 0$.
   (b) For any point M of the plane, find the g.l. of its corresponding points under the above similarities.

(c) Find the g.l. of the points M that have as image the same given point M' under the above similarities.

(d) Prove that all the s.ls that, under the above similarities, have corresponding a given fixed s.l., pass through a fixed point.

(e) Let the circle C' be the image of a given circle C under one of the above similarities. Construct such a C' in each of the following four cases:

(i) The similarity angle θ is given.

(ii) C' passes through a given point.

(iii) C' has a given radius.

(iv) C' is tangent to a given s.l.

8. Let O_1, O_2, and O_3 be the midpoints of the sides BC, CA and AB of a triangle ABC. We rotate each s.l. BC, CA and AB around the midpoints O_1, O_2, and O_3 respectively, by the same angle ϕ and in the same direction. The three new lines form another triangle $A'B'C'$.

(a) Find the g.ls of the points A', B', and C'.

(b) Prove that the triangle $A'B'C'$ is directly similar to the triangle ABC.

(c) Prove that the triangles $A'B'C'$ and ABC have the same circumcenter.

(d) Find the g.ls. of the following points of the triangles $A'B'C'$:

(i) The midpoints of the sides.

(ii) The centers of gravity.

(iii) The orthocenters.

(iv) The incenters.

(v) The circumcenters.

(e) Prove that the medians and the heights of the triangles $A'B'C'$ go through fixed points.

9. Consider two fixed points O and Q and the similarity S of center O, similarity angle $\dfrac{\pi}{2}$ and similarity ratio $k > 0$. For any point P, we let $P' = S(P)$. Find the g.l. of the points P such that P, Q and P' are collinear.

5.3 Opposite Similarities

5.3.1 General

1. Prove that the center of the negative homothety of two circles is the center of an opposite similarity between them.

2. Prove that two circles can be mapped onto each other by infinitely many opposite similarities.

3. Two polygons, each having n sides, are oppositely similar if there is a correspondence of their vertices such that the corresponding angle are opposite. Find the opposite similarity that maps one onto the other.

4. Consider any two oppositely similar ABC and $A'B'C'$ in the plane. Prove that the bisectors of the three angles of the pairs of the s.ls of the corresponding sides $\angle(BC, B'C')$, $\angle(CA, C'A')$ and $\angle(AB, A'B')$ are parallel.

5. Redraw **Figures 4.77** and **4.78** in the following cases: (a) $A = A'$. (b) $B = B'$.   (c) $A = B'$.   (d) $B = A'$.   (e) The vectors $\overrightarrow{AB}$ and $\overrightarrow{A'B'}$ intersect at interior point.   Describe the resulting opposite similarities. [Cases (a) and (b) are simple. In (c) and (d) can be done directly, or you may use a parallel translation first and then a homothety and a reflection. But, as proven in **Subsection 4.2.2**, homothety composed with a parallel translation is a new homothety.]

6. Consider two vectors $\overrightarrow{AB}$ and $\overrightarrow{A'B'}$ and let $\lambda = \dfrac{A'B'}{AB} > 0$. Let M and N be the points of the segments AA' and BB' such that $\dfrac{MA'}{AM} = \lambda = \dfrac{NB'}{BN}$. Without using the Apollonius circles give a different proof that MN is the principal axis of the opposite similarity determined by the $\overrightarrow{AB}$ and $\overrightarrow{A'B'}$.

7. (a) Let A and B be two points symmetrical in the s.l. l. Let the points A' and B' and the s.l. l' be the symmetrical of A, B and l in a s.l. k. Prove that A' and B' are points symmetrical in the s.l. l'.
   (b) Replace the points A and B with two circles or two s.ls and then state and prove the analogous results.

8. (a) Consider the reflection R_l and the homothety $H_{[Q,\lambda]}$, $\lambda \in \mathbb{R} - \{0\}$. Determine the centers, the ratios, and the two axes of the opposite similarities $R_l \circ H_{[Q,\lambda]}$ and $H_{[Q,\lambda]} \circ R_l$. (Distinguish cases for some particular values of λ and the cases $Q \notin l$ and $Q \in l$.)
   (b) Replace the reflection in the above compositions with a glide and determine the centers, the ratios, and the two axes of the new opposite similarities.

9. Consider two opposite similarities $S_1 = R_{l_1} \circ H_{[O_1,\lambda_1]}$ and $S_2 = R_{l_2} \circ H_{[O_2,\lambda_2]}$, in the form claimed by **Theorem 4.3.3**.

   (a) Prove that $S_2 \circ S_1$ is a direct similarity and describe how it can be reduced to a composition of a rotation and a homothety, as claimed in **Theorem 4.3.3**.

   (b) Find the center and the angle of the rotation and the center and the ratio of the homothety.

   (c) When this composition reduces to a rotation?

   (d) When this composition reduces to a homothety?

10. Consider an opposite similarities $S_1 = R_{l_1} \circ H_{[O_1,\lambda_1]}$ and a direct similarity $S_2 = R_{[O_2,\theta]} \circ H_{[O_2,\lambda_2]}$, in the form claimed by **Theorem 4.3.3**.

(a) Prove that $S_2 \circ S_1$ and $S_1 \circ S_2$ are opposite similarities and describe how they can be reduced to compositions of reflections and homotheties, as claimed in **Theorem 4.3.3**.

(b) Find the axes of the reflections and the centers and the ratios of the homotheties.

(c) When these compositions reduce to reflections?

(d) When these compositions reduce to homotheties?

11. Let

$$\Phi \ : \ \mathcal{P} \longrightarrow \mathcal{P}$$

be a one to one and onto mapping of the plane to itself (transformation of the plane) that maps straight lines to straight lines. Prove:

(a) If Φ preserves the absolute values of oriented angles, then it is a direct or opposite similarity (depending on whether it preserves or reverses the sign of an oriented angle). If it also preserves the length of a non-zero straight segment, then Φ is an isometry.

(b) If Φ preserves the absolute values of right oriented angles, then it is a direct or opposite similarity (depending on whether it preserves or reverses the sign of a right oriented angle). If it also preserves the length of a non-zero straight segment, then Φ is an isometry. (The proof of this claim is fairly easy if we use the dot product in the plane. But synthetically?)

5.3.2 Geometrical Constructions

1. We consider an opposite similarity of center O, similarity ratio λ, principal axis ξ and second axis ξ'. Construct a pair of corresponding points (A, A') in each of the following cases:
   (a) AA' is on a given s.l.
   (b) $\overrightarrow{AA'} = \overrightarrow{V}$ is a given vector.
   (c) A and A' are on a given circle through O.
   (d) The midpoint M of the segment AA' is given.
   (f) The perpendicular bisector on AA' is given.

2. We consider two equal circles and a fixed point on each them, A and B. Consider also a variable point on each of the circles, M and N, such that the arc $\overset{\frown}{AM}$ and $\overset{\frown}{BN}$ are oppositely equal.
   (a) Find the g.l. of the midpoints of the segments MN.
   (b) Find the g.l. of the midpoints of the segments MN, if we assume that $\overset{\frown}{AM}$ and $\overset{\frown}{BN}$ are directly equal.

5.3.3   Geometrical Loci

1. Prove:
   (a) Given a ratio $\lambda > 0$ fixed, the g.l. of the centers of all opposite similarities of ratio $\lambda > 0$ that map the s.l. l to the s.l. l' is the two straight
   lines of proportional distances.
   (b) With no restriction on the similarity ratio $\lambda > 0$, two straight line l
   and l' are mapped to each other by infinitely many opposite similarities
   whose centers cover the whole plane $\mathcal{P}$.

2. Find all the centers of the direct and / or the opposite similarities of ratio
   r that map a given s.l. to a given s.l.

Chapter 6

The Transformation of Inversion

6.1 Definitions and some properties

The transformation of inversion in the Euclidean plane $\mathcal{P}$ is defined as follows.

Definition 6.1.1 *We consider any point $O \in \mathcal{P}$ and any real number $c \neq 0$. We define the inversion with respect to the data O and c to be the mapping*

$$I_{[O,c]} \ : \ \mathcal{P} \longrightarrow \mathcal{P} \cup \{\infty\} \quad \text{such that:}$$

(a) $\forall \ P \in \mathcal{P}$ and $P \neq O$, then $I_{[O,c]}(P) = P'$, if the following two conditions are met:
(1) O, P and P' are on the same line (collinear points), i.e., P' is located on the line OP, and
(2) $\overline{OP} \cdot \overline{OP'} = c \iff \overline{OP'} = \dfrac{c}{\overline{OP}} \ \left(\iff \overline{OP} = \dfrac{c}{\overline{OP'}} \right).$

(b) If $P = O$, then $\overline{OP} = 0$ and we define $I_{[O,c]}(O) = \infty$.

In relation to this definition, we also define the following:
(a) The point O is called the **center** or the **pole of the inversion**.
(b) The number c is called the **power of the inversion**.
(c) If the power $c > 0$, then the point P' is on the ray $\overrightarrow{OP}$, and the inversion is called **positive** or **direct**. See **Figure 6.1**.

Figure 6.1: **Positive (or direct) inversion**
$$\overline{OP} \cdot \overline{OP'} = c > 0$$

(d) If the power $c < 0$, then the point P' is on the ray opposite to $\overrightarrow{OP}$, hence O is located between P and P' on the line PP', and the inversion is called **negative**, or **opposite**, or **indirect**. See **Figure 6.2**.

Figure 6.2: Negative (or Opposite, or Indirect) inversion
$$\overline{OP} \cdot \overline{OP'} = c < 0$$

As with a negative homothety, so the negative inversion is equal to the commutative composition of the corresponding positive inversion and a half-turn. That is, in general,

$$I_{[O,c]} = I_{[O,-c]} \circ R_{[O,\pi]} = R_{[O,\pi]} \circ I_{[O,-c]}.$$

We now present two **geometrical constructions of the inverse point** P' of a given point $P \neq O$. Thus, we know $OP > 0$ and the power of inversion c. Then, we must essentially find OP'.

(I): The first construction is based on **Thales' Theorem** of parallel lines intersected by two other lines. We draw an acute angle xAy and we place the segments $AR = OP$, $RS = \sqrt{|c|}$ on Ox, and $AT = \sqrt{|c|}$ on Oy, as in **Figure 6.3**.

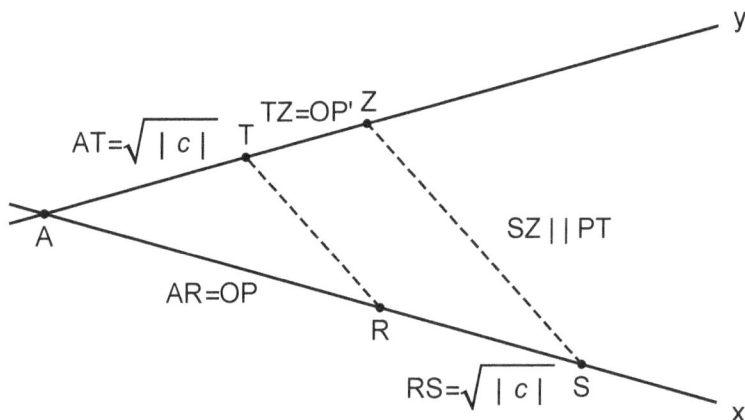

Figure 6.3: Construction of inverse points using Thales' Theorem

Then, we draw the line RT and $SZ \parallel RT$ to determine Z on Ay. By **Thales' Theorem**, we have $\dfrac{AR}{AT} = \dfrac{RS}{TZ}$. Then, $AR \cdot TZ = AT \cdot RS$, or $OP \cdot TZ = \sqrt{|c|}\sqrt{|c|} = |c|$.

Therefore, the sought length of OP' is TZ. We now place $OP' = TZ$ on the ray OP, if $c > 0$, and on the opposite ray, if $c < 0$.

(II): Take any number $c > 0$ and any points O of the plane. We will show how to find the inverses of any point $P \neq O$ under the direct and opposite inversions $I_{[O,c]}$ and $I_{[O,-c]}$, on the basis of the following well-known theorem of the geometry of a right triangle.

The square of the altitude on the hypotenuse of a right triangle is equal to the product of the two segments into which the hypotenuse is divided by the foot of the altitude. (These segments are the projections of the two perpendicular sides on the hypotenuse.)

On the basis of this Theorem, to find the inverses of a point $P \neq O$ under the inversions $I_{[O,\pm c]}$, we proceed as follows. See **Figure 6.4**.

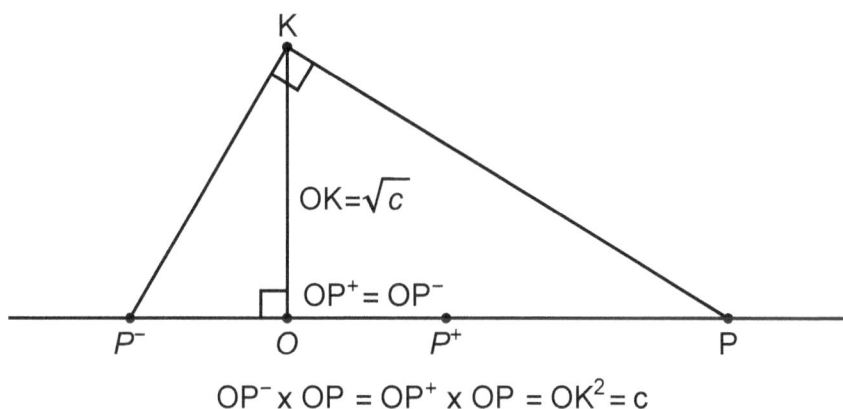

$$OP^- \times OP = OP^+ \times OP = OK^2 = c$$

Figure 6.4: Geometrical construction of inverse points for a negative or a positive inversion

(1) Draw the line OP.

(2) Draw segment $OK \perp OP$ and $OK = \sqrt{c}$.

(3) Draw KP.

(4) Draw $KP^- \perp KP$, where P^- is the common point of the line KP^- with the line OP.

(5) Find P^+ the symmetrical of P^- with respect to O.

Then:

$$\overline{OP} \cdot \overline{OP^-} = -c < 0 \qquad \text{and} \qquad \overline{OP} \cdot \overline{OP^+} = c > 0.$$

So,

$$I_{[O,-c]}(P) = P^- \qquad \text{and} \qquad I_{[O,c]}(P) = P^+.$$

[(III) Other geometrical constructions of P', the inverse point of a point $P \neq O$, are presented in **Subsection 6.4.3** and the **Appendix, Section 6.7, Part I** along with the polar lines.]

6.1.1   Two Fundamental Properties of Inversions

From the definition of inversion, we conclude two immediate and fundamental properties.

(1) By definition, if $I_{[O,c]}(P) = P'$, then $I_{[O,c]}(P') = P$
(since, $\overline{OP} \cdot \overline{OP'} = c \iff \overline{OP'} \cdot \overline{OP} = c$). Therefore,

$$I_{[O,c]} \circ I_{[O,c]} = I \quad \text{the identity.}$$

Hence, an **inversion is an involutory transformation** and as such, it is **one to one and onto**. [See **Section 2.3, Part 3, (1)** and **(2)**.]

So, any two corresponding points P and P' are called **inverse points** or **mutually inverse points,** or we say that P' is the inverse point of P and P the inverse point of P', or the pair of points (P, P') is a **pair of inverse points**. Similarly, a figure and its image under an inversion are called **(mutually) inverse figures**.

(2) If we have two different pairs of inverse points (P, P') and (Q, Q'), then, by definition,
$$\overline{OP} \cdot \overline{OP'} = c = \overline{OQ} \cdot \overline{OQ'}.$$

In basic geometry we prove that, this equality implies all four points P, P', Q and Q' are either on the same straight line (**collinear** points), or on the same circle (**concyclic** points). (So, if three of them are not of the same straight line, then the four points must be concyclic. Otherwise, that is, three points are collinear, then all four are collinear.)

The product
$$\overline{OP} \cdot \overline{OP'} = \overline{OQ} \cdot \overline{OQ'}$$

is called also **the power of the point** O **with respect to circle the** $PP'Q'Q$. Make a figure similar to **Figure 6.5**.

In general, if we consider any point O and any circle $\mathcal{C}$ in the plane, we draw any straight line through O intersecting the circle at two points P and P'. Then, the product
$$p := \overline{OP} \cdot \overline{OP'}$$

is proven (easily) to be **independent of the choice of the points** P **and** P' and is called **the power of the point** O **with respect to the circle** $\mathcal{C}$ and we symbolize it by $\mathfrak{P}_{\mathcal{C}}(O)$. (The segments $\overline{OP}$ and $\overline{OP'}$ are considered with oriented lengths on the line on which they lie.)

If O is an exterior point of the circle $\mathcal{C}$, we can draw two tangent lines from O to the circle. Let A and B be the two common points. See **Figure 6.5**. Easily, we have that **the power of the point** O **with respect to the circle** $\mathcal{C}$, as defined above, is also equal to

$$p =: \mathfrak{P}_{\mathcal{C}}(O) = OA^2 = OB^2.$$

(In **Figures 6.8** and **6.9** recognise the power of O with respect to some circle.)

All **expressions of the power of a point with respect to a circle** are very useful with inversion, and so, we mention the following:

Consider any circle $C[C, R]$, a point O outside of it and a point O' inside of it, as in **Figure 6.5**.

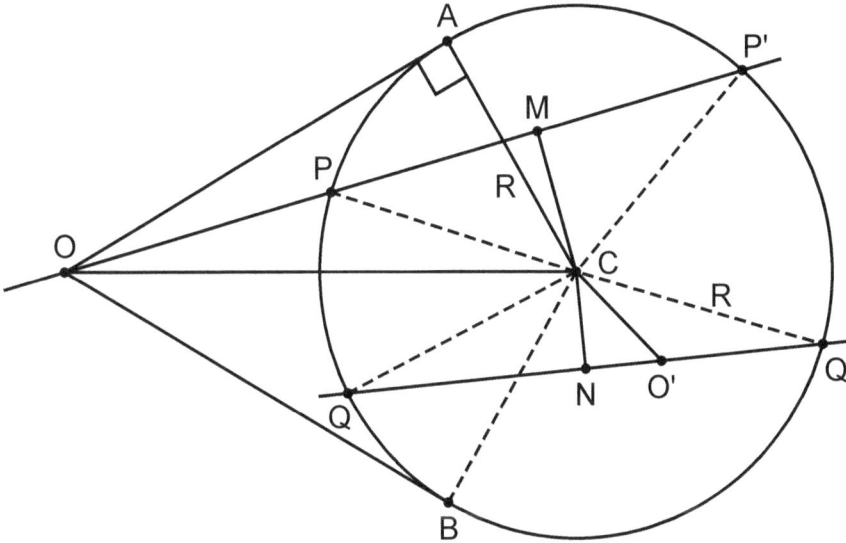

Figure 6.5: For the power of a point with respect to a circle

(a) If a point is on the circle the power is zero, since one of the segments that we multiply with, is zero. For example, the powers of A and B in the figure are zero, as the power of any point on the circle $C[C, R]$.

(b) For the point O outside the circle, we draw any secant line OPP' to the circle and the two tangent segments OA and OB. Then, the power of O with respect to $C[C, R]$ is always positive and given by the expressions

$$\overline{OP} \cdot \overline{OP'} = OA^2 = OB^2 = OC^2 - R^2 > 0.$$

(c) For the point O' inside the circle, we draw again any secant line $O'QQ'$ to the circle. Then, the power of O' with respect to $C[C, R]$ is always negative and given by the expressions

$$\overline{O'Q} \cdot \overline{O'Q'} = O'C^2 - R^2 < 0.$$

(Prove these relations or find the proofs in a geometry book. The proofs are elementary. M and N are the midpoints of the chords PP' and QQ', respectively. Now you can finish the proofs!)

6.2   Invariant or Fixed Elements

6.2.1   Fixed Points

A point A is fixed under an inversion $I_{[O,c]}$, if $A' = I_{[O,c]}(A) = A$. If a fixed point exists. then $\overline{OA} \cdot \overline{OA'} = OA^2 = c$. So, if there is a fixed point, then $c > 0$.

Consequently:

(1) A negative inversion, $(c < 0)$, has no fixed points.

(2) A positive inversion, $(c > 0)$, has fixed points all the points of the circle $C[O, \sqrt{c}]$, *since in such a case, for any point A on this circle, we have:* $\overline{OA} = \sqrt{c}$ *and* $c = \overline{OA} \cdot \overline{OA'} = \sqrt{c} \cdot \overline{OA'}$. *Then,* $\overline{OA'} = \sqrt{c} = \overline{OA}$. *Since, A' is on the ray* $\overline{OA}$, *we have $A = A'$.*

Therefore, *if $c > 0$, the circle $C[O, \sqrt{c}]$ is a point-wise fixed set, and it is called the* **circle of the positive inversion**. See **Figure 6.6**.

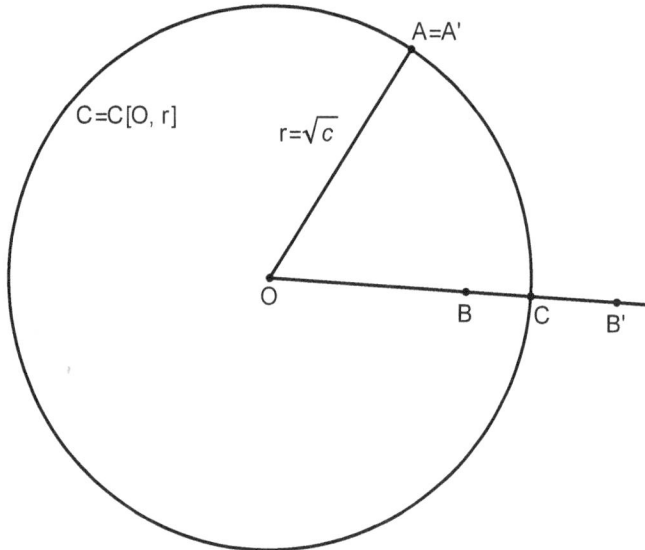

Figure 6.6: Circle of positive inversion

In case of a negative inversion, $c < 0$, the circle $C\left[O, \sqrt{|c|}\right]$ is a set-wise fixed set but not point-wise, because each point is mapped to its diametrical point. See **Figure 6.7**.

So, in general, we call the circle $C\left[O, \sqrt{|c|}\right]$ the **circle of inversion** and we will denote it with $\mathcal{C} = C\left[O, \sqrt{|c|}\right]$. Its radius is $r = \sqrt{|c|}$.

Now, consider a point $B \in \mathcal{P}$. We have:

(1) If $B \in \mathcal{C}$, then we have just seen that its image under the inversion $B' \in \mathcal{C}$. ($B' = B$, if $c > 0$ and $B' = $ diametrical point of B, if $c < 0$.)

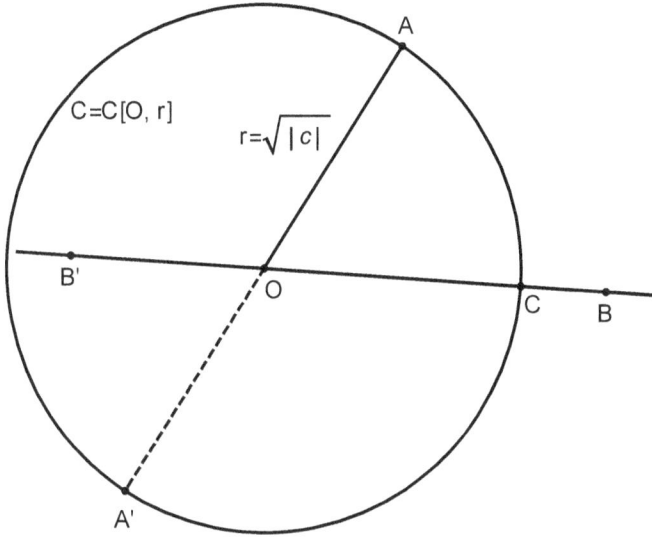

Figure 6.7: Circle of negative inversion

(2) If B is inside C, then, $\overline{OB} < r = \sqrt{|c|}$ and so $\overline{OB'} > r = \sqrt{|c|}$. That is, B' is outside C. See **Figure 6.6**.

(3) If B is outside C, then, $\overline{OB} > r = \sqrt{|c|}$ and so $\overline{OB'} < r = \sqrt{|c|}$. That is, B' is inside C. See **Figure 6.7**.

So, *if in a pair of inverse points (B, B') none of them is on C, then one of the points is inside C and the other one is outside C.*

In this context, a **useful metric relation** is the following. We let C be a point on the circle of inversion and on a segment BB' of two inverse points B and B'. With $r = \sqrt{|c|}$, we get

$$\overline{OB} \cdot \overline{OB'} = c, \text{ or, } (r + \overline{CB})(r + \overline{CB'}) = c, \text{ or, } r(\overline{CB} + \overline{CB'}) = c - r^2 - \overline{CB} \cdot \overline{CB'}.$$

Therefore, with $r = \sqrt{|c|}$, we have

$$\overline{CB} + \overline{CB'} = \frac{c - r^2 - \overline{CB} \cdot \overline{CB'}}{r} = \begin{cases} -\dfrac{\overline{CB} \cdot \overline{CB'}}{r} > 0, & \text{if } c > 0, \\[2mm] -2r - \dfrac{\overline{CB} \cdot \overline{CB'}}{r}, & \text{if } c < 0. \end{cases}$$

Thus, if $c = r^2 > 0$, then $\overline{CB} + \overline{CB'} > 0$. So, for example, as the points are arranged in **Figure 6.6**, the length of the segment BC inside the circle of inversion is less than the length of the segment CB' outside the circle.

This metric relation proves that the limiting inversion, as $0 < c \longrightarrow \infty$, is a reflection in a line. See **Figure 6.8**.

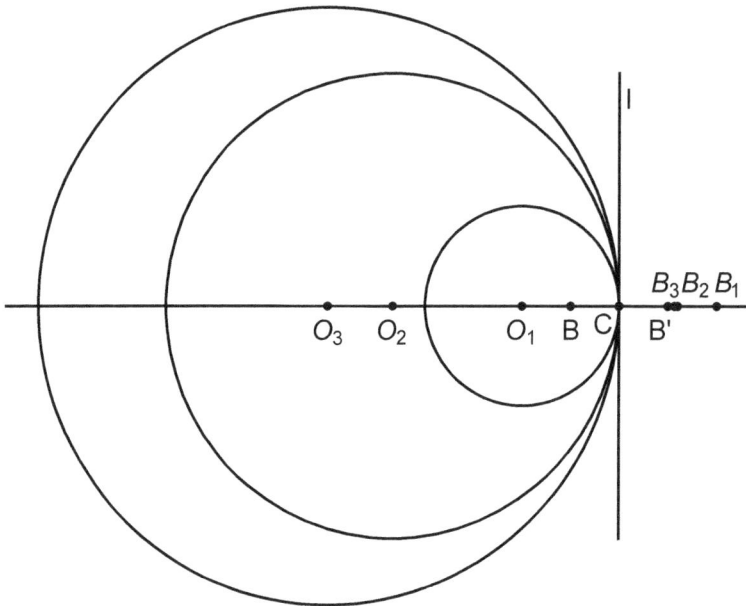

Figure 6.8: Limiting inversion is a reflection in a straight line

We pick two fixed points B and C and consider the line BC. We pick points $O_n \longrightarrow \infty$, as $n \longrightarrow \infty$, on the line BC, moving opposite to the direction $\overline{BC}$, and consider the inversions on the circles with centers O_n and radii $O_nC := \sqrt{c_n}$, for $n = 1$, 2, 3, ... Of course, $O_nC \longrightarrow \infty$, as $n \longrightarrow \infty$.

The respected inverses of B are B_n, for $n = 1$, 2, 3, ... Then, by the above metric relation, we get

$$\overline{CB} + \overline{CB_n} = -\frac{\overline{CB} \cdot \overline{CB_n}}{O_nC} \longrightarrow \frac{(bounded)}{\infty} = 0, \quad \text{as } n \longrightarrow \infty.$$

So,

$$\lim_{n \to \infty} \overline{CB_n} = -\overline{CB}.$$

Thus, the limit point of the points $\{B_n\}$, as $n \longrightarrow \infty$, is the point B' symmetrical to B in the line l, where $l \perp BC$ at C. Also, the open halves of circles $C[O_n, O_nC]$ which are tangent to l at C approach the line l, as $n \longrightarrow \infty$, point-wise. So we have the **result: "The limit of the above inversions is the reflection in the line $l \perp BC$ at C."** Therefore we have:

Two figures are mutually inverse with respect to a straight line (considered as an infinity circle) iff they are symmetrical in that line and so they are iso-metrical. The center of such an inversion (symmetry or reflection in a line) is the point at infinity of the direction perpendicular to the given straight line. This limiting inversion belongs to the set of the positive inversions.

[See also **Subsection 6.3, (4)** and **Corollary 6.5.3.**]

6.2.2 Invariant Straight Lines

For any pair of inverse points (A, A'), we have that O, A and A' are collinear. By the definition of inversion, a straight line AA' through O, the center of inversion, is set-wise invariant. Conversely, if a straight line l is invariant, then for any point $B \in l - \mathcal{C}$, its image $B' \in l - \mathcal{C}$ and $B' \neq B$. But, O is on the line $BB' = l$. See, **Figure 6.9**.

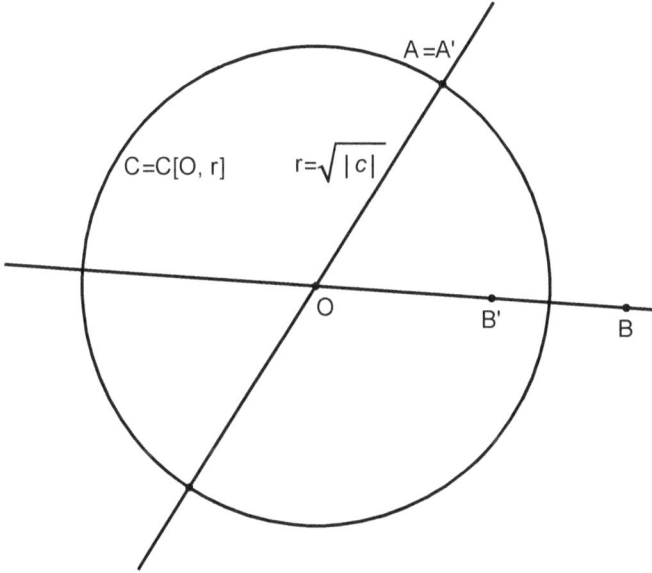

Figure 6.9: Invariant lines of an inversion

So: **A straight line is invariant under an inversion if and only if it passes through the center of inversion.**

Remark: When the inversion is positive, then all rays (half-lines) emanating at O are also invariant.

6.2.3 Invariant Circles

Apart from the circle of inversion that we saw above, we examine if there are other circles that are invariant under an inversion.

Suppose that a circle $\mathcal{D} := C[K, \rho]$ (center K, radius ρ) passes through a pair of distinct inverse points (A, A') $(A \neq A')$, under an inversion $I_{[O,c]}$. Then, $\overline{OA} \cdot \overline{OA'} = c$. Picking any point $B \in \mathcal{D}$, we draw the line OB which will intersect $\mathcal{D}$ at another or the same point B'. See **Figures 6.10** and **6.11**. Since A, A', B, and B' are concyclic, we have

$$\overline{OB} \cdot \overline{OB'} = \overline{OA} \cdot \overline{OA'} = c.$$

Therefore, B' is the image of B under the inversion. So, such a circle $\mathcal{D}$ is set-wise invariant under the inversion.

[We have already seen that the product $\overline{OB} \cdot \overline{OB'} = \overline{OA} \cdot \overline{OA'}$ is the **power of the point O with respect to the circle $\mathcal{D}$** (and is independent of the points A and A', etc.).]

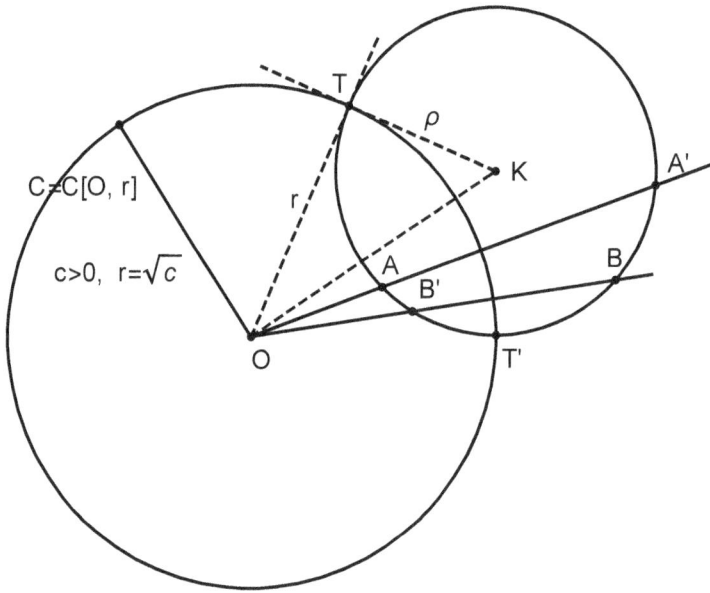

Figure 6.10: Invariant circle of a positive inversion

Conversely, if a circle $\mathcal{D} := C[K, \rho]$, other than the circle of inversion, is invariant under the inversion $I_{[O,c]}$, then for any point $A \in \mathcal{D}$ its images $A' \in \mathcal{D}$. Therefore, the circle $\mathcal{D}$ passes through a pair of inverse points. So, we have:

Theorem 6.2.1 *A circle is set-wise invariant under an inversion if and only if it passes through a pair of distinct inverse points, or,*
if and only if, the power of the center of the inversion with respect to this circle is equal to the power of the inversion.

Since one of the points of a pair of different inverse points is inside the circle of inversion and the other outside, we conclude that any invariant circle $\mathcal{D}$ has a part inside and a part outside the circle of inversion $\mathcal{C}$. So, an invariant circle intersects the circle of inversion at two points T and T'.

Now, suppose the inversion is positive, $(c > 0)$, as in **Figure 6.10**. For any point $A \in \mathcal{D}$, we have $\overline{OA} \cdot \overline{OA'} = c = r^2$ and, as we have explained before, $\overline{OA} \cdot \overline{OA'} = OK^2 - \rho^2$. Therefore,

$$OK^2 - \rho^2 = r^2 \quad \text{or} \quad OK^2 = r^2 + \rho^2 = c + \rho^2.$$

We conclude that:

(1) $OK > r$ and $OK > \rho$. So, the center of each circle, $\mathcal{C}$ or $\mathcal{D}$, is outside the other circle.

(2) The triangles OKT and OKT' are right at T and T', and so the two relevant radii are perpendicular to each other.

Thus, $OT = r = OT'$ is tangent to $\mathcal{D}$ and $KT = \rho = KT'$ is tangent to $\mathcal{C}$, and their angle at the points of intersection is right. In such a case, we say that, the two **circles intersect orthogonally** or they are **orthogonal circles**.

We can trace the steps backward to show that, if a circle intersects the circle of inversion orthogonally then the circle is set-wise invariant under the inversion.

So, we have:

Theorem 6.2.2 *Given a positive inversion, a circle, other than the circle of inversion, is set-wise invariant under the inversion if and only if this circle and the circle of inversion are orthogonal to each other.*
$(\Longleftrightarrow OK^2 = r^2 + \rho^2 = c + \rho^2.)$

Corollary 6.2.1 *If two circles intersect the circle of a positive inversion orthogonally and they intersect each other at two points, then these points are inverse of each other.*

If these circles are tangent to each other, then their point of tangency is on the circle of the positive inversion.

Proof See **Figure 6.11**.

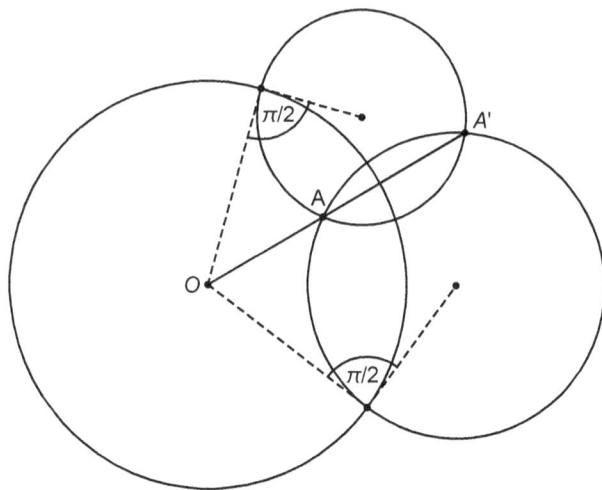

Figure 6.11: Inverse points of positive inversion

By hypothesis and the **previous Theorem**, we have that the circles are set-wise invariant under the positive inversion. Let A be one of the two points of intersection of these circles. Since the circles are invariant, the inverse point

A' of A must belong to both of them. Hence, A' is the other point of intersection of the two circles.

For the same reason, if the circles are tangent at A, then its inverse A' belong to both of them and so it is the same as A. That is, this point of tangency is fixed and so is also on the circle of the positive inversion.

■

Remarks: (1) For a reflection R_ξ in a straight line ξ, considered as limiting inversion, any circle that passes through two inverse points, i.e., points symmetrical in ξ, has ξ as diameter. Therefore, such a circle is orthogonal to and symmetrical in ξ. See **Figure 6.22** and draw a circle through A and A'.

(2) For the geometrical construction of a circle passing through a given point (or two given points, for the same matter) and being orthogonal to a given circle, as was needed in the **previous Corollary**, where the point A was given, study **Example 6.2.1** or find other solutions to this geometrical construction. See also the exercises.

Now, suppose the inversion is negative, $(c < 0)$, as in **Figure 6.12**. For any point $A \in \mathcal{D}$, we have $\overline{OA} \cdot \overline{OA'} = c = -r^2$.

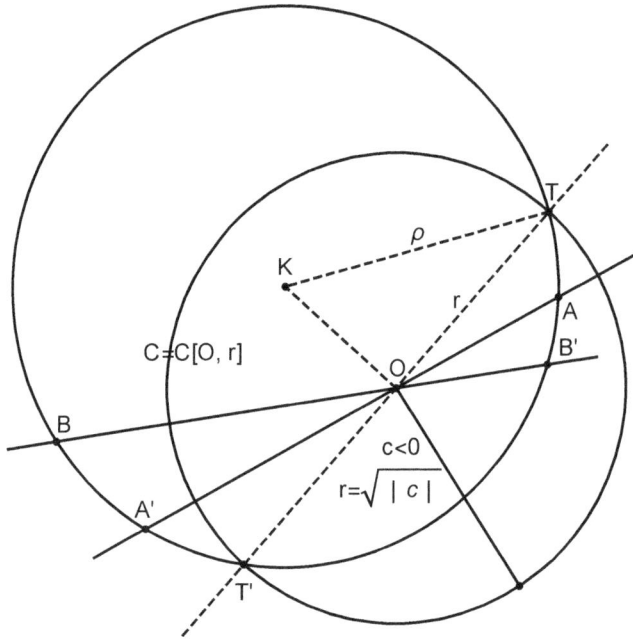

Figure 6.12: Invariant circle of a negative inversion

Again $\overline{OA} \cdot \overline{OA'} = OK^2 - \rho^2$ and so $OK^2 - \rho^2 = -r^2$.
Therefore, $OK^2 = -r^2 + \rho^2 = c + \rho^2 \iff OK^2 + r^2 = \rho^2$.
We conclude that: $OK < \rho$, $r < \rho$, the center O of $\mathcal{C}$ is inside $\mathcal{D}$, and the

triangles OKT and OKT' are right at O. So, TOT' is a diameter of C and therefore $OT = OT' = r$.

Here, we must give the following definition:

Definition 6.2.1 *If a circle D intersects a circle C at diametrical points, the we call the circle D* **pseudo-orthogonal** *to the circle C, or we say that D intersects C* **pseudo-orthogonally**.

Remark: Pseudo-orthogonality of circles is not a symmetric property, like the orthogonality of circles.

The above steps can be traced backward to prove:

Theorem 6.2.3 *Given a negative inversion, a circle, other than the circle of inversion, is set-wise invariant under the inversion if and only if this circle is pseudo-orthogonal to the circle of inversion. ($\Longleftrightarrow OK^2 = -r^2 + \rho^2 = c + \rho^2$.)*

Corollary 6.2.2 *Two circles that intersect the circle of a negative inversion pseudo-orthogonally must intersect each other at two points A and A' which are inverse of each other under the given negative inversion.*

Proof See **Figure 6.13**.

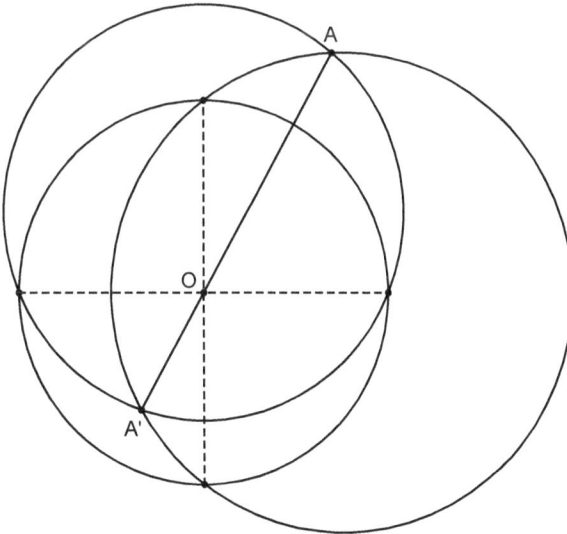

Figure 6.13: Inverse points of negative inversion

These circles intersect at two points necessarily, because any two diameters of the circle of inversion cross each other and so the center O is an interior point of the intersection of the two circles. Now, the argument of the proof is analogous to the argument of the proof of the **previous Corollary**.

■

At this point, it is appropriate to state and prove another necessary and sufficient condition for the orthogonality of two circles. But, first we need the following **property of proportions** among four numbers a, b, c, and d:

$$\frac{a}{b} = \frac{c}{d} \qquad \Longleftrightarrow \qquad \frac{a+b}{a-b} = \frac{c+d}{c-d}.$$

Proof: If $\dfrac{a}{b} = \dfrac{c}{d}$, by adding 1 and -1 to both sides of this equality, we obtain $\dfrac{a+b}{b} = \dfrac{c+d}{d}$ and $\dfrac{a-b}{b} = \dfrac{c-d}{d}$, respectively. Then, we divide these equalities sidewise and we get $\dfrac{a+b}{a-b} = \dfrac{c+d}{c-d}$ (we assume $a \neq b \iff c \neq b$).

To prove the converse, we cross-multiply the last relation and simplify to get $ad = bc$ which is equivalent to $\dfrac{a}{b} = \dfrac{c}{d}$.

Theorem 6.2.4 *Two circles are orthogonal to each other if and only if a diameter of one of the circles is divided harmonically by the other circle.*

Proof Let two circles with centers K_1 and K_2 intersect each other at the points T and T' and the diameter AK_1B of the first circle is intersected at the points C and D by the second. See **Figure 6.14**.

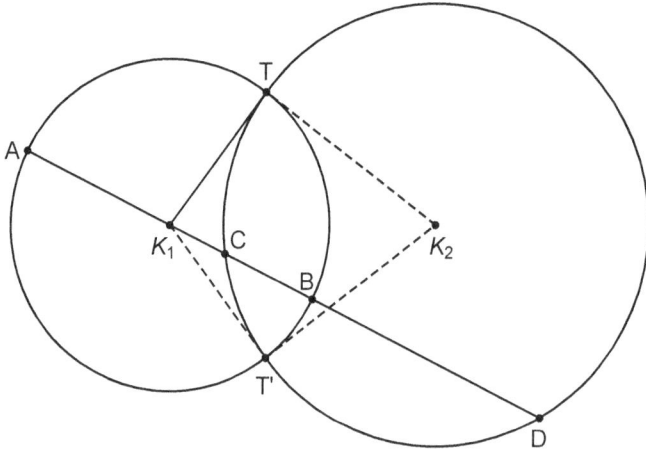

Figure 6.14: Condition for orthogonality of circles is the
harmonic quadruple condition $\dfrac{\overline{AC}}{\overline{BC}} = -\dfrac{\overline{AD}}{\overline{BD}}$

The circles are orthogonal if and only if the radius K_1T is tangent to the circle of center K_2. This happens if and only if

$$\overline{K_1C} \cdot \overline{K_1D} = K_1T^2 \qquad \Longleftrightarrow \qquad \overline{K_1C} \cdot \overline{K_1D} = K_1A^2 = K_1B^2.$$

Therefore, by **Subsection 4.5.5, (b), Remark**, this relation is equivalent to the fact that the segment AB of the diameter of the first circle is divided harmonically by the points C and D of the second circle. (So, the points A, C, B, and D in this order located on a straight line form a harmonic quadruple.) We can also prove this directly in the following way. We have

$$\overline{K_1C} \cdot \overline{K_1D} = K_1B^2 \quad \Longleftrightarrow \quad \frac{\overline{K_1D}}{\overline{K_1B}} = \frac{\overline{K_1B}}{\overline{K_1C}}.$$

By the property of proportions proven above, we obtain

$$\frac{\overline{K_1D} + \overline{K_1B}}{\overline{K_1D} - \overline{K_1B}} = \frac{\overline{K_1B} + \overline{K_1C}}{\overline{K_1B} - \overline{K_1C}}.$$

We also have $\overline{AK_1} = \overline{K_1B}$, and so we obtain

$$\frac{\overline{AD}}{\overline{BD}} = \frac{\overline{AC}}{\overline{CB}} \quad \Longleftrightarrow \quad -\frac{\overline{CA}}{\overline{CB}} = \frac{\overline{DA}}{\overline{DB}},$$

that is, the points A, C, B, and D (located on a straight line in this order) form a harmonic quadruple.

■

The **previous Theorem** and **Theorem 4.4.5** imply the following corollary:

Corollary 6.2.3 *Given two points A and B in the plane, any circle through A and B and any Apollonius circle (of any ratio) on the segment AB are orthogonal. So, any such Apollonius circle inverts point A into point B. (That is: If the center of inversion O is the center of the Apollonius circle and the power of inversion is $c = r^2$, where r is the radius of the Apollonius circle, then $\overline{OA} \cdot \overline{OB} = r^2$.)*

The above **Theorem** also prompts the following example of an important geometrical locus and geometrical construction:

Example 6.2.1 We consider a circle with center O and a point $P \neq O$ of its plane and not on the circle. Find the geometrical locus of the point P' that belongs to a circle through P and is orthogonal to the given circle, and P' is diametrical to P.

Solution See **Figure 6.15**.

(a) Consider a point P' of the sought geometrical locus, that is diametrical of P on a circle of center K and orthogonal to the given circle. Then, Q the projection of P' on the line OP belongs to the circle with center K because PP' is its diameter. Let, the diameter line OP of the given circle define the diameter segment AB on it.

Since the two circles are supposed to be orthogonal, then, by the **previous Theorem**, the point Q is the harmonic conjugate of P with respect to A and B. (See **Definition 4.1.4**.) So, Q is well determined by O and P, and the point P' is on the line l perpendicular to OP at the point Q.

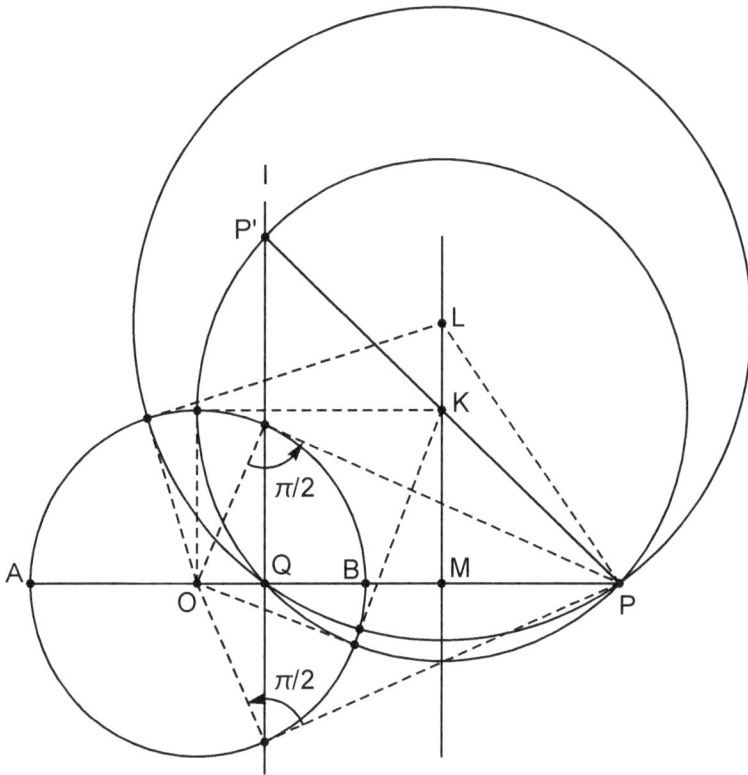

Figure 6.15: Example of a locus determined by orthogonal circles and construction of orthogonal circles

Conversely, if P' is any point of l, then, since $l \perp OP$, the circle with diameter PP' passes through Q. Also, by the choice of Q, the quadruple $\{A, B; Q, P\} = -1$ is harmonic. Therefore, by the **previous Theorem**, the circles are orthogonal.

Finally, the sought geometrical locus is the straight line l perpendicular to OP at the point Q, the harmonic conjugate of P with respect to A and B.

According to the **Definition 6.4.1** in **Subsection 6.4.1**, the line $l := QP'$ is the **polar line of the point** P and P is **the pole of** $l := QP'$ with respect to the given circle.

(b) From this example we can gather how to construct the circles orthogonal to the given circle with center O and passing through a given point $P \neq O$ that is not on the circle, in the following steps. See **Figure 6.15**.

(1) We find the points A and B, the common points of the circle and the straight line OP.

(2) Since Q is supposed to be the harmonic conjugate of P with respect to

A and B, we find Q by the proportion $-\dfrac{QA}{QB} = \dfrac{PA}{PB} := \lambda$, where $\lambda := \dfrac{PA}{PB} \neq 0$

is known, as we do, e.g., in **Lemma 4.1.1** or see the construction of a harmonic quadruple in **Subsection 4.5.5**.

(3) We draw MK the perpendicular bisector of QP.

(4) Any circle with center a point L on MK and radius $LP = LQ$, passes through P and Q and is orthogonal to the given circle with center O, by the **previous Theorem**.

The geometrical locus of the centers of all these circles is the midpoint perpendicular of the segment PQ.

(c) **Notice**: If the point P is on the given circle, then the geometrical locus l is the line perpendicular to OP at P. This line contains the centers of the circles trough P and orthogonal to the given circle. All the points P' that are diametrical to the points P are also on this line. See **Figure 6.16**.

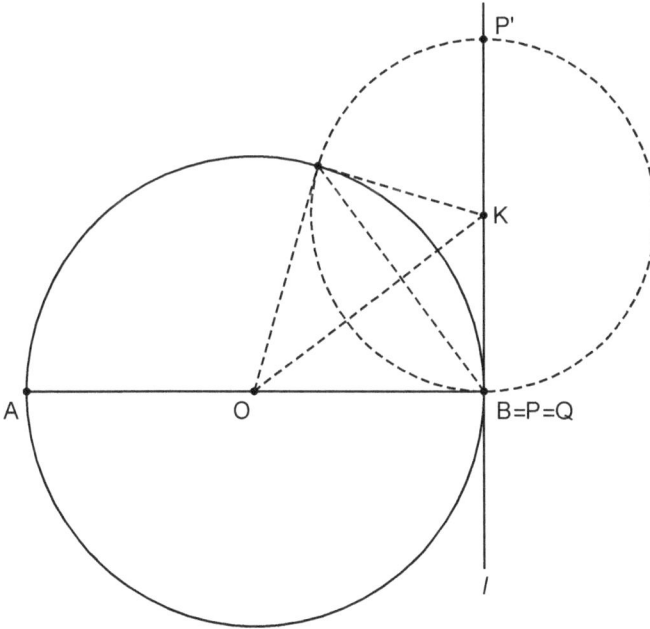

Figure 6.16: Special case of orthogonal circles

Remark: As we see in (b) and (c), the given point P can be any point of the plane [outside, inside (as the roles of P and Q can be interchanged), or on the given circle], except for the center O of the given circle. However, if $P = O$ the sought orthogonal circles become the straight lines through O, considered as circles with centers the infinity points of the directions defined by the straight lines which are perpendicular to the respective diameters, and with infinite radius. ($Q \longrightarrow \infty$, as $P \longrightarrow O$.)

(d) To construct the circle passing through two given points P and S and orthogonal to the given circle with center O, it must be either $S = Q$ [as in (b) and (c)] or the points P, S and O must not be on the same straight line. Otherwise, such a circle does not exist. So, we assume that the points P, S and O are not on the same straight line and work in the following steps. See **Figure 6.17**.

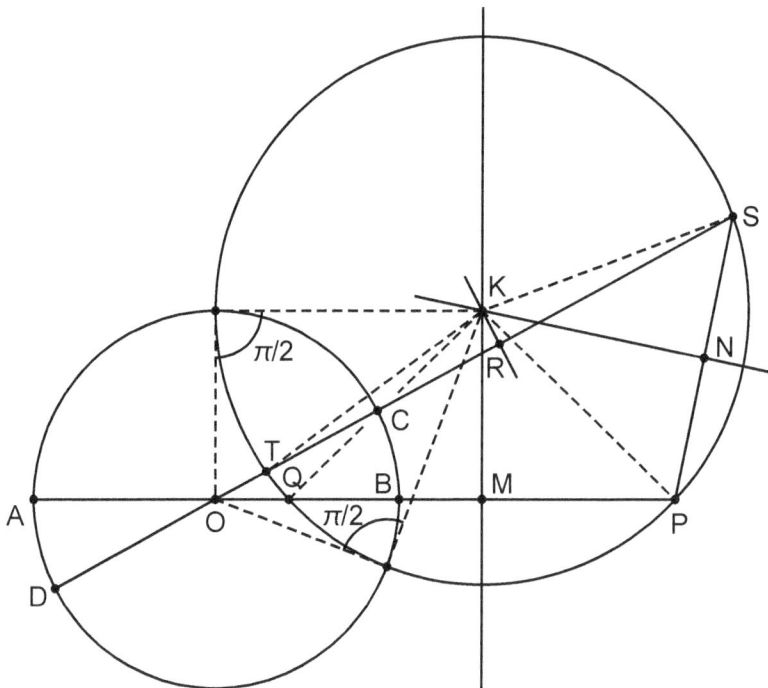

Figure 6.17: Construction of orthogonal circles

(1) We find Q the harmonic conjugate of P with respect to A and B, as in (b).

(2) We draw the straight line OS and we let C and D its points of intersection with the given circle.

(3) We find T the harmonic conjugate of S with respect to C and D.

(4) We draw the perpendicular bisectors of PQ and ST and let K be their point of intersection.

(5) The circle with center K and radius $KP = KQ = KS = KT$ is **the unique circle** passing through P and S and orthogonal to the given circle with center O.

Notice: The center K can also be determined as the intersection of the perpendicular bisectors of PQ and PS. This replaces steps (3) and (4).

▲

6.3 Determination of Inversions

An inversion is, by definition, readily known, if we know its center and its power. It can also be determined by the data of the following cases.

Case (1) *The center O and a pair of corresponding points (A, A') is given $(A \neq O$ and $A' \neq O)$ in $\mathcal{P}$.* (The center O corresponds to the ∞ point.)

In this case, it is necessary that O, A and A' are on the same straight line. Then, the power of the inversion must be $c = \overline{OA} \cdot \overline{OA'}$. So, we obtain the inversion $I_{[O,c]}$ with center$O$ and power c. See **Figure 6.18**.

Figure 6.18: Inversion determined by O and a pair of points (A, A'). Center $= O$, power $c = \overline{OA} \cdot \overline{OA'}$

Case (2) *Two pairs of corresponding points (A, A') and (B, B') are given on a straight line and $(A \neq B)$.*

(i) If $A = A'$ and $B = B'$, then the points A and B are fixed by the inversion. Hence, they belong to the circle of inversion and the inversion is positive. The center O of the inversion is any point of the midpoint perpendicular line of the segment AB and the power is

$$c = OA^2 = OB^2 > 0.$$

So, there are infinitely many inversions in this case.

(ii) Now assume that at least one of the points A or B is not fixed. So, either $A \neq A'$ or $B \neq B'$. Since $A \neq B$, we also have $A' \neq B'$. Then, the center of inversion O is located on the given straight line that contains the four given points and is determined by the following relations:

$$\overline{OA} \cdot \overline{OA'} = \overline{OB} \cdot \overline{OB'} \qquad \Longleftrightarrow \qquad (1 \neq) \; \frac{\overline{OA}}{\overline{OB}} = \frac{\overline{OB'}}{\overline{OA'}} = \frac{\overline{OB'} - \overline{OA}}{\overline{OA'} - \overline{OB}} = \frac{\overline{AB'}}{\overline{BA'}},$$

which is a know ratio. That is, the center O is the center of the homothety determined by the vectors $\overrightarrow{AB'}$ and $\overrightarrow{BA'}$ on the given line, as long as $\overline{AB'} \neq \overline{BA'}$. See **Figure 6.19** and **Lemma 4.1.1**.

After we find O, the power of the inversion is

$$c = \overline{OA} \cdot \overline{OA'} = \overline{OB} \cdot \overline{OB'}.$$

[If $\overline{AB'} = \overline{BA'}$, the inversion is special, examined in **Case (4)**, below.]

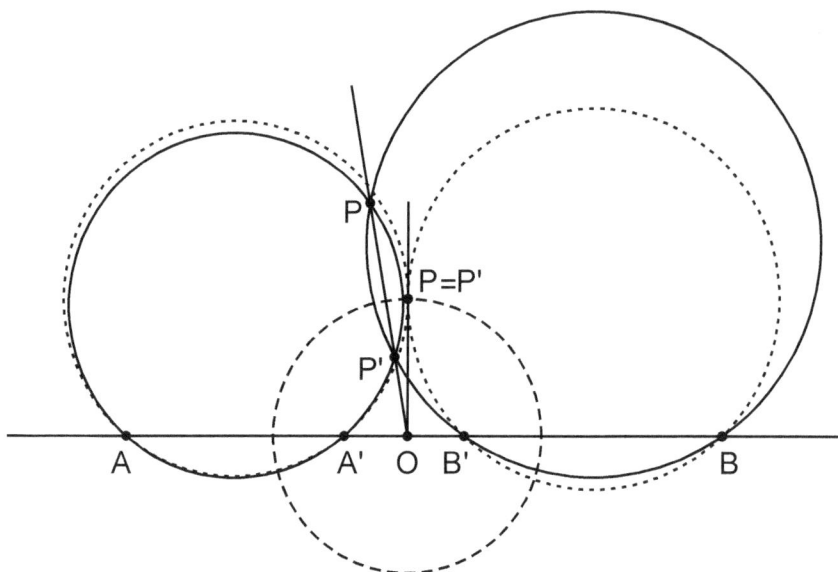

Figure 6.19: Inversion determined by two pairs of points
(A, A') and (B, B') on a straight line.
(In this Figure $0 < c$.)

Notice that if $A \neq A'$ and $B \neq B'$, the inverse of any point P not on the line AB, is the second point of intersection of the circles $(AA'P)$ and $(BB'P)$, unless the two circles are tangent to each other, in which case $P = P'$ is on the circle of inversion. From this observation, we can easily construct the center of the inversion O as the intersection of the line AB with either the line PP', or, if $P = P'$, with the tangent line to the circles $(AA'P)$ and $(BB'P)$, at $P = P'$.

But, if A is a fixed point ($A = A'$), then the inversion is positive and the circle $(AA'P)$ is not defined. We first find O and then the line OP determines the inverse point of P as either the same point ($P = P'$) or the second point of its intersection P' with the circle $(BB'P)$. In this case, the circle of inversion has radius $OA = OA'$.

Case (3) *Two pairs of corresponding points (A, A') and (B, B') are given on a circle.*

(i) We first assume that $A \neq A'$ and $B \neq B'$, and the lines AA' and BB' intersect, i.e., $AA' \nparallel BB'$. [Otherwise, see **Case (4)**, **(ii)**, below.] The point of their intersection O is the center of the sought inversion and the power is $c = \overline{OA} \cdot \overline{OA'} = \overline{OB} \cdot \overline{OB'}$. See, **Figure 6.20**.

We can construct the inverse point P' of a point P that does not belong to the straight lines AA' and BB' geometrically by drawing the circles $(AA'P)$

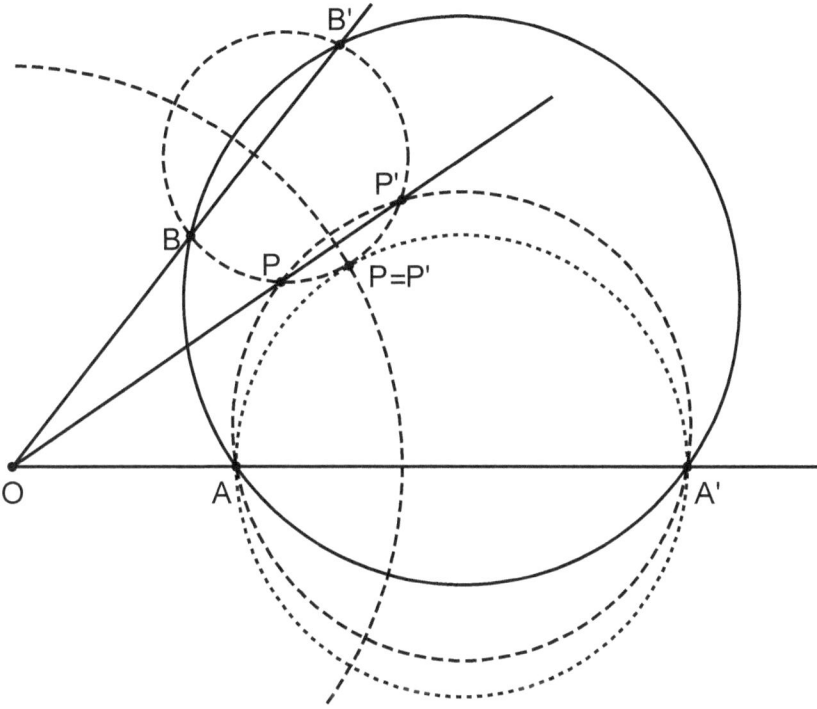

Figure 6.20: Inversion determined by two pairs of points
(A, A') and (B, B') on a circle. (Here $c > 0$.)

and $(BB'P)$. Then, either $P'(\neq P)$ is the second point of intersection of these two circles, or $P' = P$ when the two circles $AA'P$ and $BB'P$ are tangent at P. In the latter case, $P = P'$ is on the circle of inversion.

(ii) Assume $A = A'$ is a fixed point and $B \neq B'$. Then, by **Theorem 6.2.4**, the circle (ABB') intersects the circle of inversion orthogonally at $A = A'$. So, the center of inversion O is the intersection of the straight lines BB' and the tangent to the circle (ABB') at A, if these lines intersect. [Otherwise, see **Case (4), (iii)**, below.] Again, the power is $c = \overline{OA}^2 = \overline{OB} \cdot \overline{OB'}$. (Make a figure.)

Case (4) *The inversion reduces to a reflection in a straight line.*
In such a case the straight line may be viewed as a circle of center at an infinity point and infinite radius. The power of such inversion is ∞.

(i) In **Case (2)** this happens when $\overrightarrow{AB'} = \overrightarrow{BA'}$ and the vectors are on the same line. Then, the axis of reflection is the common perpendicular bisector ξ of the segments $AA' \neq 0$ and $BB' \neq 0$. See **Figure 6.21**.

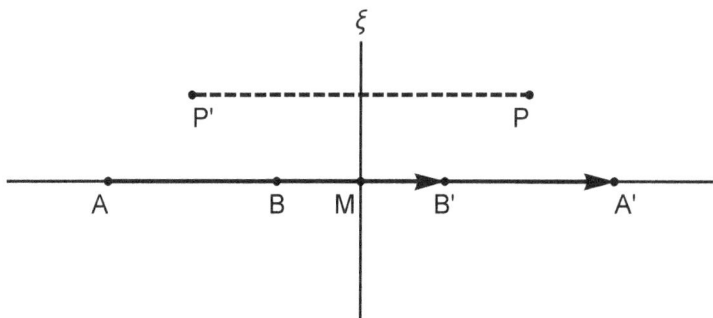

Figure 6.21: Inversion is reduced to the reflection R_ξ in the line ξ when $\overrightarrow{AB'} = \overrightarrow{BA'}$

(ii) In **Case (3), (i)**, this happens when $AA' \parallel BB'$. Since $AA'B'B$ must be inscribable, then it must be an isosceles trapezium. The axis of symmetry $\xi = MN$ is the median of the two parallel bases $AA' \parallel BB'$, which is the common perpendicular bisector for both of them. See **Figure 6.22**.

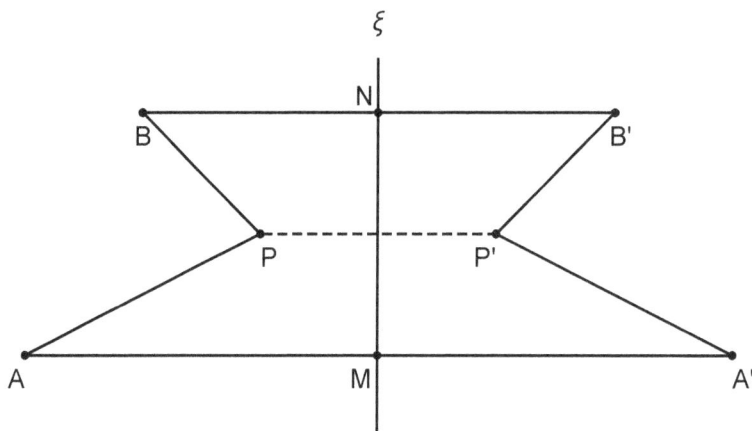

Figure 6.22: Inversion is reduced to the reflection R_ξ in the line ξ when $\overrightarrow{AA'} \parallel \overrightarrow{BB'}$

(See also the final part of **Subsection 6.2.1**.)

(iii) In **Case (3), (ii)**, this happens when the triangle ABB' is isosceles $(A = A'$, fixed point), in which case the line ξ is the axis of symmetry of ABB', i.e., the midpoint perpendicular of the base BB'. (Make a figure.)

6.3.1   Inversion Expressed Analytically

Suppose we work in the cartesian plane with an orthonormal system of axes and the points are represented by pairs of coordinates (x, y). We consider an inversion $I_{[O,c]}$, with center the point $O = (x_0, y_0)$ and power $c \neq 0$. Given any point $P = (x, y) \neq O = (x_0, y_0)$, we want to find its inverse $P' = (x', y')$. [As we know, the inverse of the point $O = (x_0, y_0)$ vanishes at infinity.]

Since $\overline{OP} \cdot \overline{OP'} = c$, we have $\overline{OP'} = \dfrac{c}{\overline{OP}}$. Then, $\|\overline{OP'}\| = \dfrac{|c|}{\|\overline{OP}\|}$, or

$$\sqrt{(x' - x_0)^2 + (y' - y_0)^2} = \frac{|c|}{\sqrt{(x - x_0)^2 + (y - y_0)^2}}.$$

Also, $\overline{OP}$ and $\overline{OP'}$ have the same or opposite direction depending on $c > 0$ or $c < 0$, respectively. Since any nonzero vector is equal to the unit vector of the same direction multiplied by the length of the given vector, we get

$$(x' - x_0, y' - y_0) = \frac{\pm|c|}{\sqrt{(x - x_0)^2 + (y - y_0)^2}} \cdot \frac{(x - x_0, y - y_0)}{\sqrt{(x - x_0)^2 + (y - y_0)^2}},$$

$$\text{or} \quad (x' - x_0, y' - y_0) = \frac{c \cdot (x - x_0, y - y_0)}{(x - x_0)^2 + (y - y_0)^2}.$$

Therefore,

$$(x', y') = (x_0, y_0) + \frac{c(x - x_0, y - y_0)}{(x - x_0)^2 + (y - y_0)^2} \Leftrightarrow \begin{cases} x' = x_0 + \dfrac{c(x - x_0)}{(x - x_0)^2 + (y - y_0)^2} \\[4mm] y' = y_0 + \dfrac{c(y - y_0)}{(x - x_0)^2 + (y - y_0)^2}. \end{cases}$$

[Also, solving this system for x and y, we find

$$\begin{cases} x = x_0 + \dfrac{c(x' - x_0)}{(x' - x_0)^2 + (y' - y_0)^2} \\[4mm] y = y_0 + \dfrac{c(y' - y_0)}{(x' - x_0)^2 + (y' - y_0)^2}, \end{cases}$$

as it should be, because P' is the inverse of P.]

If we work in the complex plane, then the above formulae are written more elegantly with complex numbers. Let $O = z_0$, $P = z$ and $P' = z'$. Then we have,

$$z' = z_0 + \frac{c(z - z_0)}{|z - z_0|^2} = z_0 + \frac{c(z - z_0)}{(z - z_0)(\overline{z} - \overline{z_0})} = z_0 + \frac{c}{\overline{z} - \overline{z_0}}$$

$$\text{and also} \quad z = z_0 + \frac{c}{\overline{z'} - \overline{z_0}}.$$

6.4   Inverses of Straight Lines and Circles

So far, we have seen the lines invariant under an inversion to be those that pass through the center of inversion. Also, we have seen that the circle of inversion and any circle that contains two inverse points is also invariant. Here, we would like to find the images of all other lines and circles under a given inversion.

6.4.1   Straight Lines Away from the Center of Inversion and Circles Passing through the Center of Inversion

We consider an inversion $I_{[O,c]}$ and a straight line l that does not pass through O. Then, the distance of O from l is the segment $0 \neq OP \perp l, \ (P \in l)$ and P' (the inverse point of P) is on the line OP and determined by the equation $\overline{OP'} = \dfrac{c}{\overline{OP}}$. See **Figures 6.23** and **6.24**. So, P and P' are completely determined by the inversion and the given line l and then the segments OP and OP' are rendered known.

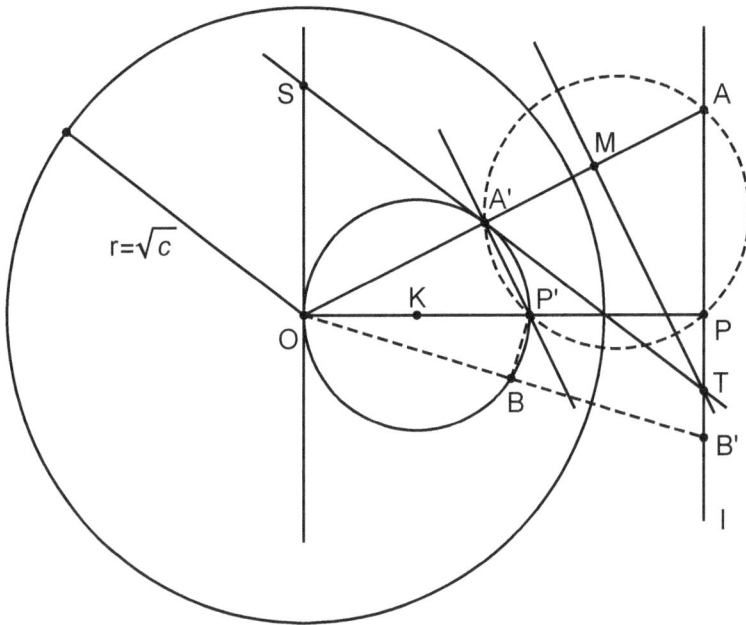

Figure 6.23: Inverse of a line not passing through O and inverse of a circle passing through O. ($c > 0$ and some properties)

Now, we pick any point $A \in l$ and consider its inverse A'. Then, by the basic property of inversions the four points P, P', A and A' are concyclic. Therefore, $\widehat{AA'P'} = \widehat{P'PA} = \dfrac{\pi}{2}$. Hence, the triangle $A'OP'$ is right at A'. This implies that A' is on the circle with diameter the known segment OP' and so its center K is the midpoint of OP' and its radius is $\rho = \left| \dfrac{\overline{OP'}}{2} \right| = \left| \dfrac{c}{2\overline{OP}} \right|$.

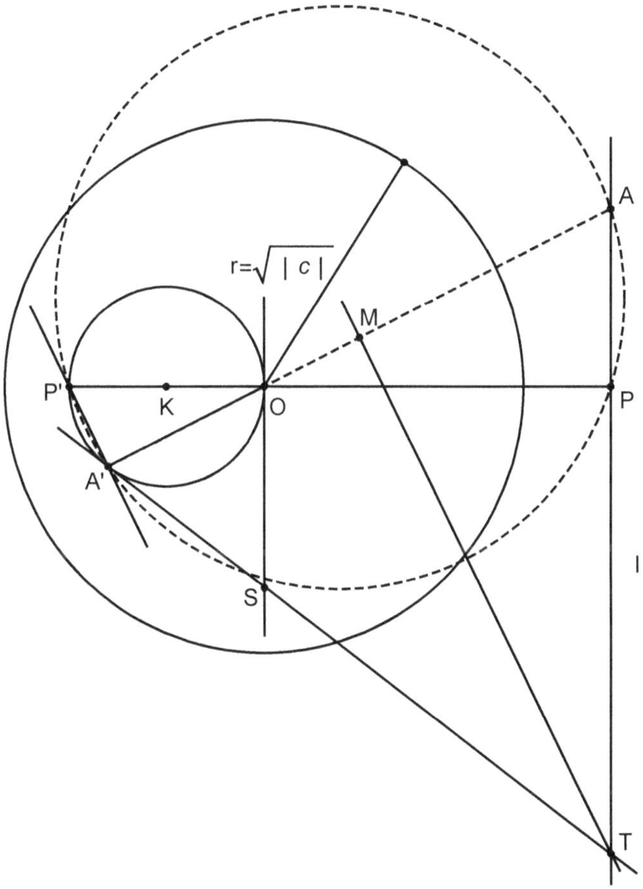

Figure 6.24: Inverse of a line not passing through O and inverse of a circle passing through O. ($c < 0$ and some properties)

These steps can be reversed to prove that any point A' of the circle with diameter OP' is inverse to a point A on the line l. That is, $I_{[O,c]}(l) = C[K, \rho]$. So, we have:

Theorem 6.4.1 *The inverse of a straight line that does not pass through the center of an inversion is a circle that passes through the center of the inversion. The diameter of this image circle (that passes through the center of inversion) is perpendicular to the given straight line and its length is equal to the absolute value of the power of the inversion divided by the distance of the center of inversion from the straight line.*

Now, using the involutory property of inversion, we have that

$$I_{[O,c]}\{C[K,\rho]\} = I_{[O,c]} \circ I_{[O,c]}(l) = I(l) = l.$$

That is, the inverse of a circle $C[K,\rho]$ that passes through the center of inversion O is the line l, constructed by reversing the steps of the previous construction. Also, if K' is the inverse point (not shown in the figures) of the center K, then

$$\overline{OK} \cdot \overline{OK'} = c = \overline{OP'} \cdot \overline{OP}.$$

This implies

$$\frac{\overline{OK'}}{\overline{OP}} = \frac{\overline{OP'}}{\overline{OK}} = \frac{2\rho}{\rho} = 2. \qquad \left(\rho = \left|\frac{\overline{OP'}}{2}\right| = \left|\frac{c}{2\overline{OP}}\right|.\right)$$

Therefore, the point $K' = I_{[O,c]}(K)$ is the symmetrical point of the point O (center of inversion) with respect to the straight line $l = I_{[O,c]}\{C[K,\rho]\}$.

Putting all these results together, we have:

Theorem 6.4.2 *Consider a **circle that passes through the center of a given inversion**. Then we have:*

(a) The inverse of such a circle under the given inversion is a straight line that does not pass through the center of inversion.

(b) The diameter that passes through the center of inversion is perpendicular to the straight line and therefore the straight line is parallel to the tangent of the given circle at the center of inversion.

(c) The distance of the straight line from the center of inversion is equal to the absolute value of the power of the inversion divided by the length of the diameter of the given circle.

(d) The inverse of the center of the given circle is the point symmetrical to the center of inversion in the straight line which is the inverse of the circle.

Remark: In **Figures 6.23** and **6.24**, we have depicted an interesting property. The tangent to the circle $C[K,\rho]$ at O is perpendicular to the line $OP'P$ and therefore parallel to l. We have also considered the tangent of $C[K,\rho]$ at A'. Suppose the two tangents (at O and A') are not parallel, which is the case as long as $A' \neq P'$, and so they intersect at the a point S. Then, the tangent $A'S$ intersects the line l at a point T and the lines $A'ST$ and l are symmetrical with respect to MT, the midpoint perpendicular of AA', which is also the bisector of the angle $\widehat{A'TA}$.

This follows from the facts: (1) the triangle $A'OS$ is isosceles, because the base angles are equal, and (2) $OS \parallel AT$. These facts imply that the triangle $AA'T$ is also isosceles. (Supply the basic details!)

We also have the following important theorem:

Theorem 6.4.3 *Consider a circle $C[K,\rho]$ and a straight line l. If they are not tangent to each other, then there are two inversions, a positive and a negative, that map one onto the other. If they are tangent, then there is only one such inversion, which is positive.*

Proof According to the what we found before, the center of an inversion that maps a line l to a circle $C[K, \rho]$ is an end-point of the diameter that is perpendicular to the line l at a point $P \in l$. The end-point of this diameter should not be on the line l, for otherwise the line is going to be invariant under the inversion and so not mapped onto the circle.

If l and $C[K, \rho]$ are not tangent to each other, there are two such end-points O_1 and O_2; and only one, O, if they are tangent.

In the first case, the two inversions are $I_{[O_1, \overline{O_1 O_2} \cdot \overline{O_1 P}]}$ and $I_{[O_2, \overline{O_2 O_1} \cdot \overline{O_2 P}]}$. One of them is positive and the other is negative. In the second case, the inversion is $I_{[O, \overline{OP}^2]}$ and is positive. (Draw figures simpler than **Figures 6.23** and **6.24**.)

∎

In view of all the above Theorems, we give the following definition.

Definition 6.4.1 *Given an inversion $I_{[O,c]}$ and a pair of inverse points (P, P'), the straight line l' perpendicular to OP' at P' is called **the polar line of** P and P is called **the pole of** l', with respect to the inversion $I_{[O,c]}$. Also, the inverse points P and P' are called **conjugate points with respect to the circle of inversion** $C[O, \sqrt{|c|}]$.*

In this definition, the line l perpendicular to OP at P is the polar line of P' and P' is the pole of l. Also, in the previous three **Theorems** and **Figures 6.23**, and **6.24**, we see that if $(P', l := PA)$ is a pair of pole and polar line, then for any point $A \in l$, the polar line of A is $A'P'$, where A' is the inverse point of A. That is, the polar line of A goes through P', the pole of $l := PA$. Analogously, the pole of the line $A'P'$ is the points $A \in l$.

Along with this context, we would like to define the **conjugate pair of points with respect to a circle** in the most general sense. As we have seen, any circle $C[O, R]$ can be considered as the circle of the positive inversion $I_{[O,c]}$, with center O and power $c = R^2 > 0$. We have also seen that any circle $C[K, \rho]$ passing through two inverse points P and P' (so, $\overline{OP} \cdot \overline{OP'} = c = R^2$) is **orthogonal to the circle of inversion** $C[O, R]$.

In such a case, the straight line OK intersects the circle $C[O, R]$ at two diametrical points A and B and the circle $C[K, \rho]$ at two diametrical points C and D which are inverse of each other in the circle $C[O, R]$ (analogous to **Figure 6.14**). Then, as we have seen the quadruple of points $\{A, B, C, D\}$ lying on the straight line OK is harmonic. That is,

$$\frac{\overline{AC}}{\overline{BC}} = -\frac{\overline{AD}}{\overline{BD}}.$$

We say the points C and D are harmonic conjugates of A and B and analogously, A and B are harmonic conjugate of C and D. By the above **Definition**, we also say that the inverse points C and D are conjugates of each other with respect to the circle $C[O, R]$. So now, we give the following general definition:

Definition 6.4.2 *Two points* M *and* N *in the plane are called* ***conjugate points with respect to the circle*** $C[O, R]$, *if the circle with diameter* MN *is orthogonal to* $C[O, R]$.

In this definition the diameter MN may pass or may not pass through O. See **Figure 6.25**. Any pair of diametrical points of the circle $C[K, \rho]$, orthogonal to $C[O, R]$, are conjugate points with respect to the circle $C[O, R]$. We also have:

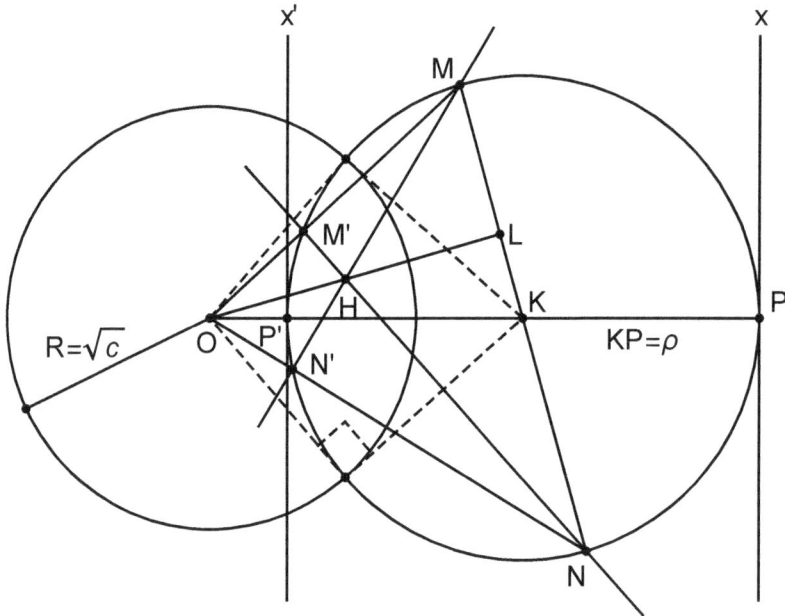

Figure 6.25: Conjugate points M and N (and P, P') with respect to the circle $C[O, R]$

(a) By **Theorems 6.2.1, 6.2.2** and **6.2.4**, besides $C[K, \rho]$, the circles with diameters MM' and NN' (not drawn in **Figure 6.25**) are also orthogonal to $C[O, R]$. (This may justify the generalization of **Definition 6.4.1** of conjugate pair of points by **Definition 6.4.2**.)

(b) The lines OM and ON intersect the circle $C[K, \rho]$ at the second points M' and N', respectively, which are the inverse points of M and N under the inversion $I_{[O, R^2]}$, since $C[K, \rho]$ is orthogonal to $C[O, R]$. ($M = M'$, if M is a point of intersection of $C[O, R]$ and $C[K, \rho]$, and similarly for N.) We again observe that the polar line of M is NM' and the polar line of N is MN'. This is so, because $NM' \perp OM$ and $MN' \perp ON$, since MN is a diameter of $C[K, \rho]$, and M' and N' are the inverse points of M and N.

(c) In **Figure 6.25**, the points P and P' are the two conjugate points with respect to $I_{[O, R^2]}$ located on the center line OK. The polar line of P is the $P'x'$ and the polar line of P' is the Px. In the case, $P'x' \parallel Px$. So, the polar line of

M passes through N and the polar line of N passes through M when the two points M and N are conjugate points with respect to the circle $C[O, R]$, except when $M = P$ and $N = P'$ are inverse points on the center line OK. In the later case the two polar lines are parallel.

(d) It is also interesting to find the pole of the line MN with respect to the circle $C[O, R]$. Let H be the common point of the lines OM and ON. This is the orthocenter of the triangle OMN, since $NM' \perp OM$ and $MN' \perp ON$ are two lines of heights of OMN. So, the line OH is perpendicular to MN at the point L. Therefore, the pole of MN is on the line OL.

But the point M (or N for the same matter) is on the line MN. So, the pole of MN is also on the polar line of M which is the straight line NM'. Therefore, the pole of MN is the point H which is the orthocenter of the triangle OMN. From this, we also get
$$\overline{OH} \cdot \overline{OL} = c = R^2.$$
(The fact $\overline{OM'} \cdot \overline{OM} = \overline{OH} \cdot \overline{OL} = \overline{ON'} \cdot \overline{ON}$ can also be obtained directly by using the similar triangles formed by the heights of a triangle.)

(e) Now, we observe that if $M \in C[K, \rho]$ approaches the point $P \in C[K, \rho]$, then N and M' approaches P' and N' approaches P. See **Figure 6.26**.

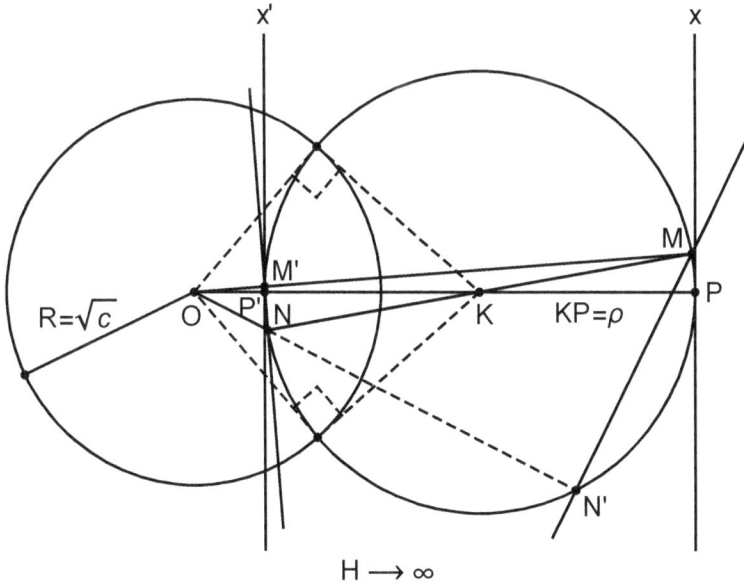

Figure 6.26: Conjugate points M and N with respect to the circle $C[O, R]$ approach P and P'

Then, the line MN' approaches the line Px and the line NM' approaches the line Px'. The triangle OMN becomes obtuse and the orthocenter H, the intersection of the heights MN' and NM', is outside the triangle OMN and approaches one of the infinity points of the parallel lines $Px \parallel Px'$.

6.4.2   Images of Circles that do not Pass through the Center of Inversion

We consider an inversion $I_{[O,c]}$ and a circle $C[K,\rho]$ that does not pass through the center O of the inversion and we want to find $I_{[O,c]}\{C[K,\rho]\}$.

In the sequel, we let $p = \mathfrak{P}_{C[K,\rho]}(O)$, the power of the center of inversion O with respect to the circle $C[K,\rho]$. Since $O \notin C[K,\rho]$, it holds $p \neq 0$.

We have already seen in **Subsections 6.2.1** and **6.2.3**: if $c > 0$, the circle of inversion and any circle orthogonal to it is mapped to itself by the inversion; if $c < 0$, the circle of inversion and any circle pseudo-orthogonal to it is mapped to itself by the inversion. So, in these cases we know the answer to the above question, and as we have seen $p = c$. Next, we want to find the inverses of the other circles that do not contain O and check that the general results we will find agree with the already known partial results. To this end, see **Figures 6.27** and **6.28**.

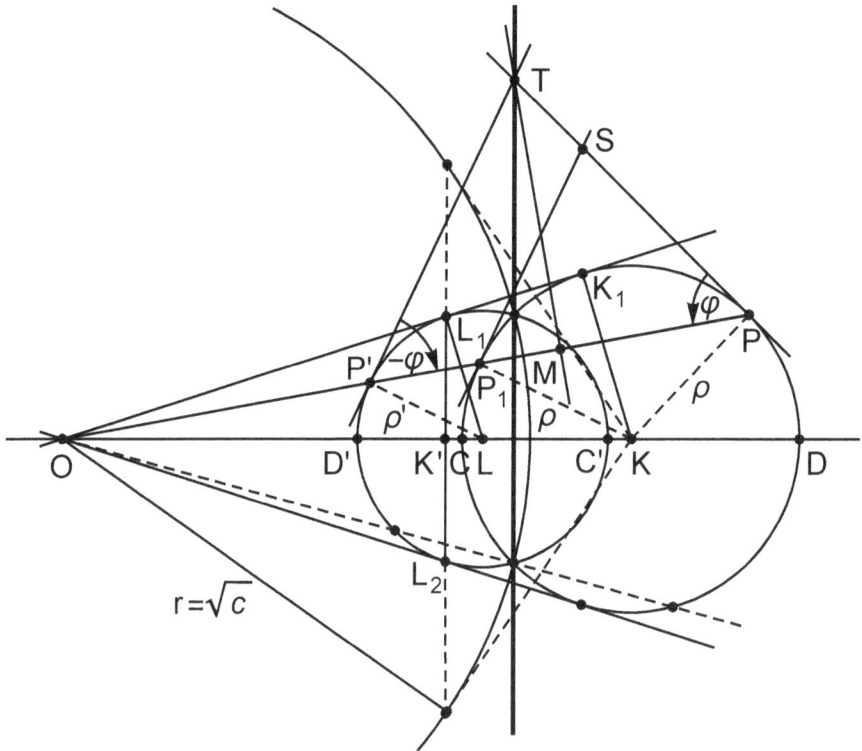

Figure 6.27: Inverse of a circle not passing through O.
($c > 0$ **and some properties)**

We pick any point $P \in C[K, \rho]$ and we consider its inverse P'. We also consider the second point P_1 at which the line OP intersects the circle, unless OP is tangent to the circle, in which case $P = P_1$. See **Figures 6.27** and **6.28**.
Then we have,

$$\overline{OP} \cdot \overline{OP'} = c \qquad \text{and} \qquad \overline{OP} \cdot \overline{OP_1} = p.$$

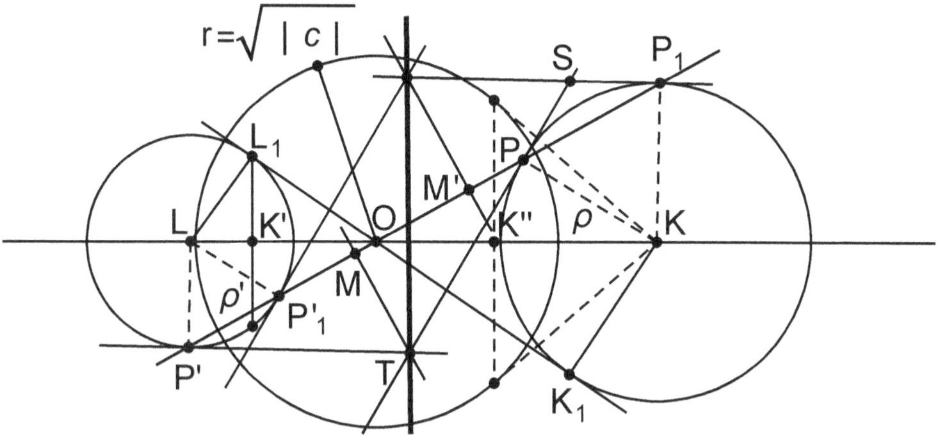

Figure 6.28: Inverse of a circle not passing through O.
($c < 0$ and some properties)

Dividing the two equations, we find

$$\frac{\overline{OP'}}{\overline{OP_1}} = \frac{c}{p}.$$

This means that P_1 corresponds to P' under the homothety

$$H_{[O, \frac{c}{p}]},$$

with homothetic center O, the center of inversion, and homothetic ratio

$$\lambda = \frac{c}{p}.$$

As P moves along the given circle $C[K, \rho]$, the point P_1 traces the same circle and the point P', which is the inverse point of P, traces the homothetic circle

$$C[L, \rho'] = H_{[O, \frac{c}{p}]}\{C[K, \rho]\}.$$

Observe that the inverse homothety $H_{[O, \frac{1}{\lambda}]} = H_{[O, \frac{p}{c}]} = H_{[O, \frac{c}{p'}]}$, where $p' = \mathfrak{P}_{C[L, \rho']}(O)$, maps $C[L, \rho']$ to $C[K, \rho]$ and we also get $pp' = c^2$.

So, we have the following Theorem:

Theorem 6.4.4 *The image of a circle $C[K, \rho]$ under an inversion $I_{[O,c]}$, such that the center of inversion O is not on this circle, is the circle $C[L, \rho']$, which is the homothetic image of $C[K, \rho]$, under the homothety $H_{[O, \frac{c}{p}]}$, where c is the power of inversion and $p = \mathfrak{P}_{C[K,\rho]}(O) = OK^2 - \rho^2$ is the power of the center of inversion, O, with respect to the given circle $C[K, \rho]$.*
{That is: the homothety $H_{[O, \frac{c}{p}]}$ has center the center O of the given inversion $I_{[O,c]}$, and ratio $\lambda = \dfrac{c}{p}$, where $p = \mathfrak{P}_{C[K,\rho]}(O) = OK^2 - \rho^2$. So, $\overline{OL} = \dfrac{c}{p} \cdot \overline{OK}$,
$\rho' = \dfrac{|c|}{|p|} \cdot \rho$, *and $H_{[O, \frac{c}{p}]} \{C[K, \rho]\} = I_{[O,c]} \{C[K, \rho]\} := C[L, \rho']$.* **Attention:**
The center L is homothetic to the center K, but not inverse.}

Important Remarks: (1) If external tangents of the given circle $C[K, \rho]$ and its inverse $C[L, \rho']$ exist, then they pass through the common center O of the inversion $I_{[O,c]}$ and the homothety $H_{[O, \frac{c}{p}]}$ on the center line KL. If internal tangents exist, then they pass through the common center of the negative inversion and negative homothety on the center line KL.

(Common tangents do not exist when one circle is inside the other one, etc.)

(2) The points P_1 and P' are homothetic, whereas P and P' are inverse. In general $P_1 \neq P$, except for the two points of the common tangents of the $C[K, \rho]$ and $C[L, \rho']$ that pass through O. According to **Definition 4.1.3**, the points P_1 and P' are homologous and the points P and P' are anti-homologous, with respect to homothetic center O.

(3) If the point P is on a common tangent from O to the circles $C[K, \rho]$ and $C[L, \rho']$, then $P_1 = P$ and $P' = P'_1$ is on the same tangent and $C[L, \rho']$. In this case, P' is both homothetic and inverse of P.

(4) By the properties of homotheties, the center L of the circle $C[L, \rho']$ is determined on the line OK by the equation $\overline{OL} = \dfrac{c}{p} \cdot \overline{OK}$ and $\rho' = \dfrac{|c|}{|p|} \cdot \rho$. {For the invariant circles of **Theorems 6.2.1, 6.2.2** and **6.2.3**, $c = \pm p = \pm \left(OK^2 - \rho^2 \right)$ and so the homothety is $H_{[O, \pm 1]}$, which is the identity or a half-turn.}

(5) Assume that the circles have common tangents. **The inverse point K' of the center K is not L, but a point $K' \neq L$ on the straight line KLO. To find K'**, we draw $L_1 K_1$, one of the common tangents to the two circles. From the properties of homotheties, we know that $L_1 K_1$ passes through O.
    Then, $OK \cdot OK' = c = OK_1 \cdot OL_1$. Therefore, the quadrilateral $K'KK_1L_1$ is inscribable and so $\widehat{KK'L_1} = \dfrac{\pi}{2}$, (because $\widehat{KK_1L_1} = \dfrac{\pi}{2}$). Hence, $L_1 K' \perp OL$. Since OL_1 is tangent to the circle $C[L, \rho']$ at L_1, the triangle $L_1 LO$ is right with hypotenuse OL and corresponding altitude $L_1 K'$. Therefore, $OK' < OL$, and

$$\text{if}\quad c > 0\quad \text{and}\quad p > 0,\quad \text{then}\quad \overline{OK'} = \frac{c}{\overline{OK}} < \overline{OL} = \frac{c}{p} \cdot \overline{OK}.$$

{More general, with any $c (\neq 0)$, if $p > 0$, then $OK' = \dfrac{|c|}{OK} < OL = \dfrac{|c|}{p} \cdot OK$.

(a) If $p > 0$, this fact could be á priori determined, since in that case we have: $0 < p = OK^2 - \rho^2 < OK^2$.

(b) If $p < 0$ ($\Longleftrightarrow O$ is inside $C[K, \rho]$), make two figures analogous to **Figures 6.27** and **6.28** and compare the differences.

(c) If $p = 0$ ($\Longleftrightarrow O \in C[K, \rho]$), see **Subsection 6.4.1**, and **Figures 6.23** and **6.24**.

(d) Analogous work must be done when one circle is inside the other one and so common tangents do not exist.}

(6) The inverse point of O with regard to the circle $C[L, \rho']$ is K'. That is, $\overline{LO} \cdot \overline{LK'} = \rho'^{\,2}$. [This follows form the fact that the triangle $LL_1 O$ is right $\left(\widehat{L_1} = \dfrac{\pi}{2} \right)$, $L_1 K'$ is the height on the hypotenuse OL and $LL_1 = \rho'$.] If $C[L, \rho']$ is replaced by a straight line l, then the inverse of O is its reflection in l.

(7) Geometrically, we can construct $C[L, \rho']$, the inverse circle of $C[K, \rho]$, in the following steps.

(a) We pick any point $P \in C[K, \rho]$ and we find its inverse point P' on OP. We also let P_1 be the other point of intersection of OP with $C[K, \rho]$.

(b) We draw the line $k \parallel K P_1$ through P'. This intersects the line OK at the point L. Now: The point L, thus determined, is the center of $C[L, \rho']$, whose radius is $\rho' = LP'$.

(8) The tangents at two inverse points P and P' of two inverse circles are symmetrical with respect to the perpendicular bisector MT of the segment PP'. See **Figures 6.27** and **6.28**.

Let T be the point of intersection of these tangents. We draw the tangent of $C[K, \rho]$ at P_1 and let S be the point of its intersection with the tangent at P. This is parallel to the tangent at P', because the corresponding radii are parallel ($KP_1 \parallel LP'$). Since the triangle SPP_1 is isosceles, so is the triangle TPP'. Now, the claim follows and moreover MT is the bisector of the angle $\widehat{PTP'}$. Therefore, these tangents form opposite angles with the straight line $OP'P$. That is, $\quad \measuredangle(PT, PP') = -\measuredangle(P'T, P'P)$.

(9) The line OK intersects both circles $C[K, \rho]$ and $C[L, \rho']$ at diameters CD and $C'D'$. As in **Figure 6.27**, we get that C' is the inverse of C and D' the inverse of D. So, by the homothety $H_{[O, \frac{c}{p}]}$, the lengths satisfy $C'D' = \left| \dfrac{c}{p} \right| CD$.

(10) We have the following useful computations of two related powers of O. The power of O with respect to $C[K, \rho]$ is
$$p = OC \cdot OD = OK^2 - \rho^2 = OK_1^2 = OP \cdot OP_1 = \text{ etc.}$$
The power $p' = \mathfrak{P}_{C[L, \rho']}(O)$, of O with respect to the circle $C[L, \rho']$, is
$$p' = \overline{OC'} \cdot \overline{OD'} = \dfrac{c}{\overline{OC}} \cdot \dfrac{c}{\overline{OD}} = \dfrac{c^2}{p} \quad \Longrightarrow \quad pp' = c^2.$$

We now have the following relative Theorem.

Theorem 6.4.5 *Consider two unequal circles C_1 and C_2.*

(a1) If the circles are not equal and not tangent, then they are mutually inverse by two inversions that, depending on the relative position of the circles, may be both positive or one positive and one negative. The negative inversion exists when the two circles have no common point. Otherwise, the circles intersect at two points and both inversions are positive.

(a2) If the circles are not equal but are tangent, then they are mutually inverse by a positive inversion, only.

(b1) If the circles are equal and have no point in common, then the positive inversion [as in (a1)] becomes the reflection in the perpendicular bisector of their center-segment. The circles are also mutually inverse by a negative inversion with center the midpoint of their center-segment and power
$$c = -\mathfrak{P}_{C_1}(M) = -\mathfrak{P}_{C_2}(M).$$

(b2) If the circles are equal and tangent (externally), then only the reflection [as in (b1)] as positive inversion between them survives.

(b3) If the circles are equal and intersecting at two points, then besides the reflection [as in (b1)], the circles are also mutually inverse by a positive inversion with center the midpoint of their center-segment and $c = \mathfrak{P}_{C_1}(M) = \mathfrak{P}_{C_2}(M)$.

(c) A special case is when the two circles C_1 and C_2 are concentric. Then, they are inverse of each other by a positive and a negative inversion whose center is the common center of the circles O and their powers are $c = \pm rR$, where $R \geq r(> 0)$ are the radii of the circles. So, the inversions are $I_{[O, \pm rR]}$.

Proof Let $C_1 = \mathbf{C}[\mathbf{K}, \rho]$ and $C_2 = \mathbf{C}[\mathbf{L}, \tau]$ be the two circles.

(a) If $\rho \neq \tau$, we know that there are two homothetic centers O_1 and O_2 between the two circles. See **Figures 6.29** and **6.30** (and also **Figures 6.27** and **6.28**).

(a1) If the circles are not tangent, then either homothetic center is not on either of the circles and so it can be used as a center of an inversion of one circle onto the other. See **Figure 6.29**.

(a2) If the circles are tangent (externally or internally) then the point of tangency is one of the homothetic centers, but it cannot be used as center of an inversion between the two given circles, since its image would be ∞. See **Figure 6.30**.

Now, we pick one of the homothetic centers that does not belong to either of the two circles, e.g., O_1. Consider any point $P_1 \in C[K, \rho]$ and its homothetic $P' \in C[L, \tau]$, with respect to O_1. The ratio of the homothety is

$$\lambda = \frac{\overline{O_1 P'}}{\overline{O_1 P_1}} = \pm \frac{\tau}{\rho},$$

with $+$, if the homothety is positive, and with $-$, if the homothety is negative. (We have seen that, the homothety is positive if O_1 is outside KL and negative if O_1 is inside KL.)

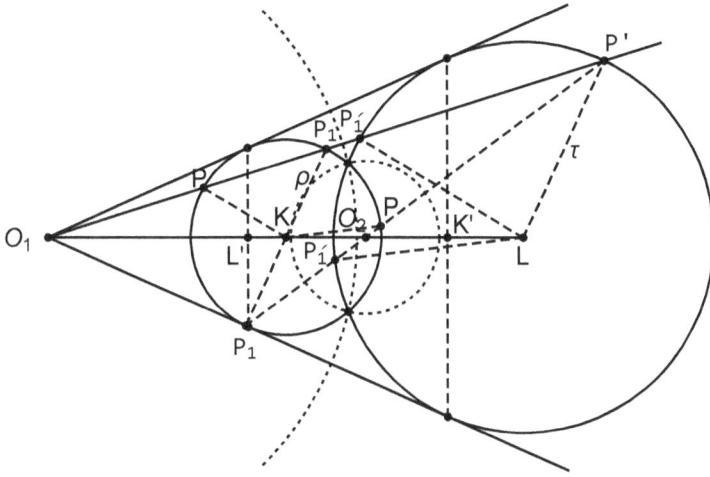

Figure 6.29: Two positive inversions of two unequal circles. The circles of the inversions are the dotted ones with centers O_1 and O_2.

Next, let line $O_1 P_1 P'$ intersect $C[K, \rho]$ at a second point P (which is equal to P_1 if $OP_1 P'$ is tangent). Then,

$$\overline{O_1 P_1} \cdot \overline{O_1 P} = p, \quad \text{which is fixed along the circle} \quad C[K, \rho].$$

$\{p > 0$ if O_1 is outside the $C[K, \rho]$, and $p < 0$ if O_1 is inside $C[K, \rho]$, as is the case of O_2 in **Figures 6.29**.$\}$

Multiplying the last two equations, we find

$$\overline{O_1 P'} \cdot \overline{O_1 P} = \lambda p.$$

This equation proves that P' is the inverse of P, under the inversion $I_{[O_1, \lambda p]}$. Since P_1 was picked arbitrarily, the points P' and P trace the two circles, $C[L, \tau]$ and $C[K, \rho]$, respectively. So, the inversion that maps $C[K, \rho]$ onto $C[L, \tau]$ is

$$I_{[O_1, c]}, \quad \text{with power} \quad c = \lambda p.$$

This inversion is positive or negative depending on λ and p being positive or negative and this in turn depends on the relative position of the two circles. We easily conclude that one negative inversion exists (i.e., $c = \lambda p < 0$), only when the two circles do not have any point in common, e.g., as in **Figure 6.28**.

Similar work is done with respect to center O_2. But, if the circles are tangent, then only one of the homothetic centers does not belong to either circle. The results of this situation follow in the same way. (Make a figure when the two circles are internally tangent. Make figures when the two circles have no common point and either the circles are outside of each other, as in **Figure 6.28**, or one circle is inside the other, and depict the circles of the positive and negative inversions.)

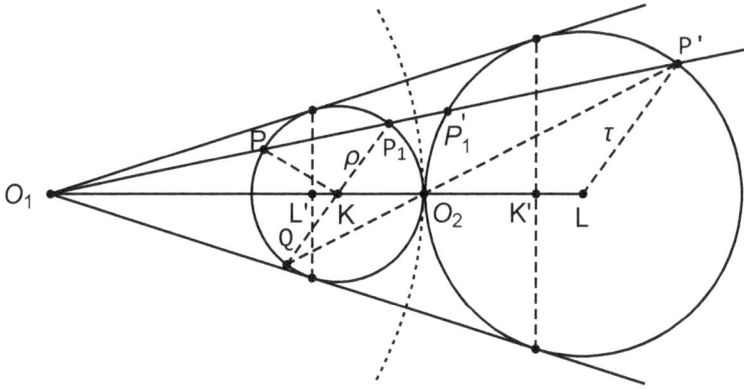

Figure 6.30: The positive inversion of two unequal tangent circles with center O_1

(b) If the two circles have equal radii, $\rho = \tau$, then we have three situations as depicted in **Figures 6.31, 6.32** and **6.33**. The results claimed follow as before.

(b1) In **Figure 6.31**, the two equal circles are non-intersecting. Then, they are symmetrical in the perpendicular bisector of the center-segment LK. The reflection in this line, is considered to be positive inversion between the two circles, with center of inversion the infinity point of the center line KL. In this situation, there is also another inversion since we have: Homothetic ratio $\lambda = -1$ and power $p = \overline{O_2 P_1} \cdot \overline{O_2 P}$. Then, the inversion is $I_{[O_2, -p]}$, which is a negative inversion.

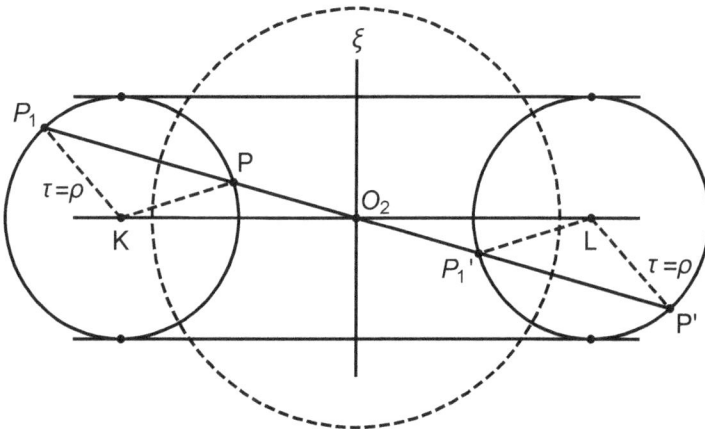

Figure 6.31: One negative inversion of two equal non-intersecting circles with center O_2. The reflection in ξ has center $O_1 = \infty$.

(b2) In **Figure 6.32**, the two equal circles are externally tangent. Here, the only inversion is the reflection in their common internal tangent, with center of inversion the infinity point of the center line KL.

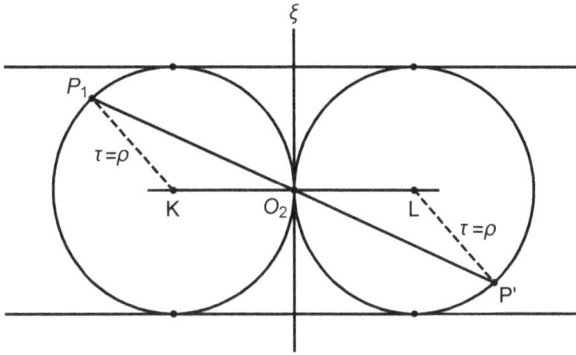

Figure 6.32: For equal tangent circles the only inversion between them is the reflection in ξ whose center is $O_1 = \infty$

(b3) In **Figure 6.33**, the two equal circles are intersecting at two points A and B. Now, besides the reflection in their common chord $AB = \xi$, there is a positive inversion between them with center O_2 the midpoint of the center-segment KL.

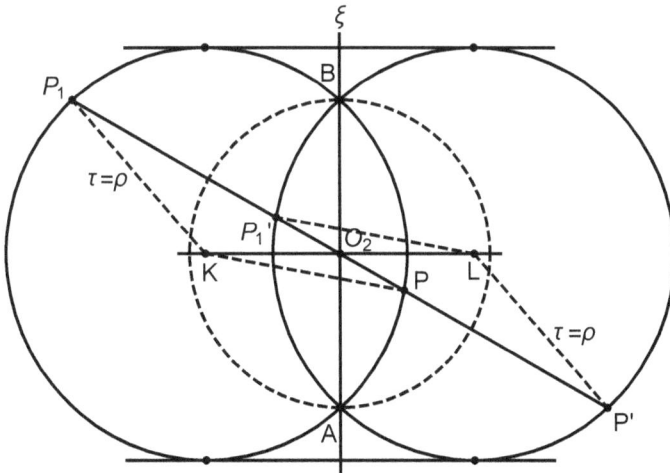

Figure 6.33: For equal intersecting circles besides the reflection in $\xi = AB$, there is a positive inversion between them with center O_2

In this situation, the circle of the positive inversion has diameter the common cord AB.

(c) In **Figure 6.34**, we have two concentric circles $C[O, r]$ and $C[O, R]$, with $R \geq r(> 0)$. These are homothetic with either the positive or negative homotheties with the same center the point O and homothetic ratios $\lambda = \pm \dfrac{R}{r}$ (or $\lambda = \pm \dfrac{r}{R}$), as we have discussed after **Theorem 4.1.2** and **Figure 4.21**. Therefore, these circles are inverse of each other under the positive inversion $I_{[O, rR]}$ or the negative inversion $I_{[O, -rR]}$. (We observe that this can be directly verified and agrees with **Theorem 6.4.4**.) The circle of inversion is $C\left[O, \sqrt{rR}\right]$ for both inversions.

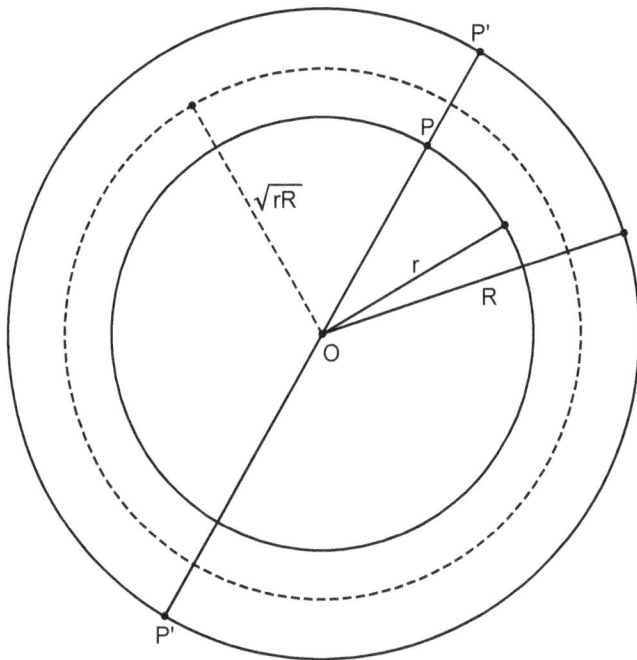

Figure 6.34: Two concentric circles are inverse of each other under the two inversions $I_{[O, \pm rR]}$

■

Remarks: (1) By **Subsection 6.4.1**, we conclude that the images of two tangent circles under an inversion with center the point of tangency are two parallel lines perpendicular to the center line of the two circle. (See also **Figure 6.77**.)

(2) For two intersecting straight lines (considered as circles of infinite radius), the two inversions that map one onto the other become the reflections in the bisectors of their angles, which are orthogonal to each other. If the lines are parallel (considered as concentric circles of infinite radius), then there is one reflection in their mid-parallel line.

In this context, we also have the following Theorem:

Theorem 6.4.6 *Given two circles, there are infinitely many pairs of inversions of the same center lying on their center line that invert the two given circles into equal circles and the centers of these equal circles may be chosen to be any points of the center line except for the points of its intersection with the two given circles. We could also choose the two center to coincide and thus make the equal circles identical.*

Proof Let $C[K_1, r_1]$ and $C[K_2, r_2]$ be the two given circles. In order for two inversions $I_{[O,c_1]}$ and $I_{[O,c_2]}$ with common center O lying on the line $K_1 K_2$, to map the two given circles into two circles of equal radius (and so to equal circles), by **Subsection 4, (j)**, and **Theorem 6.4.4**, the equal radius is

$$r := r_1 \left| \frac{c_1}{p_1} \right| = r_2 \left| \frac{c_2}{p_2} \right|,$$

where p_1 and p_2 is the power of O with respect to $C[K_1, r_1]$ and $C[K_2, r_2]$, respectively.

The radii r_1 and r_2 are known. Having picked any O on the center line $K_1 K_2$ (except for the points of intersection of the two circles and the center line), the powers p_1 and p_2 become known. Hence, we pick any c_1 and c_2 that satisfy

$$\frac{c_1}{c_2} = \pm \frac{p_1 r_2}{p_2 r_1}.$$

Then, the circles of the two inversions are $C[O, \sqrt{|c_1|}]$ and $C[O, \sqrt{|c_2|}]$. Also, the common radius of the image circles is found by the formula for r, above.

Therefore, there are infinitely many pairs of inversions of the same center O on the center line $K_1 K_2$ that can invert the two given circles into equal circles with centers on the center line $K_1 K_2$ of the given circles.

If $K_1 = K_2$, the two given circles are concentric. Then, in the place of center line, we pick any line l through $K_1 = K_2$.

To choose the centers of the two equal circles to be the same point L, then by **(4)** of the **important remarks** following **Theorem 6.4.4** we must satisfy

$$\overline{OL} = \frac{c_1}{p_1} \overline{OK_1} = \frac{c_2}{p_2} \overline{OK_2}.$$

So using this equation, we must choose O to satisfy the system of equations

$$p_1 = OK_1^2 - r_1^2, \qquad p_2 = OK_2^2 - r_2^2, \qquad \frac{\overline{OK_1}}{\overline{OK_2}} = \pm \frac{p_1 c_2}{p_2 c_1} = \pm \frac{r_1}{r_2}.$$

Then O is determined by the last equation. We may have two choices depending on the $\pm$. Next, we find p_1 and p_2 from the first two equations and subsequently we choose any compatible c_1 and c_2 (infinite choices).

If $K_1 = K_2$, then the choice is $L = K_1 = K_2$ and then, as in the proof of **Theorem 6.4.5, (c)**, we can consider the inversion $I_{[L, r_1 r_2]}$ coupled with the inversion $I_{[L,r_1^2]}$ or the inversion $I_{[L,r_2^2]}$. Of course, we have infinite choices of the powers of these inversions. (Find them all!)

∎

Remark: In the above proof, the last equation of the system was expected. As we have seen in the first part of **this Subsection**, inverse circles are homothetic. By the properties of the homotheties, we have $\dfrac{\overline{OK_1}}{\overline{OL}} = \pm\dfrac{r_1}{r}$ and $\dfrac{\overline{OK_2}}{\overline{OL}} = \pm\dfrac{r_2}{r}$. By dividing these equalities, we obtain the equation $\dfrac{\overline{OK_1}}{\overline{OK_2}} = \pm\dfrac{r_1}{r_2}$.

Definition 6.4.3 *The circle of an inversion of two given circles is called (by some authors)* **circle of antisimilarity** *(or* **antisimilitude***) of the two circles.*

6.4.3 Two Geometrical Constructions of Inverse Points and Polar Lines

In view of all things we have seen so far, we write the steps of two additional geometrical constructions of the inverse point P' of a given point $P \neq O$ under an inversion $I_{[O,c]}$, which are quite convenient.

(I) The first is done on the basis of the following well-known Theorem of the geometry of the right triangle.

The square of one perpendicular side of a right triangle is equal to the product of the hypotenuse multiplied by the projection of this side on the hypotenuse.

So, we proceed as follows.

(1) Suppose $c > 0$.
(a) If the point P is on the circle of inversion $C[O, \sqrt{c}]$, then $P = P'$.

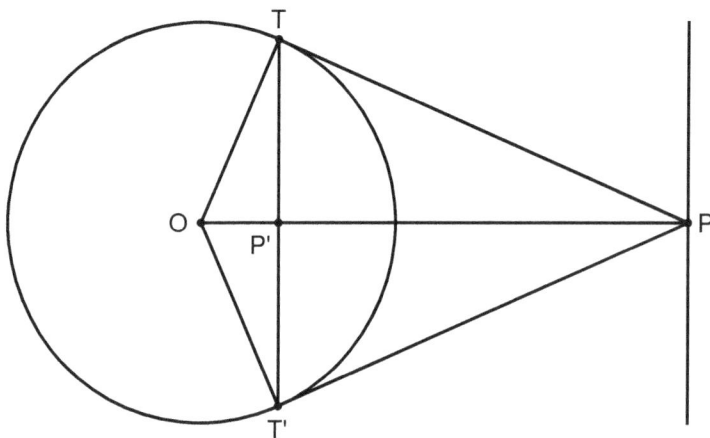

Figure 6.35: Geometrical construction of inverse points
with a positive inversion

(b) If P is outside the circle of inversion, then we draw the two tangents from P to the circle of inversion. See **Figure 6.35**. They touch the circle at points T and T'. Next, we draw the segment TT' which intersects OP at P'. Then, P' is the inverse of P.

This is proven by the fact that the right triangles $P'OT$ and TPO are similar. Then, $\dfrac{OT}{OP} = \dfrac{OP'}{OT}$ and so $OP \cdot OP' = OT^2 = c$. That is, P and P' are inverse points.

Remarks. (1) We observe that in **Figure 6.35** the circles $C[O, \sqrt{c}]$ and $C[P, PO]$ are orthogonal. In such a case (with any orthogonal circles), the inverse of P in $C[O, \sqrt{c}]$ is P' and the inverse of O in $C[P, PO]$ is P'.

(2) See also the construction of K', the inverse of the point K in **Figure 6.27**. What do you observe?

(c) If the given point is P' located inside the circle of inversion, then we reverse the steps. We draw the line OP' and the perpendicular to it at P'. This intersects the circle of inversion at two points T and T'. The tangents at these two points meet the line OP' at a common point P, which is the inverse of P.

The line TT' is the polar line of P and P is the pole of TT'. The perpendicular to OP at P is the polar line of P' and P' is its pole.

(2) Suppose $c < 0$.

Then, given a point P', we work as in the previous case with the positive inversion $I_{[O,|c|=-c]}$, to find $P_1 = I_{[O,|c|=-c]}(P')$. See **Figure 6.36**.

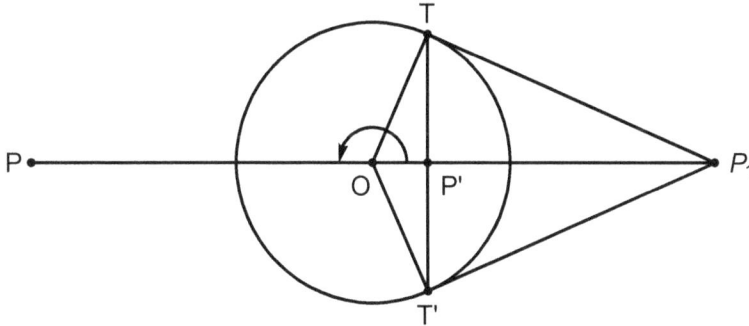

Figure 6.36: Geometrical construction of inverse points with a negative inversion

Next, we apply a half-turn to P_1 to find $P = R_{[O,\pi]}(P_1)$, the inverse of P'. That is,

$$P = R_{[O,\pi]} \circ I_{[O,-c]}(P') = I_{[O,c]}(P').$$

Also, the inverse of P is P', since inversion is involutive.

(See also the construction of K', the inverse of the point K, in **Figure 6.28**. Observe that K' and K'' are symmetrical about the point O, etc.)

(II) Another convenient geometrical construction of the inverse point P' of a point P is shown in **Figure 6.37**.

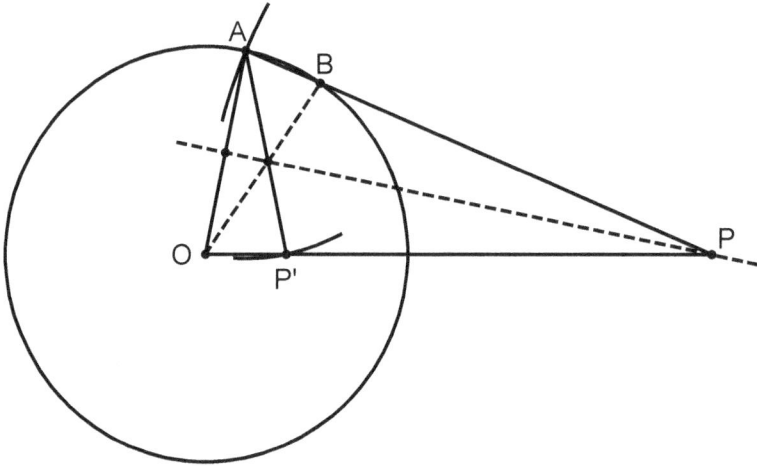

Figure 6.37: Geometrical construction of inverse points

We consider a positive inversion $I_{[O,c>0]}$ and the point P outside the circle of inversion $C[O, r = \sqrt{c}]$. Let point A be a point of intersection of the circle $C[P, PO]$ with the circle of inversion $C[O, r = \sqrt{c}]$. The circle $C[A, AO]$ intersects the straight line OP at two points. The inverse of P is the point of intersection P' lying inside the circle and between O and P.

This follows from the observation that the isosceles triangles POA and AOP' are similar because they have the angle $\widehat{AOP}$ in common. Therefore,

$$\frac{OA}{OP'} = \frac{OP}{OA} \quad \Longleftrightarrow \quad OP \cdot OP' = (OA)^2 = r^2 = c.$$

If the point P' is inside the circle, we draw the circle $C[P', r = \sqrt{c}]$ to find the point A. Then P is the intersection of the straight line OP and the perpendicular bisector of OA. (Or prove and use $OP' = AB$, etc.)

If the inversion is negative, we work first as above with the corresponding positive inversion and then we rotate P' about O by $180°$.

6.4.4   Some Short-cuts in Construction of Inverses

In **Figures 6.38-6.42**, we present some short-cuts in constructing inverses of straight lines and circles, on the basis of the previous results. Of course, we remember that, if a straight line passes through the center of inversion, its inverse is itself, as a set. Also, the circle of inversion is point-wise invariant for positive inversions and set-wise invariant for negative inversions.

Now, let us have a look at **Figure 6.38**. We observe the involutive correspondence, by a positive inversion, of a line l and a circle $\mathcal{F}$ that intersect the
circle of inversion at two points A and B. Since, these points are fixed and the
circle $\mathcal{F}$ must pass through the center of the inversion O, the construction of $\mathcal{F}$
given l, or vice-versa, is immediate, for the points O, A and B are known.

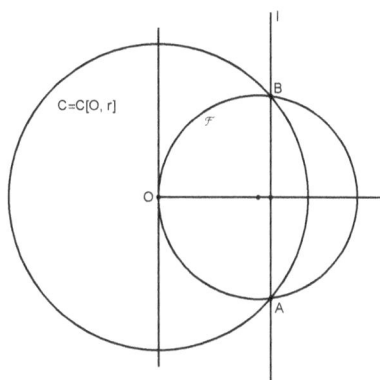

**Figure 6.38: Correspondence, by a positive inversion, of a line l and
a circle $\mathcal{F}$ that intersect the circle of inversion at two points A and B**

Now, we look at **Figure 6.39**. We observe the involutive correspondence,
by a positive inversion, of a line l and a circle $\mathcal{F}$ that are tangent the circle of
inversion at a point C. Point C is fixed and the circle $\mathcal{F}$ must pass through the
center of the inversion O and must also be internally tangent to the circle of
inversion at the point C. Then, the construction of $\mathcal{F}$ given l, or vice-versa, is
immediate, for the points O and C are known and the line l is perpendicular to
OC at C and OC is a diameter of the circle $\mathcal{F}$.

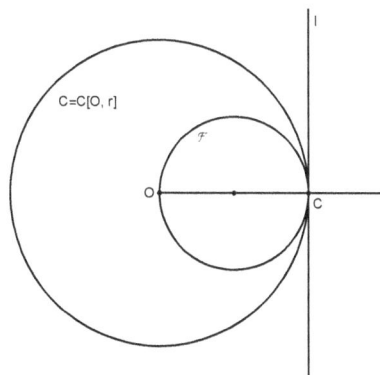

**Figure 6.39: Correspondence, by a positive inversion, of a line l and
a circle $\mathcal{F}$ that are tangent the circle of inversion at a point C**

Now, we look at **Figure 6.40**. We observe the involutive correspondence, by a positive inversion, of two circles $\mathcal{F}$ and $\mathcal{F}'$ that intersect the circle of inversion at two points A and B and do not pass through the center of inversion. The points A and B are fixed. If we have the circles, in front of us, we draw their common diameter through O, which intersects the circle $\mathcal{F}$ at point P away from O and the circle $\mathcal{F}'$ at the point P' near O. Then, P and P' are inverse points.

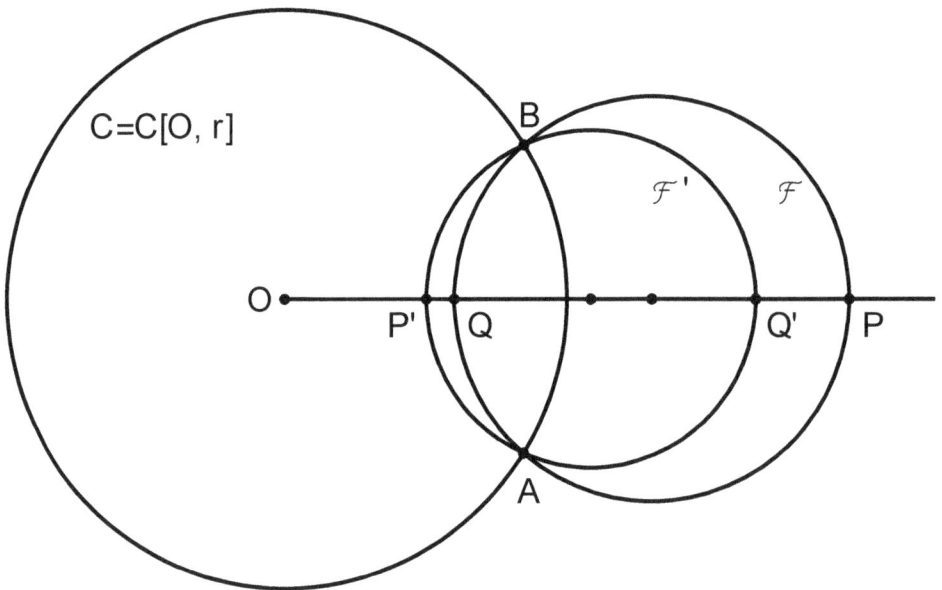

Figure 6.40: **Correspondence, by a positive inversion, of two circles**
$\mathcal{F}$ and $\mathcal{F}'$ that intersect the circle of inversion at
two points A and B and do not pass through
the center of inversion

Now, the construction of $\mathcal{F}'$ given $\mathcal{F}$, or vice-versa, is immediate. The points A, B and P (or P', as above) are known. Next, we construct the point P' (or, P, respectively) to be inverse of P (or P', respectively). Then, we draw the circle $\mathcal{F}' = (ABP')$ [or $\mathcal{F} = (ABP)$, respectively].

In this construction, we could use the points Q and Q' (as in the figure) instead of the points P and P'.

Now, we look at **Figure 6.41**. We observe the involutive correspondence, by a positive inversion, of two circles $\mathcal{F}$ and $\mathcal{F}'$ that are tangent to the circle of inversion at a point C and do not pass through the center of inversion. $\mathcal{F}$

is internally tangent and $\mathcal{F}'$ is externally tangent. The points C is fixed. If we have the circles, in front of us, we draw their common diameter through O, which, besides C, intersects the circle $\mathcal{F}$ at point P and the circle $\mathcal{F}'$ at the point P'. Then, P and P' are inverse points.

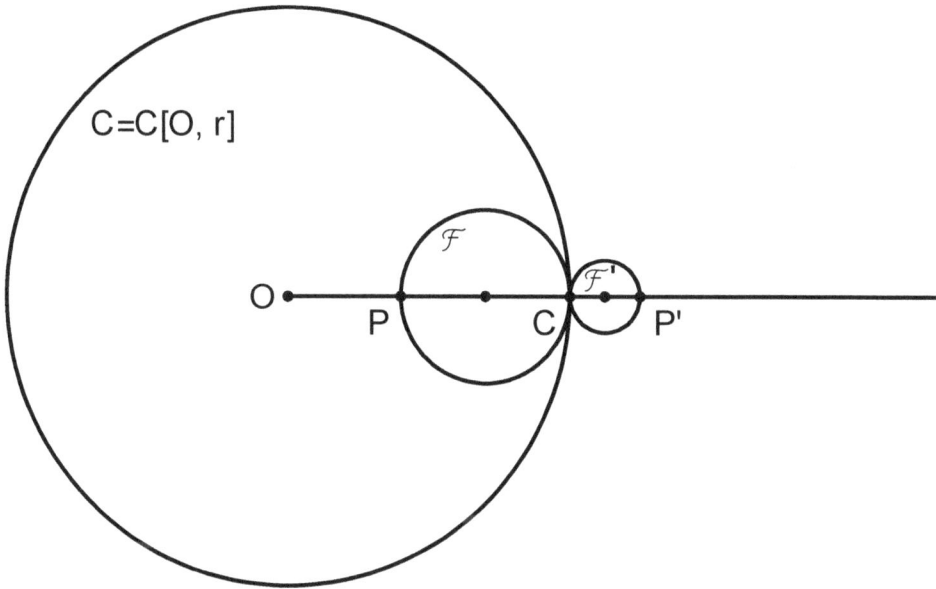

Figure 6.41: Correspondence, by a positive inversion, of two circles
$\mathcal{F}$ and $\mathcal{F}'$ that are tangent to the circle of inversion at a
point C and do not pass through the center of inversion

Now, the construction of $\mathcal{F}'$ given $\mathcal{F}$, or vice-versa, is immediate. The points C and P (or P') are known. Next, we construct the point P' (or, P, respectively) to be inverse of P (or P', respectively). Then, we draw the circle $\mathcal{F}'$ (or $\mathcal{F}$, respectively) with diameter PC (or $P'C$, respectively).

Next: To figure out the inverse of a circle $C[K, r]$ under an inversion $I_{[O,c]}$, with $c > 0$, and $C[K, r]$ not passing through O, we work as in the figure **Figure 6.42** in the following easy steps.

(1) We draw the diameter line OK, which defines the diameter CD of $C[K, r]$.

(2) We find C' and D', the inverse points of C and D (on OK), respectively.

(3) The circle $C[K', r']$ with diameter $C'D'$ is the inverse of $C[K, r]$.

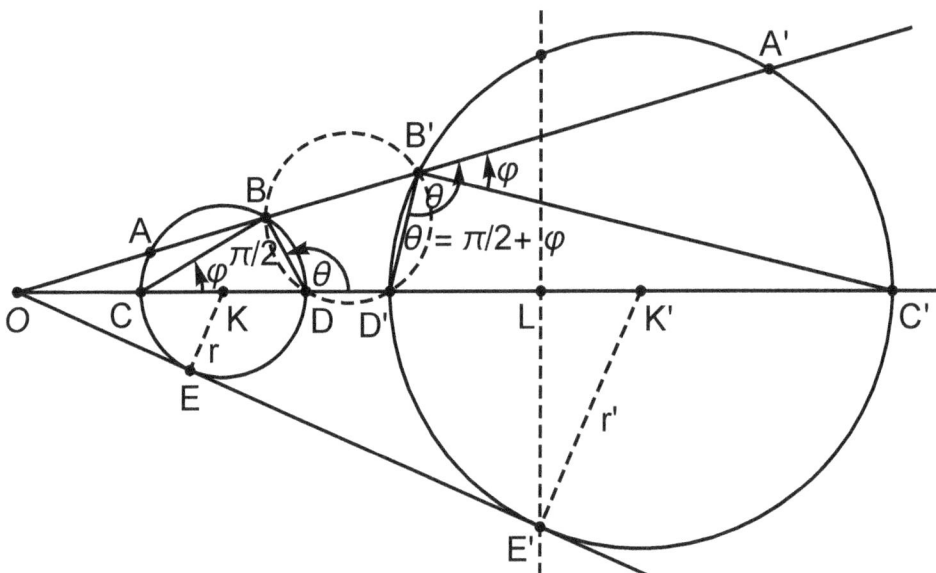

Figure 6.42: Construction of $C[K',r']$, the inverse of a circle $C[K,r]$, under an inversion $I_{[O,c]}$ such that $O \notin C[K,r]$

The justification is simple. For any line through O that intersects $C[K,r]$ at A and B (or is tangent at E), we find the inverse points A' and B' (or E'). So,

$$OA \cdot OA' = OB \cdot OB' = c = OE \cdot OE' = OC \cdot OC' = OD \cdot OD'.$$

Therefore, we have got several inscribable quadrilaterals, but we single out the $BDD'B'$ and $BCC'B'$. Also, the angle $\widehat{CBD}$ is a right angle, as inscribed in a half circle (CD is a diameter). Then, we have

$$\widehat{BCD} = \widehat{A'B'C'} := \phi,$$
$$\phi + \frac{\pi}{2} = \widehat{BDD'} = \widehat{A'D'C'} = \widehat{D'B'C'} + \phi.$$

So, $\widehat{D'B'C'} = \dfrac{\pi}{2}$ is a right angle and so B' is on the circle $C[K',r']$. (Justify these equations by elementary geometry.)

{The homothetic relation of the centers K and K', and of the points A, B with B', A', respectively, as of the points C, D, E with D', C', E', etc., was explained in **Subsection 6.4.2**. We also know that L, the inverse point of K, is the intersection of the polar line of O with respect to circle $C[K',r']$ with the line OKK' and $L \neq K'$, but L is closer to O than K'.}

Peaucellier's[1] Cell.

We consider a rhombus $ABCD$ (all sides are equal). With basis one of its diagonals, say AC, we make an isosceles triangle OAC. Then, B and D are inverse points under the inversion with center O and power $c = OA^2 - AB^2$. See **Figure 6.43**.

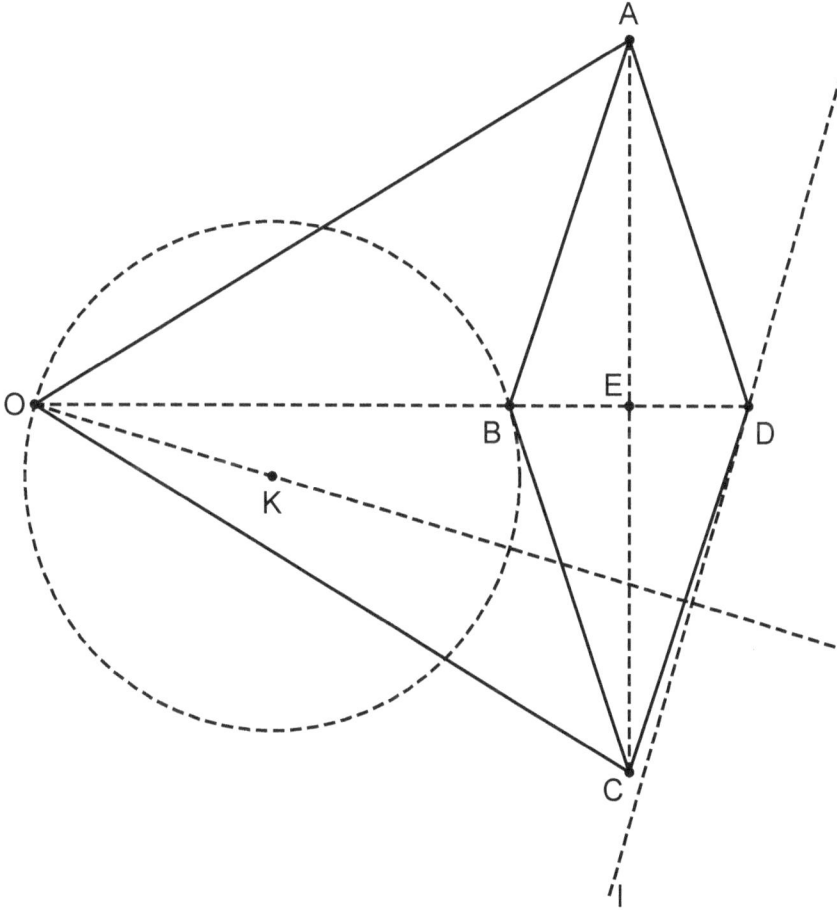

Figure 6.43: Peaucellier's Principle and Cell.

The **proof** follows from the fact that the points O, B and C are on the perpendicular bisector of the diagonal segment AC and the diagonals of the rhombus intersect each other orthogonally at their common midpoint E. Then,

---

[1]Charles-Nicolas Peaucellier, French engineer and army officer, 1832-1913. He invented the inversor in 1864.

applying the Pythagorean Theorem, we get

$$OB \cdot OD = (OE - BE)(OE + ED) = (OE - BE)(OE + BE)$$
$$= OE^2 - BE^2 = OA^2 - AE^2 - (BA^2 - AE^2) = OA^2 - BA^2 = OA^2 - AB^2,$$

which is a fixed number.

Now, if, e.g., B moves along a circle with center K and through O, then D moves along the straight line $l \perp OK$, which is the inverse of the circle, or the circle is the inverse of the straight line traced by B as D traces the straight line.

The instrument can be used in drawing inverse figures by fixing O on the paper. The circle of inversion can be drawn by flattening the rhombus so that the points B and D coincide. Then, the circle of inversion has center O and radius $OB = OD$.

6.4.5   Distance of Inverse Points

We consider an inversion $I_{[O,c]}$ and two points A and B. We let A' and B' be the inverse points of A and B, respectively. Then,

$$\overline{OA} \cdot \overline{OA'} = \overline{OB} \cdot \overline{OB'} = c \qquad \text{and so} \qquad OA \cdot OA' = OB \cdot OB' = |c|.$$

We would like to find the distance of A' and B'.

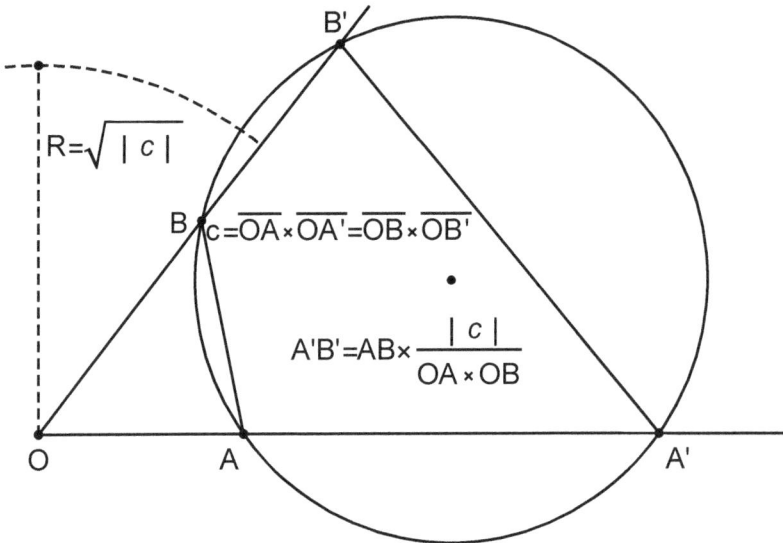

Figure 6.44: For the distance between inverse points (circle case)

(1) Assume that the points A, B, A' and B' are **concyclic**. See **Figure 6.44**. Then, the triangles OAB and $OA'B'$ are (oppositely) similar. Hence, we

have

$$\frac{OA}{OB'} = \frac{OB}{OA'} = \frac{AB}{A'B'}.$$

So,

$$A'B' = AB \cdot \frac{OA'}{OB} = AB \cdot \frac{OA' \cdot OA}{OB \cdot OA} = AB \cdot \frac{|c|}{OA \cdot OB}.$$

Finally, the distance formula for the inverse points is

$$A'B' = AB \cdot \frac{|c|}{OB \cdot OA}.$$

Remark: If OP and OP' is the distance of O from the lines AB and $A'B'$, respectively (draw them in **Figure 6.44**), then the similarity of the triangles OAB and $OA'B'$ proves that $\dfrac{A'B'}{AB} = \dfrac{OP'}{OP}$ or $\dfrac{A'B'}{OP'} = \dfrac{AB}{OP}$.

(2) Assume that the points A, B, A' and B' are **collinear**, which is the case if O, A, and B are collinear. See **Figure 6.45**.

Figure 6.45: For the distance between inverse points (straight line case)

Then,

$$\overline{A'B'} = \overline{A'O} + \overline{OB'} = -\frac{c}{\overline{OA}} + \frac{c}{\overline{OB}} = \frac{c(\overline{OA} - \overline{OB})}{\overline{OB} \cdot \overline{OA}} = \frac{c \cdot \overline{BA}}{\overline{OB} \cdot \overline{OA}} = \frac{-c \cdot \overline{AB}}{\overline{OB} \cdot \overline{OA}}.$$

So, the distance formula for the inverse points is

$$A'B' = \frac{|c| \cdot AB}{OB \cdot OA},$$

which is a formula similar to the formula of case (**1**).

6.4.6 Inversion Preserves Ratios Similar to Cross-Ratio

We consider four points A, B, C and D in the plane and their inverses A', B', C' and D' under an inversion $I_{[O,c]}$. Then, it holds

$$\frac{AC}{AD} \div \frac{BC}{BD} = \frac{AC \cdot BD}{AD \cdot BC} = \frac{A'C'}{A'D'} \div \frac{B'C'}{B'D'} = \frac{A'C' \cdot B'D'}{A'D' \cdot B'C'}.$$

This follows by the distance formula of inverse points found above. Namely, by applying the distance formula and simplifying, we get

$$\frac{A'C'}{A'D'} = \frac{AC}{AD} \cdot \frac{OD}{OC} \quad \text{and} \quad \frac{B'D'}{B'C'} = \frac{BD}{BC} \cdot \frac{OC}{OD}.$$

By multiplying these two equations side-wise, we obtain the claimed equality.

If the eight points are on the same straight line, considered as an axis going through O necessarily, we find the same relation with oriented segments. Namely:

$$\frac{\overline{AC}}{\overline{AD}} \div \frac{\overline{BC}}{\overline{BD}} = \frac{\overline{AC} \cdot \overline{BD}}{\overline{AD} \cdot \overline{BC}} = \frac{\overline{A'C'}}{\overline{A'D'}} \div \frac{\overline{B'C'}}{\overline{B'D'}} = \frac{\overline{A'C'} \cdot \overline{B'D'}}{\overline{A'D'} \cdot \overline{B'C'}}.$$

Apropos of this equation, we give the following standard definition:

Definition 6.4.4 *Given four points A, B, C and D on a straight line or better an axis, the quantity*

$$\{A, B; C, D\} := \frac{\overline{AC}}{\overline{AD}} \div \frac{\overline{BC}}{\overline{BD}} = \frac{\overline{AC} \cdot \overline{BD}}{\overline{AD} \cdot \overline{BC}} = \frac{\overline{AC}}{\overline{BC}} \div \frac{\overline{AD}}{\overline{BD}} = \frac{\overline{CA}}{\overline{CB}} \div \frac{\overline{DA}}{\overline{DB}}, \quad etc.$$

*is called the **cross-ratio (or double ratio)** of these **four points**.*[2]

So, we have proved, above, that **the cross-ratio is preserved under inversion**.

Furthermore, we see that the cross-ratio can be written in several ways. The cross-ratio has many properties and applications in several branches of mathematics. (See many properties of the cross ratio in the exercises.) The property of preservation of the cross-ratio by inversion is very important and useful. For example, see the complete **proof of Steiner's Porism, Theorem 6.6.7**.

Also, in geometry, **a quadruple of points $\{A, B; C, D\}$ on an axis is harmonic** (see **Definition 4.1.4** and **Figure 4.12**.) iff

$$\{A, B; C, D\} := -1 \qquad \Longleftrightarrow \qquad -\frac{\overline{CA}}{\overline{CB}} = \frac{\overline{DA}}{\overline{DB}}, \qquad etc.$$

Here, C must be between A and B and B between C and D. (If the order of the four points is such, then their cross ratio is negative. But, if both C and D are between A and C, then their cross ratio is positive, etc.) Hence, by what we have proved, if **the quadruple of points $\{A, B; C, D\}$ on an axis is harmonic**, then $\{A', B'; C', D'\} := -1$, and so, the quadruple of the inverse points $\{A', B'; C', D'\}$ is also harmonic (and vice-versa).

6.4.7   About Angle Preservation by Inversions

The angle of two curves at a point of their intersection is the angle between their tangent lines at that point, if of course each of these tangents exists.

The tangent line to a curve at a point A is the limit position of the secant lines AB, as B moves on the curve and approaches A. See **Figure 6.46**. If this limit exists, then the tangent exists. Otherwise the tangent does not exist. If a tangent exists, then it is unique, because the limit position is unique.

---

[2]Other authors defined the **cross-ratio (or double ratio)** in different equivalent ways. Read carefully the exact definition in the book that you study and check its equivalence with the definition we cite here.

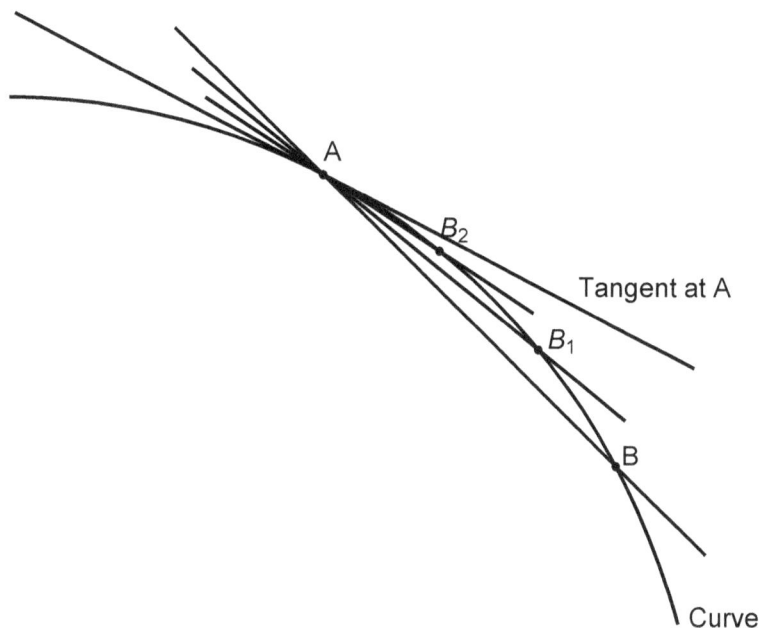

Figure 6.46: Tangent to a curve at a point A is the limit position of the secant lines AB, as B approaches A.

According to this definition a circle has tangent at any of its points, which is the usual geometrical tangent perpendicular to the radius at the point of tangency. A straight line has tangent at any point of it, which is the straight line itself. Also, this definition is the same as the definition of tangent in a course of calculus.

So, if two curves C_1 and C_2 intersect at a point A, and both have tangent AT_1 and AT_2, respectively, at A, then the angle between the two curves is the angle between their tangents. See **Figure 6.47**. In geometry, this angle may be considered non-oriented and its measure be a positive number. But, in more precise situations, we need to consider this angle to be oriented. Then, we write

$$\phi = \measuredangle(C_1, C_2) = \measuredangle(AT_1, AT_2), \quad \text{opposite to} \quad -\phi = \measuredangle(C_2, C_1) = \measuredangle(AT_2, AT_1).$$

We have seen that similarities preserve the absolute value of an angle, but the orientation is preserved by the direct similarities and is reversed by the opposite similarities. We will show that inversions preserve the absolute value of an angle and reverse the orientation. It is enough to prove this with positive inversions since any negative inversion is the commutative composition of a positive inversion and a half-turn, and the half-turns preserve oriented angles. So, in **the following Lemma and Theorem**, without loss of generality, we consider the inversion to be positive. We have already seen the result of **the following Lemma** in case that the curves involved are circles and / or straight

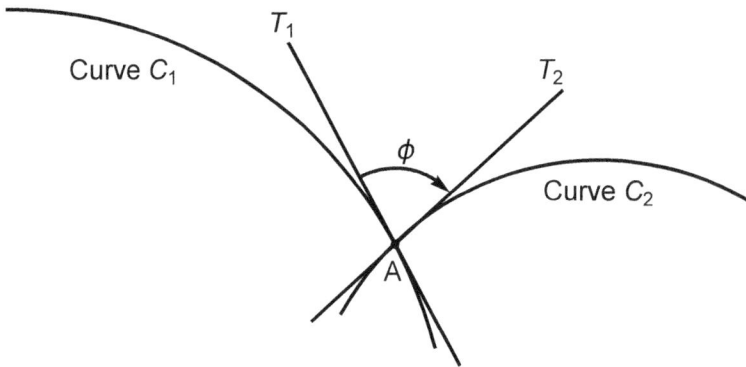

Figure 6.47: Oriented angle from curve C_1 to curve C_2

lines, in **Section 6.4**. Next, we generalize it to any curve and its inverse, as long as they admit tangents.

Lemma 6.4.1 *An inversion reverses the oriented angles between a ray through the center of inversion and two inverse curves at the corresponding inverse points of intersection.*

Proof We consider an inversion with center O and two inverse curves C_1 and C_2. We draw any ray Ox and keep it fixed. Let Ox intersect the curves C_1 and C_2 at P and P', respectively, and let PT and $P'T'$ be the corresponding tangents to the curves C_1 and C_2 at these points. See **Figure 6.48**.

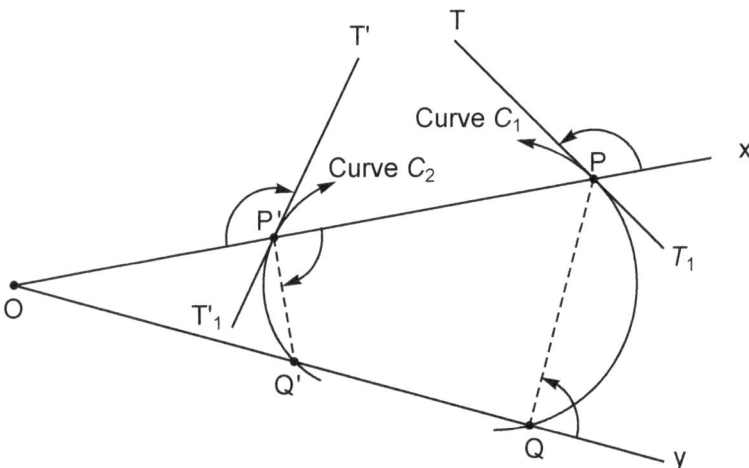

Figure 6.48: Angles from a ray through the center of inversion to two inverse curves

Now, we draw any other ray Oy that intersects the curves C_1 and C_2 at Q and Q', respectively. Since $\overline{OP} \cdot \overline{OP'} = \overline{OQ} \cdot \overline{OQ'} (= c)$, the four points P, P', Q, and Q' are concyclic. So, the angles $\angle(Qy, QP)$ and $\angle(P'x, P'Q')$ are opposite.

Next, we let Q approach P. Then, the ray Oy approaches the ray Ox, point Q' approaches P' and the lines PQ and $Q'P'$ approach the tangents PT and $P'T'$.

Hence, $\lim\limits_{Q \to P} \angle(Qy, QP) = \angle(Px, PT)$ and $\angle(Qy, QP) = -\angle(P'x, P'Q')$.

So, $\lim\limits_{Q \to P} \angle(P'x, P'Q') = \angle(P'x, P'T_1') = \angle(P'O, P'T')$. Thus we obtain,

$$\angle(Px, PT) = -\angle(P'O, P'T').$$

∎

Remark: We observe that the tangents PT and $P'T'$ are either parallel or if they intersect at a point D (not put in the **Figure 6.48**), then the triangle PDP' is isosceles. Therefore D is on the perpendicular bisector of the segment PP'. If one of the (two different) curves is a straight line (and so the other curve is a circle, and vice-versa), then D is on this straight line. (See **Figures 6.23, 6.24, 6.27, 6.28**, and the related accompanying remarks, etc.)

Now, we can prove the following important Theorem:

Theorem 6.4.7 *An inversion reverses oriented angles. That is, given two curves that intersect at a point P, their inverse curves intersect at P', the inverse of P, and the oriented angles of the original curves at P are opposite to the corresponding oriented angles of their inverse curves at P'. So, an inversion preserves the absolute values of the oriented angles and reverses their orientation.*

Proof We consider two curves C_1 and C_2 that intersect at P and their corresponding tangents PT_1 and PT_2, respectively. We also consider their inverse curves C_1' and C_2' that intersect at the point P', inverse of P and their corresponding tangents $P'T_1'$ and $P'T_2'$, respectively. See **Figure 6.49**.

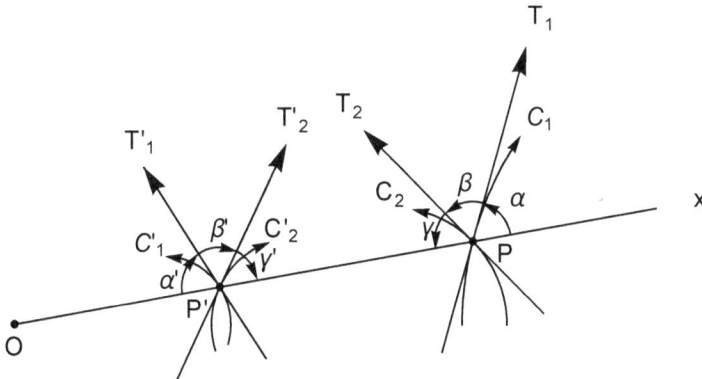

Figure 6.49: An inversion reverses the oriented angle between two curves and their inverses

We draw the ray Ox through the center of inversion O and containing the inverse points P and P'. We let

$$\alpha = \angle(Px, PT_1), \quad \beta = \angle(PT_1, PT_2), \quad \gamma = \angle(PT_2, PO),$$
$$\alpha' = \angle(P'O, P'T_1'), \quad \beta' = \angle(P'T_1', P'T_2'), \quad \gamma' = \angle(P'T_2', P'x),$$

and we have: $\quad \alpha + \beta + \gamma = \pi, \quad \alpha' + \beta' + \gamma' = -\pi, \quad \alpha = -\alpha', \quad \gamma = -\gamma'.$
(The last two equalities are obtained by the **previous Lemma**.) These four equalities imply $\beta = -\beta'$. That is,

$$\angle(C_1, C_2) = \angle(PT_1, PT_2) = -\angle(P'T_1', P'T_2') = -\angle(C_1', C_2').$$

∎

6.5   Some Compositions Involving Inversions

Case (1) We consider **two inversions of the same center** $I_{[O,c_1]}$ and $I_{[O,c_2]}$ and would like to study and investigate the two **compositions**

$$I_{[O,c_2]} \circ I_{[O,c_1]}, \quad \text{and} \quad I_{[O,c_1]} \circ I_{[O,c_2]}.$$

We firstly notice that the two compositions are inverse functions to each other. (Since inversion is involutory, we easily find that their compositions, in either order, is equal to the identity.)

We next consider any point P in the plane and let $P' = I_{[O,c_1]}(P)$ and $P'' = I_{[O,c_2]}(P')$. So, the points O, P, P' and P'' are on the same straight line. See **Figure 6.50**.

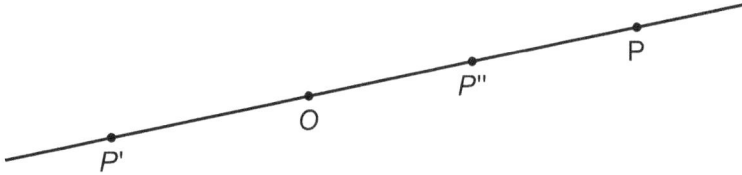

Figure 6.50: The composition of two inversions of the same center is a homothety $I_{[O,c_2]} \circ I_{[O,c_1]} = H_{[O,\lambda=\frac{c_2}{c_1}]}$

We then have the relations

$$\overline{OP} \cdot \overline{OP'} = c_1 \quad \text{and} \quad \overline{OP'} \cdot \overline{OP''} = c_2.$$

Dividing these equations, we find

$$\frac{\overline{OP''}}{\overline{OP}} = \frac{c_2}{c_1} \quad \text{or} \quad \overline{OP''} = \frac{c_2}{c_1} \cdot \overline{OP}.$$

This means that P'' is homothetic to P by the homothety $H_{[O,\frac{c_2}{c_1}]}$, with center O and ratio $\lambda = \dfrac{c_2}{c_1}$. Hence, we have:

Theorem 6.5.1 *The composition of two inversions of the same center is a homothety of the same center and homothetic ratio the ratio of the power of the second inversion divided by the power of the first inversion.*

Remark: The above composition is not commutative, since $\dfrac{c_2}{c_1} \neq \dfrac{c_1}{c_2}$, unless $c_2 = \pm c_1$, or $\lambda = \pm 1$. If $c_2 = c_1$, or $\lambda = 1$, then the composition is the identity. If $c_2 = -c_1$, $\lambda = -1$, then $I_{[O,c_2]} \circ I_{[O,c_1]} = H_{[O,-1]} = R_{[O,\pi]}$ is a half-turn or symmetry about the center point O.

From the Theorem, we draw the following corollaries:

Corollary 6.5.1 *A homothety $H_{[O,\lambda]}$ can be decomposed into a non-commutative composition of two inversions $I_{[O,c_2]}$ and $I_{[O,c_1]}$ of the same center as long as $\lambda = \dfrac{c_2}{c_1}$. So this decomposition can take place in infinitely many ways.*

Corollary 6.5.2 *The inverses $\mathcal{F}_1$ and $\mathcal{F}_2$ of a figure $\mathcal{F}$ by two inversions $I_{[O,c_1]}$ and $I_{[O,c_2]}$ of the same center, are homothetic by the homothety $H_{[O,\lambda]}$, of the same center and homothetic ratio $\lambda = \dfrac{c_2}{c_1}$.*

See **Figure 6.51**.

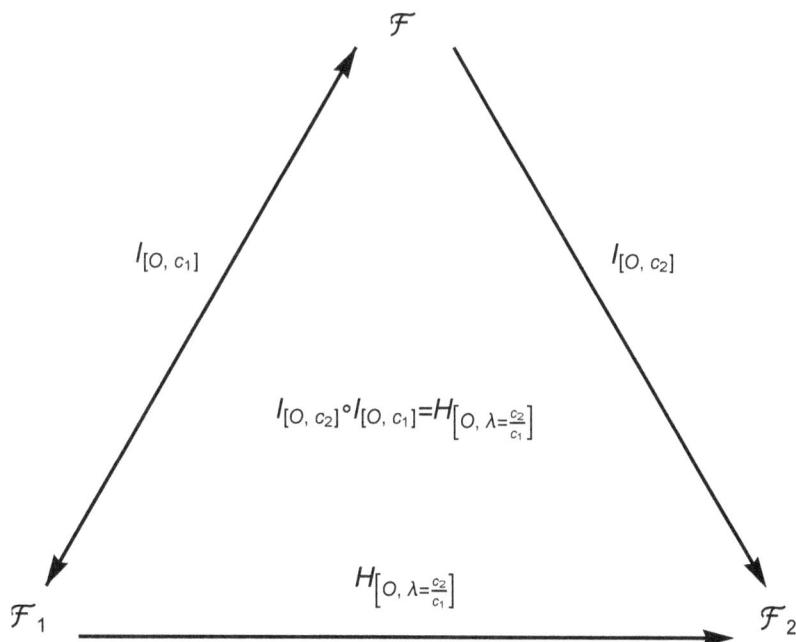

Figure 6.51: Composition of two inversions of the same center is a homothety and images $\mathcal{F}_1$ and $\mathcal{F}_2$ of a figure $\mathcal{F}$

Case (2) We want to study and simplify the **compositions of an inversion** $I_{[O,c]}$ **and a homothety** $H_{[O,\lambda]}$ **of the same center.**

$$I_{[O,c]} \circ H_{[O,\lambda]} \qquad \text{and} \qquad H_{[O,\lambda]} \circ I_{[O,c]},$$

By **Corollary 6.5.1**, we have the decompositions

$$H_{[O,\lambda]} = I_{[O,c]} \circ I_{[O,\frac{c}{\lambda}]} \qquad \text{and} \qquad H_{[O,\lambda]} = I_{[O,c\lambda]} \circ I_{[O,c]}.$$

So, using the involutory property of inversion, we get

$$I_{[O,c]} \circ H_{[O,\lambda]} = I_{[O,c]} \circ I_{[O,c]} \circ I_{[O,\frac{c}{\lambda}]} = I_{[O,\frac{c}{\lambda}]} \quad \text{and}$$

$$H_{[O,\lambda]} \circ I_{[O,c]} = I_{[O,c\lambda]} \circ I_{[O,c]} \circ I_{[O,c]} = I_{[O,c\lambda]}.$$

So, we have obtained the **Result:**

The compositions $I_{[O,c]} \circ H_{[O,\lambda]} = I_{[O,\frac{c}{\lambda}]}$ *and* $H_{[O,\lambda]} \circ I_{[O,c]} = I_{[O,c\lambda]}$ *are new inversions of the same center and powers* $\dfrac{c}{\lambda}$ *and* $c\lambda$, *respectively.*

Case (3) We want to study **the composition**

$$I_{[O_2,c_2]} \circ I_{[O_1,c_1]} \qquad \text{or} \qquad I_{[O_1,c_1]} \circ I_{[O_2,c_2]}$$

of two inversions $I_{[O_1,c_1]}$ and $I_{[O_2,c_2]}$, **of different centers,** $O_1 \neq O_2$, **and any powers,** c_1 and c_2.

As in **case (1)** above, where the centers were the same, this composition results to a new function which is not an inversion. This is expected, because inversions reverse the oriented angles and so **the composition of two inversions preserves oriented angles and therefore is not an inversion.** Also, by the involutive property of inversion, the inverse of the composition $I_{[O_2,c_2]} \circ I_{[O_1,c_1]}$ is $I_{[O_1,c_1]} \circ I_{[O_2,c_2]}$.

Besides the preservation of oriented angles, the composition of two inversions has more interesting properties and depending on the powers and the relative positions of the centers, they may also exhibit some special properties. For instance:

(a) **In such a composition, the line** O_1O_2 **is invariant**, since it is invariant by either inversion. See **Figure 6.52**.

Figure 6.52: Composition of two inversions of different centers has fixed line O_1O_2 and fixed point A as computed below, in **(c)**

(b) If the two circles of inversion intersect (at two points) or are tangent (at one point) and both powers of inversion are positive, then the points of intersection are fixed, since they are fixed by either inversion. See **Figures 6.64** and **6.63**.

(c) **A point A on the line O_1O_2 is fixed, iff $A = A''$,** where

$$I_{[O_1,c_1]}(A) = A', \qquad \text{and} \qquad I_{[O_2,c_2]}(A') = A'',$$
$$\text{and so} \qquad I_{[O_2,c_2]} \circ I_{[O_1,c_1]}(A) = A'' = A.$$

When this happens, we have

$$\overline{AO_2} = \overline{A''O_2},$$

and so,

$$\overline{O_1O_2} - \overline{O_1A} = \frac{-c_2}{\overline{O_2A'}} = \frac{-c_2}{\overline{O_1A'} - \overline{O_1O_2}} = \frac{-c_2}{\frac{c_1}{\overline{O_1A}} - \overline{O_1O_2}}.$$

Simplifying this, we find that in order for a point $A \in O_1O_2$ to be fixed, it must satisfy the **quadratic equation**, in $\overline{O_1A}$, (see **Figure 6.52**)

$$\overline{O_1O_2} \cdot \overline{O_1A}^2 - \left[c_1 - c_2 + (\overline{O_1O_2})^2\right] \cdot \overline{O_1A} + c_1\overline{O_1O_2} = 0 \qquad (\clubsuit).$$

This is the equation for determining fixed points on the line O_1O_2. Depending on the parameters involved, we may have zero, one, or two such fixed points on the line O_1O_2, found by solving this quadratic equation, as we shall see in the three cases of **Figures 6.64, 6.63** and **6.57-6.59**, respectively. Now, we have:

Theorem 6.5.2 *The fixed points of the composition*

$$I_{[O_2,c_2]} \circ I_{[O_1,c_1]}$$

are either those located on the invariant line O_1O_2 and satisfying the above quadratic equation ($\clubsuit$), or they are the common points of the two circles of positive inversions (if these circles intersect at two points).

Proof If a point $A \in \mathcal{P}$ is mapped to A' by $I_{[O_1,c_1]}$ and then A' is mapped to A'' by $I_{[O_2,c_2]}$, then A is fixed by the composition $I_{[O_2,c_2]} \circ I_{[O_1,c_1]}$ iff $A'' = A$.

But then, O_1, A and A' are collinear and also O_2, A' and $A = A''$ are collinear. Therefore:

Either, $A = A' = A''$, in which case A is a fixed point on both inversion circles and both powers are positive, $c_1 > 0$ and $c_2 > 0$, (because negative inversions do not have fixed points. Investigate this a bit more and see the **Remarks** below!).

Or, $A' \neq A = A''$ and the four points O_1, $A = A''$, A', and O_2 are collinear (on the line O_1O_2).

So, the fixed point $A = A''$ is either a common point of the circles of the two given inversions (if they are positive), or is located on the line O_1O_2. ∎

Remarks: (1) We will see or we can now prove that the quadratic equation ($\clubsuit$) has: (a) Two different real solutions if the two circles do not intersect.

(b) One double real solution if the two circles are tangent. (c) Two conjugate complex solutions if the two circles intersect at two points.

(2) Notice that, if one of the inversions is negative, then the common points of the circles of inversions are not fixed. So, if there is a fixed point, it must be on the line O_1O_2 and determined by equation ($\clubsuit$).

Now, given a point P in the plane, we want to find its image P'' by the composition

$$I_{[O_2,c_2]} \circ I_{[O_1,c_1]}.$$

We let $I_{[O_1,c_1]}(P) = P'$ and $I_{[O_2,c_2]}(P') = P''$,   and so,

$$I_{[O_2,c_2]} \circ I_{[O_1,c_1]}(P) = P''.$$

See **Figure 6.53**.

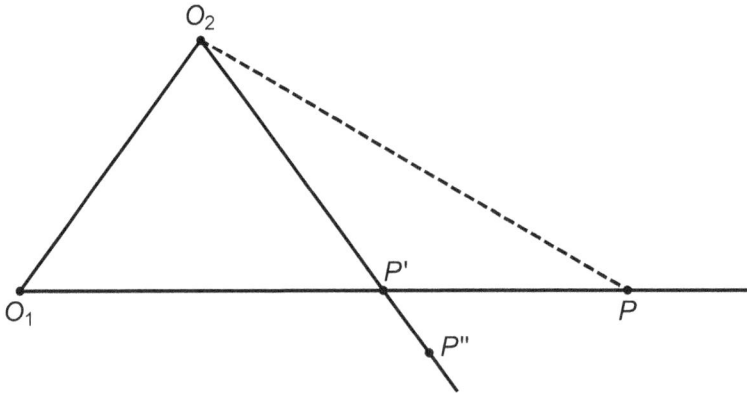

Figure 6.53: Composition of two inversions of different centers

So, we have

$$\overline{O_1P} \cdot \overline{O_1P'} = c_1 \qquad \text{and} \qquad \overline{O_2P'} \cdot \overline{O_2P''} = c_2,$$

$$\text{or:} \qquad \overline{O_1P'} = \frac{c_1}{\overline{O_1P}} \qquad \text{and} \qquad \overline{O_2P''} = \frac{c_2}{\overline{O_2P'}}.$$

So, suppose that we know c_1, c_2 and the positions of O_1, O_2, and P. We can find the position of P' on the line O_1P by

$$\overline{O_1P'} = \frac{c_1}{\overline{O_1P}}.$$

Then, the position of P'' can be found on the line O_2P' by the length of the segment O_2P' and the relation

$$\overline{O_2P''} = \frac{c_2}{\overline{O_2P'}}.$$

The length of the segment O_2P' can be computed by the known **Stewart's Theorem** for a segment from a vertex of a triangle to the opposite side. This Theorem generalizes in the plane in the following way. See **Figure 6.54**.

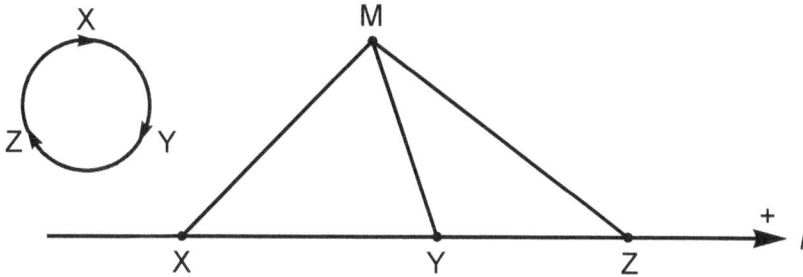

Figure 6.54: **For generalized Stewart's Theorem**

Theorem 6.5.3 (Generalized Stewart's Theorem) *Consider any axis l in the plane and three points X, Y and Z on it (in any order). Let M be any point in the plane (including the axis and the three points). Then, the following relation holds:*

$$\overline{MX}^2 \cdot \overline{YZ} + \overline{MY}^2 \cdot \overline{ZX} + \overline{MZ}^2 \cdot \overline{XY} + \overline{XY} \cdot \overline{YZ} \cdot \overline{ZX} = 0, \quad or$$

$$MX^2 \cdot \overline{YZ} + MZ^2 \cdot \overline{XY} = \overline{XZ} \cdot (MY^2 + \overline{XY} \cdot \overline{YZ}), \quad etc.$$

Find its **proof** *in the bibliography.*

∎

Comparing this generalized Stewart relation with **Figure 6.53**, we get

$$\overline{O_2O_1}^2 \cdot \overline{P'P} + \overline{O_2P'}^2 \cdot \overline{PO_1} + \overline{O_2P}^2 \cdot \overline{O_1P'} + \overline{P'P} \cdot \overline{PO_1} \cdot \overline{O_1P'} = 0.$$

Substituting $\overline{P'P} = \overline{O_1P} - \overline{O_1P'}$ and using $\overline{O_1P} \cdot \overline{O_1P'} = c_1$, we find

$$(O_2P')^2 = \frac{c_1^2 + [O_2P^2 - O_1P^2 - (O_1O_2)^2]c_1 + (O_1O_2)^2 \cdot O_1P^2}{(O_1P)^2}.$$

From this equation, we find the length of O_2P' and then compute the length $\overline{O_2P''} = \dfrac{c_2}{\overline{O_2P'}}$ to locate P'' on the line O_2P', (as P' has been already determined on the line OP).

Remark: By the formulae derived above, we see that: **The composition** $I_{[O_2,c_2]} \circ I_{[O_1,c_1]}$ **is not commutative,** in general. See **Theorem 6.5.6** for the necessary and sufficient condition for this composition to be commutative. Also note that **its inverse** is the composition $I_{[O_1,c_1]} \circ I_{[O_2,c_2]}$.

We will also need the Theorem that follows, which is very important to applications of inversion. It proves the invariance of inversiveness of inverse figures under a new inversion.

Theorem 6.5.4 (Invariance of inversiveness under inversion.) *Suppose two given points P and P_1 are inverse of each other with respect to a given circle $C[Q,r]$ and the whole figure of $C[Q,r]$ and P and P_1 is inverted with respect to another circle $C[O, \sqrt{|c|}]$. Then, in the resulting figure, the images P' and P'_1 of the given points P and P_1 are inverse with respect to $I_{[O,c]}\{C[Q,r]\}$, the image of the given circle $C[Q,r]$ under the inversion $I_{[O,c]}$.*

We must clarify that **in this Theorem** straight lines are considered circles with center a point at infinity. We know that lines and circle may be inverted to other lines and or circles interchangeably. Also, the inversion on a straight line is the reflection in this line, as we have seen in **Section 6.3, (4), (a), (b).**

Proof See **Figures 6.55** and **6.56.**

(1) Suppose the given circle is $C[Q,r]$ and two inverse points with respect to it are P and P_1, as in **Figure 6.55.**

We invert this circle and the two inverse points with respect to another circle $C[O, \sqrt{|c|}]$, for which $O \notin C[Q,r]$. The circle $C[Q,r]$ is mapped to the circle $C[K,r']$ and the points P and P_1 to P' and P'_1, respectively. We must prove that P' and P'_1 are inverses with regard to the new circle $C[K,r']$.

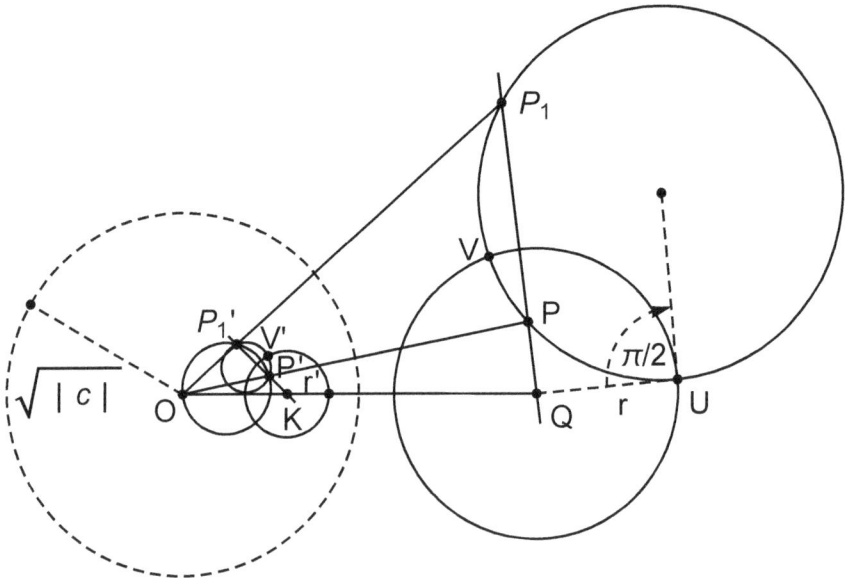

Figure 6.55: Invariance of inversiveness under inversion

We draw either the radial line QPP_1 or any two circles through the points P and P_1. The line or these circles are orthogonal to $C[Q,r]$. (See **Theorems 6.2.1, 6.2.2** and **Corollary 6.2.1.**) Since inversion preserves orthogonality the inverses of the radial line QPP_1 and the circles through the points P and P_1

are either circles or lines orthogonal to the circle $C[K, r']$ and pass through P' and P'_1. In **Figure 6.55**, we have inverted the radial line QPP_1 to the circle $(OP'P'_1)$ and the circle (PUP_1V) to the circle $(P'P'_1V')$. These new circles intersect the circle $C[K, r']$ orthogonally.

Therefore, by **Theorems 6.2.1, 6.2.2** and **Corollary 6.2.1**, the points P' and P'_1 are inverse points with respect to $C[K, r']$. (Thus, the points K, P', and P'_1 are collinear.)

(2) In **Figure 6.56**, the center O of the new circle $C[O, \sqrt{|c|}]$ is on the circle $C[Q, r]$. So, by the new inversion $I_{[O,c]}$, the circle $C[Q, r]$ becomes the straight line AB through the common points A and B of $C[Q, r]$ and $C[O, \sqrt{|c|}]$.

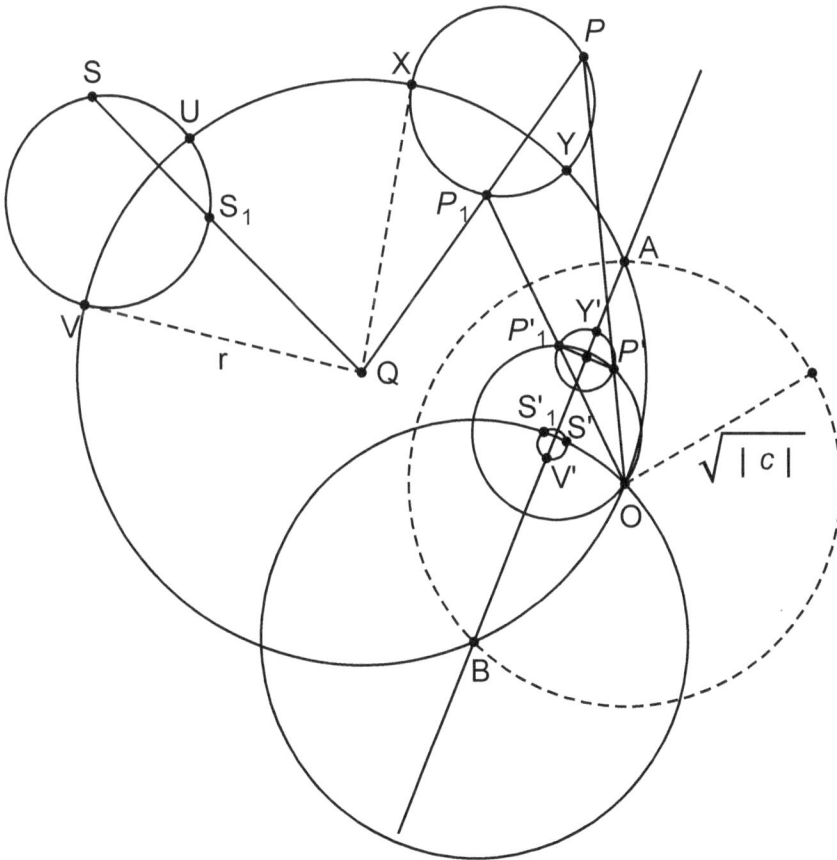

Figure 6.56: Invariance of inversiveness under inversion

The images (P', P'_1) and (S', S'_1), under the inversion $I_{[O,c]}$, with respect to $C[O, \sqrt{|c|}]$, of two pairs of inverse points (P, P_1) and (S, S_1) with regard to the circle $C[Q, r]$, are now symmetrical with respect to the line AB, as it is seen in

Figure 6.56, since the inverses of the radial lines QP_1P and QS_1S, and of the circles (PXP_1Y) and (SUS_1V), are new circles (or straight lines) orthogonal to the line AB, and so the line AB is a diameter to the new circles.

■

The following corollary is very important and supplements **Theorem 6.4.6**.

Corollary 6.5.3 *Suppose two figures $\mathfrak{F}_1$ and $\mathfrak{F}_2$ are mutually inverse under a positive inversion $I_{[O,c]}$, $(c > 0)$. The inverses of $\mathfrak{F}_1$ and $\mathfrak{F}_2$ under any inversion $I_{[Q,k]}$ with $Q \in C[O, \sqrt{c}]$ any point of the circle of the first inversion, are symmetrical in a straight line and therefore isometric. In particular, two inverse circles invert under $I_{[Q,k]}$ into equal circles.*

Proof Since $Q \in C[O, \sqrt{c}]$, the circle $C[O, \sqrt{c}]$ inverts into a straight line under $I_{[Q,k]}$. Then, by the **previous Theorem** the figures $\mathfrak{F}_1$ and $\mathfrak{F}_2$ invert into figures which are mutually inverse with respect to this straight line. Therefore, they are symmetrical in this line.

■

An interesting **family of invariant sets of the composition**

$$I_{[O_2,c_2]} \circ I_{[O_1,c_1]} \quad \textbf{with positive powers} \quad c_1 > 0 \quad \text{and} \quad c_2 > 0,$$

are the circles orthogonal to both circles of the two inversions. These circles have centers on the **radical axis** of the two non-concentric circles of the inversions. See **Figures 6.57, 6.63**, and **6.64**. These circles are invariant by the composition $I_{[O_2,c_2]} \circ I_{[O_1,c_1]}$ or its inverse $I_{[O_1,c_1]} \circ I_{[O_2,c_2]}$, because they are invariant by each inversion separately, when the powers are positive.

In **Figures 6.57, 6.63**, and **6.64**, we see the straight line which is the radical axis of two given circles. Its definition is:

Definition 6.5.1 *The radical axis of two circles, in the plane, is the geometrical locus of all points, in the plane, such that the powers of each point with respect to each of the two given circles are equal.*

Remarks: (1) Let K be a point of this locus and the given circles are $C[O_1, R_1]$ and $C[O_2, R_2]$. See **Figure 6.57**. The **condition of the above Definition is equivalent to**

$$\mathfrak{P}_{O_1}(K) = KO_1^2 - R_1^2 = KO_2^2 - R_2^2 = \mathfrak{P}_{O_2}(K).$$

(2) For points outside of both circles, the condition of this definition is **equivalent** to the fact that the tangent segments KT_1 and KT_2 drawn from any point K of the radical axis to the two given circles are equal, $KT_1 = KT_2$. This follows from $KT_1^2 = KO_1^2 - R_1^2 = \mathfrak{P}_{O_1}(K) = \mathfrak{P}_{O_2}(K) = KO_2^2 - R_2^2 = KT_2^2$. See **Figure 6.57**.

Now, with respect to the above composition, $I_{[O_2,c_2]} \circ I_{[O_1,c_1]}$, with powers $c_1 = R_1^2 > 0$ and $c_2 = R_2^2 > 0$, we have following three cases represented in **Figures 6.57, 6.63**, and **6.64**.

CASE (a): The given circles do not intersect.
So, either $O_1O_2 > R_1 + R_2$ or $O_1O_2 < |R_1 - R_2|$. See **Figures 6.57** and **6.105**.

To find the radical axis in this case, classical geometry teaches, see also **Appendix 6.7, Part II**:

(1) We find the **midpoint** M of the center-segment O_1O_2.
(2) On the center-segment O_1O_2, we determine the oriented segment $\overline{MH}$ by the relation

$$\overline{MH} = \frac{R_1^2 - R_2^2}{2\overline{O_1O_2}} = \frac{c_1 - c_2}{2\overline{O_1O_2}},$$

where $R_1 = \sqrt{c_1}$ and $R_2 = \sqrt{c_2}$ are the radii of the two inversion circles.
(3) Then, the radical axis is the line perpendicular to O_1O_2 at the point H.

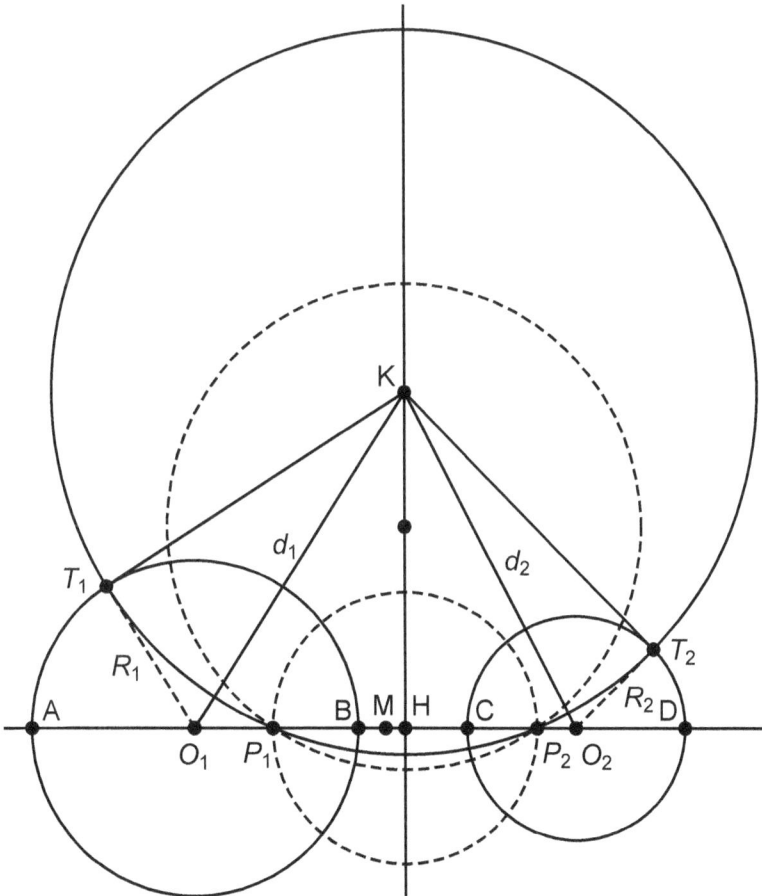

Figure 6.57: Radical axis and orthogonal circles of two given
non-intersecting circles

Remarks: (a) If the circles are outside each other, then H is located between the midpoint M of the center-segment O_1O_2 and the center of the smaller circle. This is the case iff $O_1O_2 > R_1 + R_2$. See **Figure 6.57**.

(b) If one circle is inside the other (but not concentric), then H is located outside both circles on the line O_1O_2 and on the side that the circles are closer together. This is the case iff $O_1O_2 < |R_1 - R_2|$. See **Figure 6.105**.

(c) If the circles are concentric ($R_1 \neq R_2$ and $O_1 = O_2$, so $\overline{O_1O_2} = 0$), then H becomes a point at infinity and the radical axis is the line at infinity.

(d) If the circles are equal ($R_1 = R_2$), but $O_1 \neq O_2$, then $\overline{MH} = 0$ (i.e., $H = M$) and the radical axis is the perpendicular bisector of the segment O_1O_2.

(e) The points, such as, T_1 and T_2 are anti-homologous (see **Definition 4.1.3**) with respect to the center of the corresponding homothety. The four points of any two such pairs are concyclic.

Now, when the circles are disjoint, then either $R_1 + R_2 < O_1O_2$ (the circles are outside each other), or $|R_1 - R_2| > O_1O_2$ (one circle is inside the other), H lies outside both circles (why?), and we have the following **general formulae**:

$$\overline{O_1H} = \overline{O_1M} + \overline{MH} = \frac{\overline{O_1O_2}}{2} + \frac{R_1^2 - R_2^2}{2\overline{O_1O_2}} = \frac{O_1O_2^2 + R_1^2 - R_2^2}{2\overline{O_1O_2}},$$

$$\overline{O_2H} = \overline{O_2M} + \overline{MH} = -\frac{\overline{O_1O_2}}{2} + \frac{R_1^2 - R_2^2}{2\overline{O_1O_2}} = \frac{-O_1O_2^2 + R_1^2 - R_2^2}{2\overline{O_1O_2}}.$$

[**Notice** that these formulae are valid for all positions of the two given circles, examined above or below, because we have stated them with oriented segments (lengths) and when one radius or both radii is / are zero.]

Next, we consider any circle $C[K, r]$ with center a point K on the radical axis and orthogonal to both $C[O_1, R_1]$ and $C[O_2, R_2]$. Its radius is $r := KT_1 = KT_2$, where KT_1 and KT_2 are the two equal tangent segments from K to the circles $C[O_1, R_1]$ and $C[O_2, R_2]$, respectively. See **Figures 6.57** and **Figure 6.105**. Since,

$$R_1^2 + KT_1^2 = O_1K^2 = O_1H^2 + KH^2$$

and when the circles are disjoint H lies outside both circles, i.e., $O_1H > R_1$ and $O_2H > R_2$, we have that $KT_1 = KT_2 > KH$. This inequality implies that any such circle $C[K, r]$ intersects the center line O_1O_2 at two points.

More precisely, with $C[O_1, R_1]$ and $C[O_2, R_2]$ given circles and circles $C[K, r]$ orthogonal to them and with centers on their radical axis, as above, we have the following **important facts**:

(1) *Each circle $C[K, r]$ intersects the center-line O_1O_2 at two points P_1 and P_2, which are symmetrical about H, i.e., H is the midpoint of the segment P_1P_2.*

(2) *The points P_1 and P_2 [claimed in (1)], are the same for all circles $C[K, r]$.*

(3) *When $C[O_1, R_1]$ and $C[O_2, R_2]$ are outside of each other, then P_1 is inside one of them and P_2 inside the other one. Otherwise, one of these points is inside both of them and the other outside.*

(4) *The circle with center $K := H$ and radius $\rho := HP_1 = HP_2$ is the smallest of all circles $C[K, r]$, and H is outside both circles $C[O_1, R_1]$ and $C[O_2, R_2]$.*

We **prove** these facts as follows below. (Refer to **Figure 6.57**, etc.)

Let any circle $C[K, r]$ intersect O_1O_2 at two points P_1 and P_2, as we have proved above. Since $KH \perp P_1P_2$, H is the midpoint of P_1P_2, or $P_1H = HP_2$.

The power of O_1 with respect to the circle $C[K, r]$ is $O_1T_1^2 = O_1P_1 \cdot O_1P_2$, and since $O_1T_1 = R_1$ we find,

$$R_1^2 = (O_1H - HP_1) \cdot (O_1H + HP_2)$$
$$= (O_1H - HP_1) \cdot (O_1H + HP_1) = O_1H^2 - HP_1^2.$$

Therefore, $HP_1^2 = O_1H^2 - R_1^2$, $(\iff O_1H^2 = HP_1^2 + R_1^2 > R_1^2)$.

Similarly, $HP_2^2 = O_2H^2 - R_2^2$, $(\iff O_2H^2 = HP_2^2 + R_2^2 > R_2^2)$.

So, H is outside both circles $C[O_1, R_1]$ and $C[O_2, R_2]$, and since $P_1H = HP_2$,

we have $\sqrt{O_1H^2 - R_1^2} = HP_1 = \dfrac{P_1P_2}{2} = HP_2 = \sqrt{O_2H^2 - R_2^2}$

and thus, the points P_1 and P_2 are predetermined by R_1, R_2, O_1O_2 and its midpoint M. So, P_1 and P_2 are the same for any such circle $C[K, r]$. Also, fact **(4)** follows immediately from this equation.

Using the above relations, we easily prove that: When the initial circles $C[O_1, R_1]$ and $C[O_2, R_2]$ are outside of each other, then P_1 is inside one of them and P_2 inside the other one. Otherwise, one of these points is inside both of them and the other outside. Also: the circle with center H and radius $HP_1 = HP_2$ is the smallest of all circles orthogonal to both $C[O_1, R_1]$ and $C[O_2, R_2]$ and with center on their radical axis. (See **Figures 6.57** and **6.105**.)

At this point, we must give the following related definition:

Definition 6.5.2 *We call the points P_1 and P_2 the* **Poncelet**[3] *points and the circle with center H and diameter P_1P_2 the* **Poncelet circle** *of the circles $C[O_1, R_1]$ and $C[O_2, R_2]$.*

Remarks: (1) Above, we have assumed that the two circles have no common point (iff $O_1O_2 > R_1 + R_2$, or $O_1O_2 < |R_1 - R_2|$). (See **Figures 6.57** and **6.105**.) Then, the two Poncelet points can also be found by the two solutions of the quadratic equation (♣) proven before **Theorem 6.5.2**.

(2) If the original circles are tangent to each other (iff $O_1O_2 = R_1 \pm R_2$), then the Poncelet points coincide with the point of tangency of the circles $C[O_1, R_1]$ and $C[O_2, R_2]$. ($H = P_1 = P_2$. See **Figure 6.63**.) The quadratic equation (♣) has one double solution.

(3) If the original circles intersect each other at two points A and B (iff $|R_1 - R_2| < O_1O_2 < R_1 + R_2$), see **Figure 6.64**, then the quadratic equation (♣) has two complex conjugate solutions. If we use complex coordinates with origin the point H, these solutions represent the points A and B. In this case, we say that Poncelet points are imaginary and are represented by the points A and B.

---

[3] Jean Victor Poncelet, French mathematician, 1788-1867.

In **Figure 6.58**, we have the two given non-intersecting circles of inversion circles, $C[O_1, R_1]$ and $C[O_2, R_2]$, and the Poncelet circle determined by them. Since this circle is orthogonal to both of them, it is easy to show the correspondence of points located on the Poncelet circle by the composition $I_{[O_2,c_2]} \circ I_{[O_1,c_1]}$ (or its inverse $I_{[O_1,c_1]} \circ I_{[O_2,c_2]}$, if we move in reverse order).

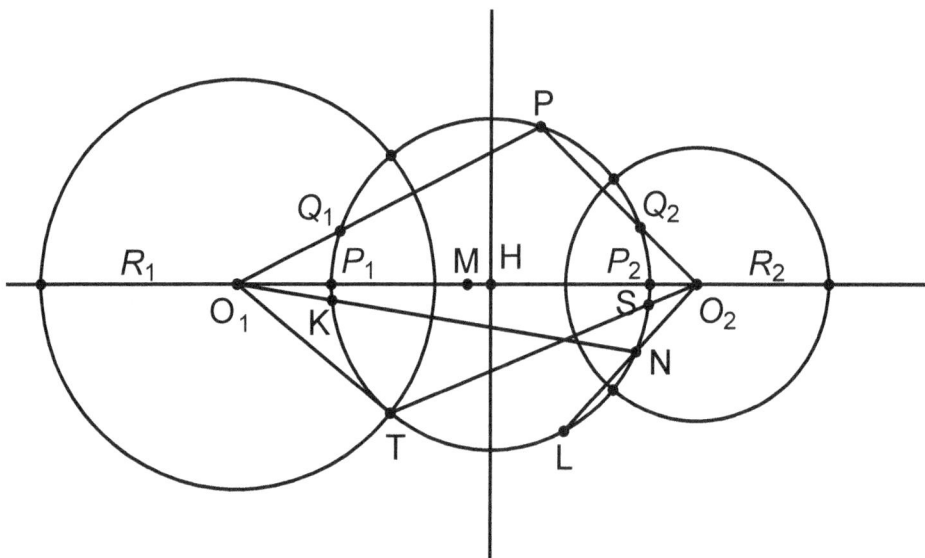

Figure 6.58: Correspondence of points on the Poncelet circle

Namely, we have depicted

$$Q_1 \longmapsto P \longmapsto Q_2, \qquad T \longmapsto T \longmapsto S, \qquad K \longmapsto N \longmapsto L.$$

We now see that *the Poncelet points are fixed by the composition*

$$I_{[O_2,c_2]} \circ I_{[O_1,c_1]} \qquad \text{(or its inverse} \quad I_{[O_1,c_1]} \circ I_{[O_2,c_2]}\text{)},$$

that is, we have,

$$P_1 \longmapsto P_2 \longmapsto P_1, \qquad P_2 \longmapsto P_1 \longmapsto P_2.$$

The points P_1 and P_2 are the only two points on the line O_1O_2 which are fixed by $I_{[O_2,c_2]} \circ I_{[O_1,c_1]}$ and / or $I_{[O_1,c_1]} \circ I_{[O_2,c_2]}$. They satisfy the quadratic equation ($\clubsuit$) of the fixed points on the center line O_1O_2, as stated and proven before **Theorem 6.5.2**. In this case, we have $O_1O_2 > R_1 + R_2$, or $O_1O_2 < |R_1 - R_2|$, and so the quadratic equation has two solutions, which represent the Poncelet points. By solving this quadratic equation, we find the positions of P_1 and P_2 on center line, in terms of $R_1 = \sqrt{c_1}$, $R_2 = \sqrt{c_2}$, and O_1O_2. (Find the two solutions and plot the Poncelet points on the center line, as an exercise!)

See **Figure 6.59**, for correspondence of points on a set-wise invariant circle orthogonal to the circles $C[O_1, R_1]$ and $C[O_2, R_2]$, of the two inversions under the composition

$$I_{[O_2,c_2]} \circ I_{[O_1,c_1]}.$$

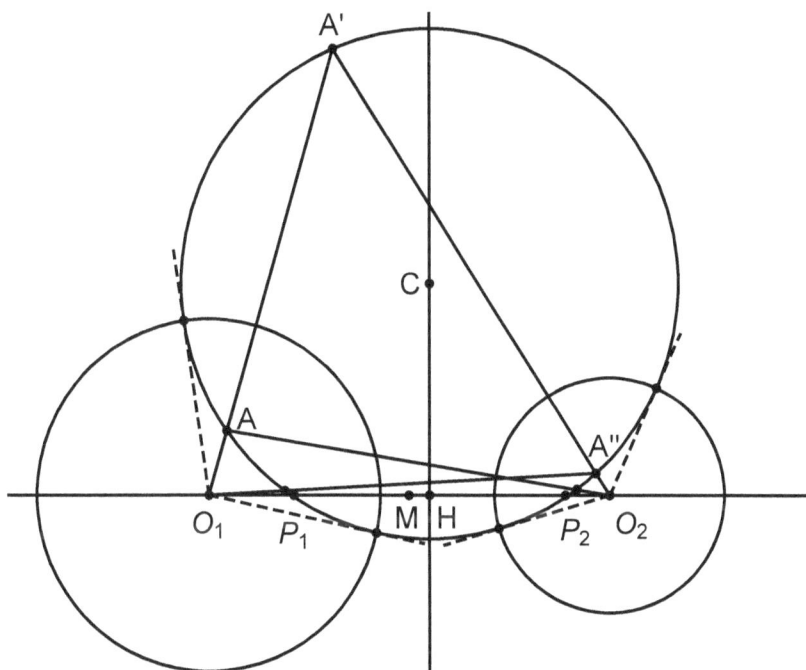

Figure 6.59: **Correspondence of points on an orthogonal circle**

We have depicted the correspondences

$$A \longmapsto A' \longmapsto A'', \qquad P_1 \longmapsto P_2 \longmapsto P_1, \qquad P_2 \longmapsto P_1 \longmapsto P_2,$$

etc. More corresponding points are placed without a label. Check the figure to notice a few additional correspondences.

6.5.1 Practical constructions of the radical axis of two circles that do not intersect

(a) Since each point of the radical axis has the same power with respect to each of the two given circles, one way to construct the radical axis is indicated in **Figure 6.60**.

We intersect the two given circles $C[O_1, r_1]$ and $C[O_2, r_2]$ by a third circle and we draw the common chord lines XY and UV, intersecting at a point C. Then, the line $CH \perp O_1O_2$ is the radical axis of $C[O_1, r_1]$ and $C[O_2, r_2]$.

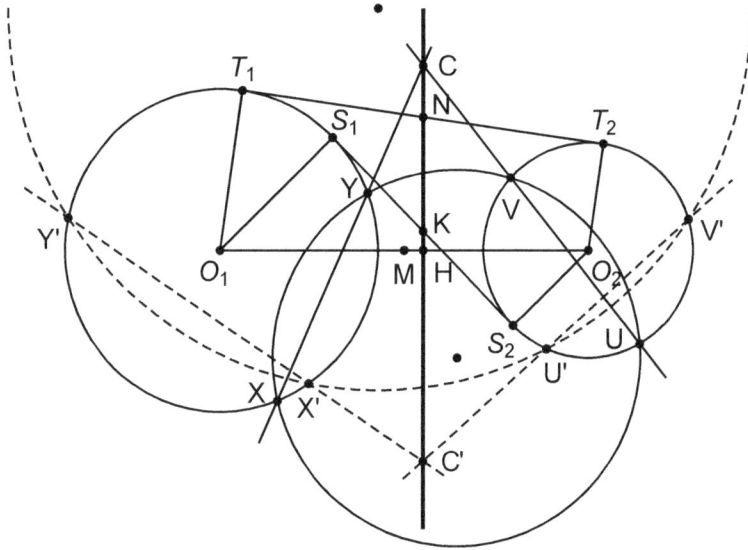

**Figure 6.60: Construction of the radical axis of
two non-intersecting circles**

Or, we intersect the two given circles $C[O_1, r_1]$ and $C[O_2, r_2]$ by another fourth circle and we draw the common chord lines $X'Y'$ and $U'V'$, intersecting at a point C'. Then, the line CC' is the radical axis of $C[O_1, r_1]$ and $C[O_2, r_2]$.

This is so, because $CX \cdot CY = CU \cdot CV$ is the common power of C with respect to all three circles. Therefore, C is on the radical axis of the given circles with centers O_1 and O_2, and the radical axis is perpendicular to the center line O_1O_2. So, knowing C, we draw $CH \perp O_1O_2$, and CH is the radical axis of the two non-intersecting circles.

(b) By the idea of this construction, we also derive the following. The two non-concentric circles are inverse of each other by two inversions, in general. Consider one of these inversions and let O be its center. *Consider now any two pairs of inverse points on them. Then, the two chords defined by these pairs of points intersect on the radical axis.* See **Figure 6.61**.

To wit: we have two non-concentric circles $C[K, r]$ and $C[K', r']$ and two inverse pairs of points (A, A') and (B, B') on them. The lines AB and $A'B'$ intersect at a point P on the radical axis of the two circles PH.

This is so because, as we have already seen with inverse pairs of points (see **Section 6.3**), the points A, A', B, and B' are concyclic, since

$$OA \cdot OA' = c = OB \cdot OB',$$

where c is the power of inversion. So, as in the previous construction, the common point P of AB and $A'B'$ has the property of equal powers with respect

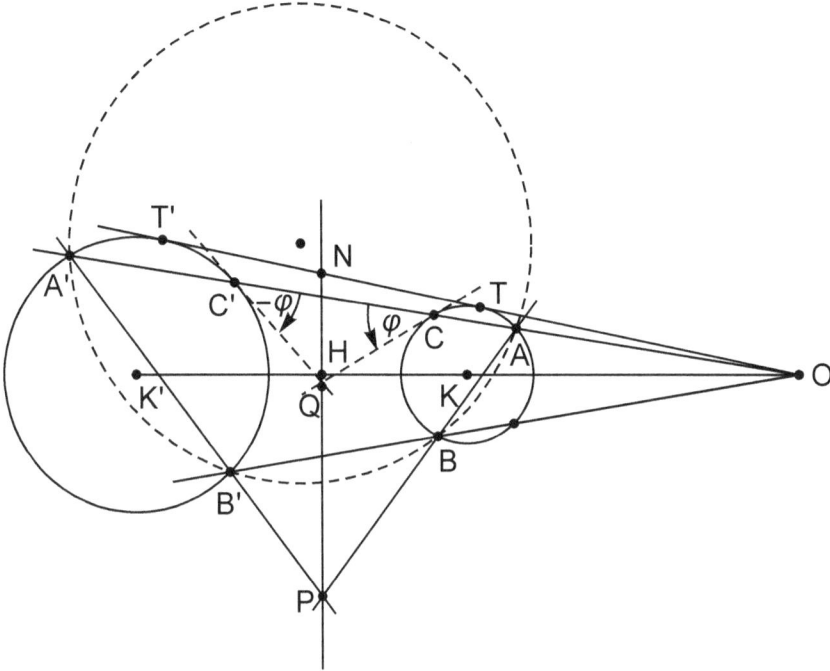

Figure 6.61: Construction of the radical axis of two non-intersecting circles by inverse pairs of points

to the two circles, i.e., $PA \cdot PB = PA' \cdot PB'$. Therefore, P is on the radical axis of the two given circles.

Also, *the tangent lines at two inverse points, let us say C and C', intersect at a point Q on the radical axis.* By **Subsection 6.4.2, Property 8**, these tangent lines form opposite angles with the corresponding ray of inversion OCC'. Hence, $QC = QC'$ and so Q is on the radical axis of the two given circles.

(c) *The radical axis passes through the midpoints of the common tangent segments (external or internal) of the two circles*, since each of these midpoints has the same power with respect to either circle. So, it can be constructed as the straight line defined by the midpoints of two common tangent segments. See points N and K in **Figure 6.60** and point N in **Figure 6.61**.

(The homothetic relation of the centers K and K' was explained in **Subsection 6.4.2**. We have also seen that the inverse point of K is its projection on the polar line of O with respect to the circle with center K'. Identify the radical axes of the circles $C[K, \rho]$ and $C[L, \rho']$ in **Figures 6.27** and **6.28**.)

6.5.2 An Important Property of the Poncelet Points

We have seen that for any two non-intersecting (and so non-tangent) circles $C[O_1, R_1]$ and $C[O_2, R_2]$, their Poncelet points P_1 and P_2 are placed on the center line O_1O_2 and determined by the formulae: $\overline{MH} = \dfrac{R_1^2 - R_2^2}{2\overline{O_1O_2}}$, where M is the midpoint of the center-segment O_1O_2 and H is the foot of their radical axis on the line O_1O_2, and $\sqrt{O_1H^2 - R_1^2} = HP_1 = HP_2 = \sqrt{O_2H^2 - R_2^2}$. Then P_1 and P_2 are placed on the line O_1O_2 appropriately, according to the relative positions of $C[O_1, R_1]$ and $C[O_2, R_2]$. (See **Figures 6.57** and **6.105**.)

The existence of the Poncelet points has an important consequence. All the circles $C[K, r]$ with center on the radical axis of $C[O_1, R_1]$ and $C[O_2, R_2]$ and orthogonal to them, pass through both Poncelet points. Since inversions preserve the absolute value of angles, if we apply an inversion with center one of the Poncelet points and any power, the circles $C[K, r]$ transform to straight lines orthogonal to the images of the circles $C[O_1, R_1]$ and $C[O_2, R_2]$ and pass through the image point of the other Poncelet point. Therefore, these infinitely many straight lines are common diameters to both image-circles. This implies that the two image-circles are concentric. See **Figure 6.62**. So, we have proven:

Theorem 6.5.5 *Given any two disjoint circles, an inversion with center one of the two Poncelet points and any power, maps the two given circles onto two concentric circles with center the image of the other Poncelet point.*

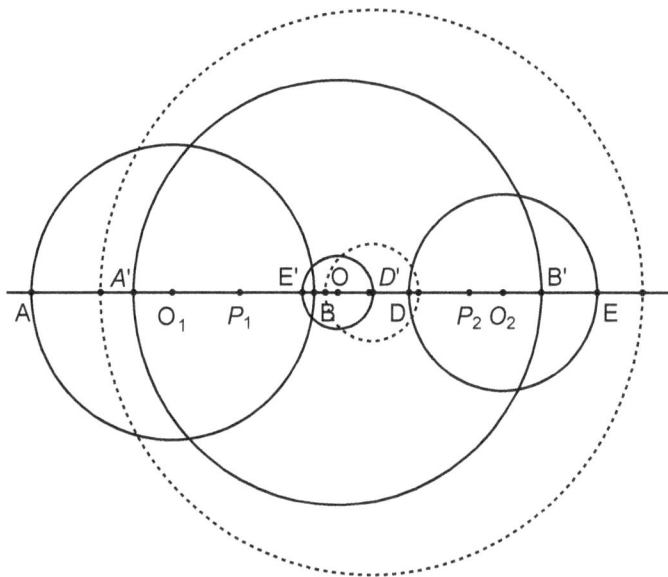

Figure 6.62: Two non-intersecting circles mapped to concentric circles by an appropriate inversion

In **Figure 6.62**, the circles $C[O_1, R_1]$ and $C[O_2, R_2]$ are mapped to two concentric circles with center O by an inversion having center the Poncelet point P_1 and some power. The circle $C[O_1, R_1]$ is mapped to the large circle and the circle $C[O_2, R_2]$ to the smaller circle. The other Poncelet point, P_2, is mapped to O. The points A, B and C are mapped to A', B' and C'. The dotted circles are obtained by using an inversion with center the Poncelet point P_2. The points D, E and F are mapped to D', E' and F'.

The particular choices of these points, in **Figure 6.62**, make the segments $A'B'$ and $D'E'$ have common midpoint the point O. Justify this!

(This result, of inverting two non-intersecting circles onto two concentric circles, can be obtained algebraically, but the system of equations obtained is rather cumbersome to solve. So, we have provided a direct synthetic geometric solution to the problem.)

Next, we continue with **cases (b) and (c)** for the given circles of inversion $C[O_1, R_1]$ and $C[O_2, R_2]$ and the composition of the corresponding inversions.

CASE (b): The given circles $C[O_1, R_1]$ and $C[O_2, R_2]$ are tangent externally or internally at the point H. So, either $O_1O_2 = R_1 + R_2$ or $O_1O_2 = |R_1 - R_2|$, respectively. See **Figure 6.63**.

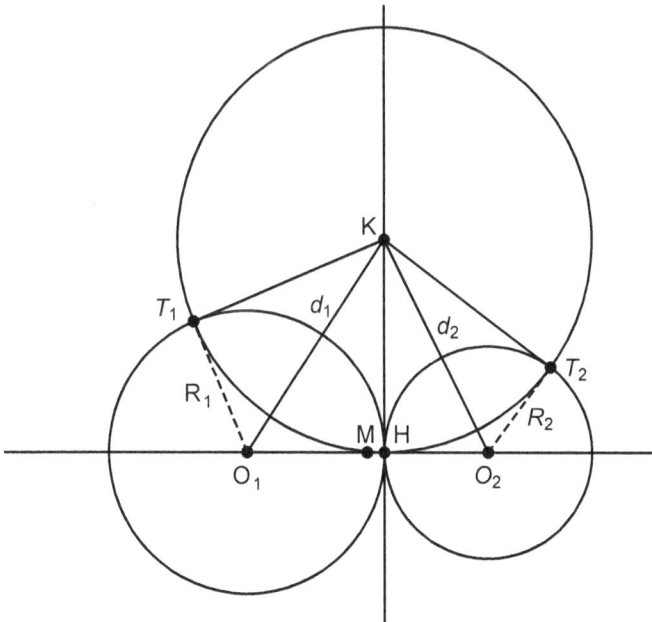

Figure 6.63: The radical axis of two tangent externally or internally circles is the common tangent

Here, the radical axis is easily found to be the common tangent and both Poncelet points coincide with the point H which is fixed by $I_{[O_1,c_1]} \circ I_{[O_2,c_2]}$

(or its inverse $I_{[O_2,c_2]} \circ I_{[O_1,c_1]}$). H satisfies the quadratic equation ($\clubsuit$) of the fixed points on the line O_1O_2, proven in **Theorem 6.5.2**. In this case, $O_1O_2 = R_1 + R_2$ and the quadratic equation with $c_1 = R_1^2$ and $c_2 = R_2^2$ has one double solution, which is the fixed point H on the line O_1O_2. (Check this!)

Remark: If with center the unique Poncelet point H we apply an inversion (as in the previous case), the tangent circles of this case become parallel lines perpendicular to the center line O_1O_2.

CASE (c): The given circles $C[O_1, R_1]$ **and** $C[O_2, R_2]$ **intersect at two points.** So, $|R_1 - R_2| < O_1O_2 < R_1 + R_2$ and the radical axis is easily found to be the straight line of the common chord. See **Figure 6.64**

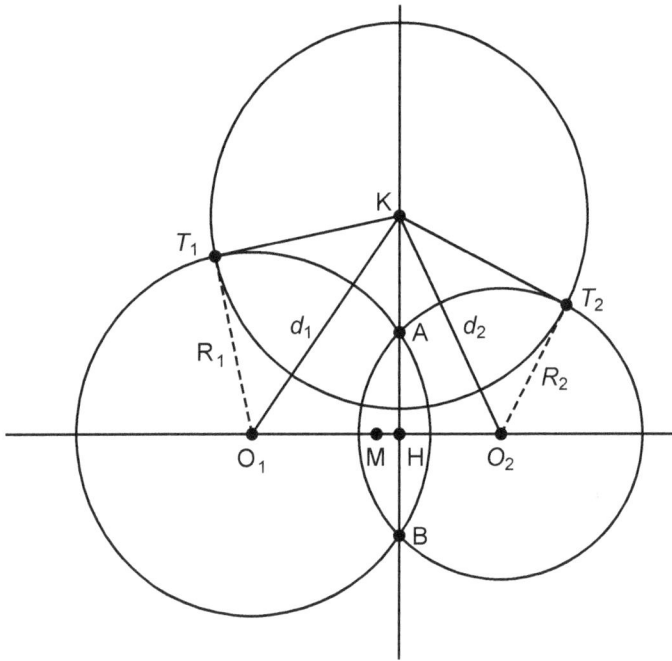

Figure 6.64: Radical axis of two intersecting circles is the line of the common chord

In this case, the common points A and B of the circles of the two inversions are fixed by each inversion and so by their composition. Both A and B are off the line O_1O_2 and symmetrical with respect to it. There are no Poncelet points and no fixed point on O_1O_2. (Here, $|R_1 - R_2| < O_1O_2 < R_1 + R_2$ and the quadratic equation ($\clubsuit$) proven and stated before **Theorem 6.5.2** has two complex solutions, which, if we use complex coordinates, represent the points A and B, complex Poncelet points. Check!) All circles orthogonal to the circles of the two inversions have centers the points of the radical axis which do not belong to the open common chord AB. An inversion with center A or B maps the circles $C[O_1, R_1]$ and $C[O_2, R_2]$ onto two intersecting straight lines (check!).

Remarks: (1) In this case, we cannot ask for common tangent to the two circles $C[O_1, R_1]$ and $C[O_2, R_2]$ from any point D of the common cord AB, but any such point has equal negative powers with respect to both circles. That is, $\mathfrak{P}_{O_1}(D) = \mathfrak{P}_{O_2}(D) = \overline{DA} \cdot \overline{DB}$. (Check!)

(2) If the circles of the two inversions are concentric, then they do not have a radical axis, except for the line at infinity. In this situation, the composition of the two respective inversions produces a homothety studied in **case (1)** at the beginning of **this Section**.

(3) Depending on the parameters involved, the composition $I_{[O_2,c_2]} \circ I_{[O_1,c_1]}$ may or may not have more invariant sets other than the line $O_1 O_2$ and the circles orthogonal to both circles of the two inversions, studied above. For example, we can easily see that:

If the circles of the two inversions are orthogonal, then each one of them is set-wise invariant under the composition. Also, each point of their intersection is a fixed point, (true for intersecting circles of any two positive inversions).

Using complex analysis, we can express the theory of inversion in terms of complex variables in the complex plane. Thus, the compositions

$$I_{[O_2,c_2]} \circ I_{[O_1,c_1]} \qquad \text{and} \qquad I_{[O_1,c_1]} \circ I_{[O_2,c_2]},$$

can be expressed as analytic functions in the complex plane, in the complex variable z. Since they are conformal (that is, they preserve the oriented angles), they are holomorphic.

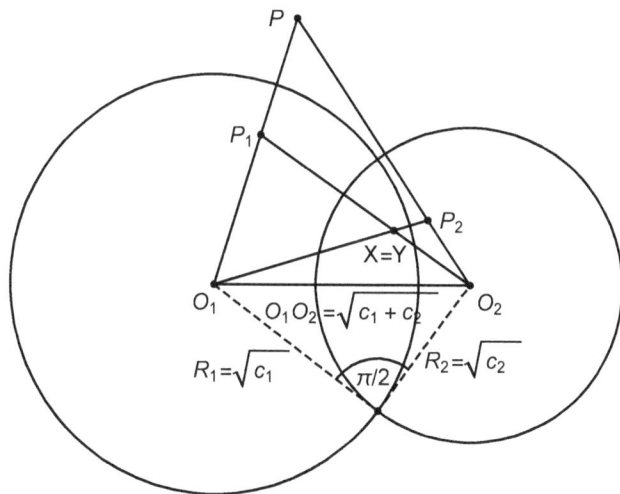

Figure 6.65: The circles of the two inversions are orthogonal

When the circles of the two inversions are orthogonal, the two compositions coincide in the union of the circles of the two inversions, which is a set with accumulation points in the complex plane. Then, as such complex functions, they coincide everywhere in the complex plane. Or, we could write both inversions as complex functions, compute the two compositions, and see that they are equal, in this case. That means, if the circles of the two inversions are orthogonal, then the composition is commutative. See **Figure 6.65**.

So, as in **Figure 6.65**, if we consider any point P in the plane and let,

$$P \longmapsto P_1 \longmapsto X \quad \text{by} \quad I_{[O_2,c_2]} \circ I_{[O_1,c_1]}, \quad \text{and}$$
$$P \longmapsto P_2 \longmapsto Y \quad \text{by} \quad I_{[O_1,c_1]} \circ I_{[O_2,c_2]},$$

then $X = Y$ (the same outcome for both compositions).

So, we have established the following interesting **Theorem**, which we are going to prove synthetically, below.

Theorem 6.5.6 *We assume that the centers of two inversions are different,* $O_1 \neq O_2$*, and both powers are positive,* $c_1 > 0$ *and* $c_2 > 0$*, then:* $I_{[O_2,c_2]} \circ I_{[O_1,c_1]} = I_{[O_1,c_1]} \circ I_{[O_2,c_2]}$ *(this composition commutes) if and only if the circles of the two inversions are orthogonal [so,* $(O_1O_2)^2 = c_1 + c_2 = R_1^2 + R_2^2$*].*

Proof. ($\Longrightarrow$) We assume that the circles of the two inversions with centers O_1 and O_2 are orthogonal and intersect at A and B and P a point in their plane. We let $I_{[O_1,c_1]}(P) = P_1$, $I_{[O_2,c_2]}(P) = P_2$, $I_{[O_1,c_1]}(P_2) = X$, and $I_{[O_2,c_2]}(P_1) = Y$. See **Figures 6.66** (and **6.69**).

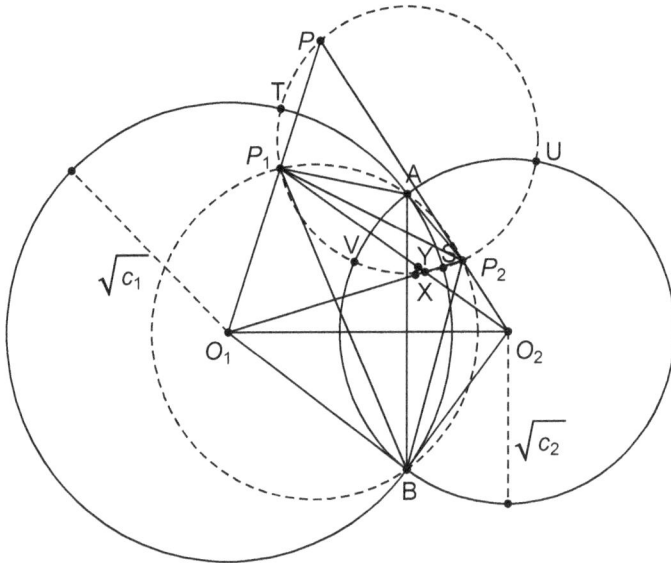

Figure 6.66: The circles of the two inversions are orthogonal
if and only if $I_{[O_2,c_2]} \circ I_{[O_1,c_1]} = I_{[O_1,c_1]} \circ I_{[O_2,c_2]}$

Way 1: We have the inverse pair of points P and P_1 under $I_{[O_1,c_1]}$. The inverses of these points under the inversion $I_{[O_2,c_2]}$, are P_2 and say X, respectively. Also, the inverse of the circle $C[O_1, c_1]$ under the second inversion is itself, since the two circles $C[O_1, c_1]$ and $C[O_2, c_2]$ are orthogonal. Therefore, by **Theorem 6.5.4**, P_2 and X are inverses with respect to the circle $C[O_2, c_2]$. That is, $X = Y$ and so, *we also obtain that* P_1, X, P_2 *and* P *are concyclic*, as (P, P_1) and (P_2, X) are pairs of inverse points in the circle $C[O_1, c_1]$.

Way 2, to be compared with **Way (1)**. Since $c_1 > 0$ and $c_2 > 0$, we consider all segments involved below non-oriented, and so of positive length. We have,

$$O_1 P \cdot O_1 P_1 = c_1 = O_1 P_2 \cdot O_1 X, \quad \text{and} \quad O_2 P \cdot O_2 P_2 = c_2 = O_2 P_1 \cdot O_2 Y.$$

So, the points P, P_1, P_2, X, and Y are on the same circle (PP_1P_2).

We apply the equation

$$(O_2 P')^2 = \frac{c_1^2 + [O_2 P^2 - O_1 P^2 - (O_1 O_2)^2]c_1 + (O_1 O_2)^2 \cdot O_1 P^2}{(O_1 P)^2},$$

proven after **Stewart's Theorem, 6.5.3**, with P_1 in place of P', and we use $(O_1 O_2)^2 = c_1 + c_2 = R_1^2 + R_2^2$, by the orthogonality of $C[O_1, c_1]$ and $C[O_2, c_2]$, and we find

$$(O_2 P_1)^2 = \frac{c_1^2 + (O_2 P^2 - O_1 P^2 - c_1 - c_2)c_1 + (c_1 + c_2) \cdot O_1 P^2}{(O_1 P)^2}$$

$$= \frac{c_1 \cdot O_2 P^2 + c_2 \cdot O_1 P^2 - c_1 \cdot c_2}{(O_1 P)^2}.$$

Similarly with P_2, we find

$$(O_1 P_2)^2 = \frac{c_2 \cdot O_1 P^2 + c_1 \cdot O_2 P^2 - c_2 \cdot c_1}{(O_2 P)^2}.$$

Dividing the last two equations and eliminating the squares, we find

$$\frac{O_2 P_1}{O_1 P_2} = \frac{O_2 P}{O_1 P}.$$

Using the equation, and **the distance formula for inverse points**, proven in **Subsection 6.4.5**, we are going to show that X and Y are equidistant from A and similarly from B. The points A and B are fixed for both inversions and applying the distance formula for inverse points, we get:

$$AP_1 = \frac{c_1 \cdot AP}{O_1 A \cdot O_1 P} = \frac{R_1 \cdot AP}{O_1 P} \quad \text{and} \quad AP_2 = \frac{c_2 \cdot AP}{O_2 A \cdot O_2 P} = \frac{R_2 \cdot AP}{O_2 P}$$

(as $c_1 = R_1^2$, $c_2 = R_2^2$, $O_1 A = R_1$ and $O_2 A = R_2$). Dividing these two relations, we find

$$\frac{AP_1}{AP_2} = \frac{R_1}{R_2} \cdot \frac{O_2 P}{O_1 P}.$$

Also,

$$AX = \frac{c_1 \cdot AP_2}{O_1 A \cdot O_1 P_2} = \frac{R_1 \cdot AP_2}{O_1 P_2} \quad \text{and} \quad AY = \frac{c_2 \cdot AP_1}{O_2 A \cdot O_2 P_1} = \frac{R_2 \cdot AP_1}{O_2 P_1}.$$

Dividing these two relations and using the two ratios found just before, we find

$$\frac{AX}{AY} = \frac{R_1}{R_2} \cdot \frac{AP_2}{AP_1} \cdot \frac{O_2P_1}{O_1P_2}$$

$$= \frac{R_1}{R_2} \cdot \frac{R_2}{R_1} \cdot \frac{O_1P}{O_2P} \cdot \frac{O_2P_1}{O_1P_2} = \frac{O_1P}{O_2P} \cdot \frac{O_2P}{O_1P} = 1.$$

So, $AX = AY := r_a$, for some $r_a > 0$, and similarly we find $BX = BY := r_b$, for some $r_b > 0$. Therefore, the points X and Y belong to the three circles (PP_1P_2), $C[A, r_a]$, and $C[B, r_b]$. Hence,

$$X = Y = O_1P_2 \cap O_2P_1.$$

[Three circles can have only one point in common and here this point is the intersection point of O_1P_2 and O_2P_1, for otherwise, we would have two different lines through two different points (impossible).]

This proves that,

$$\forall \quad P \in \mathcal{P}, \quad I_{[O_2,c_2]} \circ I_{[O_1,c_1]}(P) = I_{[O_1,c_1]} \circ I_{[O_2,c_2]}(P), \qquad \text{and so}$$

$$I_{[O_2,c_2]} \circ I_{[O_1,c_1]} = I_{[O_1,c_1]} \circ I_{[O_2,c_2]}, \quad \text{i.e., the composition commutes.}$$

($\Longleftarrow$) Now, we assume that the composition commutes and we want to prove that the circles of the two inversions are orthogonal. We use the contrapositive argument. So we suppose that the circles of the two inversions are not orthogonal. See **Figure 6.67**.

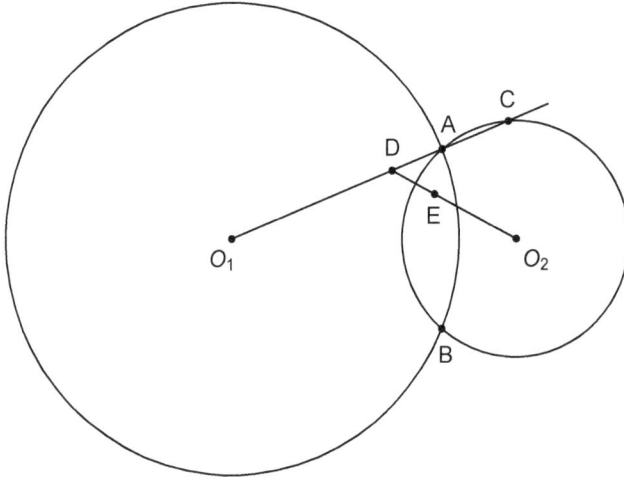

Figure 6.67: The circles of the two inversions are not orthogonal.
Then, $I_{[O_2,c_2]} \circ I_{[O_1,c_1]} \neq I_{[O_1,c_1]} \circ I_{[O_2,c_2]}$

Then, the line O_1A is not tangent to the circle $C[O_2, R_2]$ and so it intersects this circle at a second point C, which may be outside or inside the circle

$C[O_1, R_1]$. Assume that C is outside $C[O_1, R_1]$. {We do analogous work if C is inside $C[O_1, R_1]$. Also notice that $I_{[O_2,c_2]}(C) = C$.}

Let $D := I_{[O_1,c_1]}(C)$ and $E := I_{[O_2,c_2]}(D)$. Then D is inside $C[O_1, R_1]$ and outside $C[O_2, R_2]$ and so E is inside $C[O_2, R_2]$. Therefore, $D \neq A$, $E \neq D$, and $I_{[O_2,c_2]} \circ I_{[O_1,c_1]}(C) = E$ and $I_{[O_1,c_1]} \circ I_{[O_2,c_2]}(C) = I_{[O_1,c_1]}(C) = D$. Since $D \neq E$, we have that $I_{[O_2,c_2]} \circ I_{[O_1,c_1]} \neq I_{[O_1,c_1]} \circ I_{[O_2,c_2]}$. ∎

The **above Theorem** and its **proof** imply some corollaries. But, we will also need the following **Lemma**, which is a byproduct of **Theorem 6.5.4**.

Lemma 6.5.1 *We consider two inversions I_1 and I_2 in two given orthogonal circles $C[O_1, r_1]$ and $C[O_2, r_2]$, respectively. Let A and B be the two points of intersection of these circles and $P \in \mathcal{P}$ be any point of their plane. We let $P_1 = I_1(P)$ and $P_2 = I_2(P)$ be the inverses of P in the two circles, respectively. Then, the four points A, P_1, B, and P_2 are concyclic.*

Proof See **Figure 6.68**.

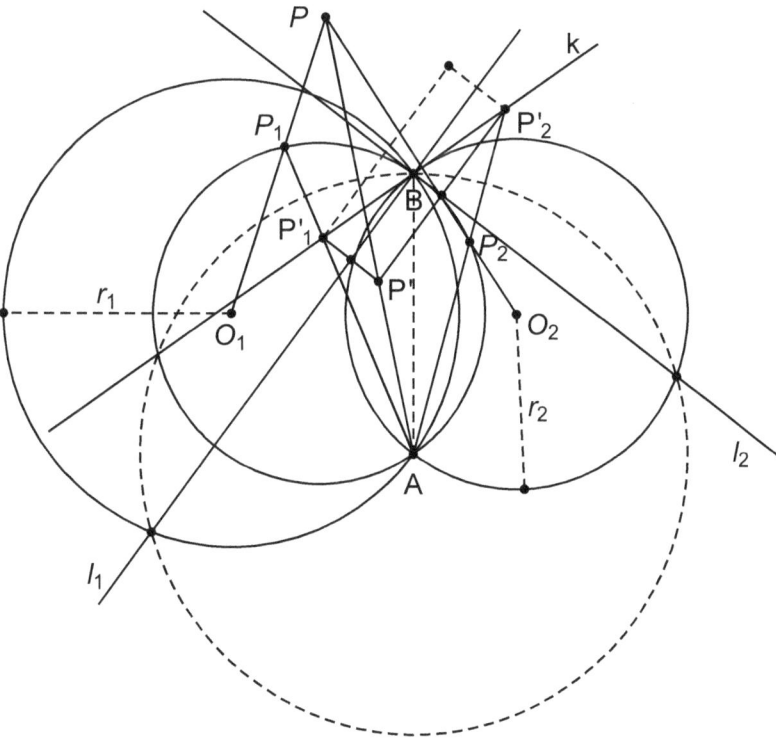

Figure 6.68: The orthogonal circles and a property of inverse points.

We consider the inversion on the circle $\mathcal{C}$ of center A and radius AB. The given circles $C[O_1, r_1]$ and $C[O_2, r_2]$, invert into two orthogonal lines l_1 and l_2, respectively, passing through B, the other common point of these circles and C.

Now, we find P', P'_1 and P'_2, the inverses of P, P_1 and P_2 with respect to C. By **Theorem 6.5.4**, P' and P'_1 are symmetrical with respect to the line l_1, and P' and P'_2 are symmetrical with respect to the line l_2. Since the lines l_1 and l_2 are orthogonal at B, the points P'_1, P'_2 and B are on the same line k (diagonal of a rectangle with center B). (Also, the composition of these two reflections amounts to a rotation about the fixed point B with angle $2 \cdot \dfrac{\pi}{2} = \pi$. This is a symmetry with respect to point B and so P'_1, P'_2 and B are on the same line k.)

Now, we invert the line k in C. Its image is a circle through both A and B and contains the inverses of P'_1 and P'_2 with respect to C, i.e., P_1 and P_2. Therefore, the four points A, P_1, B, and P_2 are concyclic.

∎

Remark: In the **proof of the above Lemma**, the choice of inversion was made by choosing a specific center (A) and a specific power (radius AB). There are problems in which the choice of the center of the inversion suffices for finding the solution, the power being any. So, it is important to be able to distinguish these cases and to make the correct choice of power when necessary. (For example, see **Example 6.6.6**.)

Now, we can continue with the following corollary of **Theorem 6.5.6**. We review the Theorem and **Figure 6.66** once more, and we obtain the following results:

Corollary 6.5.4 *Part (I)*
    (1) The circle (PP_1XP_2) is orthogonal to both circles of inversions.

    (2) The points of intersections U and V of the circle (PP_1XP_2) and the circle $C[O_2, \sqrt{c_2}]$ are inverses under the inversion $I_{[O_1,c_1]}$, and collinear with O_1.

    (3) The points of intersections S and T of the circle (PP_1XP_2) and the circle $C[O_1, \sqrt{c_1}]$ are inverses under the inversion $I_{[O_2,c_2]}$, and collinear with O_2.

    (4) The center K of the circle (PP_1XP_2) is on the radical axis of the given circles $C[O_1, \sqrt{c_1}]$ and $C[O_2, \sqrt{c_2}]$, which is the line of the common chord AB.

    *(5) The lines AB, ST, and UV intersect at the **radical center** Q of the three circles $C[O_1, \sqrt{c_1}]$, $C[O_2, \sqrt{c_2}]$, and (PP_1XP_2). (That is, Q is the unique point in the plane whose powers with regard to the three circles are equal. See **Example 6.6.4**.)*

    Part (II)
    (1) The four points P_1, A, P_2, and B are concyclic.

    (2) The centers of the circles (P_1AP_2B) and (PAB) are on the line O_1O_2.

    (3) The inverse of the circle (PAB) under either $I_{[O_1,c_1]}$, or $I_{[O_2,c_2]}$, is the circle (P_1AP_2B).

(4) The circle (PP_1XP_2) is orthogonal to the two given orthogonal circles $C[O_1, \sqrt{c_1}]$ and $C[O_2, \sqrt{c_2}]$, and to the circles (P_1AP_2B) and (PAB). Its center is on the line AB.

(5) The point X belongs to the circle (PAB).

(6) $AP_1 \cdot BP_2 = AP_2 \cdot BP_1 = \sqrt{c_1}\sqrt{c_2} \dfrac{AP \cdot BP}{O_1P \cdot O_2P}.$

(7) $AX \cdot BP = BX \cdot AP = \sqrt{c_1}\sqrt{c_2} \dfrac{AP_1 \cdot BP_1}{O_2P_1 \cdot O_1P_1} = \sqrt{c_1}\sqrt{c_2} \dfrac{AP_2 \cdot BP_2}{O_1P_2 \cdot O_2P_2}.$

(8) $\dfrac{AP_1 \cdot BP_1}{O_1P_1 \cdot O_2P_1} = \dfrac{AP_2 \cdot BP_2}{O_1P_2 \cdot O_2P_2}.$

(9) The points X, Q and P are collinear.

(10) The points P_1, Q and P_2 are collinear.

(11) $QU \cdot QV = QS \cdot QT = QA \cdot QB = QX \cdot QP = QP_1 \cdot QP_2 = \rho^2 - KQ^2$, *where ρ is the radius of the circle (PP_1XP_2).*

Proof See **Figures 6.69**, **6.68** and **6.66**.

(I) (1)-(2) The two original circles, $C[O_1, \sqrt{c_1}]$ and $C[O_2, \sqrt{c_2}]$, are given to be orthogonal. The circle (PP_1XP_2) passes through two pairs of inverse points, namely (P, P_1) and (X, P_2), under $I_{[O_1,c_1]}$. By **Theorems 6.2.1, 6.2.2** and **Corollary 6.2.1**, (PP_1XP_2) is orthogonal to the circle $C[O_1, \sqrt{c_1}]$ of the first inversion with center O_1 and so the points of intersection U and V are inverses under $I_{[O_1,c_1]}$ and collinear with O_1.

(3) Similarly the circle (PP_1XP_2) is orthogonal to the circle $C[O_2, \sqrt{c_2}]$, of the second inversion with center O_2 and the points of intersection S and T are inverses under $I_{[O_2,c_2]}$ and collinear with O_2.

(4)-(5) These claims follow from the definitions of radical axis of two circles and radical center of three circles. (See **Definition 6.5.1** and **Example 6.6.4**.)

(II) (1) The fact that the four points P_1, A, P_2, and B are concyclic follows from the previous **Lemma 6.5.1**.

(2) AB is a chord for both circles and the center line O_1O_2 is the perpendicular bisector of AB. Thus, the center line O_1O_2 contains the center of any circle having the segment AB as a chord.

(3) The inverses of the circle (PAB) under either inversion $I_{[O_1,c_1]}$ and $I_{[O_2,c_2]}$ coincide with the circle (P_1AP_2B) because: A and B are fixed points under either inversion, P_1 is the inverse of P under the inversion $I_{[O_1,c_1]}$ and P_2 is inverse

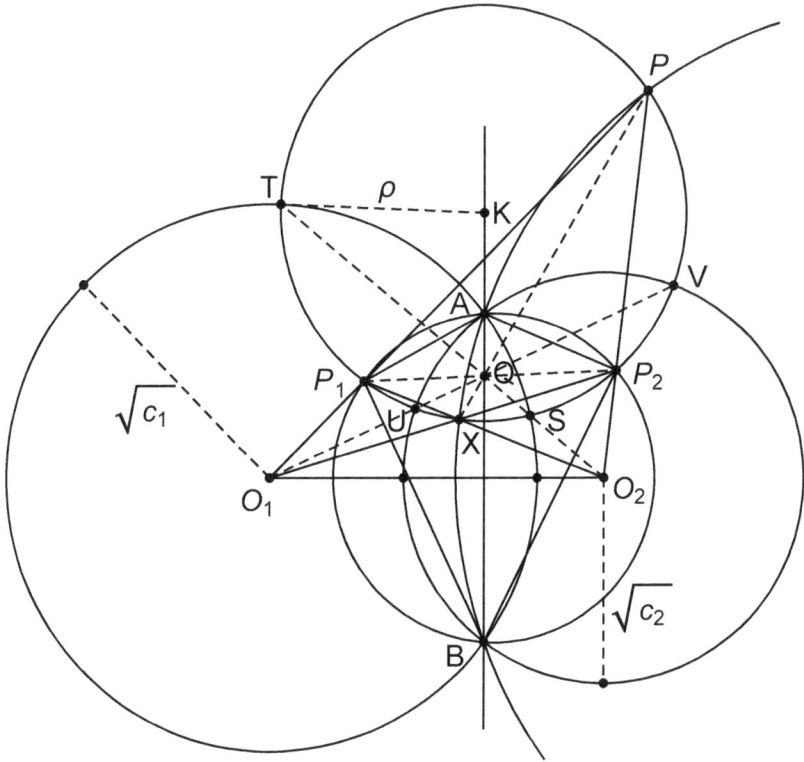

Figure 6.69: Results when the circles of two successive inversions are orthogonal

of P under the inversion $I_{[O_2,c_2]}$. Also, the four points P_1, A, P_2, and B are concyclic.

(4) At first, the circle (PP_1XP_2) is orthogonal to the circles $C[O_1, \sqrt{c_1}]$ and $C[O_2, \sqrt{c_2}]$ because it passes through pairs of inverse points with respect to either circle. Now the circles (P_1AP_2B) and (PAB) go through AB the common chord of $C[O_1, \sqrt{c_1}]$ and $C[O_2, \sqrt{c_2}]$.

Therefore, (PP_1XP_2) is also orthogonal to the circles (P_1AP_2B) and (PAB). (In fact, the same is true for any circle passing true A and B.) Hence, we also get that the center K of the circle (PP_1XP_2) is on the line AB.

(5) The point X belongs to the circle (PAB). To prove this, we will use **Ptolemy's Theorem, 6.6.1**, and **the distance formula for inverse points**, proven in **Subsection 6.4.5**.

The quadrilateral AP_1BP_2 is inscribed in the circle (AP_1BP_2). So, it satisfies

Ptolemy's relation $AP_1 \cdot BP_2 + P_1B \cdot P_2A = AB \cdot P_1P_2.$

Now, we use the **converse of Ptolemy's Theorem** to prove that the quadrilateral $AXBP$ is inscribed in the circle (ABP). The distance formula for inverse points applied on the correct inversions (which ones?) gives

$$AX \cdot BP = \frac{c_2 AP_1}{O_2 A \cdot O_2 P_1} \cdot \frac{c_2 BP_2}{O_2 B \cdot O_2 P_2} = c_2 \frac{AP_1 \cdot BP_2}{O_2 P_1 \cdot O_2 P_2},$$

$$XB \cdot AP = \frac{c_2 BP_1}{O_2 B \cdot O_2 P_1} \cdot \frac{c_2 AP_2}{O_2 B \cdot O_2 P_2} = c_2 \frac{P_1 B \cdot P_2 A}{O_2 P_1 \cdot O_2 P_2},$$

$$\text{and} \quad PX \cdot AB = \frac{c_2 P_1 P_2 \cdot AB}{O_2 P_1 \cdot O_2 P_2} = c_2 \frac{AB \cdot P_1 P_2}{O_2 P_1 \cdot O_2 P_2}.$$

The above four relations prove

$$AX \cdot BP + XB \cdot AP = PX \cdot AB,$$

which, by **Ptolemy's Theorem, 6.6.1**, proves that the quadrilateral $AXBP$ is inscribed in the circle (ABP). So, X belongs to the circle PAB.

(6) Again, applying **the distance formula for inverse points** using the appropriate inversions, we have

$$AP_1 \cdot BP_2 = \frac{c_1 AP}{O_1 A \cdot O_1 P} \cdot \frac{c_2 BP}{O_2 B \cdot O_2 P} = \sqrt{c_1}\sqrt{c_2}\frac{AP \cdot BP}{O_1 P \cdot O_2 P}$$

$$\text{and} \quad AP_2 \cdot BP_1 = \frac{c_2 AP}{O_2 A \cdot O_2 P} \cdot \frac{c_1 BP}{O_1 B \cdot O_1 P} = \sqrt{c_1}\sqrt{c_2}\frac{AP \cdot BP}{O_1 P \cdot O_2 P}.$$

(7) The proof is similar to **(6)**, by applying **the distance formula for inverse points** with two inversions (which ones?) to both $AX \cdot BP$ and $BX \cdot AP$.

(8) It follows from **(7)**.

(9)-(10) We have proved that X belongs to the circles PAB and $PP_1 X P_2$. Then, as in **(I)**, **(3)** and **(4)** above, the claims follow from the definitions of radical axis of two circles and radical center of three circles. (See **Definition 6.5.1** and **Example 6.6.4**.)

(11) Each of these products is the power of the point Q with respect to the circle $PP_1 X P_2$. Therefore, all products stated are equal to $\rho^2 - KQ^2$.

∎

In **(6)** and **(7)** above, we have seen two **inscribable** quadrilaterals, $AP_1 B P_2$ and $AXBP$, in which **the products of their opposite sides are equal**. A **quadrilateral** having these two properties is called **harmonic**. (The harmonic quadrilaterals have many interesting properties. Check bibliography and exercises.)

Before we move to the next Theorem, we must give the following definition and some remarks.

Definition 6.5.3 *A set of all the circles in the plane with the same radical axis is called a **coaxal system of circles** or a **pencil of circles**. The circles of a pencil or coaxal system are said to be **coaxal circles**.*

See **Figures 6.57, 6.63, 6.64** and **Example 6.6.5** and its **Figures**, for the coaxal figures depicted there. There are three kinds of pencils of circles in which even the radical axis may be considered as a limiting circle of infinite radius and center the infinity point of the direction of the center-line:

(1) The pencil in which all the circles pass through two points A and B. The common radical axis is the straight line of the common chord AB. See **Figures 6.64** and **6.103**. In this pencil the smallest circle is the one having diameter the segment AB. We see that two such pencils (systems) based on AB and CD are equal iff $AB = CD$. In such a case, the composition of a parallel translation and a rotation can map the center-line and the radical axis of one system to the center-line and the radical axis of the other system and all the circles of both pencils pass through two pairs of equidistant points.

(2) The pencil in which all the circles are tangent (externally and / or internally) at a point $A = H$. The common radical axis is the common tangent of the circles at the point $A = H$. See **Figures 6.63** and **6.104**. All such pencils (systems) are equal to one another, since we directly see that the composition of a parallel translation and a rotation can map the center-line and the radical axis of one system to the center-line and the radical axis of the other system and all the circles of both pencils are tangent to each other at the intersection of these two straight lines.

(3) The pencil in which no two circles intersect. See **Figures 6.57** and **Figure 6.105**. In such a pencil all the circles have the same Poncelet points and Poncelet circle. The Poncelet circle is orthogonal to all circles of the pencil and is the smallest circle with this property, as having a diameter the segment defined by the Poncelet points. Two such pencils (systems) are equal iff they have equidistant Poncelet points.

The determination of all circles in a pencil of the first and second kind, as given above, and their radical axes, by two circles that are members of this coaxal system is immediate, as we easily observe. But, for the determination of the whole system of the third kind we must say a few words. Along with this, we shall also examine some **additional properties of the Poncelet points** and results. We refer to **Figure 6.57** and we have:

(a) The foot H of the radical axis of the circles of the pencil on the center-line $O_1 O_2$ is the midpoint of the segment of the Poncelet points $P_1 P_2$, i.e., $\overline{P_1 H} = \overline{H P_2}$. The Poncelet points P_1 and P_2 are limiting circles of the pencil of radius zero.

(b) Let AB and CD be the diameters of two circles $C[O_1, R_1]$ and $C[O_2, R_2]$ of the pencil lying on the center-line. In general, the point H is different from M the midpoint of $O_1 O_2$. The two points are the same iff $R_1 = R_2$.

(b1) Since the power of H with respect to any circle of the pencil is fixed, we have:

$$\overline{HA} \cdot \overline{HB} = \overline{HP_1}^2 = \overline{HP_2}^2 \qquad \text{and} \qquad \overline{HC} \cdot \overline{HD} = \overline{HP_1}^2 = \overline{HP_2}^2.$$

Hence, by the **Remark of Case I of Subsection 4.5.5** since H is the midpoint of O_1O_2, the quadruples $\{A, P_1, B, P_2\}$ and $\{P_1, C, P_2, D\}$ are harmonic. That is, $\{P_1, P_2\}$ is common harmonic conjugate pair of the two pairs $\{A, B\}$ and $\{C, D\}$.

(b2) By the harmonicity of these quadruples and the same **Remark**, since O_1 and O_2 are the midpoints of the diameters AB and CD respectively, we also have the analogous relations

$$R_1^2 = \overline{O_1A}^2 = \overline{O_1B}^2 = \overline{O_1P_1} \cdot \overline{O_1P_2} \quad \text{and} \quad R_2^2 = \overline{O_2C}^2 = \overline{O_2D}^2 = \overline{O_2P_1} \cdot \overline{O_2P_2}.$$

Therefore the Poncelet points P_1 and P_2 are inverse of each other with respect to any circle of the pencil.

(c) Now, if we are given two non-intersecting circles $C[O_1, R_1]$ and $C[O_2, R_2]$ and we want to determine the pencil to which they belong, we follow the steps:

(c1) We mark the diameters AB and CD of the circles $C[O_1, R_1]$ and $C[O_2, R_2]$ on the line O_1O_2, respectively. Unless we are given the two Poncelet points only, and so $A = B = P_1 = O_1$ and $C = D = P_2 = O_2$, the four points A, B, C, and D are distinct one another, and the segments AB and CD either do not intersect or one is inside the other.

(c2) We find the points P_1 and P_2 so that the quadruples $\{A, P_1, B, P_2\}$ and $\{P_1, C, P_2, D\}$ are harmonic and thus determine the Poncelet points of the pencil.

The construction of P_1 and P_2 is obtained by constructing a circle orthogonal to both $C[O_1, R_1]$ and $C[O_2, R_2]$. The points of intersection of this circle with the center-line O_1O_2, P_1 and P_2, are the Poncelet points of the pencil sought.

(c3) Let H be the midpoint of the segment O_1O_2. For any point O of the center-line O_1O_2 and outside the segment O_1O_2 the radius of the respective circle of the pencil, as we have seen before, is

$$R^2 = OH^2 - HP_1^2 = OH^2 - HP_2^2.$$

This equation determines the pencil even if we are given just the Poncelet points. The radical axis of the pencil is the perpendicular bisector of the segment O_1O_2 at its midpoint H.

By the above properties, we have also obtained the following nice **Result**: *Two pairs of points $\{A, B\}$ and $\{C, D\}$ on the same straight line such that the segments AB and CD either do not intersect or one is completely inside the other, possess a unique pair of common harmonic conjugate points $\{P_1, P_2\}$.* [Prove the uniqueness as an exercise, arguing by contradiction and using the properties of harmonic quadruples. If AB and CD share the same midpoint M, then $P_1 = M$ and $P_2 = \infty$-point of the straight line $AB = CD$. Also, the absolute value of the harmonic ratio in a configuration as in **Figure 6.57** is

$$\frac{AP_1}{P_1B} = \frac{R_1 + O_1H - \sqrt{O_1H^2 - R_1^2}}{R_1 - O_1H + \sqrt{O_1H^2 - R_1^2}} = \frac{P_2D}{P_2C} = \frac{R_2 + HO_2 - \sqrt{HO_2^2 - R_2^2}}{R_2 - HO_2 + \sqrt{HO_2^2 - R_2^2}}.$$

The proof of this is straightforward and the O_1H and HO_2 are given by the formulae derived after the Remarks of Case (a).]

Next, we include the following theorem due to Casey.[4] This theorem connects the difference of powers of a point in the plane with respect to two given circles with the radical axis. Its result is convenient in many situations with coaxal circles.

Theorem 6.5.7 (Casey's Theorem) *Consider two circles $C[O_1, R_1]$ and $C[O_2, R_2]$ and a point P in the plane. Let K be the projection of P on the radical axis x of the two circles and $\overline{KP}$ the distance of P from the radical axis x of the two circles, oriented with respect to the orientation of $\overline{O_1 O_2}$. (The lines $O_1 O_2$ and KP are parallel.) We have:*

$$\text{(a) If } O_1 \neq O_2, \quad \text{then} \quad \mathfrak{P}_{O_1}(P) - \mathfrak{P}_{O_2}(P) = 2\overline{O_1 O_2} \cdot \overline{KP}.$$
$$\text{(b) If } O_1 = O_2 := O, \text{ then } \forall \ P \in \mathcal{P}, \ \mathfrak{P}_{O_1}(P) - \mathfrak{P}_{O_2}(P) = R_2^2 - R_1^2.$$

Proof (a) The two given circles are not concentric. See **Figure 6.70**. The radical axis x is perpendicular to the center-segment $O_1 O_2$ at the point H. Let M be the midpoint of $O_1 O_2$ and L the projection of P on $O_1 O_2$. Hence, $HLPK$ is a rectangle and so $\overline{HL} = \overline{KP}$. We also let H' be the symmetrical point of H about the point M. So, $\overline{H'M} = \overline{MH}$ and $\overline{H'H} = 2\overline{MH}$.

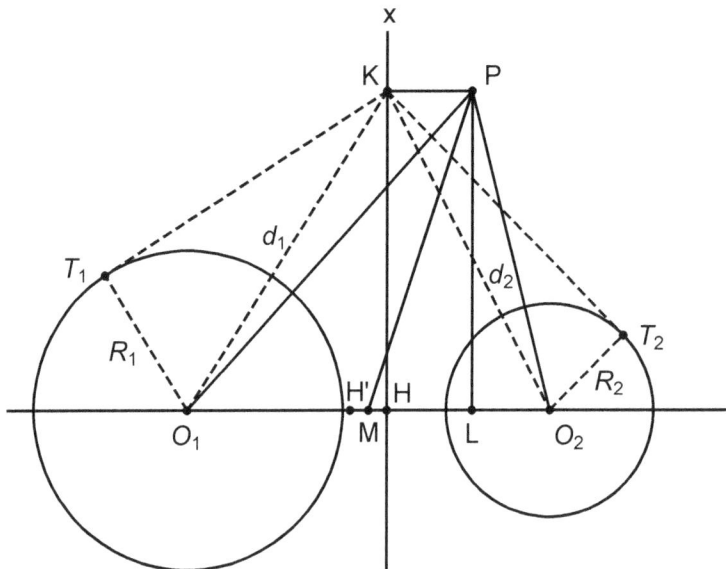

Figure 6.70: Difference of powers of a point with respect to two circles

Now, we have $\mathfrak{P}_{O_1}(P) = \overline{PO_1}^2 - R_1^2$ and $\mathfrak{P}_{O_2}(P) = \overline{PO_2}^2 - R_2^2$ and so,

$$\mathfrak{P}_{O_1}(P) - \mathfrak{P}_{O_2}(P) = \overline{PO_1}^2 - R_1^2 - \left(\overline{PO_2}^2 - R_2^2\right) = \overline{PO_1}^2 - \overline{PO_2}^2 - \left(R_1^2 - R_2^2\right).$$

---

[4] John Casey, Irish mathematician, 1820-1891.

Since H is on the radical axis, $\mathfrak{P}_{O_1}(H) = \mathfrak{P}_{O_2}(H)$ or $HO_1^2 - R_1^2 = HO_2^2 - R_2^2$. Hence,

$$R_1^2 - R_2^2 = HO_1^2 - HO_2^2 = \overline{O_1H}^2 - \overline{HO_2}^2 \quad \text{and so,}$$
$$R_1^2 - R_2^2 = (\overline{O_1H} + \overline{HO_2})(\overline{O_1H} - \overline{HO_2}) = \overline{O_1O_2} \cdot \overline{H'H} = 2\,\overline{O_1O_2} \cdot \overline{MH}.$$

By the theorem of the median of a triangle applied to triangle O_1PO_2 (followed by, e.g., the law of cosines applied to the triangles O_1PM and O_2PM), we get

$$PO_1^2 - PO_2^2 = 2\,\overline{O_1O_2} \cdot \overline{ML}.$$

Therefore, by the above relations, we obtain the final result

$$\mathfrak{P}_{O_1}(P) - \mathfrak{P}_{O_2}(P) = 2\,\overline{O_1O_2} \cdot (\overline{ML} - \overline{MH}) = 2\,\overline{O_1O_2} \cdot \overline{HL} = 2\,\overline{O_1O_2} \cdot \overline{KP}.$$

(b) The two given circles are concentric. Then, we have $O_1 = O_2 := O$ and

$$\forall\ P \in \mathcal{P}, \quad \mathfrak{P}_{O_1}(P) - \mathfrak{P}_{O_2}(P) = PO^2 - R_1^2 - (PO^2 - R_2^2) = R_2^2 - R_1^2.$$

∎

Remarks: (1) If in **(a)** we let O_1 and O_2 approach each other until they coincide, then the two circles tend to become concentric and their radical axis x tends to the straight line at infinity. So, in the limit, the product $2\,\overline{O_1O_2} \cdot \overline{KP}$ takes the indeterminate form $0 \cdot \infty$. In **(b)**, we have found that this limit value is $R_2^2 - R_1^2$, as this was expected from the determination of the foot of the radical axis on the center line O_1O_2 [see **case (a)** after **Definition 6.5.1**].
(2) We observe that $\mathfrak{P}_{O_1}(P) - \mathfrak{P}_{O_2}(P) = \mathfrak{P}_{O_1}(L) - \mathfrak{P}_{O_2}(L)$ (**Figure 6.70**).
(3) The segments $\overline{O_1H}$ and $\overline{O_2H}$ have also been computed by general formulae after **Figure 6.57**. Combine those formulae with some relations in the proof of this Theorem to derive some new relations!

Now we state five Corollaries of Casey's Theorem, which are applicable in many situations. (The missing details in the proofs are easy to provide.)

Corollary 6.5.5 *Consider two non-concentric circles $C[O_1, R_1]$ and $C[O_2, R_2]$, $(O_1 \neq O_2)$. Point P moves on the circle $C[O_2, R_2]$ and so $\mathfrak{P}_{O_2}(P) = 0$ and, by the previous Theorem, $\mathfrak{P}_{O_1}(P) = 2\,\overline{O_1O_2} \cdot \overline{d_P}$, where $\overline{d_P}$ is the oriented distance of $P \in C[O_2, R_2]$ from the radical axis of the two given circles.*

Then, the ratio of $\mathfrak{P}_{O_1}(P)$ and $\overline{d_P}$ is constant. In fact, $\dfrac{\mathfrak{P}_{O_1}(P)}{\overline{d_P}} = 2\,\overline{O_1O_2}$
(is constant).

Conversely: If $\dfrac{\mathfrak{P}_{O_1}(P)}{\overline{d_P}} = 2\,\overline{O_1O_2}$, *where $\overline{d_P}$ is the oriented distance of $P \in C[O_2, R_2]$ from a straight line x, then x is the radical axis of $C[O_1, R_1]$ and $C[O_2, R_2]$.*

Remark: If $O_1 = O_2$, then $\overline{d_P} = \infty$ and $\overline{O_1O_2} = 0$. Thus, the above equality is also true in the limiting case.

Corollary 6.5.6 *Given two circles $C[O_1, R_1]$ and $C[O_2, R_2]$ with $O_1 \neq O_2$, the locus of the points P in the plane such that $\mathfrak{P}_{O_1}(P) - \mathfrak{P}_{O_2}(P) = c$, constant, is a straight line parallel to the radical axis of the two circles at oriented distance*

$$\overline{KP} = \frac{c}{2\,\overline{O_1 O_2}}.$$

If $O_1 = O_2$ and $c = R_2^2 - R_1^2$, then the locus is the whole plane $\mathcal{P}$, but if $c \neq R_2^2 - R_1^2$, then the locus is empty $\emptyset$.

Corollary 6.5.7 *Consider three distinct circles $C[O_1, R_1]$, $C[O_2, R_2]$ and $C[O_3, R_3]$ of a coaxal system (pencil) and point P moves on $C[O_3, R_3]$. Then,*

$$\frac{\mathfrak{P}_{O_1}(P)}{\mathfrak{P}_{O_2}(P)} = \frac{\overline{O_3 O_1}}{\overline{O_3 O_2}} = constant.$$

If $O_1 = O_2$, then $O_1 = O_2 = O_3$ and the ratio $\dfrac{\mathfrak{P}_{O_1}(P)}{\mathfrak{P}_{O_2}(P)}$ is also constant.

If $R_1 = 0 = R_2$, then O_1 and O_2 are the Poncelet points of the coaxal system and the ratio $\dfrac{PO_1}{PO_2}$ is constant.

Proof Since $\mathfrak{P}_{O_3}(P)$ and the three circles have the same radical axis, by the **first Corollary**, we get

$$\frac{\mathfrak{P}_{O_1}(P)}{\mathfrak{P}_{O_2}(P)} = \frac{\frac{\mathfrak{P}_{O_1}(P)}{d_P}}{\frac{\mathfrak{P}_{O_2}(P)}{d_P}} = \frac{\overline{O_3 O_1}}{\overline{O_3 O_2}} = constant.$$

In case of concentric circles, the lengths of the tangent segments drawn from P to circles $C[O_1, R_1]$ and $C[O_2, R_2]$ have constant lengths l_1 and l_2, respectively and then

$$\frac{\mathfrak{P}_{O_1}(P)}{\mathfrak{P}_{O_2}(P)} = \frac{l_1^2}{l_2^2}.$$

The case of the Poncelet points can be proven directly or treated as the limiting case of the first case. ∎

Corollary 6.5.8 (Converse of the Previous Corollary.) *Consider two fixed circles $C[O_1, R_1]$ and $C[O_2, R_2]$. The geometrical locus of a point P such that*

$$\frac{\mathfrak{P}_{O_1}(P)}{\mathfrak{P}_{O_2}(P)} = constant$$

is a circle coaxal to $C[O_1, R_1]$ and $C[O_2, R_2]$.
    If $O_1 = O_2$ and the ratio

$$\frac{\mathfrak{P}_{O_1}(P)}{\mathfrak{P}_{O_2}(P)} = constant,$$

the geometrical locus of a point P is a circle concentric to $C[O_1, R_1]$ and $C[O_2, R_2]$.

Proof As we know (or prove it as an exercise), two circles define a coaxal system (pencil) of circles and there is a unique circle in this system passing through any given point of the plane. Consider the circle of this pencil $C[O_3, R_3]$ that passes through P. Then, O_3 is on the line O_1O_2 and, by the **previous corollary**,

$$\frac{\mathfrak{P}_{O_1}(P)}{\mathfrak{P}_{O_2}(P)} = \frac{\overline{O_3O_1}}{\overline{O_3O_2}}.$$

Since, by hypothesis,

$$\frac{\mathfrak{P}_{O_1}(P)}{\mathfrak{P}_{O_2}(P)} = \text{given constant},$$

the point O_3 is á-priori determined on the line O_1O_2. Also, P must belong to the same circle of the coaxal system and therefore the geometrical locus is the circle $C[O_3, R_3]$.

The case of concentric circles is easy.

■

Corollary 6.5.9 *(a) The circle of similarity of two given circles is coaxal with them.*

(b) An Apollonius circle on a straight segment AB determines a coaxal system (pencil) with Poncelet points A and B.

(c) Given two circles, the circle of the positive inversion under which the two circles are mutually inverse, is coaxal with the given circles.

Proof (a) See **Definition 4.4.1** (and the related **Result 2** in which this definition appears) and **Figure 4.73**. We easily prove that the ratio of the powers of any point of the circle of similarity with respect to the given circles is fixed, equal to $\frac{R^2}{r^2}$ (or $\frac{r^2}{R^2}$), where R and r are the radii of the two given circles. Then the result follows from the **previous Corollary**. (Exercise: Find this pencil and its radical axis.)

(b) It follows from the **previous Corollary**, the **Definition** of an Apollonius circle of some ratio λ on AB, and considering the points A and B as two circles with zero radii. Again the ratio of the powers of the points of any such Apollonius circle with respect to $\{A\}$ and $\{B\}$ is fixed, equal to λ^2. (Exercise: Find this pencil and its radical axis.)

(c) From the results of **Subsection 6.4.2** (and with or without the help of **Corollary 6.5.3** and the formula of the distance of inverse points, **Subsection 6.4.5**) we can prove that the ratio of the powers of the points of the circle of inversion with respect to the two initial circle is fixed, equal to $\frac{R}{r}$ (or $\frac{r}{R}$), where R and r are the radii of the two given circles. Then the **previous Corollary** applies. (Exercise: Find this pencil and its radical axis.)

■

Now, we resume the subject of the composition of two inversions. This time, we assume that **the powers of the two inversions are negative,** $c_1 < 0$ and $c_2 < 0$, and we want to study the invariant sets of the composition

$$I_{[O_2,c_2]} \circ I_{[O_1,c_1]}.$$

In this case, we have:

If $c_1 < 0$ and $c_2 < 0$, and a circle $C[K,r]$ is pseudo-orthogonal to both $C[O_1, \sqrt{|c_1|}]$ and $C[O_2, \sqrt{|c_2|}]$, then $C[K,r]$ is invariant under this composition. See **Figure 6.71**.

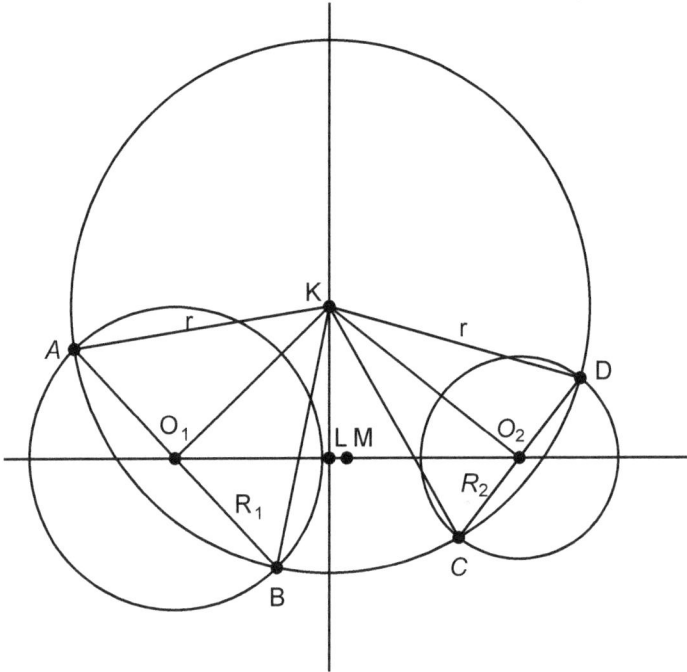

Figure 6.71: **Inversion circles intersected by another circle pseudo-orthogonally**

To find the pseudo-orthogonal circles, we compute the oriented segment

$$\overline{LM} = \frac{R_1^2 - R_2^2}{2\overline{O_1O_2}} = \frac{|c_1| - |c_2|}{2\overline{O_1O_2}},$$

where M is the midpoint of the segment O_1O_2, as before. Then, the **pseudo-radical axis of the two circles** is the line perpendicular to O_1O_2 at the point L. See **Figure 6.71**. The point L is located between the midpoint M of the center-segment and the center of the greater circle. Then, any circle with center a point K on the pseudo-radical axis and radius

$$r = \sqrt{R_1^2 + O_1K^2} = \sqrt{R_2^2 + O_2K^2},$$

is pseudo-orthogonal to both of them. If the circles are equal ($R_1 = R_2$), then $L = M$ and the pseudo-radical axis is the perpendicular bisector of O_1O_2.

Next: *If one power is positive and the other is negative, then the circles that are orthogonal to the circle of the inversion with the positive power and pseudo-orthogonal to the circle of the inversion with the negative power, are invariant under* $I_{[O_2,c_2]} \circ I_{[O_1,c_1]}$. See **Figure 6.72**.

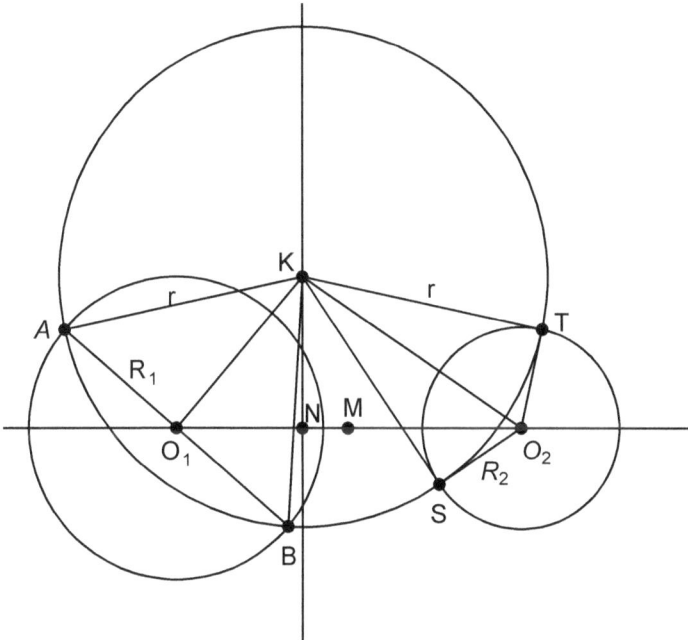

Figure 6.72: Inversion circles intersected by another circle, one pseudo-orthogonally and the other orthogonally

So: **Suppose $c_1 < 0$ and $c_2 > 0$.** To find these invariant circles, we compute the oriented segment

$$\overline{NM} = \frac{R_1^2 + R_2^2}{2O_1O_2} = \frac{|c_1| + |c_2|}{2O_1O_2}.$$

The point N is located between the midpoint M of the center-segment and the center of the circle of the inversion with the negative power. Then, we draw the line perpendicular to O_1O_2 at the point N. See **Figure 6.72**.

Now, any circle with center a point K on this line and radius

$$r = \sqrt{R_1^2 + O_1K^2} = KT,$$

is orthogonal to the circle of the inversion with positive power and pseudo-orthogonal to the circle of the inversion with the negative power.

Case (4) We examine the composition of an inversion with a homothety of different centers, in either order.

Such a composition results in a new function that may be a new inversion or something else, but still having some interesting properties.

For instance, such compositions have fixed points, invariant sets, and reverse oriented angles. Depending on the relative position of the centers, the power of inversion and the homothetic ratio, they may also exhibit some special properties.

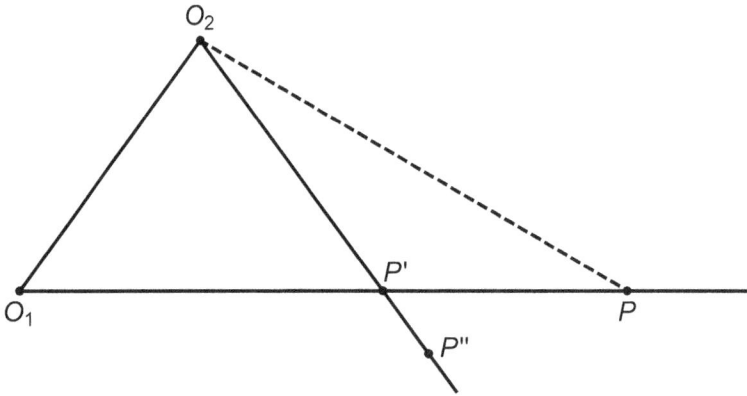

Figure 6.73: Composition of inversion and homothety of different centers

Figure 6.73 may be viewed as one of the following compositions of an inversion and a homothety of different centers,

$$H_{[O_2,\lambda]} \circ I_{[O_1,c]} \qquad \text{or} \qquad I_{[O_2,c]} \circ H_{[O_1,\lambda]}.$$

These compositions are not commutative, in general.

We can investigate and find fixed points and or invariant sets. The line O_1O_2 is again an invariant set since it is invariant by both the inversion and the homothety.

The composition

$$H_{[O_2,\lambda]} \circ I_{[O_1,c]},$$

may have **fixed point(s) on the line** O_1O_2. See **Figure 6.74**. Any such point must satisfy,

$$I_{[O_1,c]}(A) = A' \quad \text{and} \quad H_{[O_2,\lambda]}(A') = A'' \quad \text{and} \quad A = A''.$$

When this happens, we have $\overline{AO_2} = \overline{A''O_2}$. So,

$$\overline{O_1O_2} - \overline{O_1A} = -\lambda \cdot \overline{O_2A'} = -\lambda \cdot \left(\overline{O_1A'} - \overline{O_1O_2} \right) = -\lambda \cdot \frac{c}{\overline{O_1A}} + \lambda \cdot \overline{O_1O_2}.$$

Simplifying this, we find that in order for the point $A \in O_1O_2$ to be fixed, it must satisfy the quadratic equation

$$\overline{O_1A}^2 + (\lambda - 1)\overline{O_1O_2} \cdot \overline{O_1A} - \lambda c = 0.$$

This may have two or one or no real solutions. If the solutions are complex, they represent points not on the axis $\overrightarrow{O_1O_2}$ and have similar geometrical meaning [analogous to **Remark (3)** after **Definition 6.5.2**].

Figure 6.74: Composition of inversion and homothety of different centers

Similar work examining the existence of **fixed points for the composition**

$$I_{[O_2,c]} \circ H_{[O_1,\lambda]},$$

proves that in order for the point $A \in O_1O_2$ to be fixed, it must satisfy the quadratic equation

$$\lambda\overline{O_1A}^2 - (\lambda + 1)\overline{O_1O_2} \cdot \overline{O_1A} + (O_1O_2)^2 - \lambda c = 0.$$

The reader may investigate the existence or nonexistence of invariant sets and also apply **Stewart's Theorem, 6.5.3**, to find O_2P' for either composition, as we did for the composition of two inversions, earlier.

For instance, **in case of the composition**

$$H_{[O_2,\lambda]} \circ I_{[O_1,c]},$$

we find $\quad (O_2P')^2 = \dfrac{c^2 + [O_2P^2 - O_1P^2 - (O_1O_2)^2]c + (O_1O_2)^2 \cdot O_1P^2}{(O_1P)^2}.$

From this, we find O_2P' and then compute $\overline{O_2P''} = \lambda\overline{O_2P'}$ on the line O_2P', since P' is determined on the line O_1P by the equation $\quad \overline{O_1P'} = \dfrac{c}{\overline{O_1P}}.$

In case of the composition

$$I_{[O_2,c]} \circ H_{[O_1,\lambda]},$$

we find $\quad (O_2P')^2 = (1 - \lambda)O_1O_2^2 + \lambda O_2P^2 - \lambda(1 - \lambda)O_1P^2.$

From this, we find O_2P' and then compute $\overline{O_2P''} = \dfrac{c}{\overline{O_2P'}}$ on the line O_2P', since P' is determined on the line O_1P by the equation $\quad \overline{O_1P'} = \lambda\overline{O_1P}.$

Remark: Analogous work can be produced if we face a composition of an inversion and an isometry. The interested reader, by now, has enough equipment to work out each case of parallel translation, or rotation, or reflection with an inversion.

6.6   Applications

In general, inversion has found many applications in several branches of mathematics, such as complex analysis, differential equations, applied mathematics, etc. Here, we present some applications of inversion to classical geometry.

6.6.1   Seven Notable Theorems

The seven notable Theorems, we present here, were well-known for long and were proven in several ways throughout the bibliography. One of the methods of proving them is by using inversion. Here, we present this method.

Theorem 6.6.1 (Inequality of Diagonals and Ptolemy's Theorem.)
The lengths between any four points A, B, C, and D in the plane satisfy the inequality

$$AB \cdot CD + AD \cdot BC \geq AC \cdot BD.$$

The equality is valid if and only if the points are on a circle or a straight lie in the order A, B, C, and D, or an equivalent order.

Proof We consider an inversion with center A and some power $c \neq 0$. Let B', C', and D', be the inverse points of B, C, and D, respectively. Then, by the triangle inequality, we have $B'C' + C'D' \geq B'D'$ and the equality holds iff B', C', and D' are collinear and C' lies between B' and D'. See **Figure 6.75**.

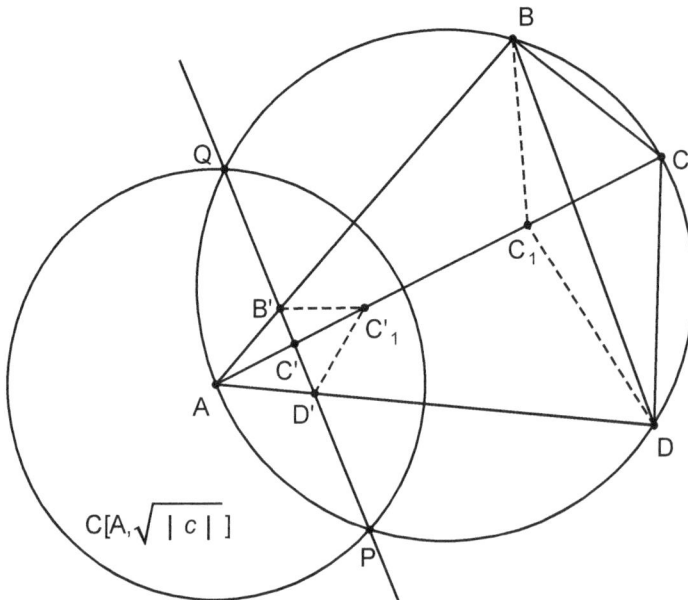

Figure 6.75: Diagonals inequality and Ptolemy's Theorem

Using the distance formula of inverse points (see **Subsection 6.4.5**), we find

$$\frac{|c|BC}{AB \cdot AC} + \frac{|c|CD}{AC \cdot AD} \geq \frac{|c|BD}{AB \cdot AD}.$$

This inequality simplifies to

$$AB \cdot CD + AD \cdot BC \geq AC \cdot BD.$$

As we have said, the equality is valid iff B', C', and D' are collinear and C' lies between B' and D'. See **Figure 6.75**. This happens iff the points B, C, and D are on a straight line through A (the center of inversion), in this order and on one side of A on this line, or on a circle through A (the center of inversion) in the order A, B, C, and D.

■

The equality part of the Theorem is the known **Ptolemy's**[5] **Theorem** of the relation between the sides and the product of the diagonals of an inscribable convex quadrilateral. See **Figure 6.75**. The converse is also true. (Prove!)

Next, we continue with Pappus's[6] Ancient Theorem and Chains.

Theorem 6.6.2 (Pappus's Ancient Theorem.) *Consider three collinear points F, G and H with G between F and H. Let C, C' and K_0 be three semicircles with diameters FH, FG and FH, respectively and lying in the same half-plane determined by the line FH. We consider a sequence of circles K_n, $n = 1, 2, 3, \ldots$, such that all of them touch C and C' and each K_n touches the circles K_{n-1} and K_{n+1}, $n = 1, 2, 3, \ldots$. (K_1 touches the semicircle K_0 and the circle K_1.) We let K_n be the center of K_n, for $n = 0, 1, 2, 3, \ldots$. Let r_n be the radius of the circle K_n and h_n the distance of its center from the line FH. Then,*

$$h_n = 2nr_n, \quad for \quad n = 0, 1, 2, 3, \ldots.$$

Proof For $n = 0$ the result is valid as $0 = 0$. Next, for each $n = 1, 2, 3, \ldots$, we let t_n be the length of the segment from the point F to the circle K_n. In **Figure 6.76**, $n = 4$. We now use the inversion with center X and power t_n^2. So, the radius of the circle of the inversion is t_n.

Then, K_n inverts to itself, since, by the choice of the inversion, it is orthogonal to the circle of inversion. The (semi-)circles C and C' invert to two straight (half-)lines l and l', since they pass through the center of inversion. Then:
(1) The line FH is invariant, since it passes through the center of inversion F.
(2) $l \parallel l'$, since the (semi-)circles C and C' are tangent to each other at F, the center of the inversion.
(3) $l \perp FH$ and $l' \perp FH$, since FH passes through the centers of C and C' and so is a diameter line for both (semi-)circles and so FH is perpendicular to both (semi-)circles.

---

[5]Claudius Ptolemy, Greek mathematician and astronomer, c.85-c.165.
[6]Pappus of Alexandria, Greek mathematician, c.290-c.350.

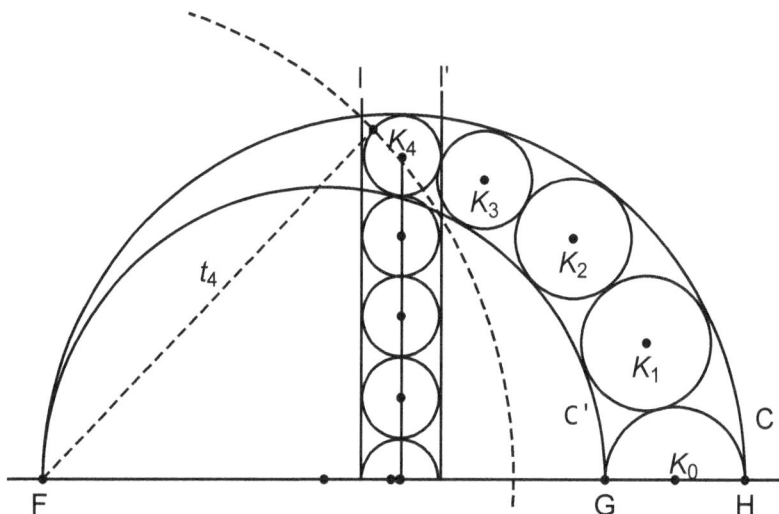

Figure 6.76: Figure for Pappus's Ancient Theorem.

(4) l and l' are tangent to the invariant circle $\mathcal{C}_n$, since $\mathcal{C}$ and $\mathcal{C}'$ are tangent to it.

The semicircle $\mathcal{C}_0$ inverts to a semicircle touching the lines l and l' and having diameter on FH. The circles $\mathcal{C}_i$, $i = 1$, 2, 3, $\ldots$, $n - 1$ invert to circles inside the strip between l and l' touching the two lines and each other successively and in the same order as the original circles. So, all the inverted circles are equal to $\mathcal{K}_n$ and all of their centers are on the mid-parallel line of l and l', which is perpendicular to FH. Therefore, the distance from the center of invariant $\mathcal{K}_n$ to the line FH is

$$h_n = 2nr_n, \quad \text{for} \quad n = 0, 1, 2, 3, \ldots .$$

■

The shape between the three semicircles $\mathcal{C}$, $\mathcal{C}'$ and $\mathcal{K}_0$, with diameters FH, FG and GH is called **arbelo**. (Arbelo = the knife of the shoemaker, in ancient Greek.) So, all the smaller circles are inside the arbelo and touching its boundary.

If we reflect **Figure 6.76** in the line FH, we get **Figure 6.77**. The tangent circles between the outer and the inner boundary of the ring form what we call **Pappus Chain of circles**. For such a chain we have the following two Theorems.

Theorem 6.6.3 *The centers K_n of the circles $\mathcal{K}_n$, and K'_n of $\mathcal{K}'_n$, the reflected circles on the center line FH, for $n = 0, 1, 2, 3, \ldots$, lie on an ellipse whose foci are the centers U and V of the boundary circles, and so its center is the midpoint O of the foci segment UV, and the major axis of the ellipse is FK_0. (So this ellipse is completely determined.)*

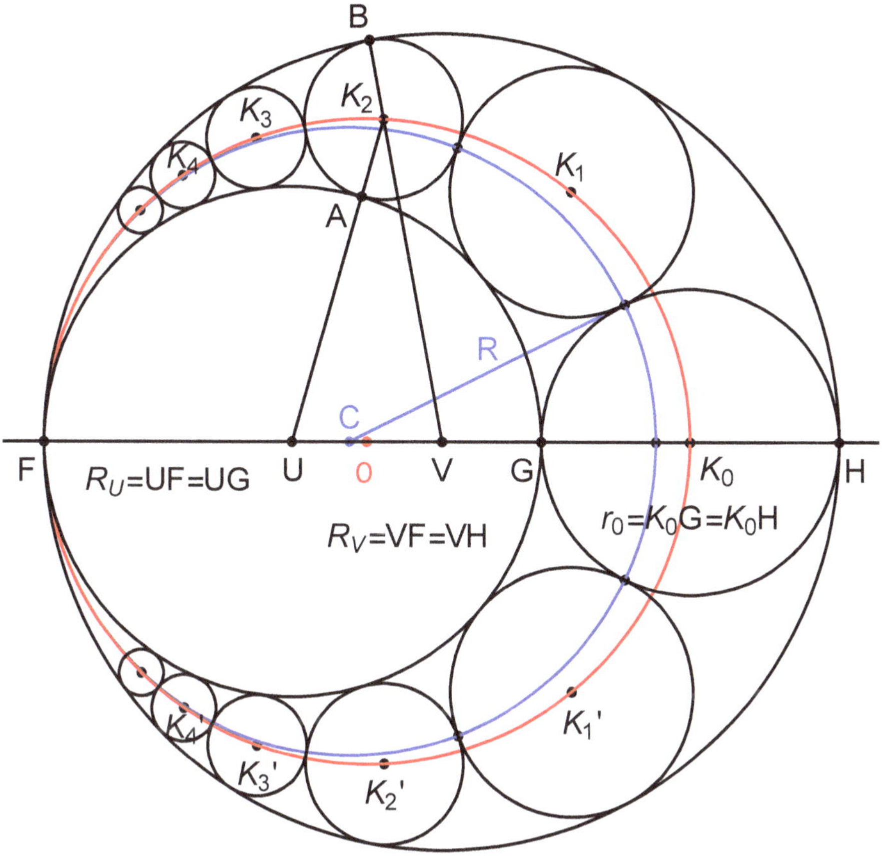

Figure 6.77: Pappus chain

Proof For any circle $\mathcal{K}_n$ (or $\mathcal{K}'_n$), (in **Figure 6.77**, we have chosen $\mathcal{K}_2$), we have

$$K_nU + K_nV = (AU + UK_2) + (VB - K_2V) = AU + VB,$$

the sum of the radii of the two given boundary circles, which is fixed. Therefore, the center K_n is on the ellipse with foci U and V. So, the center O is the midpoint of the foci segment UV and the major axis has length $AU + VB$, which is easily seen to be equal to FK_0. Thus, this ellipse is completely determined. The center O satisfies

$$FO = \frac{FV + FU}{2} = \frac{R_V + R_U}{2} = \frac{FK_0}{2}. \qquad (R_V = R_U + r_0.)$$

∎

Remark: In the next Theorem we prove that the points of tangency of the inscribed circles in a Pappus chain lie in the circumference of a circle and we determine its center and radius.

Theorem 6.6.4 *We put* $r = \dfrac{FG}{FH} = \dfrac{2R_U}{2R_V} = \dfrac{R_U}{R_V}$, *the ratio of the small radius divided by the big radius (so, $0 < r < 1$) of the two given boundary circles. Then, for $n = 0, 1, 2, 3, \ldots$, we have:*

(a) *The radius of the circle $\mathcal{K}_n$ and $\mathcal{K}'_n$ is*

$$r_n = R_V \cdot \frac{r(1-r)}{n^2(1-r)^2 + r} = \frac{FH}{2} \cdot \frac{r(1-r)}{n^2(1-r)^2 + r}. \quad [2R_V = FH = 2(R_U + r_0)].$$

(b) *The projection P_n of the center K_n of the circle $\mathcal{K}_n$ (and K'_n of $\mathcal{K}'_n$) on the center line FH, satisfies*

$$FP_n = R_V \cdot \frac{r(1+r)}{n^2(1-r)^2 + r} = \frac{FH}{2} \cdot \frac{r(1+r)}{n^2(1-r)^2 + r}.$$

(c) *In view of $h_n = 2nr_n$, proven earlier (and will be proved again below), if we let as origin $F = (0,0)$ and $K_n = (x_n, y_n)$, and so $K'_n = (x_n, -y_n)$, then*

$$K_n = (x_n, y_n) = 2R_V \cdot \left\{ \frac{r(1+r)}{2[n^2(1-r)^2 + r]}, \frac{nr(1-r)}{n^2(1-r)^2 + r} \right\},$$

$$K'_n = (x_n, -y_n) = 2R_V \cdot \left\{ \frac{r(1+r)}{2[n^2(1-r)^2 + r]}, \frac{-nr(1-r)}{n^2(1-r)^2 + r} \right\}.$$

(d) *The points of tangency of the inscribed circles in a Pappus chain lie in the circumference of a circle with diameter FM on the center line FH and such that $FM = \dfrac{4rR_V}{1+r}$. So, its center C satisfies $FC = \dfrac{2rR_V}{1+r}$ and its radius is $s = \dfrac{2rR_V}{1+r} = \dfrac{2R_U R_V}{R_U + R_V} = \dfrac{2}{\frac{1}{R_U} + \frac{1}{R_V}}.$*

We use these formulae to draw accurate Pappus chains of circles. The formula of the radius is called **Archimedes's formula**. [Archimedes, of course, gave a different complicated proof, since inversion was discovered and used a long time after him. Geometrical inversion seems to be due to Jakob Steiner ("the greatest geometer since Apollonius") who indicated a knowledge of the subject in 1824. According to Coxeter[7] {see bibliography, [2], [3], and elsewhere}, the transformation by inversion in circle was invented by L. I. Magnus[8] in 1831.]

Proof Proof As in **Theorem 6.6.2**, we are going to use an inversion with center F so that the straight line FH is invariant, the semicircle $\mathcal{C}$ with diameter FH and the semicircle $\mathcal{C}'$ with diameter FG are mapped to two straight lines perpendicular to FH, and therefore these two lines are parallel. (See **Figures 6.76** and **6.77**.) Any power $c > 0$ can be used, but for convenience we take $c = FH^2 = (2R_V)^2 = 4R_V^2$. That is, we will use the inversion $I_{[F,(2R_V)^2]}$.

Under this inversion, the point H is fixed and the semicircle $\mathcal{C}$ is mapped to the line perpendicular to FH at H. Let G' be the inverse point of G. G'

[7]Harold Scott MacDonald Coxeter, English mathematician, 1907-2003.
[8]Magnus Ludwig Immanuel, German mathematician, 1790-1861.

is on the line FH and, with the choice of c, is outside the segment FH on the side of H that F does not lie. Then, the semicircle $\mathcal{C}'$ is mapped to the line perpendicular to FH at G'. Any two successive circles of this Pappus chain are mapped to circles externally tangent, and all these image circles are tangent to these parallel lines and therefore they are equal. See **Figure 6.78**.

We have that $FG{\cdot}FG' = 4R_V^2$. Hence, $2R_U{\cdot}FG' = 4R_V^2$ and so $FG' = \dfrac{2}{r}R_V$. Therefore, the diameter of the equal image circles of the chain is

$$HG' = FG' - FH = \frac{2}{r}R_V - 2R_V = 2 \cdot \frac{1-r}{r} \cdot R_V.$$

So, the radius R of all equal image circles of the chain is $HL_0 = L_0G' = \dfrac{1}{2}HG'$, and so

$$R = \frac{1-r}{r} \cdot R_V.$$

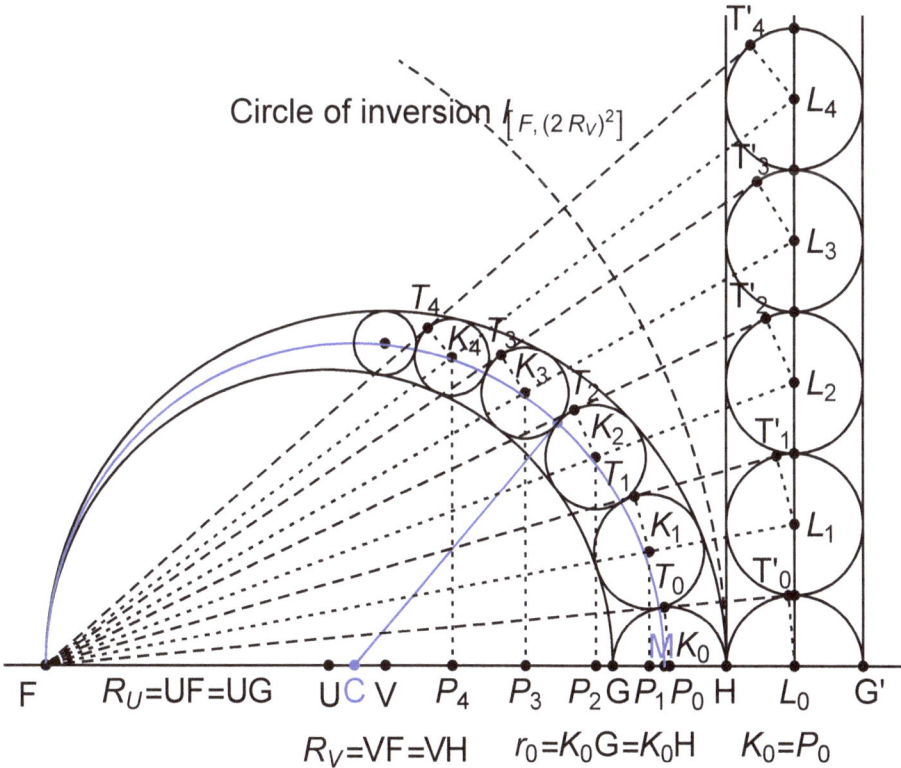

Figure 6.78: Pappus chain and calculation of r_n and FP_n

Let for any $n = 0, 1, 2, 3, 4 \ldots$, the circle of the chain $C(K_n, r_n)$ be mapped to the circle $C(L_n, R)$. Then, the distance of the center L_n from HG'

is $L_0 L_n = 2nR$ and $FL_0 = FH + HL_0 = 2R_V + R$. So,

$$FL_n^2 = FL_0^2 + L_0 K_n^2 = (2R_V + R)^2 + (2nR)^2.$$

Therefore, the power of F with respect to $C(L_n, R)$ is

$$p_n = FL_n^2 - R^2 = (2R_V + R)^2 + 4n^2 R^2 - R^2 = 4R_V^2 + 4R_V R + 4n^2 R^2.$$

By **Theorem 6.4.4**, the circle $C(K_n, r_n)$ is the image of $C(L_n, R)$ under the homothety $H_{[F, \frac{c}{p_n}]}$. That is, the center of the homothety is F and using the above formula for R and $r = \dfrac{R_U}{R_V}$, after we simplify, we find that the homothetic ratio is

$$\lambda_n = \frac{4R_V^2}{4R_V^2 + 4R_V R + 4n^2 R^2} = \frac{r^2}{r + n^2(1 - r)^2}.$$

So, the radius of $C(K_n, r_n)$ is

$$r_n = \lambda_n \cdot R = \frac{r^2}{r + n^2(1 - r)^2} \frac{1 - r}{r} R_V = \frac{r(1 - r)}{r + n^2(1 - r)^2} \cdot R_V.$$

The distance of the center K_n from the line FH (as in **Theorem 6.6.2**), is

$$h_n = \lambda_n \cdot L_0 L_n = \lambda_n (2nR) = \frac{2nr(1 - r)}{r + n^2(1 - r)^2} R_V = 2n \cdot r_n.$$

The projection of FK_n on FH is

$$FP_n = \lambda_n \cdot FL_0 = \lambda_n (2R_V + R) = \frac{r(1 + r)}{r + n^2(1 - r)^2} R_V = \frac{1 + r}{1 - r} \cdot r_n.$$

The points of tangency of the inscribed circles in a Pappus chain must lie in the circumference of a circle which is the inverse of the line $L_0 L_n$. This is a circle with diameter FM of the center line FH, where M is the inverse point of L_0. Then we compute that $FM = \dfrac{4R_v^2}{1 + r} = \dfrac{4rR_V}{1 + r}$, etc.

Finally, the remaining claims, stated in the Theorem, follow from these results immediately.

∎

Remarks. (1) The radius formula is immediately correct for $n = 0$. Plugging in $n = 0$ and $r = \dfrac{FG}{FH}$, we find: $r_0 = K_0 G = K_0 H = \dfrac{GH}{2} = R_V - R_U$.

(2) We can compute r_1 using **Stewart's Theorem, 6.5.3.** For this, examine **Figure 6.79**.

Examining **Figure 6.79** and checking the relations stated in it, we apply **Stewart's Theorem, 6.5.3,** to the triangle $UK_0 K_1$ with all segments positive, to get

$$K_1 U^2 \cdot VK_0 + K_1 K_0^2 \cdot UV = K_1 V^2 \cdot UK_0 + VK_0 \cdot UV \cdot UK_0.$$

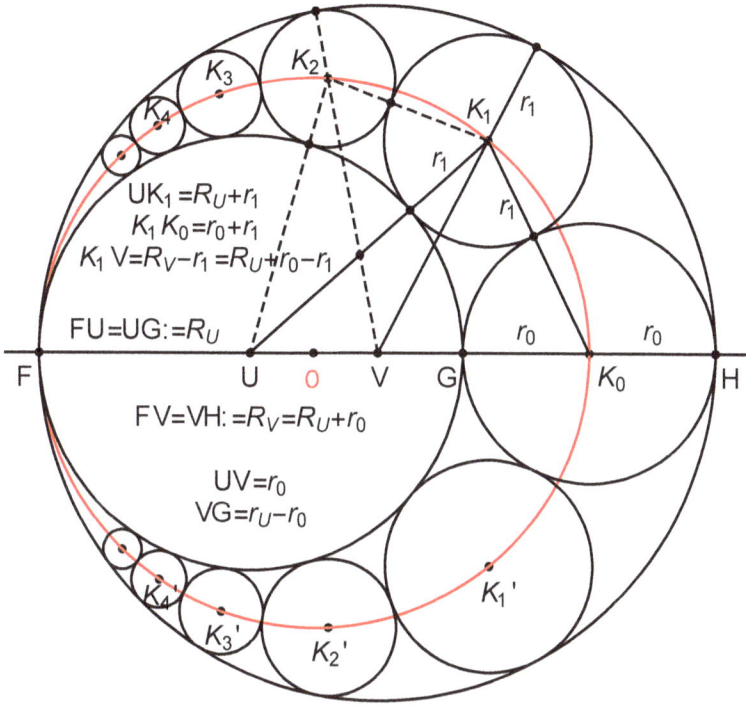

Figure 6.79: Pappus chain and calculation of r_1

We substitute in all segments in terms of the radii and solve for r_1. The solution is

$$r_1 = \frac{R_U \cdot r_0 \cdot (R_U + r_0)}{R_U^2 + R_U \cdot r_0 + r_0^2} = \frac{(R_U + r_0) \cdot R_U \cdot r_0}{r_0^2 + R_U \cdot r_0 + R_U^2},$$

which is the correct formula. (Check that it agrees with the formula found above, with $n = 1$.) (Can you find r_2 in this way?)

(3) The general formula for the radius of the circle $C[K_n, r_n]$, as in **Figure 6:77**, can also be written in the convenient form

$$r_n = \frac{(R_U + r_0) \cdot R_U \cdot r_0}{n^2 r_0^2 + R_U \cdot r_0 + R_U^2}, \quad \forall \quad n = 0, 1, 2, 3, \ldots .$$

Also, with $F = (0,0)$ and for $n = 0, 1, 2, 3, \ldots$, we have

$$x_n = \frac{R_U \cdot (R_U + r_0) \cdot (2R_U + r_0)}{n^2 r_0^2 + r_0 R_U + R_U^2} \quad \text{and} \quad y_n = \frac{2n \cdot (R_U + r_0) \cdot R_U \cdot r_0}{n^2 r_0^2 + r_0 R_U + R_U^2}.$$

(4) If we switch the roles of the circles $C[K_U, R_U]$ and $C[K_0, r_0]$, we find a new Pappus chain, as in **Figure 6.80**. (Compare this chain with the chains in

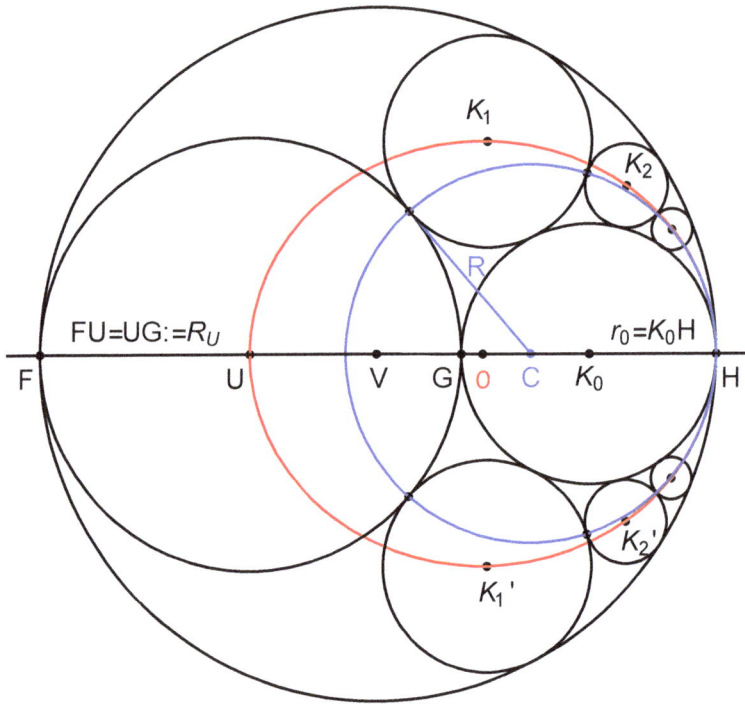

Figure 6.80: The other Pappus chain between two circles

Figures 6.77 and **6.79**.) So, in the previous formulae we interchange R_U and r_0 and we find that he radii of the tangent circles of the chain are given by the formula:

$$r_n = \frac{(r_0 + R_U) \cdot r_0 \cdot R_U}{n^2 R_U^2 + R_U \cdot r_0 + r_0^2}, \quad \forall \quad n = 1, \ 2, \ 3, \ \dots .$$

For $n = 0$, we find the radius r_U, in this case. For $n = 1$, we find again

$$r_1 = \frac{(r_0 + R_U) \cdot r_0 \cdot R_U}{R_U^2 + R_U \cdot r_0 + r_0^2} = \frac{(R_U + r_0) \cdot R_U \cdot r_0}{r_0^2 + R_U \cdot r_0 + R_U^2},$$

the same as before. In both cases the circles $C[K_1, r_1]$ and $C[K_1', r_1]$ are the same. (Expected, since they are tangent to the same three given original circles.)

{Work out and state the formulae for the centers (x_n, y_n) of the tangent circles $C[K_n, r_n]$, in this case, which are the analogous to the formulae of the original case. Determine their positions relative to point $H = (2R_U + 2r_0, 0)$.}

(5) It would be interesting to study Pappus's chains in which the initial circle of the chain $\mathcal{K}_0$ is in any general position and not just on the diameter GH, as in **Figure 6.79**.

Next, we continue with Euler's Incenter-Circumcenter Theorem.

Theorem 6.6.5 (Euler's Incenter-Circumcenter Theorem.) *Let I be the incenter of a triangle ABC and O its circumcenter. Let r be the radius of the inscribed circle and R the radius of the circumscribed circle. Then,*

$$IO^2 = R^2 - 2Rr = R(R - 2r) \geq 0 \quad \Longleftrightarrow \quad \frac{1}{R - OI} + \frac{1}{R + OI} = \frac{1}{r}.$$

Proof In **Figure 6.81**, we have let A_1, B_1 and C_1 be the points of tangency of the inscribed circle with center I (the common points of the angle bisectors) and radius r, and A', B' and C' the intersection points of the sides of the triangle $A_1B_1C_1$ with the segments IA, IB and IC, the bisectors of the angles of the triangle ABC, respectively. The segments IA, IB and IC are the perpendicular bisectors of the sides of the triangle $A_1B_1C_1$.

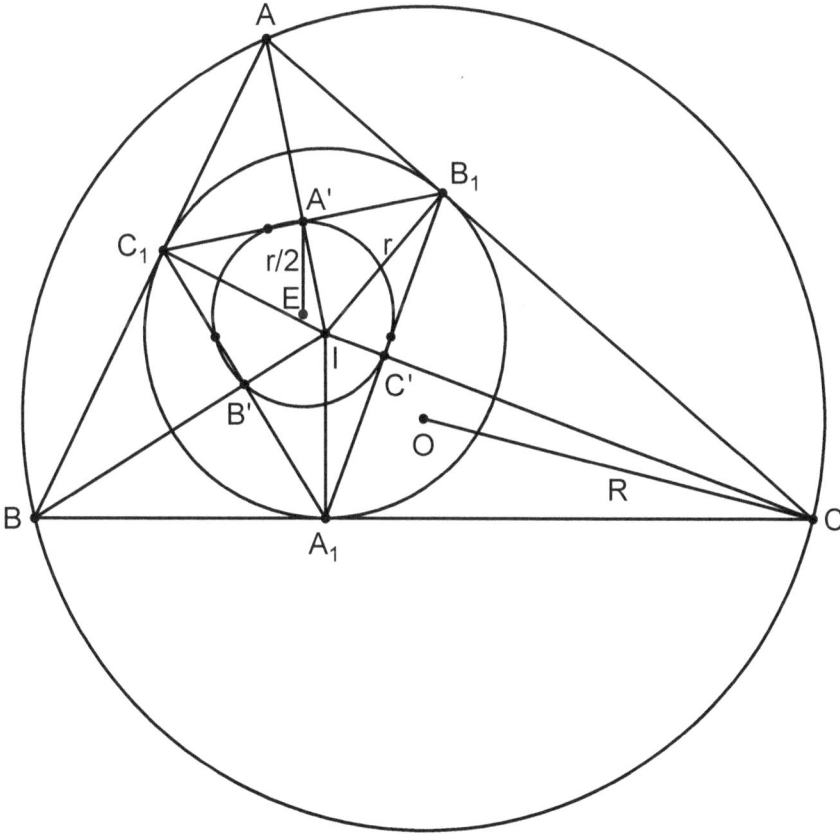

Figure 6.81: Euler's Incenter-Circumcenter Theorem

Then, as we have seen in **Subsection 6.4.3**, the line A_1B_1 is the polar line of the vertex A with respect to the inscribed circle. Therefore

$$IA' \cdot IA = r^2, \quad \text{and similarly,} \quad IB' \cdot IB = r^2 \quad \text{and} \quad IC' \cdot IC = r^2.$$

Therefore, the points A', B' and C' are the inverse points of the vertices A, B and C under the inversion

$$I_{[I, r^2]},$$

(with center I and power $c = r^2$). So, this inversion maps the circumscribed circle to the circle passing through the points A', B' and C', which is the nine-point circle of the triangle $A_1 B_1 C_1$, since A', B' and C' are the midpoints of the sides of the triangle $A_1 B_1 C_1$. Notice that the circumscribed circle to the triangle $A_1 B_1 C_1$ is the inscribed circle to the initial triangle ABC. So, the radius of the circumscribed circle of $A'B'C'$ is $\dfrac{r}{2}$.

[Remember that the radius of the nine-point circle of a triangle is equal to half of the radius of the circumscribed circle. In **Figure 6.81**, the center of the circle $(A'B'C')$ is the point E.]

As we have developed in **Subsection 4.1.1, part (j)**, the last two circles are homothetic, with homothetic ratio λ satisfying

$$|\lambda| = \frac{r}{2R}.$$

Also, the power of I with regard to the circumscribed circle is $IO^2 - R^2 < 0$. This is negative because the point I is an interior point of the triangle ABC and so an interior point of the circumscribed circle. (Also, the whole inscribed circle lies entirely inside the triangle ABC.)

Then, by **Theorem 6.4.1**, we get

$$\lambda = \frac{c}{IO^2 - R^2} \implies |\lambda| = \frac{r^2}{R^2 - IO^2}. \quad \text{So,} \quad \frac{r}{2R} = \frac{r^2}{R^2 - IO^2},$$

which simplifies to **Euler's relation for a triangle**, written as,

$$IO^2 = R^2 - 2Rr = R(R - 2r) \geq 0 \iff \frac{1}{R - OI} + \frac{1}{R + OI} = \frac{1}{r}.$$

■

Corollary 6.6.1 *In any triangle ABC, it holds $r \leq \dfrac{R}{2}$, where r is the radius of the inscribed circle and R the radius of the circumscribed circle. Equality holds if and only if the triangle is equilateral.*

■

Conversely: We consider two circles $C[O, R]$ and $C[I, r]$ such that

$$\frac{1}{R - OI} + \frac{1}{R + OI} = \frac{1}{r} \iff IO^2 = R^2 - 2Rr = R(R - 2r) \geq 0.$$

Then:

(a) The circle $C[I, r]$ lies strictly inside the circle $C[O, R]$.

(b) There are infinitely many triangles inscribed in $C[O, R]$ and circumscribed to $C[I, r]$ and one vertex, A let a say, can be picked on $C[O, R]$ arbitrarily.

Indeed: We have that $R \geq 2r > r$. Also,

$$IO^2 = R^2 - 2Rr < R^2 - 2Rr + r^2 = (R - r)^2 \text{ and so } OI < R - r.$$

This implies $C[I, r]$ lies strictly inside $C[O, R]$ and **(a)** is proved.

Now, we draw the tangent line of the circle $C[I, r]$ at A_1, any of its points, and let B and C be its points of intersection with the circle $C[O, R]$ (as in **Figure 6.81**). From the latter points, we draw the tangent lines to the circle $C[I, r]$ (other than BC), which intersect at a point A. If we show that A belongs to the circle $C[O, R]$, then we have proved assertion **(b)**.

For the proof, we notice that the circles $C[O, R]$ and (ABC) have B and C in common and, as just proved for $C[O, R]$ and by the construction of (ABC), both contain the circle $C[I, r]$. Using the inversion $I_{[I, r^2]}$, as in the direct part of this problem, the inverse of (ABC) is the circle $C\left[E, \dfrac{r}{2}\right] := (A'B'C')$ (as in **Figure 6.81**). By reversing some arguments of the proof of the direct part, and using the hypothesis

$$IO^2 = R^2 - 2Rr = R(R - 2r)$$

and the homotheties between circles, we prove that the inverse of $C[O, R]$ is also $C\left[E, \dfrac{r}{2}\right]$. So, the circles $C[O, R]$ and (ABC) must coincide and so, A is on $C[O, R]$.

Note: Similar work proves Euler's relations for the distances of the centers of the exscribed circles from the circumcenter and their radii. (Make a figure!) These relations are:

$$I_k O^2 = R^2 + 2Rr_k \iff \frac{1}{I_k O - R} - \frac{1}{I_k O + R} = \frac{1}{r_k}, \quad k = 1, 2, 3.$$

Conversely: If we consider two circles $C[O, R]$ and $C[I_k, r_k]$, $k = 1, 2, 3$, which satisfy this relation, then the circle $C[I_k, r_k]$ lies outside the circle $C[O, R]$ and there are infinitely many triangles inscribed in $C[O, R]$ and with $C[I_k, r_k]$ an exscribed circle. Again, one vertex, A let a say, can be picked on $C[O, R]$ arbitrarily.

Remark: We can also prove that: Two circles $C[O, R]$ and $C[I, r]$ are situated such that

$$\frac{1}{(R - OI)^2} + \frac{1}{(R + OI)^2} = \frac{1}{r^2},$$

if and only if, a **quadrilateral** is inscribed in $C[O, R]$ and circumscribed to $C[I, r]$ {and then infinitely many such quadrilaterals exist with one vertex picked on $C[O, R]$ arbitrarily}. (See bibliography.)

Next we continue with Feuerbach's Beautiful Theorem.

Theorem 6.6.6 (Feuerbach's Beautiful Theorem) *In every triangle ABC, the inscribed circle is internally tangent to the nine-point circle and the nine-point circle is externally tangent to the three exscribed circles.*

Feuerbach proved this Theorem by making use of several metric formulae in the triangle. The four points of tangency are called **the points of Feuerbach**. This is the reason why the nine-point circle is also called **Feuerbach's circle**. Since then, a few proofs have been derived, one of which is the one that follows and uses inversion.

Proof Let I be the incenter of ABC, K the point of intersection of the bisector of the angle $\widehat{A}$ with the side AB, and I_1 the excenter corresponding to the angle $\widehat{A}$. See **Figure 6.82**.

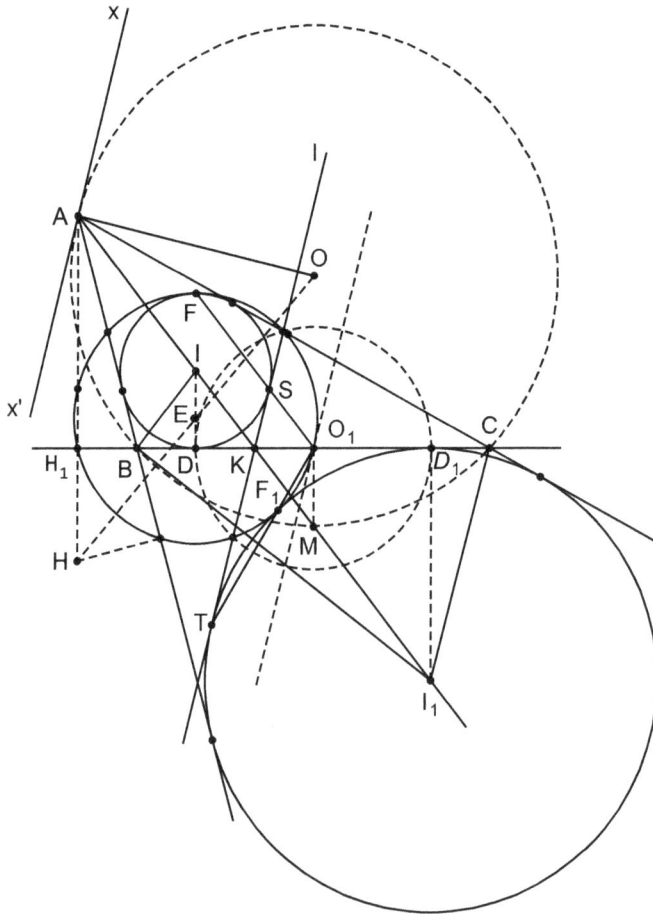

Figure 6.82: Feuerbach's Beautiful Theorem

The point I_1 is also on the bisector $\hat{A}$ (easy) and so all four points A, I, K and I_1 are collinear. They also form a harmonic quadruple. A quick proof of this fact is the following:

The lines BI and BI_1 are interior and exterior bisectors of the angle $\hat{B}$ of the triangle ABK. They meet the opposite side at I and I_1 respectively. Therefore, by the theorem of bisectors of an angle in a triangle the quadruple of points A, K, I and I_1 is a harmonic quadruple.

Then, the projections H_1, K, D and D_1 of these four points on the side BC form a harmonic quadruple (by **Thales' Theorem** of parallel lines intersected by two other lines). The midpoint O_1 of AB is also midpoint of DD_1, because, as we know in a triangle,

$$BD = D_1C = s - b,$$

where $s = \dfrac{1}{2}(a + b + c)$, is the semi-perimeter of ABC with $a = BC$, $b = AC$, and $c = AB$. (This is a known result along with many others that relate the sides of the triangle, s and the points of tangency of the inscribed and exscribed circles with the sides of the triangle. Prove it or consult bibliography!)

Then, we find

$$O_1D = O_1D_1 = \frac{1}{2}[a - 2(s - b)] = \frac{1}{2}(b - c).$$

As we have seen in **Subsection 4.5.5, (1), (b), Remark**, etc., with harmonic quadruples, we have

$$O_1D^2 = O_1D_1^2 = \overline{O_1H_1} \cdot \overline{O_1K}.$$

Now, we consider the inversion with center O_1 and power

$$O_1D^2 = O_1D_1^2 = \overline{O_1H_1} \cdot \overline{O_1K} = \left[\frac{1}{2}(b - c)\right]^2.$$

The nine-point circle passes through O_1 and therefore it is mapped onto a straight line, call it l, through K, because K is the inverse point of H_1, which is a point of the nine-point circle. By the properties of inversions of circles, the line l is parallel to the tangent of the nine-point circle at the center of the inversion O_1. (See **Example 4.2.3**.) As we know from the properties of the nine-point circle, this tangent line is parallel to the tangent xx' of the circumscribed circle at A and therefore so is l. But then,

$$\widehat{BKA} = \hat{C} + \frac{\hat{A}}{2} = \widehat{x'AK} = \widehat{AKl}.$$

This means that l is symmetrical to BC with respect to the center line II_1. Since BC is tangent to both the inscribed and the exscribed circles, then so is the line l. Hence, BC and l are the two common tangents from K to the inscribed and the exscribed circles.

We observe that the circle of inversion, with center O_1 and radius

$$O_1D = O_1D_1 = \frac{1}{2}(b - c),$$

is orthogonal to both the inscribed and the exscribed circle considered here. So, these two circles are invariant under the inversion. The inversion also maps their common tangent l to the nine-point circle of the triangle ABC. Therefore, the nine-point circle is tangent to both the inscribed and the exscribed circles.

More concretely, if S and T are the points of tangency of l with the inscribed and the exscribed circles, respectively, then the nine-point circle touches these circles at the points F and F_1, which are, of course, the second points of intersection of the lines O_1S and O_2T with these circles, respectively. ∎

Corollary 6.6.2 *In any triangle ABC, the inscribed circle is inside and touching the nine-point circle or is equal to it. Therefore,*

$$r \leq \frac{R}{2},$$

where r is the radius of the inscribed circle, $\dfrac{R}{2}$ the radius of the nine-point circle, and R the radius of the circumscribed circle.

Equality holds if and only if the triangle is equilateral, in which case the inscribed circle and the nine-point circle coincide. (Why?) ∎

Remarks: (1) If E is the center of the nine-point circle, then the points F and F_1 are on the center-lines EI and EI_1 and the two circles, respectively. They are also the second points of intersections of the lines O_1S and O_1T with the nine-point circle, respectively.

(2) If H is the orthocenter of the triangle ABC, then the triangles HBC, HCA, HAB and ABC have the same nine-point circle (Why?). Therefore the nine-point circle of the triangle ABC is tangent to $4 \times 4 = 16$ notable circles of the plane of the triangle ABC.

(3) With E the center of the nine-point circle of ABC, I the center of the inscribed circle, I_1, I_2 and I_3 the centers of the three exscribed circles, R, $\dfrac{R}{2}$, r, r_1, r_2, r_3 the radii of the circumscribed, nine-point, inscribed and exscribed circles, respectively, we have:

$$EI = \frac{R}{2} - r, \qquad EI_1 = \frac{R}{2} + r_1, \qquad EI_2 = \frac{R}{2} + r_2, \qquad EI_3 = \frac{R}{2} + r_3.$$

So,

$$EI + EI_1 + EI_2 + EI_3 = 2R + (r_1 + r_2 + r_3 - r).$$

But, $r_1 + r_2 + r_3 - r = 4R$ (prove or see bibliography), and so

$$EI + EI_1 + EI_2 + EI_3 = 6R.$$

The next theorem is called **Steiner's Porism**. For stating it, we need some definitions first. We consider **two nested non-tangent circles**, that is, one circle is inside the other and without touching. In the annulus thus formed, with boundary the two given circles, we placed a circle tangent to both boundary circles and then, we form a sequence of circles each of which is tangent to both boundary circles and to the circle preceding it. See **Figures 6.83-6.90**. If the two given circles are concentric, we say that **the annulus between them is canonical**.

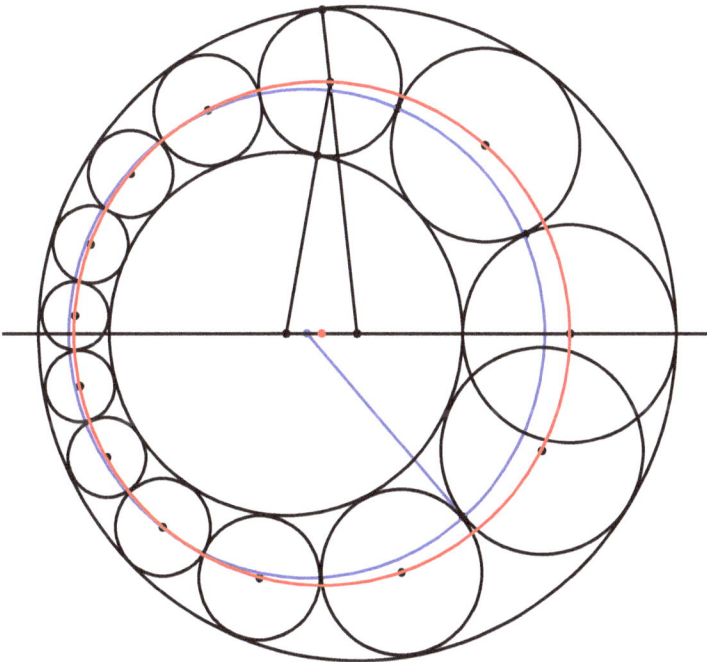

Figure 6.83: Non-closing chain 13 circles. The first and thirteenth circles overlap.

Such a sequence of circles is called **an ordinary Steiner chain (of circles)**[9] within the annulus of the two given circles. These two given circles are called **the base circles of the chain**. If a Steiner chain is inside a canonical annulus, we call it **canonical Steiner chain**. If after a finite number of circles the sequence repeats itself, that means that the last circle of the chain is also tangent to the first one, and each circle is tangent to its previously and afterward constructed

---

[9]In general the two given circles can be in any position. **A chain with base circles two given circles** is any sequence of circles [finite or infinite (as, e.g., happens in the Pappus chains)] such that each one is tangent to both base circles and the next circle of the sequence. [Thus, each circle of the chain except probably the first one is tangent to the previous circle of the sequence. Find these generalizations and many related results in the bibliography (10), (12), (13), (14), (21), etc.]

circles. If we thus continue further, the circles keep retreating themselves. In such a case, we say that **the chain closes** or we have a **closed Steiner chain**. Otherwise, we say that we have an **open chain of circles**. In **Figure 6.83** the chain has not closed after 13 circles have been constructed. In **Figure 6.85**, the chain closes when exactly 12 circles have been constructed and after that they repeat themselves, if we wish continue further. Now, we have:

Theorem 6.6.7 (Steiner's Porism) *For ordinary Steiner chains we prove:*
(1) If a chain of circles between two given nested circles closes, then every chain between the two given circles closes.
(2) So, there are infinitely many such chains and any circle tangent to the two given nested circles and lying between them belongs to a closed chain.
(3) The number of circles in every such closed chain is the same.
[(4) Therefore: if one chain does not close, then no chain closes.]

Proof Since the two base nested circles do not intersect, as we have seen in **Theorem 6.5.5**, they can be inverted to concentric circles. Then, any chain between them is inverted to a chain of image-circles between these two concentric circles, by the same inversion. So, the circles of such a chain invert to equal circles, tangent to the two new base concentric circles and vice-versa.

So, if the porism is valid for concentric base circles (they form a **canonical annulus**), then applying the same inversion to the canonical annulus and its contained circles, we transfer the result back to the original annulus.

In fact, Steiner proved that an appropriate condition between the radii and center-segment and a positive integer n exists, so that a chain closes after n circles of the chain have been constructed. It is enough to solve the problem in the canonical annulus of the two image-base circles. Then, no matter where we start in the original annulus, the chain will always close after n circles have been created, because the characteristics of the figure involving the canonical annulus are rotation invariant about its center. See **Figure 6.84**.

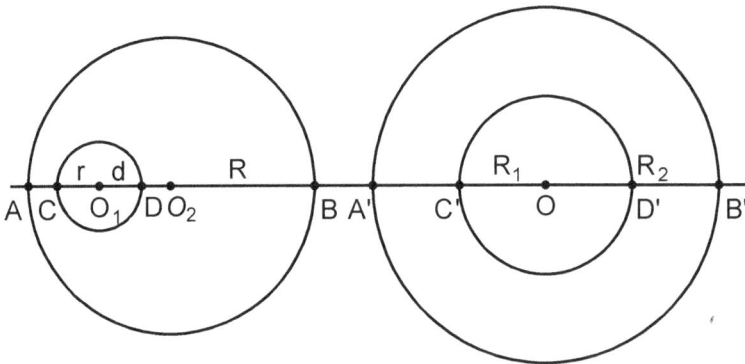

Figure 6.84:   Conditions for non-intersecting circles to invert into concentric circles that admit a closing chain

Now, to complete the proof of the Porism, we must prove the existence of closing ordinary chains. We will derive **Steiner's formula** or criterion that the two base circles must satisfy in order for any chain inside their annulus to close.

In **Figure 6.84**, we have two non-concentric circles with centers O_1 and O_2 and diameters CD and AB, respectively. The radii are $O_1C = O_1D = r$, $O_2A = O_2B = R$ and the center-segment $O_1O_2 = d$.

We consider the inversion claim by **Theorem 6.5.5**, which maps the circles $C[O_1, r]$ and $C[O_2, R]$ onto concentric circles with common center O and corresponding diameters $C'D'$ and $A'B'$. The radii of the new circles are $OC' = OD' := R_1$ and $OA' = OB' := R_2$, respectively.

Since inversions preserve the cross-ratio of four points on an axis (see **Subsection 6.4.6**), we obtain

$$[A, B; C, D] = [A', B'; C', D'], \quad \text{or,} \quad \frac{\overline{AC} \cdot \overline{BD}}{\overline{AD} \cdot \overline{BC}} = \frac{\overline{A'C'} \cdot \overline{B'D'}}{\overline{A'D'} \cdot \overline{B'C'}}.$$

Notice that the right-hand side of this equality equals

$$\frac{(R_1 - R_2)^2}{(R_1 + R_2)^2}$$

and the left-hand side

$$\frac{(R - d - r)(R + d - r)}{(R - d + r)(R + d + r)} = \frac{(R - r)^2 - d^2}{(R + r)^2 - d^2}.$$

We equate the two results and simplify to obtain **the relation of the ratios of R_1 and R_2**

$$(R - r)^2 - d^2 = Rr \left(\frac{R_1}{R_2} + \frac{R_2}{R_1} - 2 \right).$$

Now in **Figure 6.85**, we have a canonical annulus (annulus between two concentric circles) and a closing chain of n equal circles. Then, the radius of the equal circles of this chain is

$$C_1T_1 = C_2T_1 = \frac{R_2 - R_1}{2}.$$

We also have,

$$OC_1 = OC_2 = \frac{R_2 + R_1}{2} \quad \text{and} \quad OT_1 = \sqrt{OC_1^2 - C_1T_1^2} = \sqrt{R_1R_2}.$$

So, from the right triangle T_1OC_1, we get

$$\sin \left(\frac{\pi}{n} \right) = \frac{R_2 - R_1}{R_2 + R_1}.$$

Dividing numerator and denominator by R_1 and solving, we find the ratio

$$\frac{R_2}{R_1} = \frac{1 + \sin \left(\frac{\pi}{n} \right)}{1 - \sin \left(\frac{\pi}{n} \right)}.$$

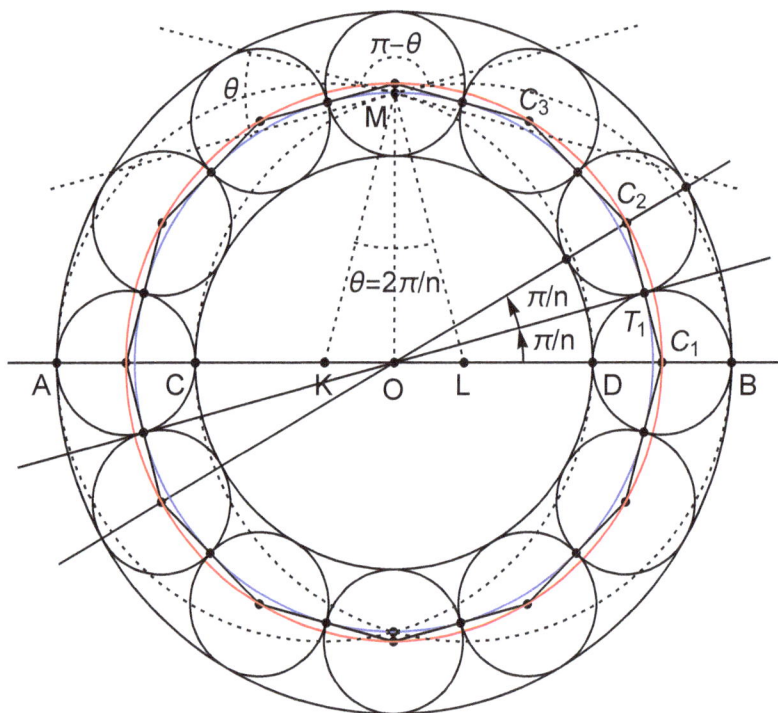

Figure 6.85: Closing simple canonical Steiner chain of circles between two concentric base circles. (The dotted parts will be used later.)

We substitute this ratio into the relation of the ratios of R_1 and R_2 above, and simplify to obtain

$$(R-r)^2 - d^2 = 4Rr\tan^2\left(\frac{\pi}{n}\right).$$

This is **Steiner's formula** that R, r, d, and n must satisfy in order for a chain inside the original annulus to close after n circles have been placed tangent in succession and tangent to the base circles.

When the base circles are concentric, $d = 0$. Then, we can solve for R, the radius of the outer circle, or for r, the radius of the inner circle, in terms of the number n and r or R, respectively, in order to construct a canonical closed chain of n circles. Notice that in this case, the radius of each circle of this canonical chain is $\rho := \dfrac{R-r}{2}$. So we have: $\tan^2\left(\dfrac{\pi}{n}\right) = \dfrac{(R-r)^2}{4Rr}\left(= \dfrac{\rho^2}{Rr}\right)$. This implies, $\sin^2\left(\dfrac{\pi}{n}\right) = \dfrac{(R-r)^2}{(R+r)^2}$, or $\sin\left(\dfrac{\pi}{n}\right) = \pm\dfrac{R-r}{R+r}$. But for $n = 2,\ 3,\ 4\ldots$,

$0 < \dfrac{\pi}{n} \le \dfrac{\pi}{2}$, and so $\sin\left(\dfrac{\pi}{n}\right) > 0$. Therefore, if $R > r$, then $R = r \cdot \dfrac{1 + \sin\left(\frac{\pi}{n}\right)}{1 - \sin\left(\frac{\pi}{n}\right)}$.

[If $R \leq r$, then $R = r \cdot \dfrac{1 - \sin\left(\frac{\pi}{n}\right)}{1 + \sin\left(\frac{\pi}{n}\right)}$. Notice that R is obtained by both $\alpha := \dfrac{\pi}{n}$ and $\pi - \alpha$. Check these formulae for R with $n = 1, 2, 3, 4, 5$ and 6, and interpret the results.]

Now, all the previously stated claims follow easily and so, this essentially finishes the proof of Steiner's Porism.

∎

Let us now illustrate some results of the **Steiner's Porism** and its proof. **Figure 6.86** is the inverse of **Figure 6.85**, under an inversion with center a point P, thus producing a Steiner chain between two non-concentric circles. The point P is one of two Poncelet points of $C[O_1, r]$ and $C[O_2, R]$, the inverses of the two given base circles, with $R = O_2 C = O_2 D > r = O_1 A = O_1 B$ and center-segment $d = O_1 O_2$. (See also **Figures 6.62** and **6.84**.) The radical axis of the circles $C[O_1, r]$ and $C[O_2, R]$, as they are situated in the figure, is located outside of them and to the left of P.

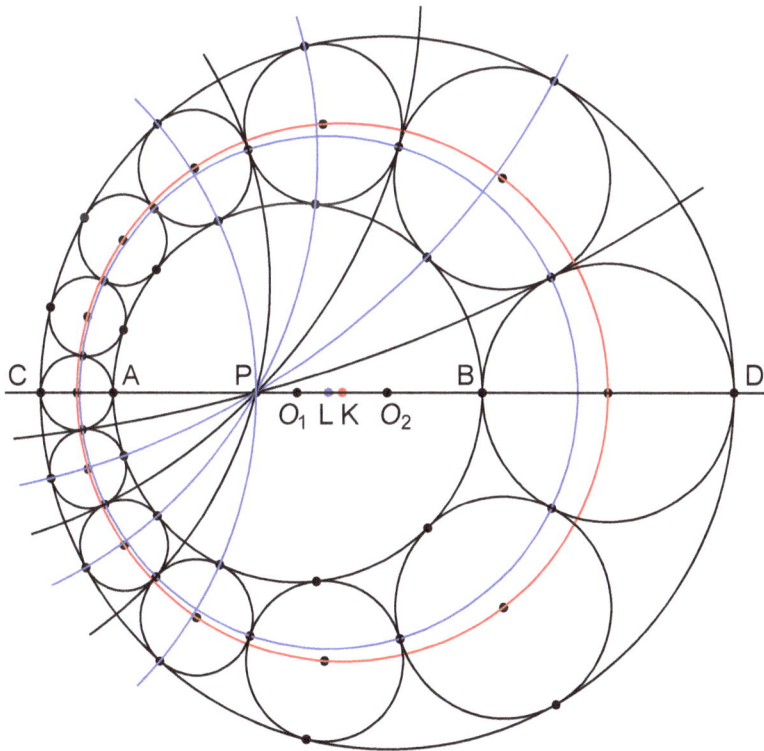

Figure 6.86: Closing simple Steiner chain of even number circles between two non-concentric circles

We must observe that the lines as OT_1, etc., in **Figure 6.85**, are inverted to circles orthogonal to the two base boundary circles, passing through the center of inversion, P, and tangent to the circles of the chain at the corresponding points. Similarly, the lines as OC_2 are inverted to circles orthogonal to both the base circles and the circles of the chain at the corresponding points, just as the line OC_1 is inverted to the line AD which passes through the center of inversion P. All the circles through P and orthogonal to the two base circles also intersect at the second Poncelet point of the two base circles. (In the figure, these circles are not drawn fully and so the second Poncelet point is not shown.)

Also as in **Figures 6.86**, **6.83**, etc., the centers of the mutually tangent circles of the chain are on an **ellipse** with:

Center K the midpoint of O_1O_2. Major semi-axis $\dfrac{r+R}{2}$. **Foci O_1 and O_2.** Their points of mutual tangency are on a **circle** orthogonal to each circle of the chain, whose center is L in the figure. This circle corresponds to the circle $C[O, OT_1]$ of **Figure 6.85**.

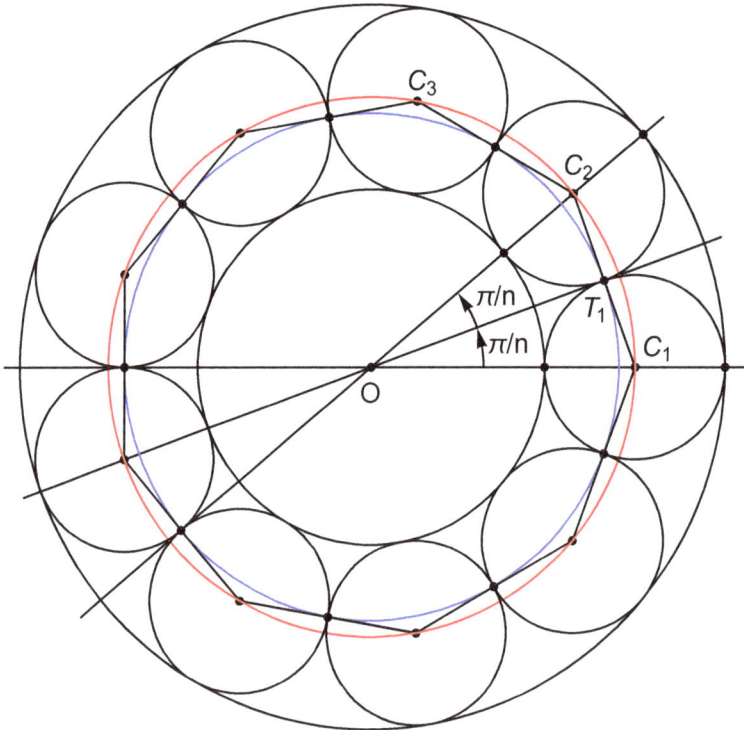

Figure 6.87: **Closing simple Steiner chain of odd number of circles between two concentric circles**

Observe the corresponding points. We have to distinguish between the two cases where the number of the circles in a closed chain is even, like depicted here, or odd. For the case of odd number, see **Figure 6.87**, and its inverse **Figure 6.88** and check the differences concerning corresponding points and lines.

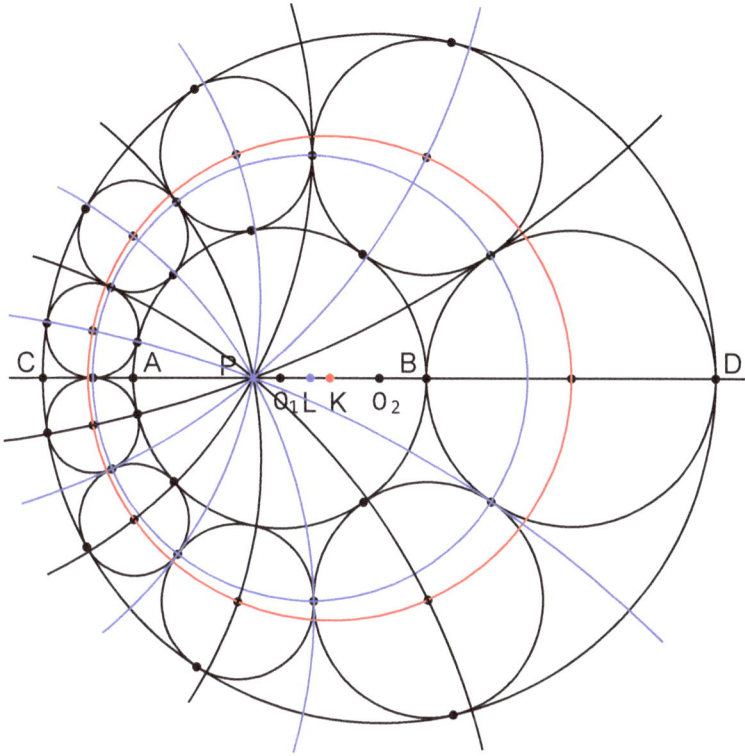

**Figure 6.88: Closing simple Steiner chain of odd number of circles
between two non-concentric circles**

Remarks: Many additional interesting results and generalizations on the
ordinary and not ordinary **Steiner chains** can be found in the bibliography.
[The **general Steiner chains, conjugate Steiner chains** and generalizations
depending on the types of tangency are not studied here. See bibliography items
(10), (12), (13) and (21).] Here we state the following:

(1) As in the Pappus chains of circles, similarly the centers of the circles of
any Steiner chain (closed or open) within two given base circles of radii $R > r$,
as specified above, lie on an ellipse with foci the center points of the two base
circles. Also, the center of the ellipse is the midpoint of the two centers of these
circles. The major axis is equal to $R + r$. If the base circles are concentric, then
the ellipse becomes a circle with center the center of the two concentric circles
and diameter $R + r$. See **Figures 6.83** and **6.85-6.88**.

(2) Chains with base circles outside one another are obtained from chains in
which one base circle is inside the other (canonical or otherwise) by an appro-

priate inversion. Or, we can use the equation $d^2 = (R + r)^2 + 4Rr \tan^2\left(\frac{\pi}{n}\right)$, which, in this case, is derived in a way analogous the way deriving the corresponding equation in the proof of **Theorem 6.6.7**. (d, R, r is the length of the center-segment and the radii of the two base circles and n is the number of circles in the chain. We would need an appropriate method to compute the radii of the circles of the chain.)

(3) In the exercises, we ask to prove that the points of tangency of the inscribed circles in a Steiner chain lie on a circle whose center and radius should be determined. The inverse points of the centers of the equal circles of a chain in a canonical annulus, as in **Figures 6.85** and **6.87** are on the same circle, but are not the centers of the circles which are the inverses of the equal circles. (Why? As we have explained the centers of the latter circles are on an ellipse.)

(4) An ordinary Steiner chain may close after wrapping around the base circles more than once. We call such a chain **multicyclic Steiner chains**. A chain that closes after one wrap is called **simple**. See **Figures 6.89** and **6.90**.

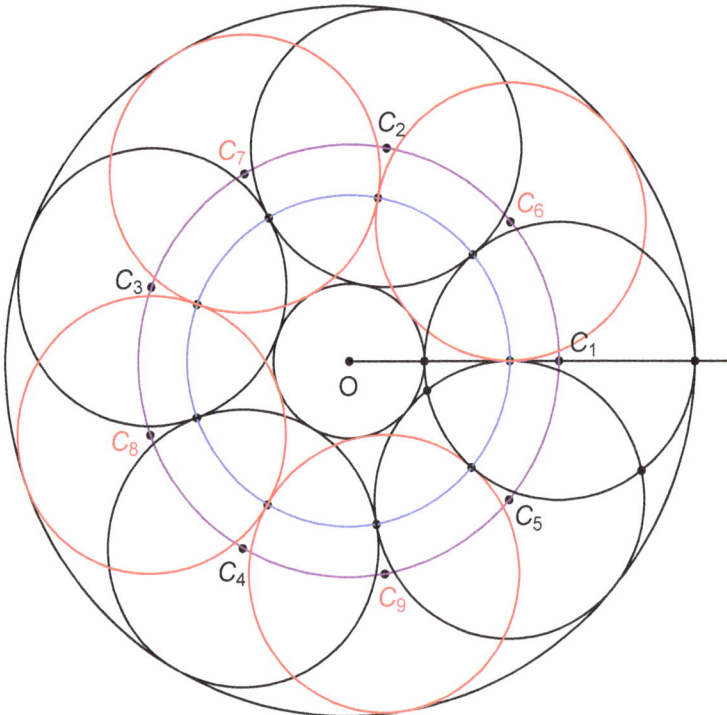

Figure 6.89: Closing multicyclic Steiner chain of odd number of circles between two concentric circles (m=2, n=9)

(5) The pairs of **Figures 6.85, 6.86** and **6.87, 6.88** suggest how to construct a **simple closing ordinary Steiner chain**. We follow the next four steps:

(a) We construct a canonical n-gon.

(b) At each vertex, we attach a circle with radius half of the side of the canonical n-gon.

(c) We fit two concentric circles with center the center of the canonical polygon so that the circles are externally tangent to the smaller and internally tangent the bigger circle.

(d) We invert the figure by an inversion whose center is not on any of the circles.

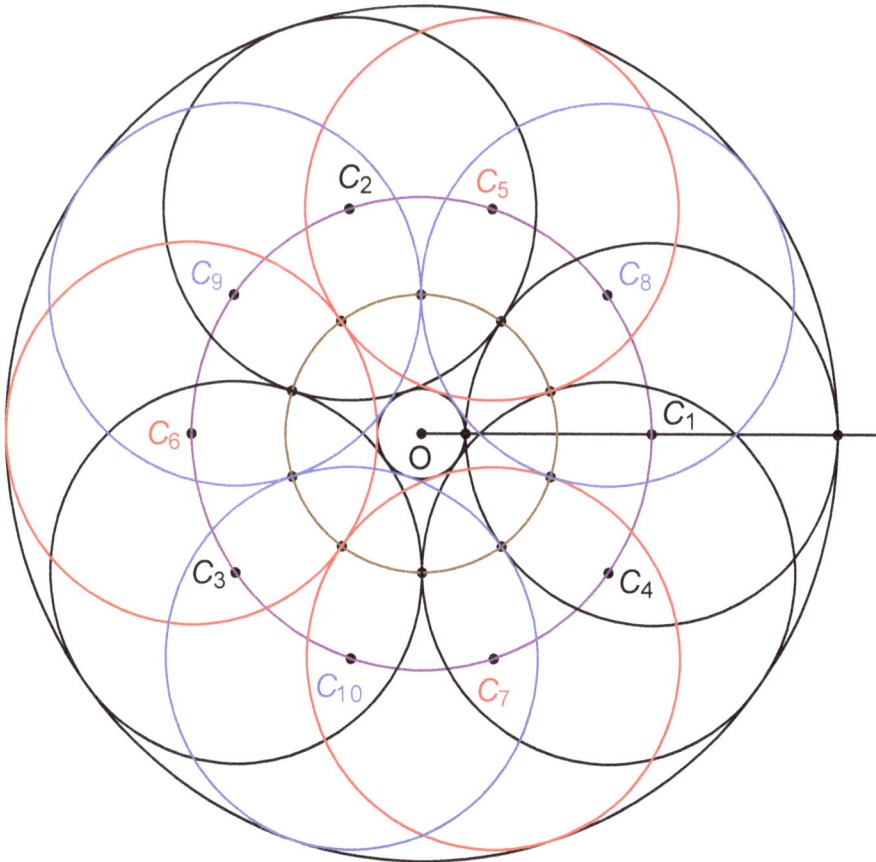

Figure 6.90: Closing multicyclic Steiner chain of even number of circles between two concentric circles (m=3, n=10)

[The tangent lines OT_1, as in **Figure 6.85**, are mapped to circles orthogonal to the two boundary base circles of the non-canonical annulus and tangent to circles of the chain, by the inversion. Also, all the lines trough the common center O (e.g., OC_1) are mapped to circles orthogonal to the two boundary base circles of the non-canonical annulus. But, the orthogonal circles corresponding to lines such as OC_1, pass through the common points of tangency of the base circles and the circles of the chain.]

(6) With a bit more work, we can construct a **multicyclic Steiner chain**. After the above exposition, it is not hard to prove that, if the central angle

$$\widehat{C_1OC_2} \quad \text{is} \quad 360° \cdot \frac{m}{n},$$

as in **Figures 6.85, 6.87, 6.89** and **6.90**, then the two concentric circles, chosen according to the Steiner's formula, admit a Steiner chain of n circles that closes after encircling the common center m times. If m and n are relatively prime, then there is no repetition of circles and the n circles are distinct. Otherwise, there are k repetitions of $\frac{n}{k}$ distinct circles, where k is the greatest common divisor of m and n.

This follows from the observation that, in this case, the tangent circles inside the canonical annulus must satisfy Steiner's Formula

$$\tan^2\left(\frac{1}{2} \cdot \frac{2m\pi}{n}\right) = \tan^2\left(\frac{m\pi}{n}\right) = \frac{\rho^2}{Rr},$$

where R is radius of the outer circle, r the radius of the inner circle and the radius of each tangent circle in the chain is

$$\rho := \frac{R - r}{2}.$$

(Explain and check this claim!)

The numbers m and n are called the **characteristic numbers of the closed Steiner chain**.

The ratio of the number of wraps m divided by the number of the circles of the chain n

$$\frac{m}{n}$$

is called **the characteristic of the chain**. [If the chain is simple, then $m = 1$ and $n \geq 2$. ($n = 2$ is a degenerate case in which the inner circle is a point, its radius is zero.)]

In any closed chain he number of wraps m is determined in the following way. We count how many times the points of tangency to one of the base circles (any one of the two) with the circles of the chain trace the base circle, if we start with a first circle of the chain and finish with it and move in one direction along this base circle, as we move successively from circle to circle of the chain.

(7) Refer to **Figure 4.85** and observe the dotted parts. Consider any or-dinary (finite) Steiner chain (simple or multicyclic) inside a canonical annulus of two concentric circles $C[O, R]$ and $C[O, r]$ with $R > r$. Let m and n be the characteristic numbers of this chain. So, we have the equality of angles

$$\widehat{C_1 O C_2} = \frac{m}{n}(2\pi).$$

When the chain contains an even number of circles, the two dotted circles are tangent to the base circles and two equal opposite circles of the chain and their center line is common diameter to these six circles. The dotted circles are equal and the length of their diameters is

$$AD = CB = R + r, \quad \text{and so their radius is} \quad s := \frac{1}{2}(R + r).$$

Call the angle of their intersection be θ. The distance of their centers from the common center O of the base circles of the chain is

$$t := \frac{1}{2}(R - r)(= \rho),$$

where ρ is the radius of the equal circles of the chain. (Also notice that, if K, L and M are the centers and a point of intersection of the dotted circles, then MO is height and median of the isosceles triangle KLM and has length $\sqrt{Rr}$.)

Consequently, the distance of their centers is $R - r$ and, by law of cosines, we find

$$(R - r)^2 = 2\left(\frac{R + r}{2}\right)^2 - 2\left(\frac{R + r}{2}\right)^2 \cos(\theta)$$

$$= 2\left(\frac{R + r}{2}\right)^2 [1 - \cos(\theta)] = (R + r)^2 \sin^2\left(\frac{\theta}{2}\right),$$

from which we get,

$$\sin^2\left(\frac{\theta}{2}\right) = \frac{(R - r)^2}{(R + r)^2}.$$

But as we have seen (in the end of the proof of **Theorem 6.6.7**), this is also the $\sin^2\left(\frac{m\pi}{n}\right)$. Therefore, with $0 \le \frac{\theta}{2} \le \pi$, we have

$$\theta = 2\pi\frac{m}{n} \quad \text{or} \quad \theta = 2\pi\left(1 - \frac{m}{n}\right) = 2\pi\frac{n - m}{n}.$$

Hence, referring to the dotted circles and their angle of intersection θ, the equality $\sin^2\left(\frac{\theta}{2}\right) = \frac{(R - r)^2}{(R + r)^2}$ ($\implies \theta = 2\pi\frac{m}{n}$ with $0 < m < n$ integers) is another necessary and sufficient condition for a Steiner chain of n circles, in a canonical annulus, to close after m wraps around the center O.

(8) Here we did not plan to exhaust all the results concerning the Steiner chains. For additional results see bibliography items: (10), pp. 113-115, (12), pp. 31-38, and (21), pp. 26-28 and 192-194. We have also stated a few results in the exercises. Look at them and verify them.

6.6.2   Examples Using Inversion

Now, we present eight examples of how to use inversion with geometrical problems. They can be solved without inversion, but here, we want to illustrate the power of the appropriate use of inversion. The interested reader can provide solutions other than the ones presented here.

Example 6.6.1 Consider a circle with center O, a diameter AB of it and a line l parallel to it $l \parallel AB$. We choose any point M on l and let A' and B' be the second points of intersections of the lines MA and MB with the circle respectively. Prove that the circle determined by the points M, A', and B' is orthogonal to the initial circle and tangent to the line l at M. See **Figure 6.91**.

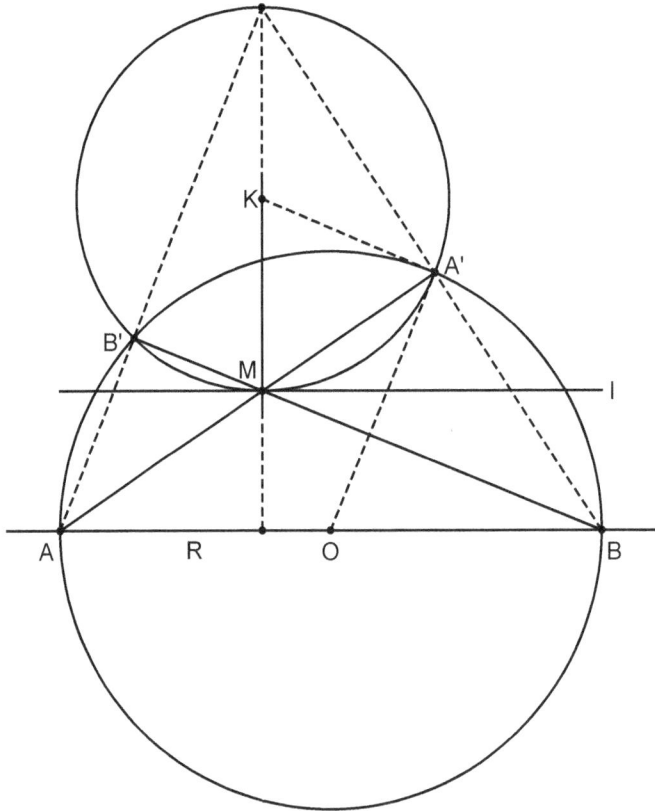

Figure 6.91: Example, E. IN. 1

A **proof** can be given without making use of inversion (provide it for the sake of comparison), but here, we present a **proof using inversion.**

We consider the inversion with center M and power the power of the point M with respect to the given circle. Then, by **Theorem 6.2.1**, the circle is invariant under this inversion, and A' and B' are the inverse points of A and B. So, the new circle determined by the points M, A', and B' is mapped to the line of the diameter AB, which intersects the initial circle orthogonally. Therefore, the given circle and the circle $(MA'B')$ are orthogonal.

Now, the center K of the circle $(MA'B')$ is on a line through the center of the inversion M and perpendicular to the diameter AB, i.e., $KM \perp AB$. Since $l \parallel AB$, the line KM is also perpendicular to l $(KM \perp l)$ and therefore l is tangent to the new circle at the point M.

▲

Example 6.6.2 We choose three fixed points A, B and C in the plane and consider all points P such that the circles (ABP) and (ACP) are orthogonal to each other at P. Find the geometrical locus of the points P. See **Figure 6.92**.

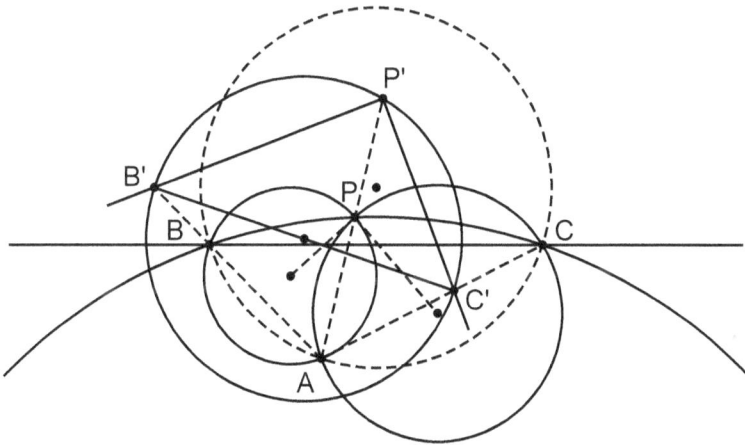

Figure 6.92: Example, E. IN. 2

We consider an inversion with center the point A and some power c. Then, the inverse points B' and C' of the given points B and C, respectively, are known. If for a point P the circles (ABP) and (ACP) are orthogonal and P' is the inverse point of P, then this inversion maps these circles onto the lines $B'P'$ and $C'P'$, respectively, and these lines must be perpendicular to each other at P'. Therefore, the geometrical locus of P' is the circle $(B'C'P')$ with known diameter $B'C'$. So, it is a known circle. Hence, the geometrical locus of P is the inverse image of this circle. Thus, we have:

(1) If A is on the circle $(B'C'P')$, then the geometrical locus of P is the known line BC.

(2) If A is not on the circle $(B'C'P')$, then the geometrical locus of P is a circle orthogonal to the circle (ABC) at B and C, and so known. This is so, because the circle (ABC) is the inverse of the line $B'C'$, which is a diameter of the circle $(B'C'P')$ and so orthogonal to it.

▲

Example 6.6.3 Construct a circle passing through two given points A and B and tangent to a given circle of center O. **Figure 6.93**.

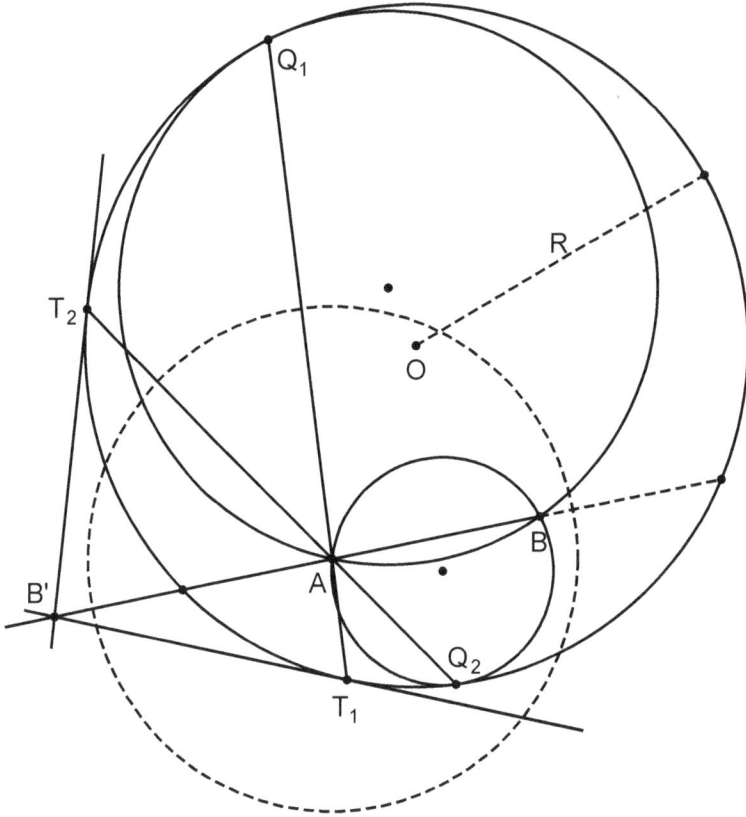

Figure 6.93: Example, E. IN. 3

We consider the inversion with center A and power the power of A with respect to the given circle with center O. Then, by **Theorem 6.2.1**, the circle is invariant under this inversion.

Now, any sought circle is inverted to a line through B' the inverse point of B and tangent to the given circle.

Hence, we find B' and we draw the tangent lines $B'T_1$ and $B'T_2$ from B' to the given circle. The inverses of these tangent lines are two circles satisfying the requirements of the problem. The points of tangency of these circles with the given circle are Q_1 and Q_2, the second points of intersection of the lines T_1A and T_2A with the given circle. So, the solution consists of the two circles (ABQ_1) and (ABQ_2).

(Investigate if there are cases in which the solution consists of one circle or there is no solution.)

▲

Example 6.6.4 Radical Center of Three Circles and an Application.

For three circles in the plane whose centers are not collinear there is a unique point C whose three powers with regard to each one of them are equal. This point is called **the radical center of the three circles**. (If the centers of the three circles are collinear, then, as we shall see below, their radical center is an infinity point and so it does not exist in the Euclidean plane.)

Therefore, *if the radical center C is an exterior point of the union of the three circular regions, then the tangent segments from it to the three circles, are all equal. Let their common length be r. In such a case, the circle $C[C,r]$ is orthogonal to the three circles.*

In **Figures 6.94, 6.95, 6.96** and **6.97**, we show how we find the radical center of three circles in the plane whose centers are not collinear in four main situations. Several additional situations, depending on the relative positions of the circles, are left for the reader to draw figures and study.

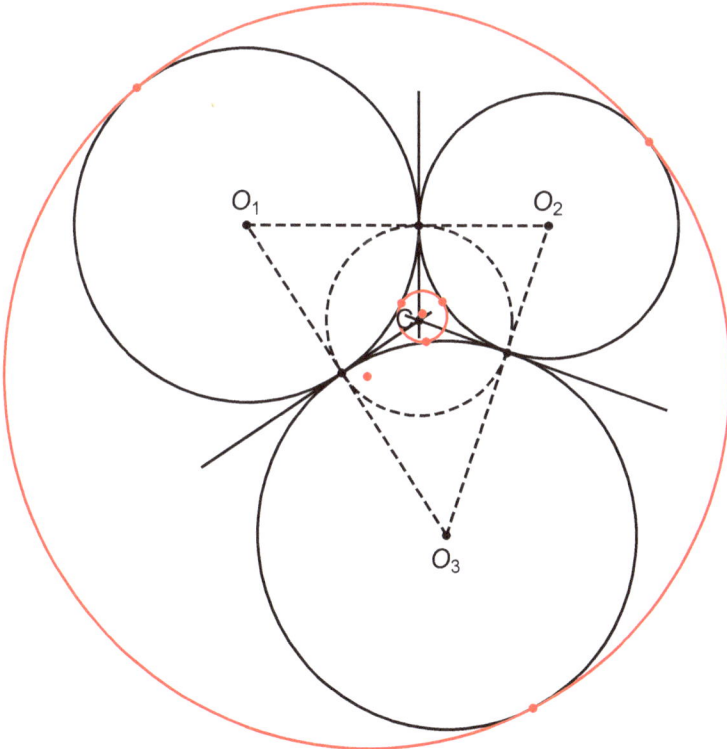

Figure 6.94: **Example, E. IN. 4, radical center of three mutually tangent circles and the two fourth tangent circles**

(a) If any two circles out of the three given circles are tangent (as in **Figure 6.94**), then the radical center is the common point of the three tangents, a fact rather immediate. In this situation, the radical center C is an exterior point to all three circles and there are 3 equal tangent segments drawn from it to the three circles. (Each tangent segment is shared by two of the circles. In this case, the radical center C is also the incenter of the triangle $O_1O_2O_3$.)

(b) If the three circles have nonempty intersection, then the radical center is the common point of the common chord lines in the three pairs of circles formed by choosing two circles at a times (**Figure 6.95**). There are two subcases:

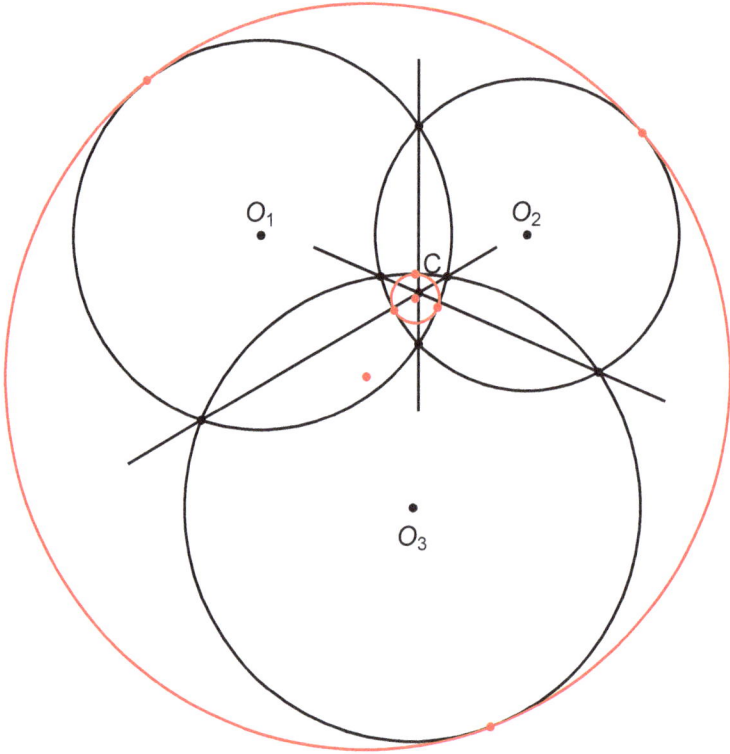

Figure 6.95: **Example, E. IN. 4, radical center of three circles and two tangent circles**

(i) If the three circles intersect at exactly one point, then this point is their radical center. Its power with respect to each circle is zero.

(ii) Otherwise, their intersection has non-empty interior and the radical center is an interior point to all three circles. Its power with respect to each circle is negative.

In these subcases, no tangent non-zero segment can be drawn from the radical center to any of the three circles. So, no circle orthogonal to the three circles exists, in this case.

(c) If the circles chosen by two are intersecting, but the three circles together have empty intersection, then the radical center is the common point of the common chord lines in the three pairs of circles formed by choosing two circles at a times (**Figure 6.96**).

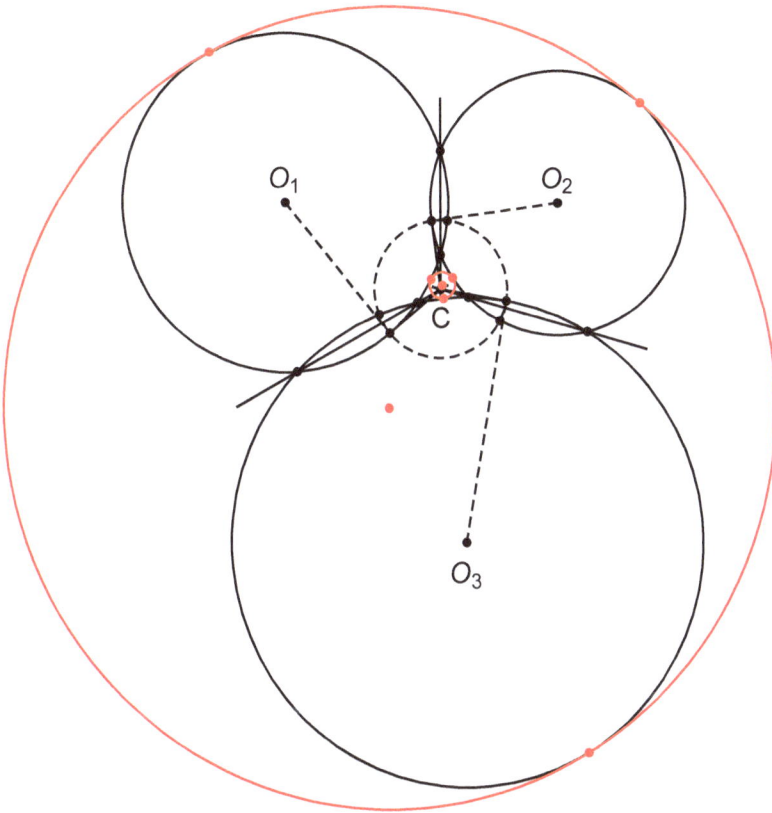

Figure 6.96: Example, E. IN. 4, radical center of three circles and two tangent circles

In this case, the radical center is an exterior point to all three circles and so a total of 6 equal tangent segments can be drawn from it to all of the three circles. (Two tangent segments for each circle.) So, in this case a circle orthogonal to the three circles exists.

(d) If no two circles intersect, and as aforementioned their centers are not collinear, then the radical center is the common point of their three radical axes (**Figure 6.97**). Here, the radical center is an exterior point to all three circles and so two tangent segments can be drawn from it to each of the three circles, for a total of 6 tangent segments. (When the centers are collinear, then the radical axes are parallel to each other and their common point is the infinity point of their common direction.)

It is straightforward to see that the three radical axes pass through the same point. Since the centers of the circles are not collinear then two of radical axes intersect at a point C. Since the powers of C with regard to these two circles are equal, then this point belongs to the third radical axis as well, forcing all

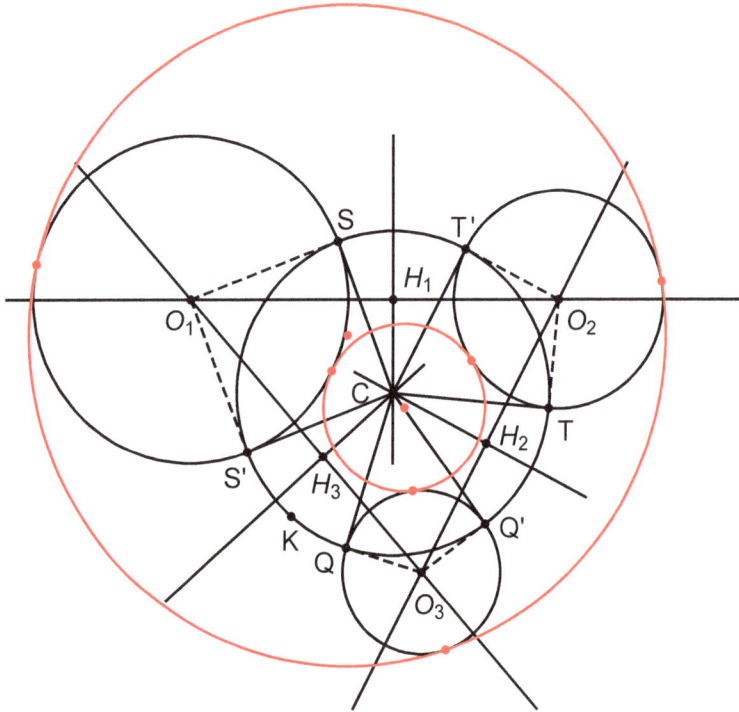

Figure 6.97: Example, E. IN. 4, radical center of three circles and two tangent circles

three to pass through the same point. **Figure 6.97** is self-explanatory! There are two tangent segments for C to each of the circles and these 6 segments have equal length.

In **Figure 6.60**, the point C is the radical center of the three circles depicted there and the same thing holds in **Figure 6.61** for the point P. Look at both figures and check why this claim is true.

Now we turn to **an application of the radical center**, when is located outside each of the three circles. In such a case, we can draw tangent segments to each of the three circles of equal length, we can draw the circle $C[C, CS]$ with center C and radius the length of the equal segments. This circle is orthogonal to all three circles, by construction.

If we pick any point $K \in C[C, CS]$ and we invert with respect to this point with some power of inversion, then the circle $C[C, CS]$, being orthogonal to the three given circles, is mapped to a straight line l and the circles are mapped to other circles of straight lines perpendicular to l. In particular, if the center of inversion K does not belong to any of the three given circles, then they are inverted to three other circles whose centers are on the straight line l. So, we have obtained the following nice **Result**:

If the radical center of three circles does not lie inside anyone of them or on all three of them, then the three circles can be inverted to other circles whose centers are collinear.

In **Figures 6.98, 6.99** and **6.100**, we see what images we expect when inverting **Figures 6.94, 6.96** and **6.97**, respectively, and the center K of inversion does not belong to any of the three given circles. **Figure 6.95** cannot admit such an inversion, as we have explained above.

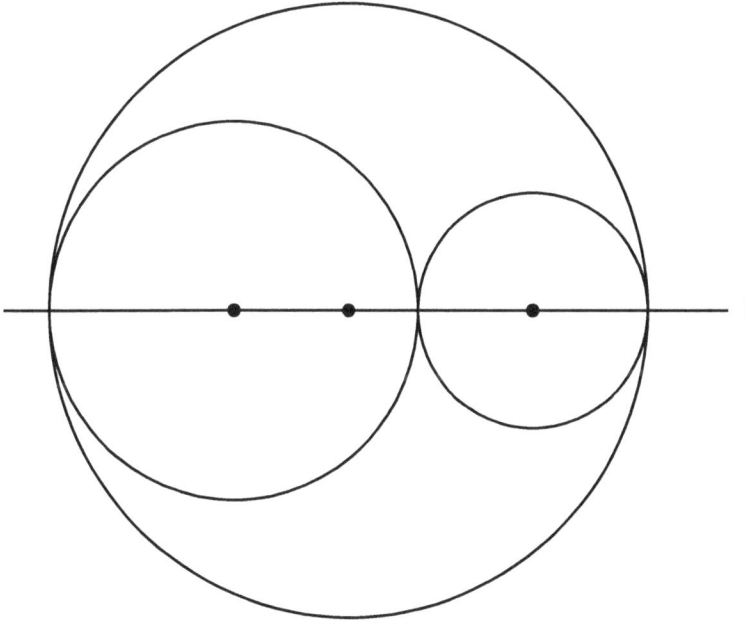

Figure 6.98: Example, E. IN. 4

The question can be asked more specific and a bit more involved: *Given three circles that have radical center that does not lie inside anyone of them or on all three of them, then find an inversion that inverts the given three circles into other circles whose centers are on a given straight line l.* (Not just collinear on any line, but on a given line.)

In such a case we must find the center of the inversion K and the power of the inversion c. At first, the center K is on the circle $C[C, CS]$ which is orthogonal to the three given circles, found above, as in **Figure 6.97**. But now, K must be especially chosen given the line l on which the new centers must lie and also the inverse of l must be the circle $C[C, CS]$. So, we work as follows.

We have the line l and the three circles $C[O_1, r_1]$, $C[O_2, r_2]$, and $C[O_3, r_3]$. Let $C[O_1', r_1']$, $C[O_2', r_2']$, and $C[O_3', r_3']$ be the sought inverses of the given circles, respectively, such that O_1', O_2', O_3', are on l.

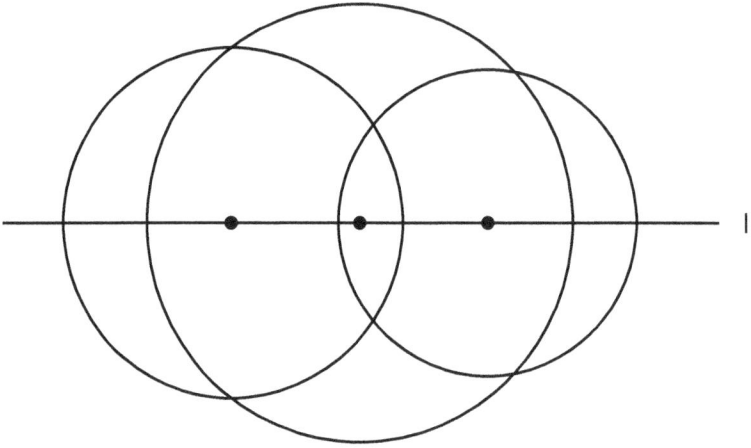

Figure 6.99: Example, E. IN. 4

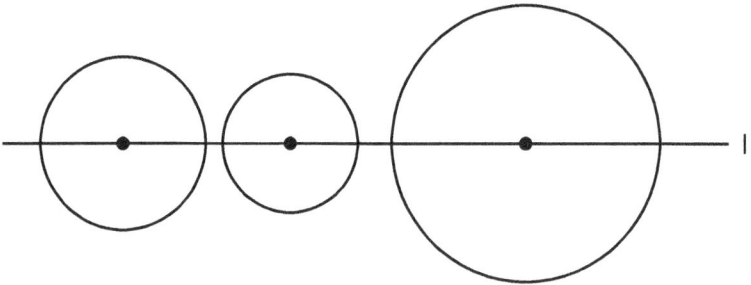

Figure 6.100: Example, E. IN. 4

The tangent to $C[C, CS]$ at K must be parallel to l, [see **Theorem 6.4.2, (b)**]. Therefore, the center K must be one of the end-points of the diameter of $C[C, CS]$ which is perpendicular to l. Next, knowing the center K, the power c is determined, as in **Subsection 6.4.1**, by the relation $KP \cdot KP' = c$, where P' is the other endpoint of the diameter $KC \perp l$, of the circle $C[C, CS]$, and P is the projection of K on l, which also lies on the line KC. See **Figure 6.101**. {The $C[O_1', r_1']$ and $C[O_3', r_3']$ are not drawn, due to space limitations, but you can make another figure with them. Remember: the points O_2 and O_2' are homothetic but not inverses (see **Subsection 6.4.2**).}

We can consider l through O_1 and keep the circle $C[O_1, r_1]$ the same by choosing the correct power. Or, we can take $l = O_1 O_2$ and keep the circles $C[O_1, r_1]$ and $C[O_2, r_2]$ the same by choosing the correct power, and so on. Solve this as an exercise.

▲

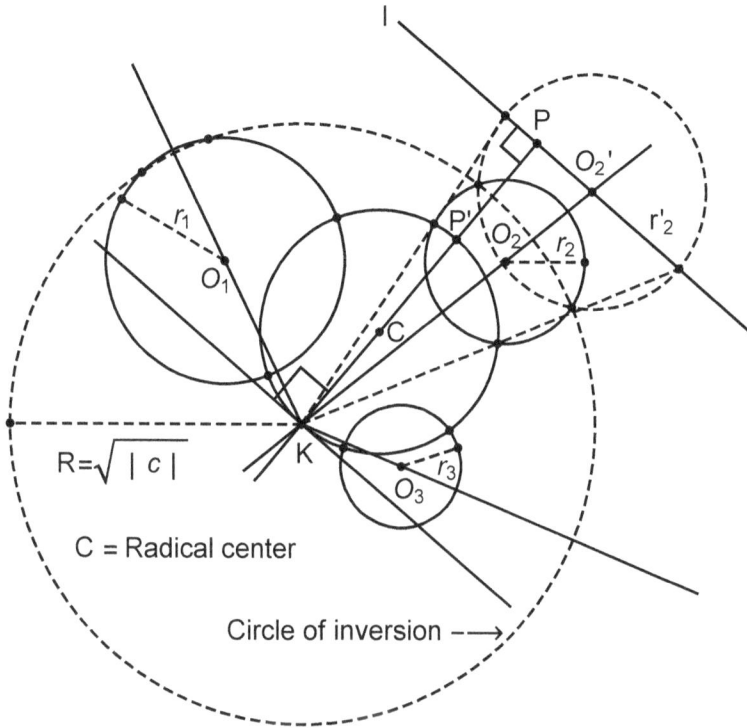

Figure 6.101: Example, E. IN. 4

Example 6.6.5 We consider two fixed circles $C[O_1, r_1]$ and $C[O_2, r_2]$. What condition the position of a point O_3 on the line O_1O_2 and a number $r_3 > 0$ must satisfy, so that, the two fixed circles and the circle $C[O_3, r_3]$ have the same radical axis.

See **Figure 6.102**. We have the three circles, M the midpoint of O_1O_2, N the midpoint of O_1O_3, H_1 the foot of the radical axis of $C[O_1, r_1]$ and $C[O_2, r_2]$ on the line $O_1O_2O_3$, and H_2 the foot of the radical axis of $C[O_1, r_1]$ and $C[O_3, r_3]$ on the line $O_1O_2O_3$.

Then, we have: $\overline{MH_1} = \dfrac{r_1^2 - r_2^2}{2\overline{O_1O_2}}$ and $\overline{O_1H_1} = \overline{O_1M} + \overline{MH_1}$, and so,

$$\overline{O_1H_1} = \frac{\overline{O_1O_2}}{2} + \frac{r_1^2 - r_2^2}{2\overline{O_1O_2}}.$$

Next: $\overline{NH_2} = \dfrac{r_1^2 - r_3^2}{2\overline{O_1O_3}}$ and $\overline{O_1H_2} = \overline{O_1N} + \overline{NH_2}$, and so,

$$\overline{O_1H_2} = \frac{\overline{O_1O_3}}{2} + \frac{r_1^2 - r_3^2}{2\overline{O_1O_3}}.$$

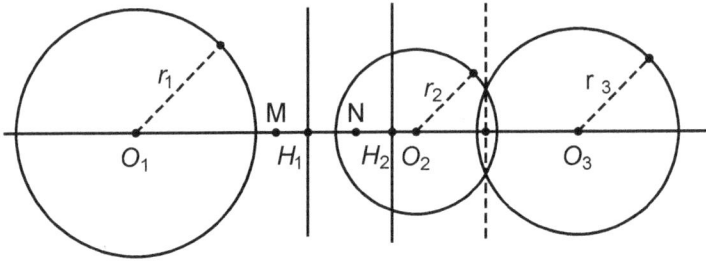

Figure 6.102: Example, E. IN.5

The three circles will have the same radical axis, iff, $\overline{O_1H_1} = \overline{O_1H_2}$. (In **Figure 6.102**, the circles do not have the same radical axis.) Therefore, it must be

$$\frac{\overline{O_1O_2}}{2} + \frac{r_1^2 - r_2^2}{2\overline{O_1O_2}} = \frac{\overline{O_1O_3}}{2} + \frac{r_1^2 - r_3^2}{2\overline{O_1O_3}} \quad \text{or,} \quad \frac{\overline{O_1O_2}}{2} - \frac{\overline{O_1O_3}}{2} = \frac{r_1^2 - r_3^2}{2\overline{O_1O_3}} - \frac{r_1^2 - r_2^2}{2\overline{O_1O_2}}.$$

Hence,

$$\overline{O_3O_2} = \frac{r_1^2 - r_3^2}{\overline{O_1O_3}} - \frac{r_1^2 - r_2^2}{\overline{O_1O_2}}.$$

After simplifying, we find the **final relation**

$$r_1^2\overline{O_2O_3} + r_2^2\overline{O_3O_1} + r_3^2\overline{O_1O_2} + \overline{O_2O_3} \cdot \overline{O_3O_1} \cdot \overline{O_1O_2} = 0.$$

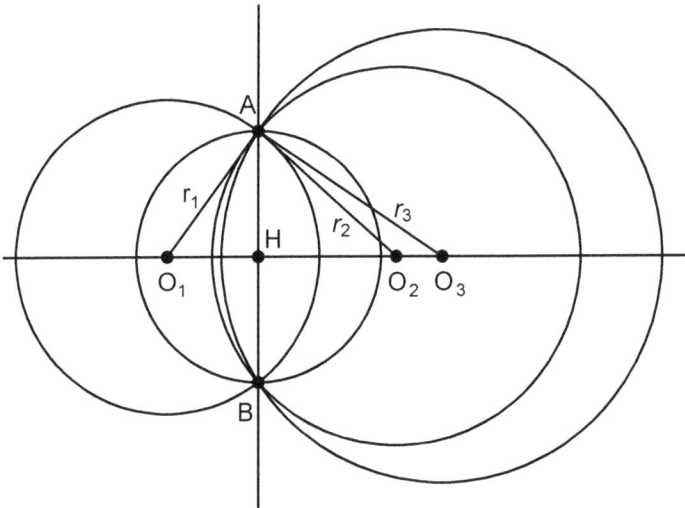

Figure 6.103: Example, E. IN.5

This is a relation similar to **Stewart's general relation** (see **Stewart's Theorem, 6.5.3**). In fact, the two relations coincide when all circles pass through two fixed points or are tangent to each other at the same point. See **Figures 6.103** and **6.104**.

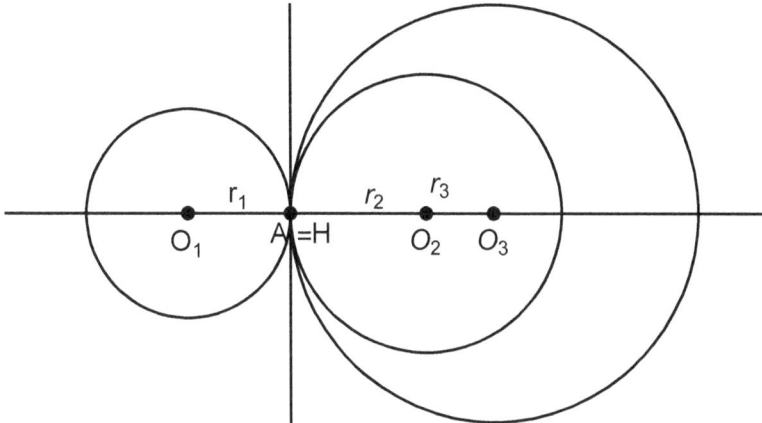

Figure 6.104: Example, E. IN.5

In these cases, we apply **Stewart's Theorem, 6.5.3,** with axis the center line, three points on this axis the three centers of the three circles and extra point a point common to all three circles. Applying **Stewart's Theorem**, we obtain the above **final relation** verbatim.

However, we can also derive this relation even when the three circles are pairwise disjoint, as in **Figure 6.102**, but have the same radical axis, using again **Stewart's Theorem, 6.5.3,** with axis the center line, three points on this axis the three centers of the three circles, extra point any point of the radical axis, and the property that characterizes the radical axis. (Exercise!)

In **Definition 6.5.3**, we have defined that a **coaxal system of circles** or a **pencil of circles** is a set of all the circles in the plane with the same radical axis is called. We have seen that the centers of all circles in a pencil are collinear on a straight line perpendicular to the common radical axis. Also, as we have noticed there are three kinds of pencils of circles:

(1) The pencil in which all the circles pass through two points A and B, depicted in **Figure 6.103**. The common radical axis is the straight line of the common chord AB. See also **Figure 6.64**.

(2) The pencil in which all the circles are tangent (externally and / or internally) at a point $A = H$, depicted in **Figure 6.104**. The common radical axis is the common tangent of the circles at the point $A = H$. See also **Figure 6.63**.

(3) The pencil in which no two circles intersect. See **Figure 6.105**. In such a pencil all the circles have the same Poncelet points and Poncelet circle. The Poncelet circle is orthogonal to all circles of the pencil and is the smallest circle

with this property. Also, compare **Figure 6.105** with **Figure 6.57**.

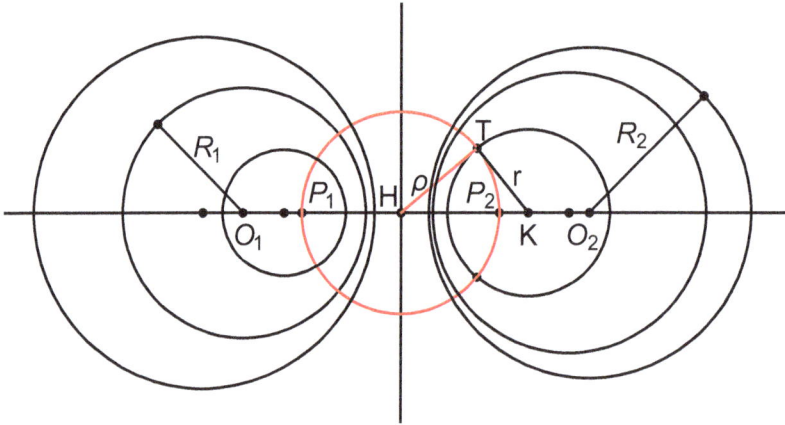

Figure 6.105: Example, E. IN.5

Given two non-intersecting circles $C[O_1, R_1]$ and $C[O_2, R_2]$, we can determine the Poncelet points P_1 and P_2 of the pencil of circles determined by $C[O_1, R_1]$ and $C[O_2, R_2]$. (This is explained in **Section 6.5** just before the **Definition 6.5.2 of the Poncelet points** and afterwards.) Then, for any point K on the straight line $O_1 O_2$ and **not between** the Poncelet points, we can find the circle of the pencil with center K, if we choose its radius r to be the tangent straight segment KT from K to the Poncelet circle. Therefore,

$$r = KT = \sqrt{HK^2 - \rho^2}, \quad \text{where} \quad \rho = HT = HP_1 = HP_2.$$

(That is, ρ is **the radius of the Poncelet circle**.)

So, we can **define**, in general, **"a radical center of three circles"** to be any point whose powers with respect to each of the three circles are equal.

We have seen that:

(1) If the centers of the three circles are not collinear, then the radical center is unique. See **Figures 6.94, 6.95, 6.96, 6.97**, etc. In this case, we talk about "the radical center" of three given circles (whose centers are not collinear).

(2) If the centers of the three circles are collinear, then: *The three circles are coaxal (belong to the same pencil) if and only if they possess two different radical centers. These two points determine the radical axis of the three circles.* In such a case, each point of the common radical axis is a radical center of these three circles, according to the above definition. For example, see **Figure 6.105**.

(3) If the centers of the three circles are collinear and the circles are not coaxal (or two centers or the three centers coincide), then the radical axes of the three pairs of circles are parallel (or the infinity line) and the radical center of the three circles is the infinity point of their parallel radical axes, (or of the infinity line). That is, in this case, the radical center is not a Euclidean point. See **Figure 6.102**.

So, we have the general **Result**: *The three radical axes of three pairs of circles formed by taking two circles out of three given circles, either they pass through the same point or are parallel, in which case all radical axes may be different or two or all three coincide.*

▲

Example 6.6.6 A problem of Archimedes

This example illustrates not only the advantages of choosing an appropriate center of an inversion, but moreover an appropriate power of the inversion in use. (Similarly with **Lemma 6.5.1**.)

We consider one of the arbelos formed by three mutually tangent circles $C[V, r_V]$, $C[U, r_U]$ and $C[K, r_K]$, such that V is between U and K (on the same line) and

$$r_V = r_U + r_K.$$

We draw the common tangent Gt of $C[U, r_U]$ and $C[K, r_K]$ at their point of tangency G and the two circles $C[K_1, r_1]$ tangent to $C[V, r_V]$, $C[K, r_K]$ and Gt, and $C[K_2, r_2]$ tangent to $C[V, r_V]$, $C[U, r_U]$ and Gt. See **Figure 6.106**.

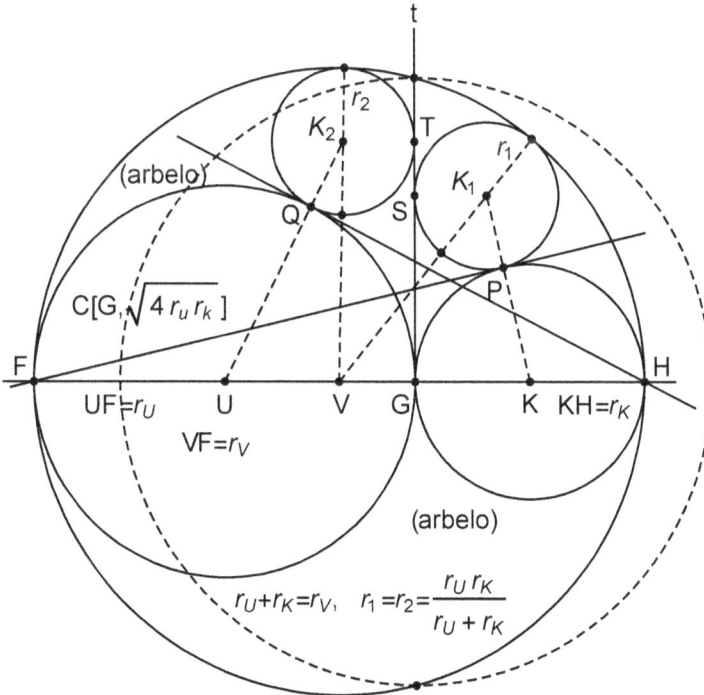

Figure 6.106: Example, E. IN.6

Then, we must prove that

$$r_1 = r_2 = \frac{r_U r_K}{r_U + r_K} = \frac{r_U r_K}{r_V}.$$

Having proved this relation, we know the radii of $C[K_1, r_1]$ and $C[K_2, r_2]$ in terms of r_U and r_K, and then, we can find the centers K_1, as a point of intersection of the circles $C[K, r_K + r_1]$ and $C[V, r_V - r_1]$, and K_2, as a point of intersection of the circles $C[U, r_U + r_2]$ and $C[V, r_V - r_2]$. Thus, we can construct these circles.

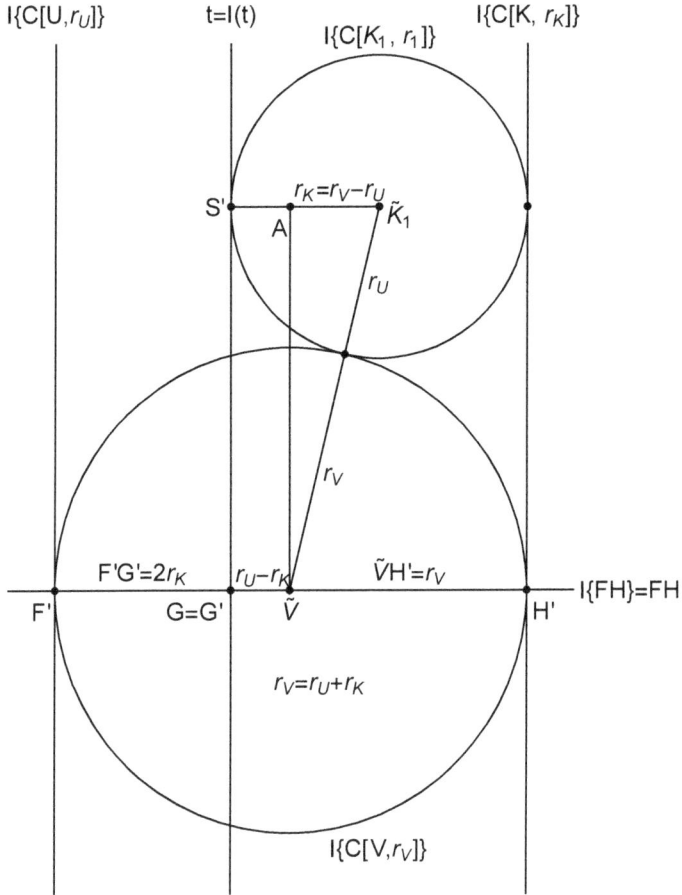

Figure 6.107: Example, E. IN.6

To prove the relation, we invert **Figure 6.106** by the inversion

$$I_{[G, \, 4 \, r_U \cdot r_K]}.$$

(The circle of inversion $C[G, \sqrt{4 r_U r_K}]$ is the dashed circle in **Figure 6.106**.) The power of inversion

$$c = 4 \, r_U \cdot r_K$$

is chosen to be the absolute value of the power of the point G with regard to the big circle $C[V, r_V]$.

In **Figure 6.107**, we have inverted **Figure 6.106**, but the circle $C[K_2, r_2]$. The work is analogous for this circle.

So, we have: The diameter line FH and the tangent Gt are invariant lines. The circles $C[U, r_U]$ and $C[K, r_K]$, are mapped to parallel lines $I\{C[U, r_U]\}$ and $I\{C[K, r_K]\}$, which are perpendicular to $F'H'$ at F' and H', respectively, where F' is the inverse point of F and H' the inverse point of H. Of course $G' = G$, because G is the center of the inversion.

Next: The circle $C[V, r_V]$, is mapped to a circle $C[\widetilde{V}, \widetilde{V}F']$, with center $\widetilde{V}$ (not inverse of V) on the line $F'G'$ and between F' and H' and tangent to the parallel lines $I\{C[U, r_U]\}$ and $I\{C[K, r_K]\}$. So, its radius is half the distance of the parallel lines $I\{C[U, r_U]\}$ and $I\{C[K, r_K]\}$. The circle $C[K_1, r_1]$, is mapped to a circle $C[\widetilde{K_1}, \widetilde{K_1}S']$ with center $\widetilde{K_1}$ (not inverse of K_1) and tangent to the parallel lines $Gt(= G't)$, at S', and $I\{C[K, r_K]\}$, and to the circle $C[\widetilde{V}, \widetilde{V}F']$. So, its radius is half the distance of the parallel lines $Gt(= G't)$ and $I\{C[K, r_K]\}$.

Now we make the following computations: $GH \cdot GH' = c = 4\, r_U \cdot r_K$ and so, $2\, r_K \cdot GH' = c = 4\, r_U \cdot r_K$. Thus, $GH' = G'H' = 2\, r_U$ and so, $\widetilde{K_1}S' = r_U$. Next, $GF \cdot GF' = c = 4\, r_U \cdot r_K$ and so, $2\, r_U \cdot GF' = c = 4\, r_U \cdot r_K$. Thus, $GF' = G'F' = 2\, r_K$. Since $r_V = r_U + r_K$, we have $G\widetilde{V} = G'\widetilde{V} = r_U - r_K$.

Then,
$$A\widetilde{K_1} = r_U - (r_U - r_K) = r_K = r_U - r_K$$
and the power of $G = G'$, with regard to the circle $C[\widetilde{K_1}, \widetilde{K_1}S']$ is
$$p = (r_V + r_U)^2 - (r_V - r_U)^2 = 4\, r_V \cdot r_U.$$

This power p refers to the inverted **Figure 6.107**. Then, applying **Theorem 6.4.2** [or **Remark (4)** that follows it] backward and keeping in mind the involutive character of inversion (i.e., inverting **Figure 6.107** back to **Figure 6.106** by the same inversion), we find
$$r_1 = \frac{c}{p} r_U = \frac{4\, r_U \cdot r_K}{4\, r_V \cdot r_U} r_U = \frac{r_U \cdot r_K}{r_V} = \frac{r_U \cdot r_K}{r_U + r_K}.$$

The formula is symmetric with respect to the indices and analogous work shows
$$r_2 = \frac{r_U \cdot r_K}{r_U + r_K} = r_1.$$

(This solution, as we have said, illustrates the role of an appropriate choice of the power of the inversion in use.)

There are several solutions of this problem that do not involve inversion, in the bibliography. (For instance, see **Example 4.2.11**.) Next, we outline another solution. Look at the next **Figure, 6.108** and give a proof of this result by completing the details of the cited steps and using the Pythagorean Theorem.

Draw line K_1W parallel to Gt. So, $K_1W \perp VK$. Then,
$$VH = r_V = r_U + r_K, \qquad\qquad KK_1 = r_K + r_1,$$
$$WK = GK - r_1 = r_K - r_1, \qquad\qquad VK_1 = r_V - r_1 = r_U + r_K - r_1,$$
$$VW = VH - WK - KH = r_U + r_K - (r_K - r_1) - r_K = r_U - r_K + r_1.$$

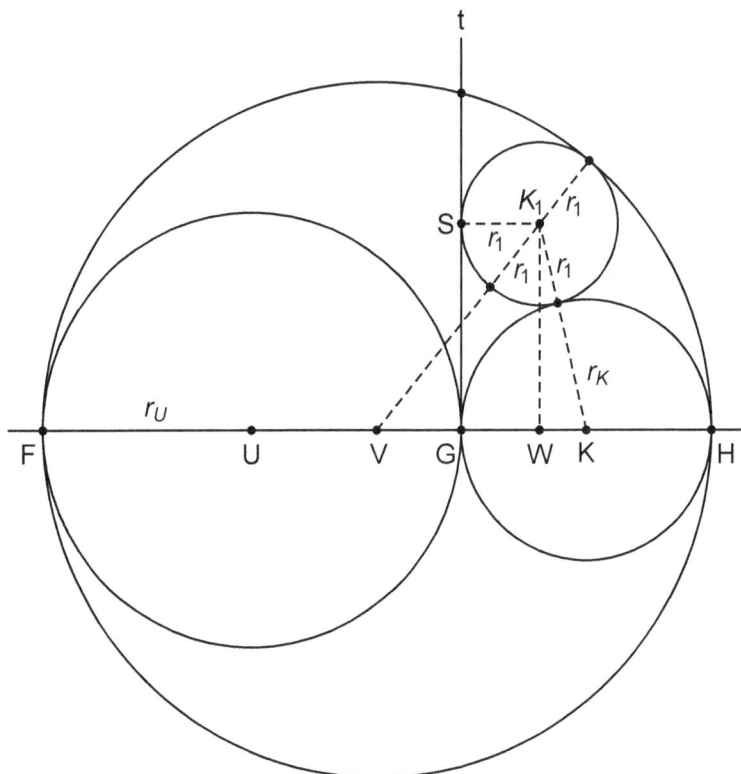

Figure 6.108: Example, E. IN.6

Substitute into the relation

$$KK_1^2 - KW^2 = K_1W^2 = VK_1^2 - VW^2,$$

simplify and obtain the result

$$r_1 = \frac{r_U \cdot r_K}{r_U + r_K}.$$

Now, make the analogous figure and work for the circle $C[K_2, r_2]$ to obtain the result for r_2.

$$r_2 = \frac{r_U \cdot r_K}{r_U + r_K} = r_1.$$

Note: In the exercises we ask to prove that in **Figure 6.106**, the s.l. PF is a common tangent of the circles $C[K, r_K]$ and $C[K_1, r_1]$, and the s.l. QH is a common tangent of the circles $C[U, r_U]$ and $C[K_2, r_2]$ and then compute the lengths of PF and QH in terms of r_K and r_U.

▲

Example 6.6.7 We consider a quadrilateral $ABCD$ such that the circumscribed circles to the triangles ABC and ADC are orthogonal. What is the relation that the sides and the diagonal must satisfy?

We let the sides be $a = AB$, $c = BC$, $c = CD$, $d = DA$, and the diagonals $x = AC$ and $y = BD$. See **Figure 6.109**.

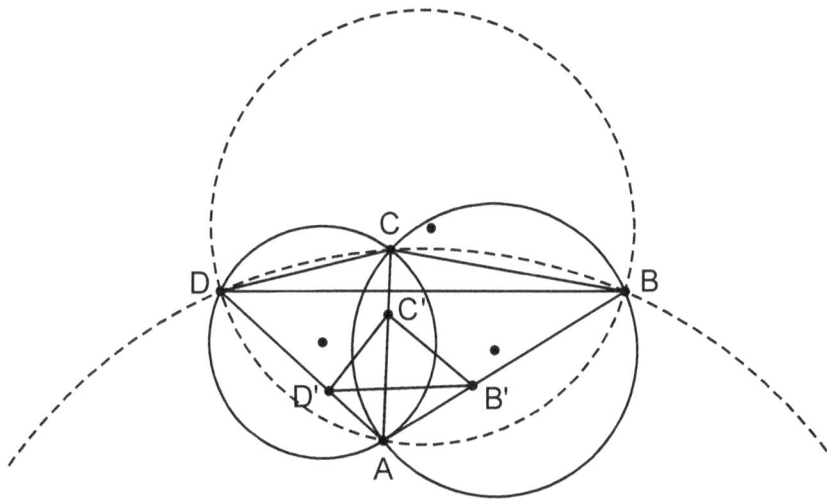

Figure 6.109: Example, E. IN.7

We consider an inversion with center A and any radius $r = \sqrt{c}$. Then, the lines AB, AC and AD are invariant and the contain B', C' and D' the inverses of B, C and D respectively. So, if the circles (ABC) and (ADC) are orthogonal the triangle $B'C'D'$ is right at C'. Hence,

$$B'D'^2 = C'B'^2 + C'D'^2.$$

Substituting these segments by the distance formula (see **Subsection 6.4.5**), we have

$$\left(\frac{c \cdot BD}{AB \cdot AD}\right)^2 = \left(\frac{c \cdot CB}{AC \cdot AB}\right)^2 + \left(\frac{c \cdot CD}{AC \cdot AD}\right)^2.$$

We substitute the letters a, b, c, d, x, and y, as set above, and simplify. Thus, we find the necessary and sufficient relation

$$x^2 y^2 = a^2 c^2 + b^2 d^2.$$

Remarks. 1) Such a quadrilateral cannot be inscribable, since an inscribable quadrilateral satisfies $xy = ac + bd$, (by **Ptolemy's Theorem, 6.6.1**). (Otherwise, one of the sides would be zero and we would not have a quadrilateral.)

2) The circle circumscribed to triangle BCD has inverse the circle circumscribed to the right triangle $C'B'D'$. The latter has diameter $B'D'$ and so the

circle $(C'B'D')$ and the line $B'D'$ intersect orthogonally. Therefore, their inverses intersect orthogonally, too. These inverses are the circles (BCD) and the circle (ABD). This means that, the circles circumscribed to triangles ABC and ADC are orthogonal if and only if the circles circumscribed to triangles BCD and the circle ABD are orthogonal. (This was expected by the symmetry of the relation found: $x^2y^2 = a^2c^2 + b^2d^2$.)

▲

Example 6.6.8 We consider a quadrilateral $ABCD$ with perpendicular diagonal $AC \perp BD$. Let Q be the point of intersection of the two diagonals. Then, the reflections of Q on the sides AB, BC, CD, DA, the points Q_1, Q_2, Q_3, Q_4, respectively, are concyclic. See **Figure 6.110**.

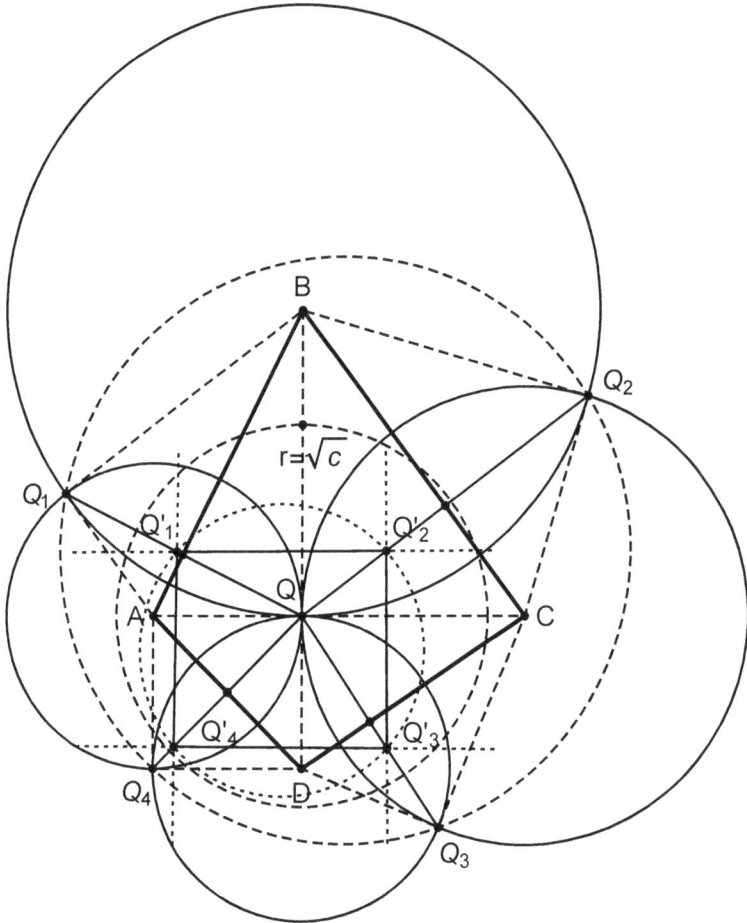

Figure 6.110: Example, E. IN.7

By the conditions stated, the circle $C[A, QA]$ passes through Q_4 and Q_1, and is tangent to the diagonal BD. Similar results for the circles $C[B, QB]$, $C[C, QC]$, $C[D, QD]$. So, $C[A, QA]$ is orthogonal to $C[B, QB]$, $C[D, QD]$ and tangent to $C[C, QC]$.

We use inversion with center Q and radius any $r = \sqrt{c}$. The above four circles pass through Q. Then, the tangent circles become parallel lines and the orthogonal circles perpendicular lines. Also, the image of these circles contains Q'_1, Q'_2, Q'_3, Q'_4 the inverse points of Q_1, Q_2, Q_3, Q_4. Hence, $Q'_1 Q'_2 Q'_3 Q'_4$ is a rectangle, and therefore inscribable in a circle, which does not pass through Q.

Now, applying the inversion to the circle $(Q'_1 Q'_2 Q'_3 Q'_4)$, we get a circle passing through Q_1, Q_2, Q_3, and Q_4, proving the claim.

This result is also true when the quadrilateral $ABCD$ is not convex and the diagonal AC and BD are perpendicular. See **Figure 6.111**.

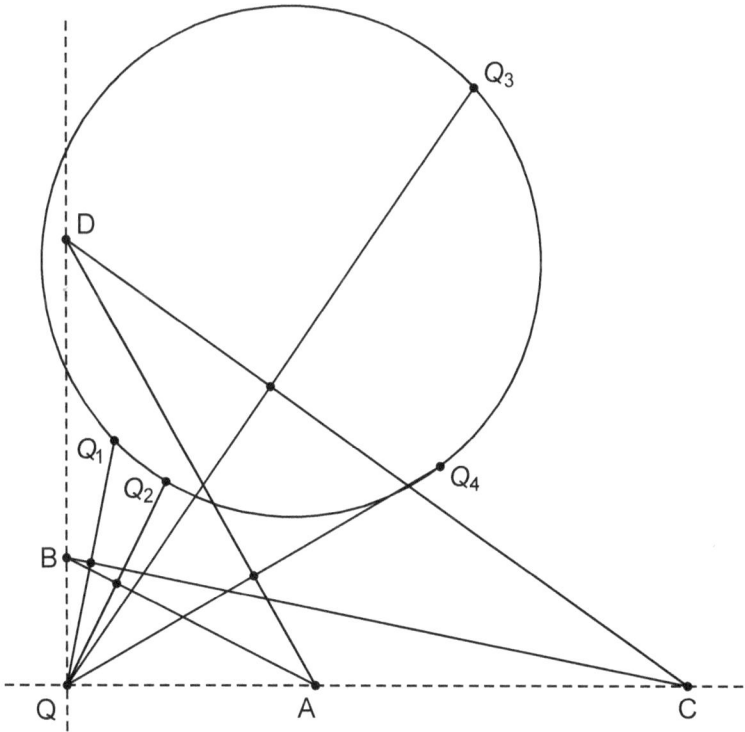

Figure 6.111: Example, E. IN.7

(This example can be solved analytically, if we introduce coordinates with axes the perpendicular diagonals AC and DB and origin the point Q.)

▲

For further results, applications, and examples study and solve the exercises listed in the following chapters.

6.7   Appendix

Part I: Double Ratio, Polars and Poles

There are many results and exercises with regard to double ratio, polar lines and poles. Here and in the exercises, we present only a few in order to briefly show a different treatment of the connection of the double ratio with polar lines and poles. Study this large chapter of geometry from pertinent bibliography. We start with the following fundamental Theorem:

Theorem 6.7.1 (Pappus of Alexandria.) *We consider four straight lines* $\{k,\ m,\ l,\ n\}$, *in this order from left to right, and passing through a point* Z *finite or infinite. Let any two straight lines* u *and* u' *intersect these lines at points* $\{A,\ C,\ B,\ D\}$ *and* $\{A',\ C',\ B',\ D'\}$, *respectively. Then, we have the equality of the double ratios*

$$\{A, B; C, D\} = \{A', B'; C', D'\}.$$

Proof See **Figure 6.112** and pay attention to the order of the lines and points and the point Z is a finite point. Through the points B and B' we draw lines parallel to k. Let E and F and E' and F' be the points of their intersections with m and n, respectively. The triangles EBC and ACZ are similar and so are the triangles BFD and ADZ, since they have equal corresponding angles.

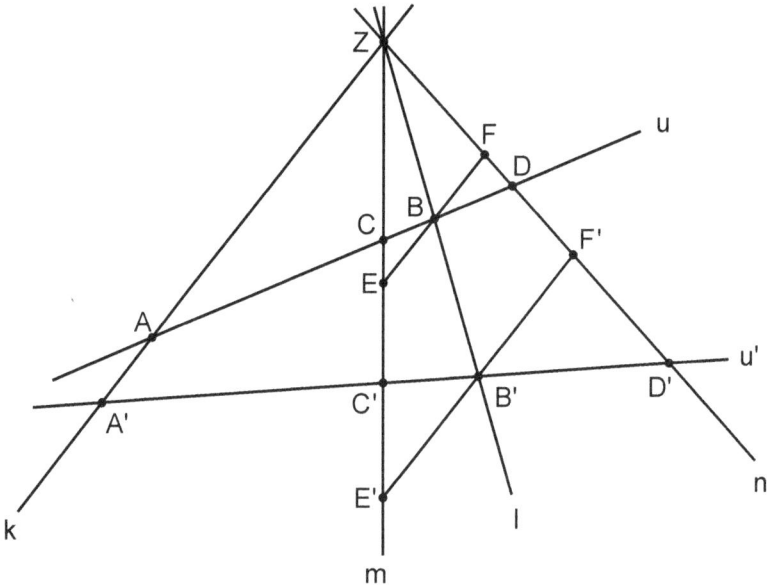

Figure 6.112: An Invariance of the Double Ratio

So, we have

$$\frac{CA}{CB} = \frac{AZ}{BE} \quad \text{and} \quad \frac{DA}{DB} = \frac{AZ}{BF}.$$

Hence,

$$\frac{CA}{CB} \div \frac{DA}{DB} = \frac{AZ}{BE} \div \frac{AZ}{BZ} = \frac{BF}{BE}.$$

Similarly we prove

$$\frac{C'A'}{C'B'} \div \frac{D'A'}{D'B'} = \frac{B'F'}{B'E'}.$$

Now, since $u \parallel k \parallel u'$ by Thales' Theorem, we have

$$\frac{BF}{BE} = \frac{B'F'}{B'E'}.$$

The three above equalities prove the claim of the Theorem (written with or without oriented segments)!

If $Z = \infty \Longleftrightarrow k \parallel m \parallel l \parallel n$, the Theorem follows immediately from Thales's Theorem. (Draw figure and check.)

■

We have two immediate corollaries.

Corollary 6.7.1 *Under the conditions of the above Theorem, B is the midpoint of EF (and so is B' for E'F', i.e., EB = BF and E'B' = B'F') if and only if $\{A, B; C, D\} = -1$ (the quadruple is harmonic).*

Corollary 6.7.2 *Under the conditions of the above Theorem, if for a line u $\{A, B; C, D\} = -1$ (the quadruple is harmonic), then for any other line u' $\{A', B'; C', D'\} = -1$.*

Definition 6.7.1 *A pencil of four straight lines is called harmonic if the quadruple of the points of intersection with any straight line intersecting them is harmonic.*

In **Figure 6.113** the pairs of straight lines (k, l) and (m, n) are called **harmonic conjugate pairs** of straight lines.

Note: It may be that a point of intersection is ∞, that is, the intersecting line is parallel to one of the lines of the pencil. The same result is still valid, by Thales' Theorem, or as a limiting case. (Draw figure and elaborate!)

Given three lines going through a point, **Figure 6.112** and **Corollary 6.7.1** tell how to find the fourth line so that all four lines make a harmonic pencil. (Elaborate and list the steps. Also, see **the sine condition for a harmonic pencil** in the exercises which is convenient in many situations.)

Corollary 6.7.3 *For any two given lines the two bisectors of their angles are perpendicular and the set of all these four lines form a harmonic pencil of lines. **Conversely**: If the lines in a pair of two conjugate pairs of lines of a harmonic pencil are perpendicular, then they are the angle bisectors of the angles formed by the other pair.*

(Draw figure, use **Corollary 6.7.1** and elaborate!)

Next, we have the following theorem:

Theorem 6.7.2 *We consider two straight lines Zx and Zy and a point A not on either one. Let any two straight lines through A intersect the two lines at the points $\{P, Q\}$ and $\{R, S\}$, respectively. Let Y be the point of intersection of PS and QR and the line ZY intersect RQ and RS at M and N, respectively. Then, the quadruples $\{S, N, R, A\}$, and $\{Q, M, P, A\}$ are harmonic.*

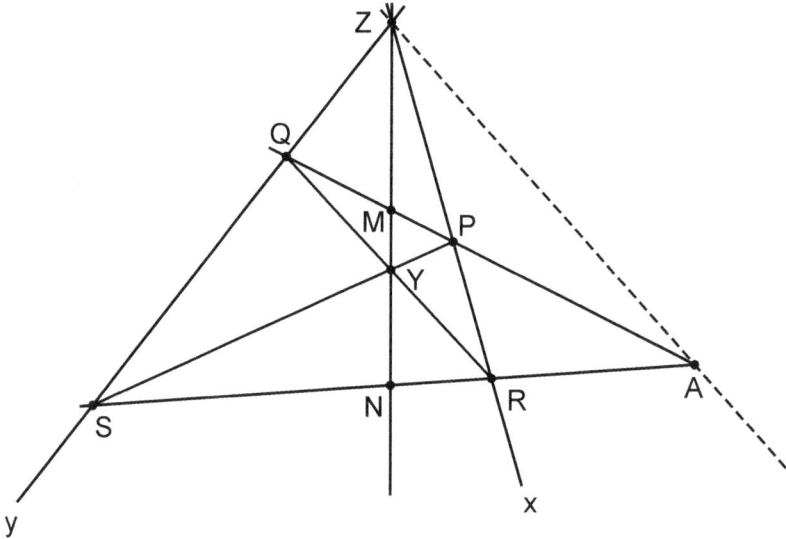

Figure 6.113: Construction of Polar Line w.r.t. Two Lines

Proof See **Figure 6.113**. We apply **Ceva's Theorem, 4.2.6**, to triangle ZSR and point Y and **Menelaus's Theorem, 4.2.5**, to the same triangle and the intersecting straight line APQ. We have:

$$\frac{ZQ}{QS} \cdot \frac{SN}{NR} \cdot \frac{RP}{PZ} = 1$$

and

$$\frac{ZQ}{QS} \cdot \frac{SA}{RA} \cdot \frac{RP}{PZ} = 1.$$

Dividing these equalities, we find

$$\frac{SN}{NR} = \frac{AS}{AR}.$$

By **Definition 4.1.4** of harmonic quadruple (careful with the arrangement of the points on the straight line here and in the definition), this implies that the quadruple $\{S, N, R, A\}$, is harmonic and, by **Corollary 6.7.2**, $\{Q, M, P, A\}$ is also harmonic. [That is, $\{S, R; N, A\} = -1 \iff \{S, N; R, A\} = 1 - (-1) = 2$.]

Again Z can be ∞, i.e., $Zx \parallel Zy$. (Draw figure and elaborate!) ■

The above results yield the following corollary.

Corollary 6.7.4 *If in the previous Theorem* Zx, Zy, A *and* N *are fixed and* $\{S, R; N, A\} = -1$, *then letting the line* APQ *move, the points* M *and* Y *describe the fixed straight line* ZN, *which is the locus of all points* M, *such that through* $\{Q, P; M, A\} = -1$.

Also, the pair of lines (ZY, ZA) *is the harmonic conjugate pair of the pair of lines* $(ZS = Zx, ZR = Zy)$, *and the pencil of the four lines* $\{ZS, ZN, ZR, ZA\} = \{ZQ, ZM, ZP, ZA\}$, *is harmonic.*

Remark: Notice that in **Figure 113**, $SRAPZQ$ is a complete quadrilateral and the lines AZ, PS and QR are its three diagonals. Y is the point of intersection of the two diagonals PS and QR. So, some of the above results can be rephrased in terms of complete quadrilaterals, as we encounter in many books.

Definition 6.7.2 *The line* ZN *of the above* **Corollary** *is called the* **polar line of** A **with respect to the lines** Zx **and** Zy. *The point* A *is called the* **pole of** ZN **with respect to the lines** Zx **and** Zy.

The points $\{P, Q, S, R\}$ may be on a circle $C[O, r]$ (concyclic points). See **Figure 6.114**. Then, by **Theorem 6.2.4** and **Example 6.2.1**, we have that the points M and N are on the polar line of A with respect to the given circle $C[O, r]$ (since the diameter FGA is divided harmonically by A and A', the

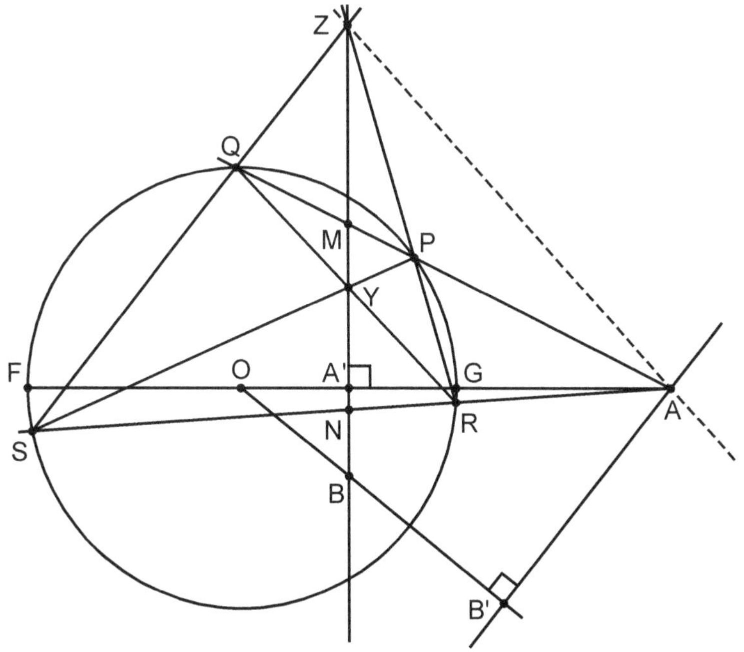

Figure 6.114: Construction of Polar Line w.r.t. a Circle

inverse of A in the circle $C[O, r]$ and this diameter could be in the place of the intersecting line SRA). Therefore, the polar line of A with respect to the given circle $C[O, r]$ is ZY (or MN) and A is the pole of ZY. The point Y is á-priori determined and so YZ is also á-priori determined. Hence, $ZY \perp AO$ at A', the inverse of A in $C[O, r]$.

Also, if B' is the inverse of a point $B \in ZY$, then the polar line of B with respect to the given circle $C[O, r]$ is the line $B'A$, and $B'A \perp OB$.

Remark: The exposition of this Appendix constitutes another treatment of the polar lines and poles studied in **Example 6.2.1**, **Subsection 6.4.1**, **Definition 6.4.1**, and **Subsection 6.4.3**.

Note: In the above exposition, given two straight lines or a circle and a point not on them, we can construct the inverse point and the polar line of this point with respect to the straight lines or the circle, by using the straight edge only. (List the steps of these interesting one-instrument constructions!)

Part II: Theorem of Medians and Radical Axis

The following theorem of medians of a triangle is very important in classical geometry and can be used in many exercises if the book.

Theorem 6.7.3 (Theorem of Medians) *We consider any triangle ABC its median $m_a = AD$ (D is the midpoint of BC), ED the projection of this median on the side $a = BC$ (E is the foot of the altitude on $a = BC$ from the vertex A) and, without loss of generality, we may assume that $b = AC \geq c = AB$. Then we have:*

$$(a) \quad b^2 + c^2 = 2 \cdot m_a^2 + \frac{a^2}{2},$$

$$(b) \quad b^2 - c^2 = 2 \cdot a \cdot ED,$$

$$(c) \quad a^2 = 4[m_a^2 - bc\cos(\widehat{A})].$$

Analogous results for the other medians are obtained by a cyclic permutation of the letter a, b, and c.

Proof See **Figure 6.115**.

If $b = c$, i.e., the triangle ABC is isosceles, then $E = D$, i.e., $ED = 0$ and height $AE = AD = m_a$ median. Then, the relations **(a)** and **(b)** follow immediately form the Pythagorean Theorem applied to the right triangles ACE and ABE.

If $b > c$, then E and B are on the same side of D along the line BC. We use the Pythagorean Theorem again and from the right triangles ACE and ABE and the relation $BD = CD = \dfrac{a}{2}$, we get

$$b^2 = AE^2 + CE^2 = AE^2 + \left(\frac{a}{2} + ED\right)^2 = AE^2 + \frac{a^2}{4} + a \cdot ED + ED^2,$$

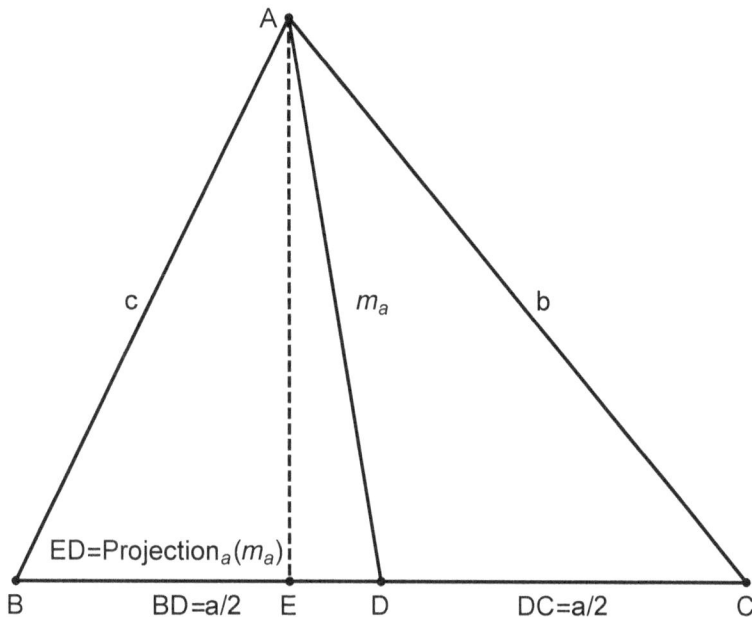

Figure 6.115: Theorem of Medians

$$c^2 = AE^2 + BE^2 = AE^2 + \left(\frac{a}{2} - ED\right)^2 = AE^2 + \frac{a^2}{4} - a \cdot ED + ED^2.$$

By adding and subtracting the two equalities sidewise, and using the right triangle AED ($\widehat{E} = 90°$, $AE^2 + ED^2 = AD^2 = m_a^2$), we find the two results

(a) $b^2 + c^2 = 2\left(AE^2 + ED^2\right) + \frac{a^2}{2} = 2 \cdot m_a^2 + \frac{a^2}{2}$,　**(b)** $b^2 - c^2 = 2 \cdot a \cdot ED.$

Result **(c)** follows from **(a)** and the law of cosines.

∎

Corollary 6.7.5 *In any triangle ABC we have*

$$\text{(a)}\quad m_a^2 = \frac{2(b^2 + c^2) - a^2}{4} \quad\Longleftrightarrow\quad m_a = \frac{\sqrt{2(b^2 + c^2) - a^2}}{2},$$

$$\text{(b)}\quad b^2 - c^2 = 2a \cdot \overline{ED} \quad\Longleftrightarrow\quad \overline{ED} = \frac{b^2 - c^2}{2a},$$

$$\text{(c)}\quad \overline{BE} = \frac{a}{2} - \overline{ED} = \frac{a}{2} - \frac{b^2 - c^2}{2a} = \frac{a^2 - b^2 + c^2}{2a},$$

$$\text{(d)}\quad \overline{EC} = \frac{a}{2} + \overline{ED} = \frac{a}{2} + \frac{b^2 - c^2}{2a} = \frac{a^2 + b^2 - c^2}{2a}.$$

Analogous results for the other medians are obtained by a cyclic permutation of the letter a, b, and c.

Remarks: (1) Formula **(a)** can also be derived by **Stewart's Theorem, 6.5.3**.

(2) Using **(c)** and simplifying $AE^2 = c^2 - BE^2 = c^2 - \left(\dfrac{a^2 - b^2 + c^2}{2a} \right)^2$, we find **Heron's formula** for the height $h_a := AE$,

$$h_a = \frac{2}{a}\sqrt{s(s-a)(s-b)(s-c)}, \quad \text{where} \quad s = \frac{a+b+c}{2}$$

(half of the perimeter of ABC), and so the area of ABC is

$$\text{Area}(ABC) = \sqrt{s(s-a)(s-b)(s-c)}.$$

Next, we will use the **Theorem of the Medians** to find the **radical axis of two circles** $\mathcal{C}_1 = C[O_1, R_1]$ and $\mathcal{C}_2 = C[O_2, R_2]$. We have used the radical axis several times in **Section 6.5**, and so let us explain how we obtain it.

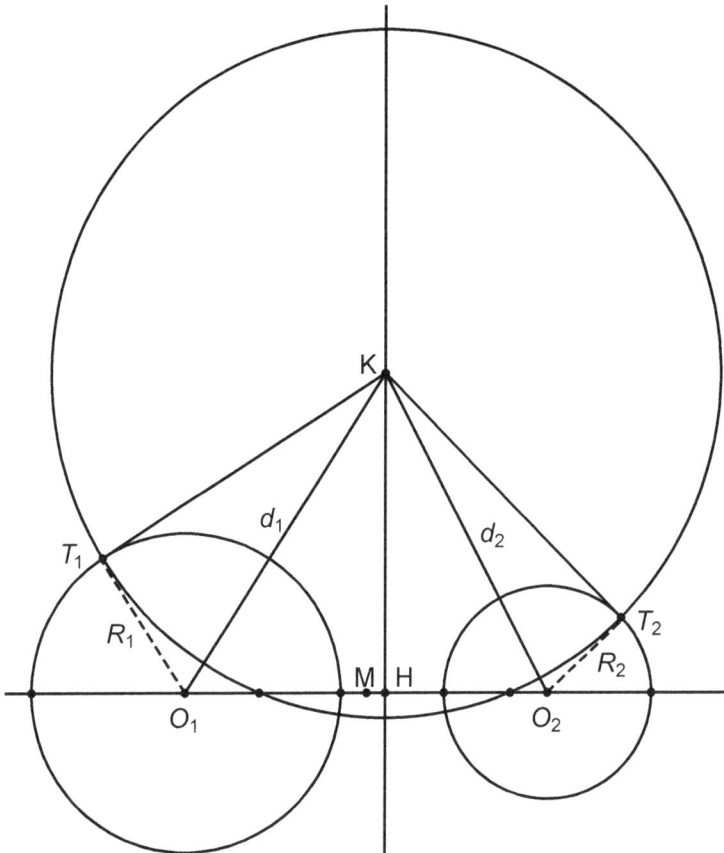

Figure 6.116: Radical Axis of Two Circles

By definition, *"the **radical axis** of two given coplanar circles is the geometrical locus of the points of their plane that have equal powers with respect to these circles".*

We assume that $O_1 \neq O_2$ (the circles are not concentric). Suppose $K \in \mathcal{P}$, such that

$$\mathfrak{P}_{C_1}(K) = \mathfrak{P}_{C_2}(K)$$

and consider $d_1 = KO_1$ and $d_2 = KO_2$. See **Figure 6.116**. Since,

$$\mathfrak{P}_{C_1}(K) = KO_1^2 - R_1^2 = d_1^2 - R_1^2$$

and

$$\mathfrak{P}_{C_2}(K) = KO_2^2 - R_2^2 = d_2^2 - R_2^2,$$

K must satisfy

$$d_1^2 - d_2^2 = R_1^2 - R_2^2.$$

Then, by relation **(b)** of the **Theorem of the Medians, 6.7.3** [or **(b)** of the **previous Corollary**], to the triangle KO_1O_2, to obtain

$$2\overline{O_1O_2} \cdot \overline{MH} = R_1^2 - R_2^2 \quad \Longleftrightarrow \quad \overline{MH} = \frac{R_1^2 - R_2^2}{2\overline{O_1O_2}},$$

where M is the midpoint of O_1O_2 (known point) and H is the projection of K onto O_1O_2. Therefore, the point K is on the straight line perpendicular to O_1O_2 at the point H.

Now, using the **Theorem of the Medians, 6.7.3, (b)**, we directly prove that any point of the straight line HK satisfies the definition of radical axis and no other point of the plane does. (Provide details.) Therefore:

*"The **radical axis** of $C_1 = C[O_1, R_1]$ and $C_2 = C[O_2, R_2]$ is the whole straight line perpendicular to O_1O_2 at the point H",*

found by the above equation.

Remarks: Now we can justify the following:

(a) If $O_1 \neq O_2$ and $R_1 = R_2$, then $H = M$ (the midpoint of the center segment O_1O_2).

(b) If $O_1 = O_2$ and $R_1 \neq R_2$, then H is the infinity point of the straight line O_1O_2 and the radical axis is the line at infinity.

(c) If $O_1 = O_2$ and $R_1 = R_2$, then the two circles coincide and it makes no sense to ask for the radical axis of one circle. (In this case, every point of the plane satisfies the definition of the radical axis.)

(d) If the circles are tangent, then $O_1O_2 = |R_1 \pm R_2|$ and the radical axis is the common tangent.

(e) If the circles intersect at two points, then $|R_1 - R_2| < O_1O_2 < R_1 + R_2$ and the radical axis is the common chord.

(f) If the circles are outside each other, then $O_1O_2 > R_1 + R_2$ and the radical axis is located between them and not intersecting them, as in **Figure 6.116**.

(g) If the circles are inside each other and not concentric, then $0 < O_1O_2 < |R_1 - R_2|$ and the radical axis is located outside of them and on the side that the circles are closer to each other.

(h) The tangent segments to the two circles from any point K of the part of radical axis outside the intersection of the two circles are equal.

(i) The circles with centers the points of the part of radical axis outside the intersection of the two circles and radii the tangent segments to the circles from these points are orthogonal to the two given circle.

Part III: Three General Results

Following **Coxeter [4], pp. 90-91, on circle-preserving transformations**, we state the following three general results. Let

$$\Phi \ : \ \mathcal{P} \longrightarrow \mathcal{P}$$

be a one to one and onto mapping of the plane to itself (transformation of the plane).

(a) If Φ maps straight lines to straight lines and preserves the absolute values of oriented angles, then it is a direct or opposite similarity (depending on whether it preserves or reverses the sign of an oriented angle). If it also preserves the length of a non-zero straight segment, then Φ is an isometry. (This result follows form the material of **Chapters 2 and 4**.)

(b) If Φ maps straight lines to straight lines and circles to circles correspondingly, then it is either a direct or an opposite similarity. If it also preserves the length of a non-zero straight segment, then Φ is an isometry. (This results needs some proof. You may clinch a proof by starting with proving that the center and any diameter of a circle is mapped to the center and a diameter of the corresponding circle and that the right angles are mapped to right angles.)

(c) If Φ maps straight lines and circles to straight lines and circles (not necessarily correspondingly) and takes an ordinary point to an infinity point, then

$$\Phi = J \circ S,$$

where J is an inversion and S is an isometry. {See, **Coxeter [4], pp. 90-91**. To prove this we firstly need to prove that: *A one to one function of the plane into itself that maps straight lines and circles to straight lines and circles and has a fixed point is a similarity.* Then we use **Theorem 4.3.2** and **Cases (1) and (2) of Section 6.5** of this text.}

Chapter 7

Exercises on Circles

Powers of Points with respect to Circles
Orthogonal and Pseudo-orthogonal Circles
Steiner Chains

Notation: The symbol $\mathfrak{P}_O(P)$ means the power of the point P with respect to a given circle with center O and a certain radius. If the circle is $C[O,R]$ and the radius R matters, we can also write $\mathfrak{P}_{C[O,R]}(P)$.

We abbreviate "straight line", "-es" with "s.l.", "s.ls"; "geometrical locus", "-es" with "g.l.", "g.ls"; and "if and only if" with "iff".

7.1 General

1. Let $a = BC$, $b = CA$ and $c = AB$ be the sides of a triangle ABC. Let also G and H be the center of gravity (centroid) and the orthocenter of the triangle. Prove that the power of G with respect to the circumcircle $(ABC) := C[O,R]$ is equal to

$$\mathfrak{P}_{(ABC)}(G) = GO^2 - R^2 = \frac{-1}{9}\left(a^2 + b^2 + c^2\right)$$

$$\text{and so,} \quad GO^2 = R^2 - \frac{1}{9}\left(a^2 + b^2 + c^2\right).$$

Next, use the properties of the segment OH of Euler's line to prove that

$$OH^2 = 9R^2 - \left(a^2 + b^2 + c^2\right) \quad \text{and} \quad GH^2 = 4R^2 - \frac{4}{9}\left(a^2 + b^2 + c^2\right).\,[1]$$

---

[1]The quantity $P := \frac{1}{2}\left(a^2 + b^2 + c^2\right)$ was named as **the power of the triangle** ABC by the Spanish geometer mathematician Juan Jacobo Durán Loriga (1854-1911). Also, the quantities $P_a := \frac{1}{2}\left(-a^2 + b^2 + c^2\right)$, $P_b := \frac{1}{2}\left(a^2 - b^2 + c^2\right)$ and $P_c := \frac{1}{2}\left(a^2 + b^2 - c^2\right)$ were named as **the partial powers of the triangle**. These terms were invented due to the fact that all these quantities, P, P_a, P_b, and P_c, appear in many important relations for the triangle. Some of these relations will be encountered in the exercises that follow.

2. Let $a = BC$, $b = CA$ and $c = AB$ be the sides of a triangle ABC and let R be the radius of its circumcircle. Prove:

   (a) For any triangle $9R^2 \geq a^2 + b^2 + c^2$. Equality holds iff ABC is equilateral. (Hint: Use the previous problem.)

   (b) ABC is acute   iff $a^2 + b^2 + c^2 > 8R^2$.

   (c) ABC is right   iff $a^2 + b^2 + c^2 = 8R^2$.

   (d) ABC is obtuse   iff $a^2 + b^2 + c^2 < 8R^2$.

3. Let $a = BC$, $b = CA$ and $c = AB$ be the sides of a triangle ABC. Let A', B', and C' be the symmetrical points of the vertices A, B and C of a triangle with all angles acute, with respect to the opposite sides. Prove that the sum of the powers of the points A', B', and C' with respect to the circumcircle of the triangle is equal to
$$\mathfrak{P}_{(ABC)}(A') + \mathfrak{P}_{(ABC)}(B') + \mathfrak{P}_{(ABC)}(C') = a^2 + b^2 + c^2.$$

4. Let $a = BC$, $b = CA$ and $c = AB$ be the sides of a triangle ABC. Let G be the center of gravity (centroid) of a triangle ABC. Prove that the powers of the vertices A, B, and C with respect to circles (GBC), (GCA), and (GAB), respectively, are equal to
$$\mathfrak{P}_{(GBC)}(A) = \mathfrak{P}_{(GCA)}(B) = \mathfrak{P}_{(GAB)}(C) = \frac{1}{3}\left(a^2 + b^2 + c^2\right).$$

5. Let $a = BC$, $b = CA$ and $c = AB$ be the sides of a triangle ABC and O_1, O_2, and O_3 their midpoints respectively. Prove that the sum of the powers of the vertices of a triangle ABC with respect to the nine-point circle is
$$\mathfrak{P}_{(O_1O_2O_3)}(A) + \mathfrak{P}_{(O_1O_2O_3)}(B) + \mathfrak{P}_{(O_1O_2O_3)}(C) = \frac{1}{4}\left(a^2 + b^2 + c^2\right).$$

6. Consider a convex quadrilateral $ABCD$ which is both inscribable and circumscribable. Prove that the power of the point of intersection of two opposite sides with respect to the circumscribed circle is equal to the square of the distance from this point to the center of the inscribed circle.

7. Consider a circle $C[O, R]$ and a point $A \neq O$ inside of it. Prove:

   (a) Among all the chords through A the shortest one is the chord perpendicular to OA. (What happens if $A = O$?) This chord is called the **minimum chord of $C[O, R]$ at the point** A.

   (b) If the minimum chord in **(a)** is XY, then
$$\mathfrak{P}_O(A) = -\left(\frac{AX}{2}\right)^2 = -\left(\frac{AY}{2}\right)^2.$$

8. (a) Consider two circles $C[O_1, R_1]$ and $C[O_2, R_2]$ that have a common chord UV. (If $U = V$, the chord is just a point). Prove that for any point A of UV the minimum chords of $C[O_1, R_1]$ and $C[O_2, R_2]$ at the point A are equal.

   (b) Consider three circles $C[O_1, R_1]$, $C[O_2, R_2]$ and $C[O_3, R_3]$ such that each pair of them has a common chord or a common point. Prove that these three chords intersect at the same point and if this common point is on the circles or is an interior point to all three of them, then the minimum chords of $C[O_1, R_1]$, $C[O_2, R_2]$ and $C[O_3, R_3]$ at this common point are equal. (This common point is the **radical center** of these three circles.)

9. Let two circles $C[O_1, R_1]$ and $C[O_2, R_2]$ intersect at two points A and B or externally tangent at a point $A(= B)$. Let the common external tangent segment closer to the point A be CD with $C \in C[O_1, R_1]$ and $D \in C[O_2, R_2]$.

(a) Prove that the circles (ACD) and (BCD) are equal and their radius is $r = \sqrt{R_1 R_2}$.

(b) Prove that $\dfrac{AC}{AD} = \sqrt{\dfrac{R_1}{R_2}}$.   (c) Find $\dfrac{BC}{BD}$ in terms of R_1 and R_2.

(d) Compute AC, AD, BC, and BD in terms of R_1, R_2 and $O_1 O_2$ (or equivalently AB).

10. Let two circles $C[O_1, R_1]$ and $C[O_2, R_2]$ be externally tangent at a point A. Let one of the equal common external tangent segments be CD with $C \in C[O_1, R_1]$ and $D \in C[O_2, R_2]$ and d the distance of A from CD. We also consider the circle $C[O, R]$ which is tangent to CD and externally tangent to the circles $C[O_1, R_1]$ and $C[O_2, R_2]$. Prove:

$$\text{(a)}\quad CD = 2\sqrt{R_1 R_2}.$$

$$\text{(b)}\quad \frac{2}{d} = \frac{1}{R_1} + \frac{1}{R_2} \quad \text{or} \quad d = \frac{2R_1 R_2}{R_1 + R_2}.$$

$$\text{(c)}\quad AC = 2R_1 \sqrt{\frac{R_2}{R_1 + R_2}} \quad \text{and} \quad AD = 2R_2 \sqrt{\frac{R_1}{R_1 + R_2}}.$$

$$\text{(d)}\quad R = \frac{R_1 R_2}{(\sqrt{R_1} + \sqrt{R_2})^2}.$$

$$\text{(e)}\quad \frac{1}{R_1^2} + \frac{1}{R_2^2} + \frac{1}{R^2} = 2\left(\frac{1}{R_1 R} + \frac{1}{R_2 R} + \frac{1}{R_1 R_2}\right).$$

11. Consider two orthogonal circles $C[O_1, R_1]$ and $C[O_2, R_2]$. Let A and B be their common points, CD their external common tangent segment and EF its projection on the center s.l. of the circles, $O_1 O_2$. Prove that

$$EF = AB = \frac{2R_1 R_2}{\sqrt{R_1^2 + R_2^2}}.$$

12. From the foot of the height AH_1 of a triangle ABC draw s.ls parallel to the sides AB and AC that intersect AB at B_1 and AC at C_1, and also s.ls perpendicular to the sides AB and AC intersecting AB at B_2 and AC at C_2. Prove that the s.ls BC, $B_1 C_1$ and $B_2 C_2$ pass through the same point.

13. Consider a circle $C[O, R]$ and two points A and B in the plane. Consider all circles through A and B that intersect $C[O, R]$ at two points or are tangent to $C[O, R]$ at one point. Prove that all the common chords or

the common tangent of $C[O, R]$ with these circles pass through the same point of the s.l. AB or all are parallel to AB.

14. Consider two orthogonal circles $C[O_1, R_1]$ and $C[O_2, R_2]$. Prove:
    (a) If $R_1 > 0$ and $R_2 > 0$, then O_1 is outside $C[O_2, R_2]$ and O_2 is outside $C[O_1, R_1]$.
    (b) If $O_2 \in C[O_1, R_1]$, then either $R_2 = 0$, that is, $C[O_2, R_2] = \{O_2\}$ is a point, or $R_1 = \infty$, that is, $C[O_1, R_1]$ is a straight line.
    (c) If three circles are not coaxal and have nonempty intersection, then there is no non-trivial circle orthogonal to all three of them.

15. Two circles are orthogonal iff the square of the radius of one of them is equal to the power of its center with respect to the other circle.

16. Two circles are orthogonal iff the diameter of one of them is divided harmonically by the other circle.

17. Two circles intersecting at two points are orthogonal iff any segment through one of the common points and with endpoints on the two circles subtend a right angle with the other common point.

18. Two circles are orthogonal iff the common exterior tangent segment subtends angle $= \dfrac{\pi}{4}$ with one of their common points and an angle $= \dfrac{3\pi}{4}$ with the other one.

19. Two circles $C[O_1, R_1]$ and $C[O_2, R_2]$ are orthogonal at their common points A and B iff for any point P of one of the circles the chords PA and PB intersect the other circle at diametrical points C and D.

20. Two circles $C[O_1, R_1]$ and $C[O_2, R_2]$ are orthogonal iff any point P of the circle of diameter $O_1 O_2$ has opposite powers with respect to $C[O_1, R_1]$ and $C[O_2, R_2]$.

21. Two circles $C[O_1, R_1]$ and $C[O_2, R_2]$ are orthogonal at their common points A and B iff through A (or B) there exist two perpendicular lines intersecting the circles at points P and S on $C[O_1, R_1]$ and Q and T on $C[O_2, R_2]$, and all four points are distinct.

22. Consider a triangle ABC and with centers the vertices A, B, and C we draw circles pairwise orthogonal. If where $a = BC$, $b = CA$, $c = AB$ are the sides of the triangle, prove that the radii of these three circles are

$$R_A^2 = \frac{b^2 + c^2 - a^2}{2}, \qquad R_B^2 = \frac{c^2 + a^2 - b^2}{2}, \qquad R_C^2 = \frac{a^2 + b^2 - c^2}{2}.$$

23. Two circles $C[O_1, R_1]$ and $C[O_2, R_2]$ are orthogonal iff they have two diameters that are segments defined in an orthocentric quadruple. (**Definition:** Four points A, B, C, D form an **orthocentric quadruple**, if each point is the orthocenter of the triangle with vertices the other three.)

24. Consider two circles $C[O_1, R_1]$ and $C[O_2, R_2]$ and any circle $C[K, r]$ that intersects them orthogonally. Prove:
(a) If $C[O_1, R_1]$ and $C[O_2, R_2]$ intersect each other, then $C[K, r]$ does not intersect $O_1 O_2$.
(b) If $C[O_1, R_1]$ and $C[O_2, R_2]$ are tangent to each other (on $O_1 O_2$), then $C[K, r]$ passes through the point of tangency on $O_1 O_2$.
(c) If $C[O_1, R_1]$ and $C[O_2, R_2]$ do not intersect, then $C[K, r]$ intersects $O_1 O_2$ at two fixed points {the same points for any such $C[K, r]$, called the **Poncelet points**}.

25. **A circle $C[O, R]$ is called pseudo-orthogonal to a circle $C[K, r]$, if** $C[O, R]$ intersects $C[K, r]$ at two diametrical points A and B {i.e., AB is a diameter of $C[K, r]$}. Prove:
(a) $C[O, R]$ is pseudo-orthogonal to a circle $C[K, r]$ iff the power of K with respect to $C[O, R]$ is $-r^2$. [$\mathfrak{P}_O(K) = -r^2$.]
(b) The property of pseudo-orthogonality is not symmetric.

26. Suppose $C[O, R]$ is pseudo-orthogonal to a circle $C[K, r]$. Prove:
(a) The radical axis of $C[O, R]$ and $C[K, r]$ passes through the center K of $C[K, r]$.
(b) If the circle $C[O, R]$ passes through a fixed point C, then it also passes through another fixed point D of KC, such that

$$\overline{KC} \cdot \overline{KD} = -r^2.$$

27. Consider the circle $C[O, R]$ and $t > 0$. Then prove:
(a) If $\mathfrak{P}_O(K) = t^2$, then $C[O, R]$ is orthogonal to the circle $C[K, t]$.
(b) If $\mathfrak{P}_O(K) = -t^2$, then $C[O, R]$ is pseudo-orthogonal to the circle $C[K, t]$.
(c) If $\mathfrak{P}_O(K) = 0$, then $C[O, R]$ is orthogonal or pseudo-orthogonal to the point-circle $C[K, 0] = \{K\}$.

28. Consider a right triangle ABC, with $\widehat{A} = 90°$ and any point P on the hypotenuse BC. Prove:
(a) The circles (ABP) and (ACP) are orthogonal.
(b) The circles through P and tangent to AB at B and to AC at C are orthogonal.

29. Consider a circle $C[K, R]$ and point A outside of it. Let B be a point outside of $C[K, R]$ and on the s.l. defined by the two points of tangency of the tangent lines from A to $C[K, R]$. Prove:

(a) The circle of diameter AB intersects the circle $C[K, R]$ orthogonally.

(b) The circles with centers A and B and intersecting $C[K, R]$ orthogonally, intersect each other orthogonally.

30. Consider two circles $C[O_1, R_1]$ and $C[O_2, R_2]$. Prove that all the circles intersecting the two given circles pseudo-orthogonally pass through two fixed points of the s.l. $O_1 O_2$.

31. The circle $C[O, r]$ is an Apollonius circle on a segment AB, of some ratio λ, iff $\overline{OA} \cdot \overline{OB} = r^2 (> 0)$.

Now justify: (a) r^2 is the power of O with respect to any circle passing through A and B. (b) O is outside the straight segment AB. (c) A and B are mutually inverse in the Apollonius circle $C[O, r]$.

32. Consider a triangle ABC and suppose $a = BC \geq b = CA \geq c = AB$, without loss of generality. We call the **Apollonius circle of the triangle with respect to a side**, the circle with diameter the segment on the side in consideration and with endpoints the common points of this side with the internal and external bisectors of the opposite angle. That is, the internal and external angle bisectors of the angles of a triangle ABC define on the opposite sides the pairs of points (D, D'), (E, E') and (F, F'). The three Apollonius circles of the triangle with respect to its sides are the circles with diameters the segments DD', EE' and FF'. Prove:

(a) Prove that the three Apollonius circles of the triangle ABC intersect at two common points (common to all three) M and N. (So the three circles share MN as the common chord.) So, their centers are on the same straight line, called the **Lemoine[2] line or axis of the triangle**. These points are always distinct and equidistant from the Lemoine line. When one of these points is an infinity point?

(b) If P and Q are the points at which the s.l. MN intersects the circle (ABC), then the lines AP and AQ are the bisectors of the two angles $\widehat{MAN}$.

(c) Prove that each of the three Apollonius circles of ABC intersects the circumcircle of the triangle ABC orthogonally. Therefore, their common points [proven in **(a)**] are on the same s.l. with the circumcenter and they are inverse of themselves with regard to the circumcircle of the given triangle ABC.[3]

(d) Prove that the radii of the three Apollonius circles of ABC are equal to

$$r_a = \frac{abc}{b^2 - c^2}, \qquad r_b = \frac{abc}{a^2 - c^2}, \qquad r_c = \frac{abc}{a^2 - b^2},$$

and so satisfy

$$\frac{1}{r_b} = \frac{1}{r_a} + \frac{1}{r_c}.$$

(e) The Apollonius circle of a triangle ABC with respect to the side BC has radius equal to the radius of the circumcircle (ABC) iff the angle between the median and the height corresponding to the vertex A is $\dfrac{\pi}{4}$.

---

[2]Émile Michel Hyacinthe Lemoine, French civil engineer and mathematician, 1840-1912.

[3]This s.l., called the **Brocard line or axis of the triangle**, also passes through the **symmedian point**. [Pierre René Jean Baptiste Henri Brocard, French meteorologist and mathematician, 1845-1922.] The symmedian point has many important properties. See, bibliography, e.g., R. A. Johnson, etc. It is also called **Lemoine's point** or **Grebe's point**. [Ernst Wilhelm Grebe, German mathematician, 1804-1874.]

33. Consider a convex quadrilateral $ABCD$ inscribed on a circle $C[O, r]$. Let AB and CD intersect at E, and BC and AD intersect at F. Prove:

    (a) There exists point P on the segment EF satisfying the two relations

    $$EP \cdot EF = EB \cdot EA \qquad \text{and} \qquad FP \cdot FE = FA \cdot FD.$$

    (b) $$EF^2 = \mathfrak{P}_O(E) + \mathfrak{P}_O(F).$$

34. Consider a square $ABCD$ and let l be the length of its side. On the sides AB and BC pick points E and F, respectively, such that $AE = a$ and $CF = b$, given lengths.

    (a) Prove that the necessary and sufficient condition that a, b and l must satisfy in order the circles $C[E, a]$ and $C[F, b]$ intersect orthogonally is $a + b = l$.

    (b) If the condition in (a) is fulfilled, prove that the perpendicular bisector of EF passes through the center the square $ABCD$.

    (c) Prove that the orthogonal circles $C[E, a]$ and $C[F, b]$ have one of the common points on the diagonal AC and the other one on the circumscribed circle of the square $ABCD$.

35. Consider a circle $C[O, R]$ and two of its radii, Ox and Oy, making an angle $\angle(Ox, Oy) = \alpha$, where $0 < \alpha \le \pi$.

    (a) Prove that the circle internally tangent to $C[O, R]$ and the two radii Ox and Oy, has center located on the bisector of the angle $\angle(Ox, Oy)$ and its radius is

    $$\rho = R \cdot \frac{\sin\left(\frac{\alpha}{2}\right)}{1 + \sin\left(\frac{\alpha}{2}\right)}.$$

    Therefore, the (smallest) distance of O from such a circle is

    $$r = R \cdot \frac{1 - \sin\left(\frac{\alpha}{2}\right)}{1 + \sin\left(\frac{\alpha}{2}\right)}.$$

    (b) If $\alpha = \dfrac{2\pi}{n}$, where $n = 2, 3, 4 \ldots$, then we can construct n equal circles as in (a) which are successively externally tangent and the last circle is externally tangent to the first one.

    (c) Prove that the circle $C[O, r]$ is externally tangent to all the n circles constructed in (b). {Therefore, the n circles constructed in (b) form an ordinary, simple and closed Steiner chain with base circles the two concentric circles $C[O, R]$ and $C[O, r]$.}

36. This problem must be considered as a **project** and you may need to consult bibliography. This is a **theorem proven by J. Steiner**, and surprised him so much that he considered it one of the most remarkable theorems in all geometry.

In this problem we consider closed Steiner Chains of **even number of circles** S_1, S_2, ... S_{2k} with base circles B_1 and B_2 in any position. In such a chain, the pairs of circles S_1 and S_{k+1}, S_2 and S_{k+2}, and so on, are called **opposite** or **antipodal circles** of the chain.

(a) In **Figure 6.85** observe the dotted circles between the base circles and two opposite circles. Find their centers and radii, in general. Then observe that the two base circles and the two dotted circles always form a special closed chain (not canonical) of characteristic $\dfrac{1}{4}$ with base circles any two opposite circles. Prove that given two base circles that do not intersect (one is either inside or outside the other) such a chain is always possible and find the centers and radii of the four circles in terms of the radii of the base circles and the distance of their centers. (What happens to this special chain if the base circles are tangent or intersect?)

(b) Prove that one angle of the dotted circles between any two apposite circles of a chain is equal to $\dfrac{m}{n} \cdot (2\pi)$ and the other angle is its supplementary $\left[\pi - \dfrac{m}{n} \cdot (2\pi) \right]$.

(c) Prove that any two antipodal circles in a closed Steiner chain with $n = 2k$ and characteristic $\dfrac{m}{n}$, can serve as the base circles of another closed Steiner chain with characteristic $\dfrac{m'}{n'}$, such that $\dfrac{m}{n} + \dfrac{m'}{n'} = \dfrac{1}{2}$. This new chain is called **conjugate chain** of the given chain. The original chain and the new chain are called **conjugate chains**.

(d) Check that with n even, n' may be even or odd, but if n' is odd then n is even. So, we can find chains which are conjugate to chains with odd number of circles. Also, a finite chain is possible, if $2m < n$. What happens if $2m = n$?

(e) Depending on the positions and the sizes of the base circles of the original chain, the new chain may contain both B_1 and B_2, or one of them, or none. If the original chain is canonical and n' is even, then the conjugate chain contains both B_1 and B_2, but if n' is odd, then it contains either B_1 or B_2, but not both (there are two chains, one contains B_1 and the other B_2). Also, keep in mind that, some circles may have infinite radius, i.e., they are straight lines. Such straight lines are considered as tangent circles if and only if they are parallel. In this case, the point of tangency is the point at infinity that belongs to their direction.

[You may want to draw some examples of the conjugate chains of some Steiner chains, as above. For big chains this becomes cumbersome. You can work with canonical Steiner closed chains, since all Steiner closed chains are images of such chains under some inversion. Also, the canonical Steiner closed chains have symmetrical behavior and this helps to discover the conjugate chain. Try with characteristics: $\dfrac{1}{4} + \dfrac{1}{4} = \dfrac{1}{2}$, $\dfrac{1}{3} + \dfrac{1}{6} = \dfrac{1}{2}$ (find the two cases), $\dfrac{1}{8} + \dfrac{3}{8} = \dfrac{1}{2}$, $\dfrac{1}{2} + \dfrac{1}{\infty} = \dfrac{1}{2}$ (degenerate case), etc.]

7.2 Geometrical Constructions

1. Given a circle construct another circle orthogonal to it.

2. Construct a circle orthogonal to a given circle and passing through a given point. There are infinitely many such circles. If the radius of the sought circle(s) is given, there may be 0, 1, or 2 such circles.

3. Construct a circle orthogonal to a given circle and passing through two given points. Depending on the relative positions of the data there may exist 1 or ∞ such circles, the straight line included and considered as a circle of infinite radius. When the unique circle is an infinite straight line? (∞ occurs if the two given points can play the role of Poncelet points.)

4. Construct a circle orthogonal to two given circles and passing through a given point.

5. Construct 3 circles orthogonal to each other with given radii a, b and c.

6. Construct 3 circles orthogonal to each other with given centers A, B & C.

7. Construct a circle orthogonal to a given circle and having center on a given s.l. or on a given circle.

8. Construct a circle such that the tangent segments drawn from three fixed points A, B, and C to the circle have given lengths x, y, and z, respectively.

9. Construct a circle pseudo-orthogonal to three given circles.

10. Construct a circle intersected pseudo-orthogonally by three given circles.

11. Construct a circle pseudo-orthogonal to a given circle and orthogonal to two given circles.

12. Construct a circle orthogonal to a given circle and pseudo-orthogonal to two given circles.

13. Construct a circle pseudo-orthogonal to two given circles and intersected pseudo-orthogonally by a given circle.

14. Construct a circle of radius R, tangent to a given circle and pseudo-orthogonal to a given circle.

15. Consider a circle $C[O, R]$ and s.l. l whose distance from O is $OH = 2R$. Construct:
    (a) Circle with given radius r, tangent to l and orthogonal to $C[O, R]$. Check the conditions for such a circle to exist.
    (b) Circle tangent to l at a given point $P \in l$ and orthogonal to $C[O, R]$.

16. Consider a circle $\mathcal{C}$ and a point A. A s.l. through A intersects $\mathcal{C}$ at the points B and C. Construct another s.l. through A intersecting $\mathcal{C}$ at B' and C' such that the circles with diameters BB' and CC' are:
(a) Tangent.   (b) Orthogonal.

17. On the s.l. BC of the basis of a triangle ABC find point(s) P such that
$$\frac{AP^2}{BP \cdot CP} = r,$$ given ratio. [The point(s) P may be interior or exterior of the interval AB.]

Find the minimum possible value of $r > 0$ and investigate the conditions under which solution(s) exist(s). An interesting case is $r = 1$.

What happens to r as $P \longrightarrow \pm\infty$, or $P \longrightarrow B$, or $P \longrightarrow C$?

7.3   Geometrical Loci

1. Consider two circles $C[O, R]$ and $C[Q, R']$ that intersect at the points A and B. Find the g.l. of the points P for which it holds $\mathfrak{P}_O(P) + \mathfrak{P}_Q(P) = 0$.

2. Prove that the g.l. of the centers of the circles that are intersected pseudo-orthogonally by two given circles with non-empty intersection is a segment of the part of their radical axis that is located inside their intersection.

3. Consider a fixed circle $C[O, R]$ and a fixed point P.
(a) Prove that any circle passing through P and intersecting $C[O, R]$ orthogonally passes through a second fixed point.
(b) Find the g.l. of the centers of the circles in (a).

4. Consider a circle $C[O, R]$, a fixed point A and circles $\mathcal{C}$ passing through A and intersecting $C[O, R]$ at points M and N. Find the g.l. of the points of intersection of the s.ls MN and the tangent lines of $\mathcal{C}$ at A.

5. Consider two circles $C[O_1, R_1]$ and $C[O_2, R_2]$ and a variable circle $C[K, r]$. Find the g.l. of the centers K if:
(a) $C[K, r]$ is pseudo-orthogonal to both $C[O_1, R_1]$ and $C[O_2, R_2]$. {Answer: The g.l. is the symmetrical of the radical axis of $C[O_1, R_1]$ and $C[O_2, R_2]$ with respect to the midpoint of $O_1 O_2$.}
(b) $C[K, r]$ is orthogonal to $C[O_1, R_1]$ and pseudo-orthogonal to $C[O_2, R_2]$.
(c) $C[K, r]$ is orthogonal to $C[O_1, R_1]$ and is intersected pseudo-orthogonally by $C[O_2, R_2]$.
(d) $C[K, r]$ is pseudo-orthogonal to $C[O_1, R_1]$ and is intersected pseudo-orthogonally by $C[O_2, R_2]$.

6. Find the g.l. of the points whose power with respect to a given circle is equal to λ^2 (given).

7. Find the g.l. of the points whose powers with respect to two given circles have sum κ^2 (given).

Chapter 8

Exercises on Radical Axis

Radical Axis, Radical Center and Pencils of Circles

We abbreviate "straight line", "-es" with "s.l.", "s.ls"; "geometrical locus", "-es" with "g.l.", "g.ls"; and "if and only if" with "iff".

8.1 General

1. Consider three circles with centers on a s.l. and the three radical axes of the three pairs of circles formed out of these three circles. Prove that either 2 or the 3 radical axes coincide or those that do not coincide are parallel.

2. There is point on the radical axis of two circles which has the same negative power with respect to each of the circles if and only if the two circles intersect.

3. Consider two circles $C[O, r]$ and $C[O', r']$, and one homothety between them with center K. A homothety ray Kx intersects them at two corresponding pairs of points (A, A') and (B, B'), in this order, and another homothety ray Ky intersects them at two corresponding pairs of points (M, M') and (N, N'), in this order.

   Prove:

   (a) The tangent lines at N and M' intersect on the radical axis of the two circles $C[O, r]$ and $C[O', r']$.

   (b) The quadrilateral $BNM'A'$ is inscribable.

   (c) The chord s.ls BN and $A'N'$ intersect on the radical axis of the two circles $C[O, r]$ and $C[O', r']$.

451

[In the case of homothetic circles, the points that make up a pair of corresponding points are usually called **homologous points**. Here, A and A'
are homologous, as B and B', or M and M', and N and N'.

But, the points B and A' (on the same ray Ox) or N and M' (on the
same ray Oy) , as given above, are called **anti-homologous points**. (So
are the points M and N', or A and B'. Make figure and see their relative
positions.)]

4. Given two circles C_1 and C_2, the plane P is partitioned into three pairwise
disjoint subsets with respect to the relation of the powers of the points
with respect to the two given circles. These subsets are:

$$\{A \in P \mid \mathfrak{P}_1(A) = \mathfrak{P}_2(A)\},$$
$$\{A \in P \mid \mathfrak{P}_1(A) > \mathfrak{P}_2(A)\},$$
$$\{A \in P \mid \mathfrak{P}_1(A) < \mathfrak{P}_2(A)\}.$$

5. Three circles have centers O, P and Q on the same s.l. and Q is the
midpoint of OP. Prove that: if the three circles have the same radical
axis, then for any point A of the plane it holds

$$\mathfrak{P}_O(A) + \mathfrak{P}_P(A) = 2\,\mathfrak{P}_Q(A).$$

6. You can use **Casey's Theorem, 6.5.7**, to prove the Euler relations that
compute the distances of the circumcenter O from the incenter I and
the excenters I_1, I_2 and I_3 of the triangle, in terms of the radii of the
circumcircle, R, incircle, ρ, and excircles ρ_1, ρ_2 and ρ_3.

$$
\begin{aligned}
OI^2 &= R^2 - 2R\rho, &\quad \text{or} \quad& \mathfrak{P}_O(I) = -2R\rho, \\
OI_1^2 &= R^2 + 2R\rho_1, &\quad \text{or} \quad& \mathfrak{P}_O(I_1) = 2R\rho_1, \\
OI_2^2 &= R^2 + 2R\rho_2, &\quad \text{or} \quad& \mathfrak{P}_O(I_2) = 2R\rho_2, \\
OI_3^2 &= R^2 + 2R\rho_3, &\quad \text{or} \quad& \mathfrak{P}_O(I_3) = 2R\rho_3.
\end{aligned}
$$

7. Consider two circles $C[O_1, r_1]$ and $C[O_2, r_2]$ and let $d = \overline{O_1 O_2}$.

(a) Suppose the two given circles intersect or are tangent at a point A.
Let $\theta = \widehat{O_1 A O_2}$ (in absolute value).
Prove that
$$\cos(\theta) = \frac{r_1^2 + r_2^2 - d^2}{2\,r_1 r_2}.$$

(b) Suppose the circles are not concentric and are intersected by a third
circle $C[O, r]$ and the angles as in (a) are θ_1 and θ_2, respectively. Let h
be the distance of O to the radical axis of the two original circles.
Prove
$$r[r_1 \cos(\theta_1) - r_2 \cos(\theta_2)] = d \cdot h.$$

But, if the original circles are concentric, then

$$r[r_1 \cos(\theta_1) - r_2 \cos(\theta_2)] = \frac{1}{2} \left(r_1^2 - r_2^2 \right).$$

8. Prove that if the radical center of three circles is inside (outside) one of the them, then it is inside (outside) all three of them.

9. Prove that three points are collinear if and only if each one of them has the same power with respect to two circles.

10. If a circle is completely inside another circle, then the radical axis of the two circles is completely outside both of them.

11. (a) If each of two given points has the same power with respect to three circles, then these circles have radical axis the s.l. determined by the two points and the centers of the three circles are collinear.

    (b) If no two circles intersect, prove that either one or two of the circles is/are inside a circle of the three given circles.

    (c) Generalize to $n \geq 3$ circles.

12. Consider two points A and B symmetrical with respect to the center of a given circle. We draw parallel segments AP and BQ of the same direction and with P and Q points of the circle. Prove that the product $AP \cdot BQ$ is constant.

13. Consider a triangle ABC, its heights AH_1, BH_2, and CH_3 and its ortho-center H. Prove:

    (a) The radical center of the circles with diameters BC, BH and CH is the vertex A, etc.

    (b) The radical center of the circles with diameters BC, CA and AB is the orthocenter H.

    (c) The radical center of the circles with diameters AH_1, BH_2 and CH_3 is the orthocenter H. That is

$$\overline{HA} \cdot \overline{HH_1} = \overline{HB} \cdot \overline{HH_2} = \overline{HC} \cdot \overline{HH_3}$$

$$= 4R^2 \cos(\widehat{A}) \cos(\widehat{B}) \cos(\widehat{C}) = \frac{1}{2} \left(a^2 + b^2 + c^2 \right) - 4R^2,$$

    where R is the circumradius, $a = BC$, $b = CA$ and $c = AB$ are the sides, and $\widehat{A}$, $\widehat{B}$ and $\widehat{C}$ the angles of the triangle.

    (d)
$$\overline{AH_3} = \frac{b^2 + c^2 - a^2}{2c}, \quad \text{etc.,}$$

    and
$$\overline{AH} \cdot \overline{AH_1} = \frac{b^2 + c^2 - a^2}{2}, \quad \text{etc.}$$

(e) (1) If any points P_1, P_2, P_3 are chosen on the s.ls BC, CA and AB, respectively, and are not collinear, then the circles with diameters AP_1, BP_2, CP_3 have the orthocenter H as radical center. So, if two of the circles intersect, then the common chord goes though the orthocenter H. (2) If P_1, P_2 and P_3 are collinear, then the three circles are coaxal and H is on their radical axis. So, the centers of the three circles are on the same s.l. and the radical axis is the s.l. through H and perpendicular to the s.l. of the centers.

[In both cases, make figures for a triangle with: (1) all angles acute, (2) one angle right, and (3) one angle obtuse, to see the noticeable differences in the figures.]

(f) Prove that the points of intersection of the circles with diameters the medians are on the heights.

14. If there are two points A and B that have equal powers with respect to each of n given circles, then these circles belong to the same pencil. (The centers of the n circles are on the same s.l. and the radical axis is the s.l. AB which must be perpendicular to the center line. See that, to prove the claim of this exercise, it suffices to prove it for $n = 3$.)

15. Consider a complete quadrilateral $ABCA'B'C'$ with collinear ordered triples of points

$$\{A,\ B,\ C\}, \quad \{A,\ B',\ C'\}, \quad \{B,\ A',\ C'\} \quad \{C,\ A',\ B'\}$$

and diagonals the straight segments AA', BB' and CC'. Prove:

(a) The midpoints of the diagonals AA', BB' and CC' are collinear. (**Newton's**[1] **Theorem.**)

(Hint: One way is to use Menelaus's Theorem appropriately. Or, find a different way.)

(b) The circles with diameters AA', BB' and CC' are coaxal. (**Gauß**[2]-**Bodenmiller**[3] **Theorem.**)

(c) The orthocenter of the triangle ABC', is the radical center of the three above circles.

(d) The orthocenters of the four triangles ABC', $A'B'C'$, $AB'C$, $A'BC$, formed by the complete quadrilateral $ABCA'B'C'$, are on the common axis of the circles in (b).

(Hint: You may use the results of the previous two exercises.)

16. Consider a circle $C[O, R]$, a point A and a circle C passing through A and with center any point of $C[O, R]$. Prove that the radical axis of the two circles $C[O, R]$ and C is tangent to a fixed circle.

---

[1]Sir Isaac Newton, English mathematician and physicist. One of the greatest mathematician and scientists of all times. 1643-1727.

[2]Carl Friedrich Gauß, German mathematician. One of the greatest mathematician and scientists of all times. 1777-1855.

[3]Antonius Bodenmiller, German mathematician, 1744-1838.

17. Consider a triangle ABC and a s.l. parallel to BC intersecting AB and AC at P and Q, respectively. Prove that the radical axis of the circles with diameters BQ and CP is the s.l. of the height AH_1 of the triangle ABC.

18. Consider a triangle ABC with sides $a = BC$, $b = CA$ and $c = AB$. Prove that the radical centers of the circles $C[A, a]$, $C[B, b]$, and $C[C, c]$ is the symmetrical point of the orthocenter H with respect to the circumcenter O of the triangle ABC.

19. Prove that the radical center of the three exscribed circles of a triangle ABC is the incenter of the triangle.

20. Prove that the sides of the orthic triangle of a scalene triangle ABC intersect the opposite sides of ABC at three collinear points on the radical axis l of the circumscribed circle (ABC) and the nine-point circle of the triangle ABC. Hence, Euler's line of the triangle ABC is perpendicular to l. What happens if ABC is isosceles and / or equilateral?

21. Consider two non-concentric circles $C[O, r]$ and $C[Q, R]$ such that $C[O, r]$ is inside $C[Q, R]$. ($r < R$ and $O \neq Q$.) Draw a tangent l to $C[O, r]$ at a point T, non-perpendicular to OQ, and let A and B the point of intersection of l with $C[Q, R]$. Let S be the harmonic conjugate of T with respect to A and B. Prove that the midpoint of ST belong to the radical axis of $C[O, r]$ and $C[Q, R]$.

22. For any point P of the plane not on the common radical axis of the circles of a pencil (of any kind) there exist exactly one circle of the pencil that passes through P.

23. The set of all circles orthogonal to the circles of a pencil that are tangent at a point A is another pencil of circles tangent at the point A.

24. Prove that there are infinitely many circles orthogonal to each of the circles of a pencil. These circles define a new pencil. (Any such two pencils are called **orthogonal or conjugate pencils.**)

25. Consider a right triangle ABC with $\widehat{A} = 90°$ and constant height AH_1 but vertices B and C variable. Draw $H_1D \perp AB$ and $H_1E \perp AC$. Prove that the quadrilaterals $CBDE$ are inscribable and their circumscribed circles form a pencil of circles.

26. Consider three circles with common radical axis (coaxal) and the set $\mathcal{S}$ of the points of one of the circles which are exterior to the other two circles. Prove that the ratio of the tangent straight segments drawn from any point of $\mathcal{S}$ to the other two circles is constant.

27. Prove that the set of all circles each of which intersects two given circles pseudo-orthogonally is a pencil. Determine this pencil. The set of the

centers of these circles is the straight line symmetrical to the radical axis of the two given circles about the midpoint of their center-segment. This straight line is called **pseudo-radical axis** of the two given circles.

28. Prove that the set of all circles each of which intersects a given circle orthogonally and another circle pseudo-orthogonally is a pencil. Determine this pencil and the set of the centers of these circles.

29. Prove that the radical center of three circles each of which intersects the other two orthogonally, is the orthocenter of the triangle with vertices the centers of the circles.

30. Prove that the set of all Apollonius circles on a straight segment AB ($A \neq B$ and the ratio is any $\lambda > 0$) is a pencil with Poncelet points A and B. Prove that the common radical axis is the perpendicular bisector of AB and the orthogonal pencil is the set of all circles passing through A and B.

31. Consider three circles $C[O_1, r_1]$, $C[O_2, r_2]$, and $C[O_3, r_3]$ pairwise disjoint. Prove that the limit points (Poncelet points) of the two pencils determined by $C[O_1, r_1]$ and $C[O_2, r_2]$, and by $C[O_1, r_1]$ and $C[O_3, r_3]$, are concyclic.

32. Explain why any two points P_1 and P_2 in the plane completely define a pencil whose Poncelet points are the P_1 and P_2. (What is: the Poncelet circle, the line of the centers of the circles of the pencil, the common radical axis, etc.)

33. Let in **Figure 6.97** the radii of the circles with centers O_1, O_2, and O_3 be R_1, R_2, R_3, respectively, $d_1 = O_2 O_3$, $d_2 = O_3 O_1$, $d_3 = O_1 O_2$, and R the radius of the circle commonly orthogonal to them with center the radical center of the three circles C. We want to find the distances of C from O_1, O_2, and O_3, and from the sides of the triangle $O_1 O_2 O_3$ and the radius R. Without loss of generality, assume that $R_1 \geq R_2 \geq R_3 \geq 0$. We have proven in the text:

$$O_1 H_1 = \frac{d_3^2 + R_1^2 - R_2^2}{2d_3}, \qquad O_1 H_3 = \frac{d_2^2 + R_1^2 - R_3^2}{2d_2}, \qquad \text{etc.}$$

(a) Notice that $O_1 H_1 C H_3$ is inscribable ($\widehat{H_1} = \widehat{H_3} = 90^o$) with diameter OC and use the law of sines with the diameter and the cosine of $\widehat{O_1}$ in terms of d_1, d_2, and d_3, to prove

$$O_1 C^2 = \frac{d_1^2(d_3^2 + R_1^2 - R_2^2)(d_2^2 + R_1^2 - R_3^2)}{(d_1 + d_2 + d_3)(d_1 - d_2 + d_3)(d_1 + d_2 - d_3)(-d_1 + d_2 + d_3)}$$
$$+ \frac{(d_2^2 - d_3^2 + R_2^2 - R_3^2)[d_2^2(-R_1^2 + R_2^2) + d_3^2(R_1^2 - R_3^2)]}{(d_1 + d_2 + d_3)(d_1 - d_2 + d_3)(d_1 + d_2 - d_3)(-d_1 + d_2 + d_3)}, \qquad \text{etc.,}$$

and then

$$R^2 = O_1 C^2 - R_1^2, \qquad \text{etc.,}$$

and

$$H_1C^2 = O_1C^2 - O_1H_1^2, \quad \text{etc.}$$

(b) These formulae apply to **Figures 6.94, 6.95**, and **6.96**. Simplify them for **Figure 6.94**.

(They simplify nicely since $d_1 = R_2 + R_3$, etc. For instance,

$$H_1C = H_2C = H_3C = R = \sqrt{\frac{R_1 R_2 R_3}{R_1 + R_2 + R_3}},$$

$$O_1C = \sqrt{R_1^2 + R^2} = \sqrt{\frac{R_1(R_1 + R_2)(R_1 + R_3)}{R_1 + R_2 + R_3}}, \quad \text{etc.})$$

(c) Explain what happens in **Figures 6.95**.

(d) When the three circles pass through the same point?

(e) Examine the cases where some or all $R_i = 0$, $i = 1$, 2, 3 or equal to each other.

(f) Examine the case where only two circles are tangent.

(g) Examine the case where one / two circles is / are inside another circle.

34. Consider four circles each of which is externally tangent to the other three. Thus, there are six points of tangency and so we can consider $\binom{6}{3} = 20$ different subsets containing three of these points. Prove that four of these subsets define four different mutually tangent circles, such that, three of them are mutually externally tangent and the fourth circle contains these three circles and therefore they are internally tangent to it. Also, each new circle is orthogonal to three of the fourth original circles. Characterize the centers of these circles and find their radii in terms of the radii of the initial four circles. [Use (b) of the previous exercise. Look at **Figure 6.94** and expand it.]

Observe and justify that the dual question is also true. That is, given four different mutually tangent circles, such that, three of them are mutually externally tangent and the four circles containing the three circles which are internally tangent to it, by working backward, we can find four circles each of which is externally tangent to the other three and orthogonal to three of the initial circles.

[This result is related to **Descartes' problem of four mutually kissing (tangent) circles**. The result of this problem can solve Apollonius' problem of tangencies in some special cases. Find related bibliography and study the problem. For example: Coxeter, H. S. M., *Introduction to Geometry*, pp.13-15. Pedoe, D., *Geometry a Comprehensive Course*, pp. 157-158, etc.]

8.2   Geometrical Constructions

1. Consider a circle $C[O, r]$ and two points A and B. A variable circle $C[K, \rho]$ passing through A and B intersects $C[O, r]$ at P and Q.

   (a) Prove that PQ passes through a fixed point.

   (b) Construct a circle $C[K, \rho]$ such that $PQ = s$, a given segment.

2. Consider a s.l. l and two points A and B such that l is not parallel to AB. Let C be the intersection of l and the s.l. AB. A variable circle $C[K, \rho]$ passing through A and B intersects l at P and Q.

   (a) Prove that the product $CP \cdot CQ$ is constant.

   (b) Construct a circle $C[K, \rho]$ such that $PQ = s$, a given segment.

3. Construct a triangle ABC if the side AB, the angle $\widehat{A}$, and the bisector b_a of the angle $\widehat{A}$ are given.

4. Construct a circle belonging to a given pencil and intersecting a given circle orthogonally.

5. Construct a triangle ABC if the side BC, the height on BC, and the angle bisector of $\widehat{A}$, are given.

6. Construct a circle belonging to a given pencil and passing through a given point.

7. Construct a circle belonging to a given pencil and tangent to a given circle.

8. Construct a circle belonging to a given pencil and tangent to a given s.l.

9. Construct a circle of given center O and belonging to the pencil defined by:

   (a) The two points A and B common to all the circles of the pencil.

   (b) The two Poncelet points.

   (c) The common point of tangency and the radical axis.

   (d) A circle and the radical axis of the pencil such that this circle and the radical axis do not intersect.

10. Given any three circles examine the construction of a circle that intersects the three given circles orthogonally. The cases are:

   (a) There is no such a circle.

   (b) There is only one circle.

   (c) There are infinitely many circles.

   Explain when each of these cases occurs and how we construct a circle that intersects the three given circles orthogonally.

   Also, examine the case in which the radical center exists, but a circle that intersects the three given circles orthogonally does not exist.

11. Given three circles and three points construct a circle such that each radical axis of this circle with each one of the three given circles passes through one of the three given points, respectively.

8.3 Geometrical Loci

1. Prove that the g.l. of the points whose difference of its powers with respect to two given circles $C[O_1, R_1]$ and $C[O_2, R_2]$ is a real number p, is a s.l. perpendicular to the center line of the two circles. Determine this s.l. from p and the straight segment O_1O_2. What happens if with $O_1 = O_2$.

2. Consider three concentric circles $C[O, R]$, $C[O, r]$ and $C[K, \rho]$. Find the g.l. of the centers of the circles that intersect $C[O, r]$ orthogonally and the radical axis of any such a circle with $C[K, \rho]$ is tangent to $C[O, R]$.

3. Consider a circle and two fixed points A and B. Draw s.ls through A intersecting the circle at P and Q. Find the g.l. of the centers of the circles (BPQ).

4. Prove that the g.l. of the points of the plane whose ratio of the powers with respect to two given circles is a constant ρ, is a circle of the pencil determined by the two given circles, and whose center divides, internally or externally, the center-segment of the two given circles in ratio ρ. Find the radical axis of this pencil.

5. Consider a circle $C[O, R]$, and a point A in the plane. We rotate a diameter BOC of $C[O, R]$. Find the g.l. of the centers of the circles (ABC). Investigate various cases depending on the position of A relative to $C[O, R]$.

6. (a) A corollary of **Casey's Theorem, 6.5.7,** states: The g.l. of a point P such that the ratio of its powers with respect to two given circles $C[C_1, R_1]$ and $C[C_2, R_2]$ is fixed, ρ let us say, is a circle $C[C, R]$ which is coaxal (has the same radical axis) with $C[C_1, R_1]$ and $C[C_2, R_2]$ (so the center C is on the s.l. C_1C_2), and

$$\frac{\mathfrak{P}_{C_1}(P)}{\mathfrak{P}_{C_2}(P)} = \frac{\overline{CC_1}}{\overline{CC_2}} = \rho \, (= \text{fixed}).$$

Describe $C[C, R]$ by the data of this situation. Notice that in this situation, we can also allow $R_1 = 0$ and / or $R_2 = 0$.

(b) Let $C[C, R]$ be the circle of similarity of two given circles $C[C_1, R_1]$ and $C[C_2, R_2]$, (see **Definition 4.4.1**). Prove that $C[C, R]$ is coaxal with $C[C_1, R_1]$ and $C[C_2, R_2]$. Describe the circle $C[C, R]$ (find its radius R and center C) by the data of this situation ($\overline{C_1C_2}$, R_1, R_2, etc.). What is the fixed ratio of the powers of its points with respect to the two given circles?

7. The ratio of the radii of two variable circles is given fixed and both circles are tangent to a given s.l. and located in the same closed half plane defined by this s.l. Find the g.l. of the intersection of the other exterior tangent of the circles with their radical axis.

Chapter 9

Exercises on Inversion

Inversion, Cross-Ratio (Double), Stewart's Theorem and Ptolemy's Theorem

We abbreviate "straight line", "-es" with "s.l.", "s.ls"; "geometrical locus", "-es" with "g.l.", "g.ls"; and "if and only if" with "iff".

9.1 General

1. Consider a circle $C[O, r]$, a constant point P not on it and a variable diameter AB of $C[O, r]$.
   (a) Prove that the circles (PAB) pass through a second fixed point. (The first is P.)
   (b) Let M and N be the second points of intersections of $C[O, r]$ with the s.ls PA and PB. Prove that the s.ls MN pass through a fixed point.
   (c) Prove that the circles (PMN) pass through a second fixed point. (The first is P.)

2. Consider four inverse pairs of points (A, A'), (B, B'), (C, C') and (D, D') under an inversion $I_{[O,c]}$.
   (a) Prove that $\sphericalangle(CA, CB) + \sphericalangle(C'A', C'B') = \sphericalangle(OA, OB)$.
   (b) As the point C traces a circle passing through A and B find the set of points C'.
   (c) Prove that $\sphericalangle(CA, CB) + \sphericalangle(DB, DA) = \sphericalangle(C'A', C'B') + \sphericalangle(D'B', D'A')$.

3. Prove that two circles can be mapped onto each other by one or two inversions.

4. Find subsets of the plane which are set-wise fixed by an inversion (not listed in the text).

5. Prove that the images of two circles under an inversion with center on their radical axis have radii proportional to the radii of the original circles. This is true with three circles if the inversion has center their radical center.

6. In Euclidean plane geometry, the **Problem of Apollonius (problem of tangencies)** is to construct circles or s.ls that are tangent to three given circles in a plane, where some or all of the circles may be points or s.ls. (Find pertinent bibliography and study this problem and its solution. For example: Pedoe, D., *Geometry a Comprehensive Course*, pp. 101-104, 151, etc.) Depending on the situation, this problem has zero or one or two or three or four or six or eight solutions or infinite. For example, there is no solution, when the three circles are nested without touching or the three s.l. are parallel. There are more situations with no solutions. Find them all.
   On the basis of this result, prove that any three circles, some or all of which may be points or s.ls, for which the Apollonius' problem has a solution, can be inverted to three circles tangent to a s.l. or to a given s.l. (S.ls are considered as tangent circles if and only if they are parallel. The point of tangency is the point at infinity that belongs to their direction.)

7. Let $\mathcal{F}$ and $\mathcal{F}'$ be two inverse figures under an inversion $I_{[O,c]}$ and let $\mathcal{G}$ and $\mathcal{G}'$ be their images under another inversion $I_{[O',c']}$, respectively. Prove that $\mathcal{G}$ and $\mathcal{G}'$ are inverse of each other under a third inversion.

8. We consider a circle $\mathcal{C}$, a diameter AB of it, a point C on the s.l. AB, and the s.l. $l \perp AB$ at C. Let a s.l. through C intersect $\mathcal{C}$ at P and P' and the s.ls AP and AP' intersect l at D and D'. Prove that $\overline{CD} \cdot \overline{CD'}$ is constant.

9. Prove that a set of circles passing through two points A and B is transformed, under an inversion with center A, to a set of s.ls through B', the inverse of B.

10. Consider a circle $\mathcal{C}$ and a point A exterior to it. A s.l. passing through a fixed point P intersect $\mathcal{C}$ at the points B and C.
    (a) Prove that the circles (ABC) pass through a fixed point.
    (b) If the s.ls AB and AC intersect the circle at the second points D and E, prove that the s.l. DE and the circle (ADE) pass through two fixed points F and G, respectively.
    (c) Find the g.ls of M and N, the points of intersections of the s.ls BC and DE and the circles (ABC) and (ADE), respectively.

11. Find the set of all inversions of a given power c that transform two concentric circles into two equal circles.

12. Consider a pencil of circles tangent to each other at a point A. Prove that an inversion $I_{[O,c]}$ with $O \neq A$ transforms these to other circles tangent to each other at A', the inverse point of A. If $O = A$, then the circles are transformed, under $I_{[O,c]}$, to parallel lines perpendicular to s.l. OA.

13. A point Q on a circle $C[O, r]$ is projected on two perpendicular diameters d_1 and d_2 of $C[O, r]$ at the points A and B, respectively. Let P be the pole of AB with respect to $C[O, r]$ and P_1 and P_2 its projections on d_1 and d_2.
    (a) Prove that $P_1 P_2$ is tangent to $C[O, r]$ at Q.
    (b) Let the circle of diameter $P_1 P_2$ intersect $C[O, r]$ at C and D. Prove that A, B, C and D are collinear.

14. Let circle C be tangent to two given circles, or to a given circle and a given s.l. Prove that the points of tangency are inverses under an inversion that maps the two given circles, or the given circle and the given s.l.

15. (a) If two circles or one circle and a s.l. intersect at two points, then any inversion with center one point of intersection maps the two circles or the circle and the s.l. onto two intersecting straight lines.
    (b) If two circles or one circle and a s.l. are tangent, then any inversion with center the point of tangency maps the two circles or the circle and the s.l. onto two parallel straight lines.
    (c) A circle and a s.l. do not intersect. Find the inversion that maps the circle and the s.l. onto two concentric circles.

16. We are given four rods, such that $AB = CD = a$ are two of the sides and $AC = BD = b$ are the diagonals of a convex quadrilateral $ABCD$. (The sides BC and DA are not given.) The rods can rotate freely at the joint points A, B, C and D.
    (a) Prove that in any position that variable quadrilateral $ABCD$ is an isosceles trapezium.
    (b) From the midpoint M of AB, we draw a s.l. parallel to BC which intersects the s.ls AC and BD at the points E and F, respectively. Prove that $ME \cdot MF$ is constant.
    (c) Find the g.l. of F, when E traces a circle passing through M.

17. On an axis we consider the points $A = -4$, $B = 1$, $C = 3$, and $D = -2$. Compute the cross-ratio $\{A, B; C, D\}$.

18. If $\{A, B; C, D\} = r$, prove the following six results which prove that *"there are at most 6 different values of the cross-ratio for a configuration of four points on an axis, instead of $4! = 24$."*
    (a) $\{A, B; C, D\} = \{B, A; D, C\} = \{C, D; A, B\} = \{D, C; B, A\} = r$.

    So, we have the **rule: The cross-ratio does not alter, if we switch two of the points and at the same time we switch the other two.**

    (b) $\{B, A; C, D\} = \dfrac{1}{r} = \{A, B; D, C\}$.

    (c) $\{A, C; B, D\} = 1 - r$. So $\{A, B; C, D\} + \{A, C; B, D\} = (\text{etc.}) = 1$.

    (d) $\{A, D; B, C\} = 1 - \dfrac{1}{r} = \dfrac{r-1}{r}$.
    So, $\{B, A; C, D\} + \{A, D; B, C\} = (\text{etc.}) = 1$.

(e) Let A, B, C and D be four points on a s.l. Prove: B is between A and C and C between B and D iff $\{A, B; C, D\} + \{A, C; B, D\} = 1$.

(f) $\{A, D; C, B\} = \dfrac{r}{r-1} = 1 + \dfrac{1}{r-1}$.

(g) Let A, B, C and D be four points on a s.l. Prove: B is between A and C and C between B and D iff $\{A, B; C, D\} + \{A, C; B, D\} = 1$.

19. Prove:

(a) $\{A, B; C, D\} = \dfrac{\dfrac{1}{\overline{AB}} - \dfrac{1}{\overline{AD}}}{\dfrac{1}{\overline{AB}} - \dfrac{1}{\overline{AC}}}$.

(b) $\{A, B; C, D\}$ is a harmonic quadruple iff
$$\frac{1}{\overline{AC}} + \frac{1}{\overline{BD}} + \frac{1}{\overline{AD}} + \frac{1}{\overline{BC}} = 0.$$

(c) $\{A, B; C, D\}$ is a harmonic quadruple iff
$$\overline{AB} \cdot \overline{CD} + 2\overline{AD} \cdot \overline{BC} = 0.$$

20. On an axis we consider the points $A = 1$, $B = 4$, and $C = 7$. Find the point D such that $\{A, B; C, D\} = -2$.

21. If $\{A, B; C, D\} = \dfrac{1}{2}$, compute $\{A, C; B, D\}$, $\{A, D; B, C\}$ and $\{A, C; D, B\}$.

22. Consider a rhombus $ABCD$. Let O be its center and Q the center of gravity (centroid) of the triangle ABD. Find the cross-ratio $\{A, O; Q, C\}$.

23. Consider a triangle ABC and a point P. Let A' be the point of intersection of AP and BC and P_1 be the point for which $\{A, A'; P, P_1\} = t$ (given). Also let B' be the point of intersection of BP_1 and AC and P_2 the point for which $\{B, B'; P_1, P_2\} = t$. Prove:
(a) C, P, and P_1 are of the same s.l.
(b) If the s.l. determined at (a) intersects AB at a point C', then prove $\{C, C'; P_2, P\} = t$.

24. Consider a triangle ABC, a s.l. l and a point $O \in l$. Let the symmetrical s.ls of OA, OB, OC in l intersect the s.ls BC, CA, and AB at the points A', B' and C', respectively. Prove that A', B' and C' are collinear.

25. Consider a triangle ABC, G its center of gravity (centroid) and P any point of the plane. Prove the **Leibniz**[1] **relation**

$$PA^2 + PB^2 + PC^2 = 3PG^2 + \frac{1}{3}\left(a^2 + b^2 + c^2\right),$$

---

[1] Gottfried Wilhelm (von) Leibniz, German polymath, philosopher and mathematician, 1646-1716.

where $a = BC$, $b = CA$, $c = AB$ are the sides of the triangle.
For which point P, is the sum of the squares of the distances of the point from the vertices

$$PA^2 + PB^2 + PC^2$$

minimum? How much is this minimum?

26. A circle $C[B, r]$ is completely inside a circle $C[A, R]$ and passes through A. Prove that the radical axis of these circles is the inverse of $C[B, r]$ under the inversion in the circle $C[A, R]$.

27. Let two circles $C[O_1, R_1]$ and $C[O_2, R_2]$ be tangent at a point F and a third circle $C[O, R]$ tangent to both of them at the points A and B, respectively. Let A' and B' be the second points of intersection of $C[O, R]$ with the s.ls. FA and FB, respectively. Prove that $A'B'$ is a diameter of $C[O, R]$ parallel to the center s.l. $O_1 O_2$.

28. Consider four points A, B, C and D in the plane. Prove that the quantity

$$\frac{AB \cdot CD + BC \cdot AD}{AC \cdot BD}$$

is invariant under any inversion.
Prove that if these points are on a circle or a s.l., then this quantity is equal to 1.

29. Let t be the length of the tangent segment of two circles $C[O_1, r_1]$ and $C[O_2, r_2]$. We invert these circles by an inversion $I[O, c]$, where the center O is not on either circle.
Prove that the length of the corresponding tangent segment of the inverse circles is

$$t' = \frac{c}{\sqrt{p_1 p_2}} t$$

where $p_1 = OO_1^2 - r_1^2$ and $p_2 = OO_2^2 - r_2^2$ {the powers of O with respect to the circles $C[O_1, r_1]$ and $C[O_2, r_2]$}.

30. Consider $ABCD$ an **inscribable quadrilateral** with sides $a = AB$, $b = BC$, $c = CD$, and $d = DA$, and diagonals $u = AC$ and $v = BD$.

   (a) Prove

$$\frac{u}{v} = \frac{ad + bc}{ab + cd}.$$

   Also prove that if this relation is true, then the quadrilateral is inscribable.

   (b) Combine **(a)** with **Ptolemy's Theorem, 6.6.1**, $(uv = ac + bd)$, to prove

$$u = \sqrt{\frac{(ac + bd)(ad + bc)}{ab + cd}} \qquad \text{and} \qquad v = \sqrt{\frac{(ac + bd)(ab + cd)}{ad + bc}}.$$

(c) Let K be the common point of the diagonals of $ABCD$. Prove that

$$\frac{KA}{KC} = \frac{ad}{bc}.$$

(d) Let P be any point of the plane. Prove

$$PA^2 \cdot \text{Area of triangle}(BCD) + PC^2 \cdot \text{Area of triangle}(DAB)$$
$$= PB^2 \cdot \text{Area of triangle}(CDA) + PD^2 \cdot \text{Area of triangle}(ABC).$$

31. Consider $ABCD$ a convex quadrilateral with sides $a = AB$, $b = BC$, $c = CD$, and $d = DA$, and diagonals $u = AC$ and $v = BD$.

(a) Prove that the **area formula of the convex quadrilateral** $ABCD$

$$\text{Area}(ABCD) = \frac{1}{4}\sqrt{4u^2v^2 - (a^2 + c^2 - b^2 - d^2)^2}.$$

(b) If $ABCD$ is an inscribable quadrilateral, then use **(a)**, **Ptolemy's Theorem, 6.6.1**, and the previous exercise to prove:

(b1) **Brahmagupta's**[2] **formula** for the area of an inscribable quadrilateral

$$\text{Area}(ABCD) = \sqrt{(s-a)(s-b)(s-c)(s-d)}\,,$$

where $s = \dfrac{a+b+c+d}{2}$ is the semi-perimeter of $ABCD$.

(b2) The radius of the circumscribed circle is given by

$$R = \frac{1}{4}\sqrt{\frac{(ab+cd)(ac+bd)(ad+bc)}{(s-a)(s-b)(s-c)(s-d)}}.$$

(c) If $ABCD$ is both inscribable and circumscribable quadrilateral, then

$$a + c = b + d \qquad \text{and} \qquad \text{Area}(ABCD) = \sqrt{abcd}$$

and then simplify the formula for R in **(b2)**, in this case.

32. A **quadrilateral** $ABCD$ is called **harmonic** if it is inscribable and

$$ac = bd \quad \Longleftrightarrow \quad \frac{a}{b} = \frac{d}{c} := \lambda,$$

where, $a = AB$, $b = BC$, $c = CD$, and $d = DA$.

(a) A square is obviously harmonic quadrilateral. Prove that the inverse of the vertices of a square under an inversion whose center is not on the circle of the square are the vertices of a harmonic quadrilateral.

---

[2]Brahmagupta, Indian mathematician and astronomer, 598-670.

(b) Consider a square and a point not on its circle. Prove that the s.ls that join this point with the vertices of the square intersect its circle at four points which are vertices of a harmonic quadrilateral.

(c) A quadrilateral is harmonic if and only if its vertices are obtained as inverse of the vertices of a square.

(d) We consider a harmonic quadruple $[A, B; C, D]$ on an axis l and we consider its inverse $[A', B'; C', D']$ under an inversion $I_{[O,c]}$ with $O \notin l$. Prove that the quadrilateral $A'B'C'D'$ is inscribable and the products of its opposite sides are equal.

(e) Draw a circle and pick an exterior point X. Let the two tangents from X touch the circle at the points A and C. Let a straight line from X intersect the circle at the points B and D. Prove that the quadrilateral $ABCD$ is harmonic. Then prove:

(e1) Let the straight line AC intersect the side BD at the point S. Prove that $\{X, S; B, D\}$ is a harmonic quadruple.

(e2) The straight line AC passes through the common point of the tangents to the circle at the points B and C.

(e3) If N is the midpoint of the diagonal BD, then

$$\frac{NA}{NC} = \frac{a^2}{b^2} = \frac{d^2}{c^2}.$$

(f) Consider the harmonic quadrilateral $ABCD$ and the midpoints M and N of the diagonals AC and BD, respectively. Compute all the segments involved in terms of the sides $a = AB$, $b = BC$, $c = CD$, and $d = DA$ and prove

$$MB \cdot MD = MA^2 = MC^2 \quad \text{and} \quad NA \cdot NC = NB^2 = ND^2.$$

Also prove that AC is the bisector of the angle $\widehat{BMD}$ and BD the bisector of angle $\widehat{ANC}$.

[Use the results concerning the interior and exterior bisectors in a triangle.]

(g) Find the simplified formulae for the area and the circum-radius for a harmonic quadrilateral (using two sides and λ).

33. Look at **Figure 6.81** and let $a = BC$, $b = CA$, $c = AB$ and $s = \dfrac{a+b+c}{2}$. Prove

$$\frac{a(s-a)}{(B_1C_1)^2} = \frac{b(s-b)}{(C_1B_1)^2} = \frac{c(s-c)}{(B_1A_1)^2} = \frac{R}{r}.$$

(You need **Ptolemy's Theorem** and some metric relations in a triangle.)

34. In **Figure 6.44** compute AB' and $A'B$ in terms of AB, OA, OB and c.

35. Suppose that for a triangle ABC there are two points in one of the open arcs that the vertices A, B and C divide the circumscribed circle, such that the distance of each of them from the opposite vertex of ABC is equal to the sum of its distances from the two vertices which are the endpoints of this arc. Prove that ABC is equilateral.

36. A constant angle is rotated around its constant vertex A and its sides intersect a fixed s.l. l at B and C. Prove that the circle (ABC) is tangent to fixed circle.

37. When the triangles ATA' and OSA' in **Figure 6.23**, etc., are right isosceles?

38. Consider two circles C and C' outside each other and two fixed points $A \in C$ and $A' \in C'$ on them, respectively. Two circles D and D' are tangent to C and C' at the points A and A' and to each other at the point T. Prove that by an appropriate inversion the image of set of the points T consists of two circles orthogonal to each other.

39. Examine **Figures 6.77, 6.78, 6.79, 6.83, 6.85, 6.86, 6.87, 6.88** and prove that the points of tangency of the circles inscribed between the two base circles in a Pappus chain or a Steiner chain lie in a circumference of a circle. Find the center and radius of this circle.

40. Consider two intersecting circles and inscribe circles between them, such that any two successive inscribed circles are (externally) tangent. Prove that the points of tangency of the inscribed circles lie in a circumference of a circle.

41. Consider two concentric circles $C[O, R]$ and $C[O, r]$ with $R > r$ to be the base circles of an ordinary Steiner Chain of n equal circles. Let ρ be the radius of each circle of the chain and $\phi = \dfrac{\theta}{2} = \dfrac{\pi}{n}$ (see **Figure 6.85**). Prove that:

$$\sin(\phi) = \frac{\rho}{r + \rho} \quad \Longleftrightarrow \quad \rho = \frac{r \sin(\phi)}{1 - \sin(\phi)}, \quad \text{and} \quad \frac{R}{r} = [\sec(\phi) + \tan(\phi)]^2.$$

42. Consider a circle $C[O, R]$ and two diameters d_1 and d_2 perpendicular to each other. A point $M \in C[O, R]$ is projected on d_1 and d_2 to points A and B, respectively.

(a) If P is the pole of AB with respect to $C[O, R]$ and P_1 and P_2 the projections of P onto d_1 and d_2, prove that $P_1 P_2$ is tangent to $C[O, R]$ at M.

(b) The circle with diameter $P_1 P_2$ intersects $C[O, R]$ at C and D. Prove that the points A, B, C and D are collinear, i.e., the s.l. CD is the s.l. AB.

43. Consider a circle $C[O, r]$ and a point A in its plane. Let a be the polar s.l. of A with respect to the circle and take a point B on it. Consider b the polar s.l. of B with respect to the circle. Let also C be the point of intersection of a and b. Prove:

(a) In the triangle ABC, each side is the polar s.l. of the opposite vertex, with respect to the circle $C[O, r]$. Such a triangle is called **polar triangle with respect to the circle** $C[O, r]$.

(b) The orthocenter of ABC is the point O.

(c) Consider a polygon $A_1 A_2 \ldots A_n$ and the s.ls a_1, $a_2, \ldots$, a_n the polar s.ls of the vertices, with respect to the circle $C[O, r]$, respectively. These intersect at the points A'_1, $A'_2, \ldots, A'_n$ which are vertices of a new polygon whose vertices are the poles of the sides of the first polygon.
If three vertices of the polygon $A_1 A_2 \ldots A_n$ are on the same s.l. then their polar s.ls pass through the same point.

If three of the s.ls a_1, a_2, $\ldots$ a_n pass through the same point, then their poles with respect to $C[O, r]$, lie on the same s.l.

(d) Two opposite vertices and the common point of the diagonal of a quadrilateral inscribed in the circle $C[O, r]$, are the vertices of a polar triangle of the circle $C[O, r]$.

44. Suppose the circles $C[M, r]$ and $C[N, R]$, $r < R$, are mutually inverse under the inversion $I_{[O,c]}$, $c > 0$. For any point $Q \in C[O, \sqrt{c}]$ prove that

$$r \cdot QN^2 - R \cdot QM^2 = Rr(R - r).$$

45. See **Example 6.6.6** and **Figure 6.106**, **Application 11**, and prove that the s.l. PF is a common tangent of the circles $C[K, r_K]$ and $C[K_1, r_1]$, and the s.l. QH is a common tangent of the circles $C[U, r_U]$ and $C[K_2, r_2]$. Then compute the lengths of PF and QH in terms of r_K and r_U.

46. (a) Prove that any four points on a circle can be inverted into the vertices of a rectangle.

(b) Two equal circles intersect at two points and two other equal circles (not necessarily equal to the first two) intersect the first two equal circles such that there are four points of intersection, two of which are the points of intersection of the first two equal circles, and on every point of intersection there are three of the four circles. Prove that the four points of intersection are vertices of a parallelogram.

(c) Prove that any four points in the plane can be inverted into the vertices of a parallelogram.

47. Consider three equal circles with centers K, L and M and passing through the same point H. Let A, B and C be the other points of intersection of the three pairs of the circles. Prove:

(a) H is the orthocenter of the triangle ABC.

(b) The triangles ABC and KLM are equal and their equal sides are parallel.

(c) The circle defined by A, B and C is equal to the initial circles. If O is its center, then O is the orthocenter of the triangle KLM.

(d) Any four points **not** on a circle can be inverted into the vertices and orthocenter of a triangle.

48. (a) Consider any three points A, B, C and their corresponding inverses A', B', C' under an inversion of center O. Find the angles of the triangle $A'B'C'$ in terms of the corresponding angles of triangle ABC and the angles $\widehat{AOB}$ $\widehat{BOC}$ and $\widehat{COA}$. [The answers depend on the relative positions of A, B, C and O and different configurations must be examined or find general formulae with oriented angles (mod π).]
    (b) Any three points in the plane can be inverted to vertices of a triangle similar to a given triangle.
    (c) Two triangles can be placed in such a way that their vertices are mutually inverse.

49. Prove that there are at most eight inversions each of which transforms three given circles into equal circles, or circles of given radii. The case that no such inversion exists is also possible.

50. Consider a circle $C[O_1, r_1]$ to the felt of another circle $C[O_2, r_2]$. Let $d = O_1 O_2$ be the length of the center-segment, and t_e and t_i the lengths of the external and internal tangent segments, respectively, if they exist. Let the line $O_1 O_2$ intersect $C[O_1, r_1]$ and $C[O_2, r_2]$ at P_1, Q_1 and P_2, Q_2, respectively, from left to right.
    (a) Prove that

$$t_e^2 = d^2 - (r_1 - r_2)^2 = (d + r_1 - r_2)(d - r_1 + r_2),$$
$$t_i^2 = d^2 - (r_1 + r_2)^2 = (d + r_1 + r_2)(d - r_1 - r_2).$$

[So, it must be $d^2 - (r_1 - r_2)^2 > 0$ and / or $d^2 - (r_1 + r_2)^2 > 0$. Otherwise, the respective tangent segments do not exist.]
    (b) Prove that

$$\frac{\overline{P_1 Q_2} \cdot \overline{Q_1 P_2}}{\overline{P_1 Q_1} \cdot \overline{P_2 Q_2}} = \frac{d^2 - (r_1 + r_2)^2}{4 r_1 r_2} \quad \text{and} \quad \frac{\overline{P_1 P_2} \cdot \overline{Q_1 Q_2}}{\overline{P_1 Q_1} \cdot \overline{P_2 Q_2}} = \frac{d^2 - (r_1 - r_2)^2}{4 r_1 r_2}.$$

(c) Let a circle orthogonal to both $C[O_1, r_1]$ and $C[O_2, r_2]$ intersect them at the points R_1, S_1 and R_2, S_2, respectively, as we move on this circle in the positive direction. Prove that

$$4 \cdot \frac{\overline{R_1 S_2} \cdot \overline{S_1 R_2}}{\overline{R_1 S_1} \cdot \overline{R_2 S_2}} = 4 \cdot \frac{\overline{P_1 Q_2} \cdot \overline{Q_1 P_2}}{\overline{P_1 Q_1} \cdot \overline{P_2 Q_2}} = \frac{t_e^2}{r_1 r_2},$$
$$4 \cdot \frac{\overline{R_1 R_2} \cdot \overline{S_1 S_2}}{\overline{R_1 S_1} \cdot \overline{R_2 S_2}} = 4 \cdot \frac{\overline{P_1 P_2} \cdot \overline{Q_1 Q_2}}{\overline{P_1 Q_1} \cdot \overline{P_2 Q_2}} = \frac{t_i^2}{r_1 r_2},$$

and that these quantities are invariant under any inversion of the figure of the circles $C[O_1, r_1]$ and $C[O_2, r_2]$, with center of inversion lying either inside both or outside both circles $C[O_1, r_1]$ and $C[O_2, r_2]$. (If the center of the inversion is inside one of the circles and outside the other, then this result is not valid. Examine why!)

(Hint: Use the fact that by an appropriate inversion the circle $R_1 S_1 R_2 S_2$ can be exchanged with the line $P_1 Q_1 P_2 Q_2$, **Theorem 6.4.3**, and also use **Subsection 6.4.6**, etc.)

51. Refer to **Theorems 6.6.4** and **6.6.2** and **Figures 6.78, 6.77, 6.76** and **6.79**. Replace the semicircle $C[K_0, r_0]$ by the whole circle that is tangent to the two boundary base circles and located in a general position. We want to construct the corresponding **general Pappus chain** starting with $C[K_0, r_0]$ and placing chain circles in both sides of it. We can work in the following way:

We let $0 < r := \dfrac{R_U}{R_V} < 1$, $d' = 2sR_V$ the signed distance of the center L_0 or $C[L_0, R]$ from the line HG', and $d = tR_V$ the signed distance of K_0 from the line FH, where s and t are appropriate real numbers. Then prove:

(a) $R = \dfrac{1-r}{r} \cdot R_V$ (the same as in **Theorems 6.6.4**), and for

$$n = 0,\ 1,\ 2,\ 3,\ \dots . \quad \text{we get:} \quad r_n = \frac{r(1-r)}{r + [sr + 2n(1-r)]^2} \cdot R_V,$$

$$h_n = \frac{2[sr + n(1-r)]}{1-r} \cdot r_n, \quad \text{and} \quad FP_n = \frac{1+r}{1-r} \cdot r_n.$$

(b) $t = \dfrac{2rs}{1 + rs^2}$ (so, $t = 0 \Longleftrightarrow s = 0$ and $-\sqrt{r} \le t \le \sqrt{r}$),

and so

$$-\sqrt{r} R_V = -\sqrt{R_U R_V} \le d \le \sqrt{R_U R_V} = \sqrt{r} R_V.$$

That is,

$$\max d = \sqrt{R_U R_V} \quad \text{and} \quad \min d = -\sqrt{R_U R_V}.$$

We then have,

$$d = 0 \Longleftrightarrow t = 0 \Longleftrightarrow s = 0 \Longleftrightarrow d' = 0, \quad \text{and}$$

$$d \ne 0 \Longrightarrow s = \frac{rR_V \pm \sqrt{r^2 R_V^2 - rd^2}}{rd}, \quad \text{and } d' = 2sR_V.$$

Explain the two values of s when $d \ne 0$ and $d \ne \pm\sqrt{R_U R_V} = \pm\sqrt{r} R_V$.

(c) Prove that the center of the circle of the chain with the largest / smallest possible signed distance $d = \pm\sqrt{R_U R_V} = \pm\sqrt{r} R_V$ from the s.l.

FH is located at distance $\sqrt{R_U R_V}$ on the perpendicular bisector of the segment UV (whose midpoint is O). The radius of this circle is $\dfrac{R_V - R_U}{2}$.

(d) Study, state and prove the analogous results for an ordinary Steiner chain.

52. Consider two intersecting circles at points A and B. Consider any two circles $C[O_1, r_1]$ and $C[O_2, r_2]$ in their intersection and tangent to the two circles. Let d_1 and d_2 be the distances of the centers O_1 and O_2 from the common chord AB of the two circles. Prove that $\dfrac{d_1}{r_1} = \dfrac{d_2}{r_2}$.

53. Two intersecting circles partition the plane into four disjoint parts in each of which we can inscribe an infinite chain of circles. Study these chains and their properties.

54. *Inversion preserves separations*: Let A, B, C and D be four points on a s.l. or a circle and such that B is between A and C and C between B and D. Let A', B', C' and D' be their inverses by an inversion $I_{[O,c]}$. Prove that B' is between A' and C' and C' between B' and D' on the image s.l. or circle.

55. (a) Consider an inversion $I_{[O,c]}$ and two pairs of inverse points (A, A') and (B, B'). Prove that the triangles AOB and $A'OB'$ are similar.
(b) Let A', B', C' be the inverse points of the vertices of a triangle ABC under an inversion $I_{[O,c]}$ with O the circumcenter of ABC. Prove that the triangles $A'B'C'$ and ABC are homothetic and find the center and the ratio of their homothety.

56. *Inversion of any three distinct points (three vertices of a triangle, possibly degenerate) in the plane to vertices of a triangle congruent to a given triangle.*

Consider any three distinct points X, Y, and Z in the plane and any given triangle ABC. Let $\widehat{X}$, $\widehat{Y}$, and $\widehat{Z}$ be the angles of the triangle XYZ, which may be degenerate when X, Y, and Z are collinear, and $\widehat{A}$, $\widehat{B}$, and $\widehat{C}$ the angles of the triangle ABC. Let O_1 be the vertex of the isosceles triangle YO_1Z with base YZ and equal angles $\widehat{X} + \widehat{A} - \dfrac{\pi}{2}$ and O_2 the vertex of the isosceles triangle ZO_2X with base ZX and equal angles $\widehat{Y} + \widehat{B} - \dfrac{\pi}{2}$.
(O_1 and O_2 must lie in the correct closed half-planes of the s.ls YZ and ZX depending on the equal angles being acute or obtuse.) Let O be the second point of intersection (other than Z) of the circles $C[O_1, O_1Z]$ and $C[O_2, O_2Z]$ and let

$$c = \frac{OX \cdot OY \cdot AB}{XY}.$$

Let X', Y', Z' be the inverse points of X, Y, Z under the inversion $I_{[O,c]}$. Prove that the triangles $X'Y'Z'$ and ABC are congruent.

What can you say about the special case in which the triangles XYZ and ABC are equal?

[This is valid in both cases: (1) XYZ is an honest triangle, and (2) XYZ is a degenerate triangle and so a straight segment (with X between Y and Z). So any three distinct points of the plane can be mapped to any triangle by the composition of an inversion and an isometry.]

[The case in which A, B, C are collinear needs investigation. If such an inversion exists, then its center must be on the circle or the straight line (X, Y, Z).]

9.2 Geometrical Constructions

1. Consider two circles C_1 and C_2 and two points A and B on them, respectively. Construct a point P on the radical axis of C_1 and C_2, such that if the lines PA and PB intersect C_1 and C_2 at second points A' and B', the line $A'B'$ is perpendicular to the radical axis.

2. Construct a circle through a given point and tangent to a given circle and a given s.l.

3. Construct a triangle ABC, if we are given $BC = a$, the altitude $AH_1 = h$, and the product $CA \cdot CD$ where CD is the projection of BC on CA.

4. Construct a circle passing through a given point and being tangent to a given circle and a s.l.

5. Construct a circle passing through a given point and cutting two given circles at two given angles.

6. Construct a circle passing through a given point and cutting two given circles orthogonally.

7. Construct a circle passing through two given points and cutting a given circle orthogonally.

8. On a s.l., we consider two constant points A and B, a variable point P, and the inversion $I_{[P, \overline{PA} \cdot \overline{PB}]}$.

   (a) For given P and Q construct $Q' = I_{[P, \overline{PA} \cdot \overline{PB}]}(Q)$.

   (b) As Q moves on a s.l. l and the s.l. QQ' has constant direction, find the g.l. of Q'.

9. Construct a circle through a given point and tangent to two given circles.

10. Through a point A inside an angle $\widehat{xOy}$, construct a segment BAC with $B \in Ox$, $C \in Oy$ and $AB \cdot AC = k^2$.

11. Through the point of intersection A of two circles C and C' construct a segment BAC with $B \in C$, $C \in C'$ and $AB \cdot AC = k^2$.

12. From two points A and B of a circle $\mathcal{C}$ construct two parallel chords AD and BC, such that $AD \cdot BC = k^2$.

13. Construct a circle passing through a given point and being tangent to two given circles.

14. Construct a circle passing through a given point, being tangent to a given s.l. and intersecting a given circle at a given angle.

15. Construct a circle tangent to three given circles that have exactly one point in common.

16. Consider a circle, a point P, and a chord of it. Find a segment through P that intersects the chord and the circle at A and B, respectively, and $PA \cdot PB = k^2$.

17. From a point X outside an angle construct a segment XAB intersecting the sides of the angle and $XA \cdot XB = k^2$.

18. Construct a triangle ABC given the side of the inscribed square with one side on BC, the angle $\widehat{A}$ and the product of the segments into which one vertex of the square divides the side AB (or AC).

19. Construct a circle tangent to two given circles and passing through a point of their radical axis.

20. Consider three positive inversions. Find a point concyclic with its three inverse points under the given inversions.

21. Construct a convex inscribable quadrilateral when its four sides are given and each one of them is less than the sum of the other three. There are three such inscribable quadrilaterals, in general.

22. In a given parallelogram inscribe a rhombus of given area. Investigate the conditions under which this construction is possible and the number of solutions.

9.3 Geometrical Loci

1. Let $\mathcal{C}$ be a given circle and AB a constant chord of it. For any point $M \in \mathcal{C}$, we consider the circles $\mathcal{D}$ and $\mathcal{D}'$ through M and tangent to AB at A and B, respectively. Find the g.l. of the second point of intersection of $\mathcal{D}$ and $\mathcal{D}'$.

2. Two variable orthogonal circles $\mathcal{C}$ and $\mathcal{D}$ intersect each other at points P and P' and are tangent to a s.l. at two fixed points A and B, respectively. Find:

(a) The g.ls of the points Q and Q', the inverses of P and P' under the inversion $I_{[A,AB^2]}$.

(b) The g.ls of P and P'.

3. On a s.l., we pick three points A, B, and C. Let a point P move along the perpendicular bisector of AB and the s.l. CP intersect the circle (PAB) at a second point Q. Find the g.l. of Q.

4. We are given a circle $C[O, r]$ and a s.l. l that does not intersect it. From a point $M \in l$ we draw the segments MA and MB tangent to $C[O, r]$. Find the g.l. of the orthocenters of the triangles MAB, as M traces l.

5. We have an isosceles triangle ABC, with vertex A. On a s.l. through A, we consider two points M and N such that

$$\frac{MB}{MC} = \frac{NB}{NC}.$$

If N traces a s.l., or a circle, find the g.l. of M.

6. Consider a circle $C[O, r]$, a fixed point A inside of it, and a right angle $\angle(AB, AC) = 90°$, with vertex A and B and C on the circumference of $C[O, r]$. Prove:

(a) The g.l. of the common point of the tangents to $C[O, r]$ at B and C, as the angle $\angle(AB, AC) = 90°$ rotates around A, is a circle.

{First prove that the midpoint M of BC satisfies

$$MO^2 + MA^2 = r^2,$$

as the angle $\angle(AB, AC) = 90°$ rotates around A. So, the g.l. of M is the circle $C[N, \rho]$ contained inside $C[O, r]$, with center the midpoint N of OA and radius

$$\rho = \sqrt{\frac{r^2}{2} - \frac{OA^2}{4}}.$$

Then prove that the circle which is the inverse of $C[N, \rho]$ under the inversion $I_{[O, r^2]}$, is the g.l. sought.}

(b) The g.l. of the projections of A on the variable BC's is the circle $C[N, \rho]$ found above.

(c) The three circles $C[N, \rho]$, $C[O, r]$, and the circle found in (a) are coaxal (have the same radical axis) and the point A is one of the Poncelet points of the coaxal system they determine.

7. Let C be a given circle and AB a constant chord of it. For any point P of the line AB we draw the two circles $\mathcal{D}$ and $\mathcal{D}'$ through P and tangent to C at A and B, respectively.

(a) Prove that $\mathcal{D}$ and $\mathcal{D}'$ intersect at a constant angle.

(b) If F is the second point of intersection of $\mathcal{D}$ and $\mathcal{D}'$, prove that the s.l. PF passes through a fixed point. Now, find the g.l. of F.

(c) Let $\mathcal{G}$ and $\mathcal{G}'$ be the circles tangent to the three circles C, $\mathcal{D}$ and $\mathcal{D}'$. Then, each s.l. through P and one of the points of tangency of $\mathcal{G}$, $\mathcal{G}'$ and $\mathcal{D}$, $\mathcal{D}'$ passes through a fixed point.

(d) Suppose $\mathcal{D}$ and $\mathcal{D}'$ are orthogonal to a fixed circle and are tangent to a fixed circle different from $\mathcal{C}$. Find the g.ls of the point of tangency of $\mathcal{G}$, $\mathcal{G}'$ with the circles $\mathcal{D}$, $\mathcal{D}'$.

8. We take three different collinear points A, B and C, with B between A and C. A variable circle $\mathcal{C}$ is tangent to the line ABC at C. We let T be the point of tangency of the tangent drawn from A to $\mathcal{C}$ and different than AC. Let P be the common point of BT with $\mathcal{C}$. Find the g.l. of P.

9. We take three different collinear fixed points F, A and B, with A between F and C and a s.l. l intersecting FAB at F. On l, we pick any point T and consider the inversion $I_{[T,TB^2]}$. Let P be the inverse of A under this inversion. Find the g.l. of P, as T moves along l.

10. A variable circle passing through two fixed points A and D intersects two constant s.ls l_1 and l_2 at the points B and C, respectively. Find the g.l. of the orthocenter of the triangle ABC.

11. We consider a circle of center O, a point A inside of it and a point B outside of it. We draw a s.l. through B intersecting the circle at points C and D and we draw the circles (CAE) and (DAF). Find the g.l. of the center of the circle circumscribed to the triangle AEF.

12. We consider two circles $\mathcal{C}$ and $\mathcal{C}'$ outside each other, and two fixed points A and A' on them, respectively. Two circles $\mathcal{D}$ and $\mathcal{D}'$ are tangent to $\mathcal{C}$ and $\mathcal{C}'$ at A and A' and they are tangent to each other at T. By using an appropriate inversion prove that the set of all points T is the union of two orthogonal circles.

13. Take three points A, B and C on a s.l. and any circle passing through B and C. Join A with the midpoint M of the arc $\overarc{BC}$. Find the g.l. of the intersection of this circle with the s.l. AM.

14. Consider a circle $\mathcal{C}$ and two fixed points A and B and any point $M \in \mathcal{C}$. The s.ls MA and MB intersect the circle $\mathcal{C}$ at second points C and D and the s.l. through C and parallel to AB intersects $\mathcal{C}$ at a point E.

(a) Prove that the s.l. DE intersects the s.l. AB at a fixed point P.

(b) If we suppose that $\mathcal{C}$ and B are fixed, but A traces a circle, find the g.l. of P.

15. We take a circle with center O and a point P. For any circle through O and P, we let P' be the point of intersection of the s.l. OP with the common chord. Find the g.l. of P', as P moves on some given s.l.

16. Consider three collinear points A, B and C, variable circle tangent to the s.l. ABC at C, and the other tangent s.l. from A to the circle touching the circle at T. The s.l. BT intersects the circle at another point P. Find the g.l. of P.

17. Consider three fixed collinear points F, A and B, in this order, and a s.l. l going through F. Pick a point S on l and consider the inversion $I_{[S,SB^2]}$ with center S and power SB^2. Let $P = I_{[S,SB^2]}(A)$ be the inverse of A under this inversion. Find the g.l. of P, when the S traces the s.l. l.

18. Consider four points A, B, C, and O in the plane, and an inversion with center O. Let A', B' and C' be the inverses of A, B, and C under this inversion. If O is such that $A'B' = A'C'$, find the g.l. of O.

19. Consider two parallel lines $l_1 \parallel l_2$. Consider points A and C on l_1 and B and D on l_2 such that

$$AB = DC = a > AD = BC = b$$

are given segments. Take a point O between A and D and let $AO = \lambda$ a given length. Draw through O a line parallel to $l_1 \parallel l_2$ and let E and F be its common points with AB and CD, respectively. (a) Prove that the product $OE \cdot OF := k^2$ is constant. (b) Now, keep the point O fixed. Prove that the g.l. of the point F as the point E moves on a given circle of center W and passing through O is the s.l. which is the inverse of the circle under the inversion $I_{[O,k^2]}$.

[With this we make the so called **Hart's**[3] **inversor**, if we replace all straight segments involved with rods that are assembled with rivets at the vertices (joins) and can rotate with certain tension around the rivets.]

9.4 Exercises on the Appendix

1. **Sine condition for a harmonic pencil of four rays.** Consider four distinct rays Px, Py, Pw, and Pz in this order and consider, without loss of generality, the non-oriented angle $\widehat{xPy} := \alpha$, $\widehat{yPw} := \beta$, $\widehat{wPz} := \gamma$, and $0 < \widehat{zPx} = \alpha + \beta + \gamma := \delta < \pi$. Make a figure and use the law of sines appropriately to prove:

   The pencil of these four rays (Px, Py, Pw, Pz) is harmonic iff

   $$\frac{\sin(\alpha)}{\sin(\beta)} = \frac{\sin(\delta)}{\sin(\gamma)}.$$

2. Consider a quadrilateral $ABCD$, without loss of generality convex. Let P be the common point of its diagonals AC and BD, Q the common point of the sides AB and CD (extended) and R the common point of the sides BC and AD (extended). Prove:

   (a) P and Q are harmonic conjugate points with respect to the pair of the s.ls BC and AD.

---

[3] Harry Hart, English mathematician, 1848-1920. He published this inversor in 1874.

(b) P and R are harmonic conjugate points with respect to the pair of the s.ls AB and CD.

(c) Q and R are harmonic conjugate points with respect to the pair of the s.ls AC and BD.

(d) The point P is the only point with the same polar line with respect to the pair of the s.ls BC and AD and the pair of the s.ls AB and CD, which is the s.l. QR.

3. Let l and l' be two s.ls and A, B, C any three points of l and A', B', C' any three points of l'. Prove that the points $D = BC' \cap B'C$, $E = CA' \cap C'A$, $F = AB' \cup A'B$ are collinear (on the same s.l.). (Pappus of Alexandria.)

Bibliography

[1] Baloglou, G. 2007. *Isometrica*. SUNY Oswego.

[2] Baxter, J., Fristedt B., Rogness J., 2017. *College Geometry I, Using Vectors and Calculus*. School of Mathematics, University of Minnesota.

[3] Coolidge, L. J., 2012. *A Treatise on the Circle and the Sphere*, classic reprint series. Forgotten Books.

[4] Coxeter, H. S. M., F. R. S. 1969. *Introduction to Geometry*, second edition. Wiley.

[5] Coxeter, H. S. M., F. R. S., Greitzer S.L. 1967. *Geometry Revisited*. The Mathematical Association of America, New mathematical Library.

[6] Dodge, C. W. 1972. *Euclidean Geometry and Transformations*. Dover.

[7] Frère, G.-M. 1926. *Exercises de Géométrie*, 7me, éd. Librairie Générale, Paris 1926.

[8] Greenberg, M. J. 1993. *Euclidean and Non-Euclidean Geometries, Development and History*, third edition. Freeman.

[9] Guggenheimer, H. W. 1967. *Plane Geometry and Its Groups*. Holden-Day.

[10] Isaacs, I. M., 2001. *Geometry for College Students*. Brooks / Cole.

[11] Johnson, R. A. 2007. *Advanced Euclidean Geometry*. Dover.

[12] Lemaire, G. 1926. *Méthodes de Résolution et de Discussion des Problèmes de Géométrie*, onzième édition. Paris Librairie Vuibert.

[13] Lowell, C. J. 2012. *A Treatise on the Circle and the Sphere*. Classic reprint series (1916). Forgotten Books.

[14] Ogilvy, C. S. 1969. *Excursions in Geometry*. Reprint. Dover.

[15] Pedoe, D. 1970. *Geometry a Comprehensive Course*. Dover.

[16] Posamentier, A. 2002. *Advanced Euclidean Geometry*. Key College Publishing.

[17] Roussos, I. M. 2010-2016. *Lecture Notes in Euclidean Geometry.*

[18] Roussos, I. M. 2012. "On the Steiner Minimizing Point and the Corresponding Algebraic System". The College Mathematics Journal Vol. 43, No. 4 (September 2012), pp. 305-308.

[19] Smart, R. J. 1993. *Modern Geometries*, fifth edition. Books / Cole.

[20] Yale, P. B. 1968. *Geometry and Symmetry*, Dover.

[21] Yaglom, I. M. 1973. *Geometric Transformations I, II, III*, translated from the Russian by Allen Shields and Abe Shenitzer. The MAA, New Mathematical Library.

[22] Yaglom, I. M. 2009. *Geometric Transformations IV, Circular Transformations*, translated from the Russian by Abe Shenitzer. The MAA, Anneli Lax New Mathematical Library, #44.

Index

www.ingramcontent.com/pod-product-compliance
Lightning Source LLC
Chambersburg PA
CBHW081220220326
41598CB00037B/6840